Foundations of Cellular Neurophysiology

Daniel Johnston and Samuel Miao-Sin Wu

Foundations of Cellular Neurophysiology

with illustrations and simulations by Richard Gray

A Bradford Book

The MIT Press
Cambridge, Massachusetts
London, England

This book was formatted in LaTeX at Baylor College of Medicine by Richard Gray. The typeface is Lucida Bright and Lucida New Math, created by Charles Bigelow and Kris Holmes. This book was printed and bound in the United States of America.

Library of Congress Cataloging-in-Publication Data

Johnston, Daniel, 1947–
    Foundations of cellular neurophysiology / Daniel Johnston and Samuel Miao-Sin Wu : simulations and illustrations by Richard Gray.
        p.   cm.
    "A Bradford book."
    Includes bibliographical references and index.
    ISBN 0-262-10053-3
    1. Neurophysiology   2. Neurons   I. Wu, Samuel Miao-Sin.
II. Title.
    [DNLM: 1. Neurons—physiology.   2. Ion channels—physiology
3. Synaptic Transmission—physiology.   WL 102.5 J72f 1994]
QP355.2.J64 1994
591.1'88—dc20
DNLM/DLC
for Library of Congress   93-49890
                                                                        CIP

To our parents, professors, and colleagues.

Contents in Brief

Contents in Detail

Preface

This book was prepared from the authors' lecture notes in a course on cellular neurophysiology that has been given to first-year graduate students in neuroscience since 1983. Dan Johnston is primarily responsible for chapters 4, 7, 11–15, and appendixes A and B, while Sam Wu wrote chapters 1–3, 5, 6, and 8–10. Rick Gray taught the laboratory portion of the course, which consisted of computer simulations of neurophysiological experiments. Students with widely different backgrounds have taken this course successfully. The only prerequisite we have found necessary is for the student to have had (fairly recently) a calculus course that includes some differential equations. We would therefore expect that this book could be used as a text for any course intended for graduate students and advanced undergraduates provided that they have the necessary background in math.

Because this book is intended primarily as a textbook and not as a general review of neurophysiology, we have tried to avoid breaking up the text with numerous citations and have instead included a list of suggested readings for each chapter at the end of the book. The lists are alphabetical and in two parts, one for books and reviews and one for original articles. Some of these references will provide background reading for the material in the chapter, while others are actually cited within the text. Even though the references are grouped by chapter, we have put them all at the end of the book so that it will be easier to scan the reference lists for all the chapters and find a particular citation. The included references are heavily weighted toward books and reviews rather than the original publications, because we feel these are more appropriate for students. We hope that this practice does not offend our colleagues who may find that their work is mentioned but not cited as an original publication. We have also used many of these original articles, reviews, and books as source material for equations, figures, and general explanations of concepts throughout this book. We are therefore grateful for the efforts of our colleagues for without these sources this book would not have been possible.

One of the major reasons that the course from which this book is derived has been so successful is our emphasis on problem solving. We have found that students get out of the course what they put into it. If they are compulsive about working the homework problems every week, then they usually can master the material successfully. If, instead, they tend to focus on getting the correct answer to a problem (either from a colleague or from an old answer sheet) rather than being sure that they understand the material behind the problem, then they are usually much less successful. It is difficult to learn the concepts in this book without working through the problems and seeing how the principles of neurophysiology are utilized and applied in practice. It is for these reasons that the answers to the homework are put into two appendixes—one with short answers that the student can consult while trying to work a problem, and the second with the complete solutions, which should be consulted only as a last resort. The homework problems should also serve as a rich source of exam questions since in many cases they were created for that purpose.

A major problem we encountered when writing this book was in choosing the proper nomenclature. We found to our distress that there were few "standard" practices in neurophysiology. We tried where possible to use the practice that is common in physics and engineering and let capital letters represent constant or peak values of variables and small letters represent variables that are functions of time. We nevertheless were forced to make a few exceptions to this rule so that the symbols in this book would not differ wildly from those in the original papers or from those used in other books.

The authors and reviewers have made a considerable effort to minimize mistakes in the equations and homework problems. Nevertheless, it is inevitable that in a quantitative book of this sort with so many homework problems and solutions to homework problems that many errors are still to be found. We want to, first, apologize for any inconvenience these errors may cause and, second, encourage readers of this book to notify us (or the publisher) of these errors so that they can be corrected in time for the next edition.

We have many people to thank for helping to make this book a reality. First and above all, we thank Dr. Richard Gray for his heroic effort in preparing the figures, many of which resulted from computer simulations of the underlying neurophysiology. His efforts in this regard are recognized by including his name on the front cover even though he is not an author. We also want to thank Diane Jensen for typing and editing much of the book. Special thanks is reserved for Dr. Nelson Spruston, who gave us detailed comments on most of the chapters and solicited help with the rest of them from Drs. David Colquhoun, Michael Häusser, Johannes Helm, Keiji Imoto, Greg Stuart, Alfredo Villarroel, and Lonnie Wollmuth. Their critical feedback has greatly improved the final product and for this we are most grateful. Drs. Ted Carnevale and Zach Mainen also gave us useful comments on the book. Drs. Jimmy Zhou, Enrico Stefani, and King-Wai Yau proofread some chapters for us, Ken Tsai made a figure for us from unpublished work, and Drs. Peter Saggau and Bill Ross advised us on the contents of appendix B. We thank Erik Cook for helping to design the front cover. A number of our students have helped a great deal in proofreading the text and homework in this book. These students are (in alphabetical order): Bob Avery, Erik Cook, David Egelman, David Leopold, Jean-Baptiste Le Pichon, Craig Powell, Kris Radcliffe, Saurabh Sinha, and Ling-Gang Wu. We also want to express our sincere appreciation to Dr. Jim Patrick, the Head of the Division of Neuroscience, for his encouragement during the writing of this book.

This book was written using LaTeX and other TeX utilities running under the NEXTSTEP operating system. We are deeply indebted to Donald Knuth for developing TeX and to Leslie Lamport and Tomas Rokicki for writing their wonderful utilities. The TeX typesetting language made our job as authors enormously easier. We also want to thank Fiona Stevens and the staff at The MIT Press for their professional help and enthusiasm in the process of making this book a reality. Finally, we want to thank all of our professors who taught us neurophysiology and gave us the proper *foundations* in this immense and complicated field. We are only too happy to try and pass these foundations along to others.

We also wish to thank our wives and children for their support and patience during the long gestation of this book.

Daniel Johnston
Samuel M. Wu

April 10, 1994
Houston, Texas
FCN@mossy.bcm.tmc.edu.

List of Symbols, Units, and Physical Constants

Symbols (units)

(The symbols are listed by the chapter in which they first appear.)

Chapter 1

I = current (A)

V = voltage or potential difference (V)

Na^+ = sodium ion

K^+ = potassium ion

Cl^- = chloride ion

Ca^{2+} = calcium ion

Chapter 2

J = ion flux (molecules/sec-cm^2)

$\mathbf{J}$ = molar flux (mol/sec-cm^2)

[X] = concentration of substance X

D = diffusion coefficient (cm^2/sec)

∂_{el} = electrical conductivity (molecules/V-sec-cm)

μ = mobility (cm^2/V-sec)

$u = \mu/N_A$ = molar mobility (cm^2/V-sec-mol)

E_i = equilibrium potential of ion species i

P = permeability (cm/sec)

β = partition coefficient (unitless)

$\log_{10}$ = log = logarithm (base 10)

ln = natural logarithm

Chapter 3

R_m = specific membrane resistance (Ω-cm^2)

G_m = specific membrane conductance (S/cm^2)

C_m = specific membrane capacitance (F/cm^2)

V_m = membrane potential (mV)

I_m = membrane current (A)

I_i = ionic current (A)

I_C = capacitance current (A)

$E_r = E_{\text{rest}}$ = resting potential (mV)

V_{rev} = reversal potential (mV)

$G_r = G_{\text{rest}}$ = resting conductance (S)

$\frac{dI}{dV}$ = slope condutance (S)

I^* = instantaneous current (A)

I^∞ = steady-state current (A)

ΔG = free energy of activation (cal/mol)

$k_1, k_2, k_{-1}, k_{-2}, \alpha, \beta\ldots$ = rate coefficients (sec^{-1})

δ = factor of asymmetry of the energy barrier (unitless)

Chapter 4

r_i = internal resistance/length of cylinder (Ω/cm)

r_m = membrane resistance/length of cylinder (Ω-cm)

c_m = capacitance/length of cylinder (F/cm)

r_o = external resistance/length of cylinder (Ω/cm)

R_i = internal resistivity (Ω-cm)

$I_0 = I_{\text{in}}$ = injected current (A)

i_m = membrane current/length of cylinder (A/cm)

i_i = intracellular current/length of cylinder (A/cm)

i_C = capacitance current/length of cylinder (A/cm)

G_N = input conductance (S)

R_N = input resistance (Ω)

λ = length or space constant (cm)

ρ = dendritic to somatic conductance ratio (unitless)

f = frequency (Hz)

ω = radial frequency = $2\pi f$ (radians)

a = radius (cm)

d = diameter (cm)

x = distance (cm)

X = electrotonic distance (unitless)

L = electrotonic length (unitless)

l = length (cm)

t = time (msec)

τ_m = membrane time constant (msec)

$T = t/\tau_m$

q = electrical charge (C)

θ = conduction velocity (cm/sec)

Chapters 5, 6, and 7

y, n, m, and h = probability of gating particles (unitless)

τ = time constant (sec)

V_H = holding voltage (mV)

V_c = command voltage (mV)

V_{th} = threshold voltage (mV)

γ = single-channel conductance (S)

Q = gating charge (C)

g_{Na} = Na$^+$ conductance as function of time and voltage (S)

g_K = K$^+$ conductance as function of time and voltage (S)

g_L = leakage conductance (S)

g_{Ca} = Ca^{2+} conductance as function of time and voltage (S)

g_{Cl} = Cl$^-$ conductance (S)

$\overline{g}$ = maximum conductance (S)

G = a constant conductance (S)

Refer to Table 7.1, page 208 for list of symbols for different ionic currents.

Chapter 9

$x(t)$ = random variable (arbitrary unit)

μ_x = mean value of random variable $x(t)$ (x)

$\sigma_x{}^2$ = variance (x^2)

$|\sigma_x|$ = standard deviation (x)

$p(x), f(x)$ = probability density function, pdf

$P(x), F(x)$ = probability distribution function or cumulative distribution function (unitless)

$E[x]$ = expectation value of $x(t)$ (x)

$R_x(\tau)$ = correlation function of $x(t)$ (x^2)

$C_x(\tau)$ = covariance function of $x(t)$ (x^2)

$S_x(f)$ = power spectral density function (x^2-sec)

C_k^n = number of combinations of k events, choosing from a collection of n events

$\mathcal{F}$ = Fourier transform

$\mathfrak{R}$ = real part of a complex variable

Chapter 10

$\text{prob}[A|B]$ = conditional probability

$P_{ij}(t)$ = transition probability from state i to state j

$\mathbf{P}(t)$ = transition probability matrix

$\mathbf{Q}$ = infinitesimal matrix

$\mathbf{I}$ = identity matrix

$\det \mathbf{Q}$ = determinant of matrix $\mathbf{Q}$

λ = eigenvalue (sec^{-1})

$\overline{\tau}_o$ = mean open time (sec)

$\overline{\tau}_c$ = mean closed time (sec)

Chapters 11–15

EPP = end-plate potential (mV)

mEPP = miniature end-plate potential (mV)

EPSP = excitatory postsynaptic potential (mV)

IPSP = inhibitory postsynaptic potential (mV)

EPSC = excitatory postsynaptic current (A)

IPSC = inhibitory postsynaptic current (A)

I_s = synaptic current (A)

I_{cl} = clamp current (A)

I_H = holding current (A)

G_s = peak synaptic conductance (S)

g_s = synaptic conductance as function of time (S)

E_s = synaptic equilibrium potential (mV)

p = probability of release

m = quantal content

$\overline{q}$ = mean quantal size (mV)

$\overline{V}$ = mean EPSP size (mV)

n = number of release sites

CV = coefficient of variation

σ^2 = variance

$\mathcal{L}$ = Laplace Transform

LTP = long-term potentiation

PTP = post-tetanic potentiation

R_{int} = internal resistance of a cylindrical compartment

Appendix A

Y = admittance (S)

Z = impedance (Ω)

R_s = series resistance (Ω)

R_a = access resistance for voltage clamp (Ω)

R_e = microelectrode tip resistance (Ω)

R_f = feedback resistance (Ω)

A = amplifier gain

e = elementary charge (C)

f = frequency (Hz)

Appendix B

c = speed of light in a vacuum (m/sec)

λ = wavelength of light (nm)

v = speed of light in a medium (m/sec)

n = refractive index

I = image size (m)

i = image distance (m)

O = object size (m)

o = object distance (m)

f = focal length (m)

M = magnification (unitless)

NA = numerical aperture of a lens

R = resolution (μm)

F = relative fluorescence emission intensity (unitless)

Other symbols are defined in the text as needed.

Units

Å = angstrom = 10^{-10} m

c = centi = 10^{-2}

m = milli = 10^{-3}

μ = micro = 10^{-6}

n = nano = 10^{-9}

p = pico = 10^{-12}

f = femto = 10^{-15}

k = kilo = 10^{3}

M = mega = 10^6

G = giga = 10^9

nt = newton (kg-m/sec^2)

joule = nt-sec

V = volt

A = ampere (C/sec)

C = coulomb

Ω = ohms (V/A)

S = $1/\Omega$ = siemens (A/V)

F = farad (sec-A/V)

db = decibels

m = meter

cm = centimeter

L = liter

ml = 10^{-3} L

μl = 10^{-6} L

M = molarity

mol = mole

sec = second

Hz = hertz (cycles/sec)

cal = calories

Physical Constants

N_A = Avogadro's number = 6.023×10^{23} molecules/mol

F = Faraday's constant = 9.648×10^4 C/mol

ϵ_0 = the permittivity constant = 8.85×10^{-12} F/m

k = Boltzmann's constant = 1.381×10^{-23} joule/°K

R = gas constant = 1.987 cal/mol-°K

e = elementary electrical charge = 1.602×10^{-19} C

°K = absolute temperature = °C + 273.16

Foundations of Cellular Neurophysiology

1   Introduction

The fundamental task of the nervous system is to communicate and process information. Animals, including humans, perceive, learn, think, deliver motion instructions, and are aware of themselves and the outside world through their nervous systems. The basic structural units of the nervous system are individual neurons, and neurons convey neural information by virtue of electrical and chemical signals.

In the human nervous system there are about 10^{12} neurons. A typical 1 mm^3 cortical tissue contains about 10^5 neurons. Cells in the nervous system exhibit extraordinary morphological and functional diversities. Some neurons are as small as a few micrometers (μm), but some bear axons as long as one to two meters. Some neurons have large, flamboyant dendrites, whereas others have no dendrites or axons. The number of different morphological classes of neurons in the vertebrate brain is estimated to be near 10,000. Figure 1.1 illustrates a few examples of morphologically distinct classes of neurons. The number of functionally different classes of neurons in the brain is probably even higher, because neurons of similar morphology (e.g., pyramidal cells in figure 1.1) may have different functions.

Neurons communicate with one another through specialized contact zones, the synapses. There are about 10^{15} synapses in a human brain. Synapses can be either electrical (synapses that contain intercellular bridging pores that allow current flow) or chemical (synapses that release chemical messengers or neurotransmitters). Electrical and chemical synapses can be observed with the electron microscope. Electrical synapses are normally located at the *gap junctions*, which consist of specialized proteins that form channels bridging the interiors of two neurons and allow current flow from one neuron to the other. The typical chemical synapse consists of synaptic vesicles in the presynaptic neuron and membrane thickening in the presynaptic and postsynaptic membranes (active zones). The pat-

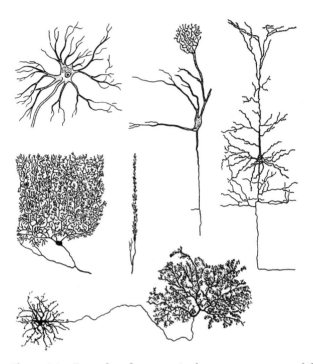

Figure 1.1 Examples of neurons in the nervous system exhibiting various morphology. From the upper left in clockwise order: motor neuron from the spinal cord, mitral cell from olfactory bulb, pyramidal cell from cortex, horizontal cell from retina, and Purkinje cell (front and side views) from cerebellum. (From Nicholls et al. 1992 and Fisher and Boycott 1974.)

terns of synaptic connections in the nervous system are extremely complex. Some neurons make synaptic contacts with other neurons nearby, while others send long axons that make synapses up to a meter away. Most neurons are polarized: They receive synaptic input at the dendritic end and make output synapses at the axonal end (figure 1.2). The number of synaptic connections made by a neuron can be extraordinarily large. The dendrite of a mammalian motor neuron, for example, receives inputs from about 10^4 synapses. Neuronal polarization does not imply, however, that synapses are always formed in one direction. A large number of neurons make reciprocal (feedback) synapses onto their presynaptic cells, others make lateral synapses with parallel neurons, and many neurons make serial synapses that form local loops among several neuronal processes.

The primary difference between neurons and most other cells in the body (e.g., liver cells) is that neurons can generate and transmit neural signals. Neural signals, either electrical or chemical, are the messengers used by the nervous system for all its functions. It is of paramount importance to understand the principles and mechanisms of neural signals. Despite the extraordinary diversity and complexity of neuronal morphology and synaptic connectivity, the nervous system adopts a number of basic principles of signaling for all neurons and synapses. The principles of neural signaling and the physical laws and mechanisms underlying them are the focus of this book.

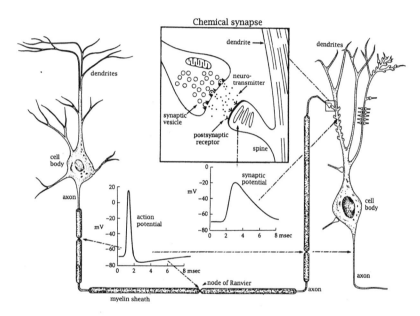

Figure 1.2 Neurons convey information by electrical and chemical signals. Electrical signals travel from the cell body of a neuron (left) to its axon terminal in the form of action potentials. Action potentials trigger the secretion of neurotransmitters from synaptic terminals (upper insert). Neurotransmitters bind to postsynaptic receptors and cause electric signals (synaptic potential) in the postsynaptic neuron (right). Synaptic potentials trigger action potentials, which propagate to the axon terminal and trigger secretion of neurotransmitters to the next neuron. (Adapted from Kandel et al. 1991 and from L.L. Iversen, copyright © 1979 by Scientific American, Inc. All rights reserved.)

In neurons or other excitable cells, electrical signals are carried primarily by transmembrane ion currents, and result in changes in transmembrane voltage. In the nervous system four ion species are involved in transmembrane currents: sodium (Na^+), potassium (K^+), calcium (Ca^{2+}), and chloride (Cl^-), with the first three carrying positive charges (cations) and the fourth carrying negative charges (anions). The movement of these ions is governed by physical laws (which are discussed in detail in chapter 2). The energy source of ion movement is the ionic concentration gradient across the membrane, which is maintained by ion pumps whose energy is derived from hydrolysis of ATP molecules. These concentration gradients set up the electrochemical potential across the membrane, which drives ion flow in accordance with the laws of diffusion and drift (Ohm's law).

Although the energy sources and ion species involved in electrical signals are relatively simple, the control mechanisms for the passage of ions across the membrane are quite complicated. Ions flow across the membrane through aqueous pores formed by transmembrane protein molecules, the ion channels. These channel protein molecules undergo conformational changes that under certain conditions allow ion passage (*gate* in the open state), but under other conditions prohibit ion passage (*gate* in the closed state). The principles of ion permeability and channel gating have been the main focus of neurophysiology for more than 50 years, and they are the cornerstones of our understanding of neural signaling. We shall spend a large portion of this book describing these principles. In chapters 2 and 3, we provide formal definitions of various electrical parameters of excitable cells. Membrane permeability and conductances are described in terms of current-voltage (*I-V*) relations. Under certain conditions the *I-V* relations of the membrane are approximately linear (constant membrane conductance), whereas under other conditions the *I-V* relations are nonlinear. The linear membrane properties will be described in chapters 3 and 4, and the nonlinear properties will be described in chapters 5 and 6.

As noted earlier in this chapter, most neurons are not of regular shapes, but exhibit complex morphology. A signal generated at the tip of a dendrite will excite the cell body differently from a signal generated at the cell body itself, because the dendritic signal must travel along the thin dendritic processes to the cell body. It is vitally important, therefore, to incorporate the morphology and shape of the neuron into any analysis of neural signaling. Under conditions where the membrane is linear, electrical signals in one part of a neuron diffuse passively (down the electrochemical gradient) to other parts of the neuron. Principles dealing with such signal spreading are embodied in what is called *linear cable theory*, and are described in chapter 4.

Under the conditions where *I-V* relations are nonlinear, membrane conductances vary with respect to transmembrane voltage and/or time. These time- or voltage-dependent nonlinearities mediate complex signaling behaviors of neurons. Stimulation of a neuron may trigger a sequence of these voltage- and time-varying conductance changes, and this may lead to the generation and propagation of action potentials, the primary form of electrical signal in the nervous system. Mechanisms of voltage- and/or time-dependent nonlinearities are described in chapter 5, and principles and experiments concerning the generation and propagation of action potentials in the squid giant axon (Hodgkin and Huxley's analysis) are described in chapter 6. Chapter 7 gives a brief account of the properties of ion conductances other than Hodgkin and Huxley's Na^+ and K^+ conductances in the squid axon. Although a large number of functionally diverse channels exist in the nervous system, the principles used for analyzing the squid axon are largely applicable to the analysis of other voltage- and time-dependent conductances.

Our understanding of ion channels up to the mid-seventies was restricted to population analysis: measurements of gross current in the whole cell that contains populations of individual channels. In 1976, Neher and Sakmann developed the *patch-clamp* technique, which allows recording of current flowing through a single channel. Molecular cloning techniques in recent years have facilitated our understanding of the molecular structure of single ion channel proteins. A brief account of the molecular structure and patch-clamp recording of single ion channels is given in chapter 8.

Patch-clamp records of single channel currents reveal that ion channels open and close randomly. This means that the behavior of individual ion channels cannot be predicted or described by deterministic mathematical equations. Rather, stochastic analysis must be employed. Chapter 9 gives a concise description of the basic mathematical tools for stochastic analysis.

The statistical properties of single channels and ensembles of single channels are discussed. In chapter 10, a stochastic formulation of channel mechanisms is described. To avoid tedious mathematical manipulations, we derive the basic differential equation for transition probabilities (Chapman-Kolmogorov equation) with the simplest two-state scheme and give the generalized rules of this equation for the general n-state scheme for any channels. This formulation is intended to help readers to acquire the mathematical background for single-channel and whole-cell current data analysis.

Mechanisms of synaptic transmission are described in chapters 11–15. Most of what we know about the physiology of synaptic transmission was obtained from studying two classical preparations: the frog neuromuscular junction and the squid giant synapse. The frog neuromuscular junction was the preparation used by Bernard Katz and his colleagues when they formulated the so-called *quantum hypothesis* for transmitter release. This hypothesis states that transmitter is released in uniform packets or *quanta* from presynaptic terminals. Chapter 11 is devoted to a discussion of this quantum hypothesis. The theory and the associated equations for analyzing the spontaneous and evoked release of transmitter from a typical presynaptic ending are derived in this chapter. The methods formulated from this hypothesis are called *Quantal Analysis* and have been used in various ways for almost 40 years to investigate mechanisms of synaptic transmission. There are also *changes* in transmitter release that are dependent on the prior history of activity at a synapse. These changes in synaptic functioning are called use-dependent or activity-dependent *synaptic plasticities* and are discussed briefly in chapters 11, 12, and 15.

Neurotransmitter release requires a rise in $[Ca^{2+}]$ in the presynaptic terminal. Chapter 12 discusses the data, derived primarily from the squid giant synapse, for another fundamental hypothesis of neurophysiology, the Ca^{2+} *hypothesis* for transmitter release. Although Ca^{2+} is known to be required for transmitter release, very little is known about the biochemical steps between Ca^{2+} entry and release. The chapter outlines the physiology related to Ca^{2+} entry, transmitter release, and the action of the transmitter on the postsynaptic cell, and thus provides a framework for whatever these biochemical steps might be. A few of the current ideas for the biochemistry of release are also discussed briefly.

Chapter 13 moves to the postsynaptic side of the synapse. Neurotransmitter molecules released from presynaptic terminals bind to specialized receptors and cause the opening of channels. There are many types of receptors, some of which produce excitation and some, inhibition. The analysis of excitation and inhibition from a macroscopic perspective is discussed in this chapter. Reversal potentials, *I-V* curves, conductances, and kinetics are important parameters for investigating and understanding the mechanisms of synaptic transmission. Moreover, most synapses occur on dendrites and spines at some distance from the cell body. The problems associated with the study of remotely located synapses and the effects of spines on synaptic inputs are also discussed in this chapter. Although most of the emphasis in chapter 13 is on fast, chemical transmission, some of the mechanisms associated with slower, conductance-decrease synaptic events and electrical synapses are also presented.

The recording of electrical events from intact nervous systems is often done using extracellular recordings. These recordings are possible because the electrical events in single neurons produce electrical fields in the extracellular space that can be detected with suitable electrodes. Chapter 14 briefly outlines a semiquantitative analysis of extracellular field potentials and their measurements.

The last chapter outlines some of the theories and basic principles involved in trying to understand behaviors such as learning and memory from cellular neurophysiology. A basic tenet of learning and memory research is that memories are stored as changes in the strength of synaptic connections. Alteration in the structure of dendritic spines has been one attractive candidate for a mechanism of such changes in synaptic strength. A quantitative analysis of whether morphological changes in spines could be a substrate for memory is presented in chapter 15. Also, long-term potentiation (LTP), first presented in chapter 11, is discussed at some length in chapter 15. LTP is attractive as a possible mechanism for a change at a synapse that could underlie memory. Computational modeling is a powerful method for trying to understand the behavior of ensembles of neurons and how they process and store information. We end chapter 15 with an analysis of one particular type of model that has some structural similarities with the hippocampus, an area of the brain that is important for memory.

This book also includes four appendixes. The first reviews the operation and analysis of simple electrical circuits. If the student using this book has no background in basic electrical circuits and circuit elements, then this appendix may be useful. Appendix B summarizes some of the optical methods in common use in cellular neurophysiology. Most of the chapters in this book include an extensive list of suggested homework problems; appendix C gives short answers for most problems, and appendix D has the complete solutions for the problems. We have separated the solutions into two appendixes, so that the student can check his or her answer quickly by referring to appendix C, consulting appendix D only as a last resort.

It is not the intention of this book to give a comprehensive account of the physiology of the nervous system. We spend little time describing specific neurons, synapses, or pathways in the brain. The objective of this book is to describe the basic principles that are not neuron or pathway specific but are generally applicable to most, if not all, neurons. We also recognize that neurophysiology is a fast-growing field, and that new information and concepts emerge on a daily basis. For this reason, we have usually avoided describing the most recent findings in certain areas (e.g., the structure and function of various ion channels or the mechanisms of LTP), and have concentrated instead on the basic analytical principles and concepts that are applicable in many situations. The only two exceptions to this approach are the squid giant axon and the frog neuromuscular junction. We describe these two preparations in great detail not only because they are historically important, but also because they provide general concepts and analytical tools (e.g., the Hodgkin and Huxley gate model and the Katz quantal analysis of synaptic transmission) that can be applied to other neurons and synapses.

We believe that one cannot avoid mathematics if one wants to learn neurophysiology thoroughly. This is why we attempt to describe most physiological principles quantitatively in this book. Additionally, we give worked examples in the text, and problems at the end of each chapter. Our own and our students' experiences have taught us that problem solving is an indispensable part of learning neurophysiology.

2   Ion Movement in Excitable Cells

2.1   Introduction

Electric signals in humans and animals are carried by dissociated ions: K^+, Na^+, Cl^-, and Ca^{2+}. These ions carry positive or negative charges and flow from one part of the body to another. In excitable cells, movement of ions across the plasma membrane results in changes of electrical potential across the membrane, and these potential changes are the primary signals that convey biological messages from one part of the cell to another part of the cell, from one cell to another cell, and from one part of the body to another part of the body.

Ions in biological systems are not uniformly distributed. Ion concentrations in one compartment are quite different from those in other compartments. The concentration of K^+ ions inside most animal cells, for example, is much higher than that in the extracellular space. Such differences in ion distribution result in concentration gradients or chemical potentials in biological systems. Based on thermodynamic principles, ions tend to flow from regions of high concentrations to regions of low concentrations—a phenomenon known as diffusion. Physical principles of ion diffusion will be described in this chapter.

Because dissociated ions carry electric charges, their movement is influenced not only by concentration gradients but also by electric fields. In most parts of the body, the net electric charge of biological molecules is zero. In other words, the number of positive charges in a given volume equals the number of negative charges (known as space-charge neutrality). An important exception to this space-charge neutrality occurs within the plasma membrane of individual cells. Since most plasma membranes are permeable to some ion species but not to others, a separation of charge normally occurs across the membrane. This ion separation results in an electric field across the cell membrane, which profoundly influences the movement of ions through pores or channels situated in the plasma membrane. Physical principles dictating ion movement in electric fields will be discussed in this chapter.

2.2   Physical laws that dictate ion movement

In this section, four fundamental physical laws essential for describing the movements of ions in biological systems will be discussed. For simplicity, we shall consider only the one-dimensional system; that is, we assume ions move only along the x-axis. This simplification is adequate for most biological systems we will encounter because our primary concern is ion movement across the cell membrane.

The first two laws concern two processes: diffusion of particles caused by concentration differences and drift of ions caused by potential differences. The third law concerns the relationship between the proportional coefficients of the first two processes, the diffusion coefficient D and the drift mobility μ. The fourth law states the basic principle of separation of charges in biological systems. As we will discuss later in this chapter, these four physical laws are the foundations and basic mathematical tools for deriving fundamental equations in neurophysiology. These include the Nernst-Planck, Nernst, Goldman-Hodgkin-Katz, and Donnan equilibrium equations.

2.2.1   Fick's law for diffusion

$$J_{\text{diff}} = -D\frac{\partial[C]}{\partial x}, \tag{2.2.1}$$

where J is diffusion flux (molecules/sec-cm^2); D is the diffusion coefficient (cm^2/sec); and [C] is the concentration of ion (molecules/cm^3). The negative sign indicates that J flows from high to low concentration. Concentrations are used for dilute solutions; otherwise the activity of solutes should be used.

Equation 2.2.1 is Fick's first law. It is an empirical law that states that diffusion takes place *down* the concentration gradient and is *everywhere* directly proportional to the magnitude of that gradient, with proportionality constant D.

2.2.2 Ohm's law for drift

Charged particles (e.g., ions) in a biological system will experience an additional force, resulting from the interaction of their electric charges and the electric field in the biological environment. The flow of charged particles in an electric field can be described by

$$
\begin{aligned}
J_{\text{drift}} &= \partial_{el}E \\
&= -\mu z[C]\frac{\partial V}{\partial x},
\end{aligned}
\tag{2.2.2}
$$

where J_{drift} is the drift flux (molecules/sec-cm^2), ∂_{el} is electrical conductivity (molecules/V-sec-cm), E is electric field (V/cm) = $-\frac{\partial V}{\partial X}$, V is electric potential (V), μ is mobility (cm^2/V-sec), z is the valence of the ion (dimensionless), and [C] is the concentration.

Equation 2.2.2 states that drift of positively charged particles takes place *down* the electric potential gradient and is *everywhere* directly proportional to the magnitude of that gradient, with the proportionality constant equal to $\mu z[C]$.

2.2.3 The Einstein relation between diffusion and mobility

Einstein (1905) described diffusion as a random walk process. He demonstrated that the frictional resistance exerted by the fluid medium is the same for drift as it is for diffusion at thermal equilibrium, and diffusion coefficient and mobility can be related by

$$
D = \frac{kT}{q}\mu,
\tag{2.2.3}
$$

where k is Boltzmann's constant (1.38×10^{-23} joule/°K), T is absolute temperature (°K), and q is the charge of the molecule (C).

This relationship formally states that diffusion and drift processes in the same medium are additive, because the resistances presented by the medium to the two processes are the same. This relationship greatly simplifies our quantitative descriptions of ion movement in biological systems, since ions in living cells usually are influenced by both concentration and electric potential gradients.

2.2.4 Space-charge neutrality

In a given volume, the total charges of cations is *approximately* equal to the total charge of anions, i.e.,

$$\sum_i z_i^C e[C_i] = \sum_j z_j^A e[C_j], \tag{2.2.4}$$

where z_i^C is the valence of cation species i; z_j^A is the valence of anion species j; e is the charge of a monovalent ion; and $[C_i]$ and $[C_j]$ are concentrations of ion species.

Space-charge neutrality holds for most parts of living bodies. The only exception is within the cell membrane due to separation of charges. An example of the amount of charge separation to establish a membrane voltage is given in example 2.1.

Example 2.1

The membrane capacitance of a typical cell is 1 μF/cm^2 (i.e., 10^{-6} uncompensated coulombs of charge on each side of the 1 cm^2 membrane are needed to produce 1 V across the membrane), and the concentration of ions inside and outside of the cell is about 0.5 M. Calculate the fraction of uncompensated ions on each side of the membrane required to produce 100 mV in a spherical cell with a radius of 25 μm.

Answer to example 2.1

Surface area $= 4\pi a^2 = 4\pi (0.0025 \text{ cm})^2 = 7.85 \times 10^{-5} \text{ cm}^2$.

Total volume $= \dfrac{4}{3}\pi a^3 = \dfrac{4}{3}\pi (0.0025 \text{ cm})^3 = 6.5 \times 10^{-8} \text{ cm}^3$.

Number of ions needed to charge up 1 cm^2 membrane to 100 mV:

$$
\begin{aligned}
n &= \frac{q \times 1 \text{ cm}^2}{e} = \frac{C \cdot V \times 1 \text{ cm}^2}{1.6 \times 10^{-19} \text{ C}} \\
&= \frac{10^{-6}\left(\frac{C}{\text{V-cm}^2}\right)(10^{-1} \text{ V})(\text{cm}^2)}{1.6 \times 10^{-19} \text{ C}} = 6 \times 10^{11}.
\end{aligned}
$$

Therefore, the number of uncompensated ions needed for the cell

$= 6 \times 10^{11} \ (\text{cm}^{-2}) \times 7.85 \times 10^{-5} \text{ cm}^2 = 4.7 \times 10^7$.

Total number of ions in the spherical cell

$= \left[(0.5 \times 6.02 \times 10^{23})/1000 \text{ ml}\right] \times 6.5 \times 10^{-8} \text{ ml} = 2 \times 10^{13}$.

Fractional uncompensated ions $= \dfrac{4.7 \times 10^7}{2 \times 10^{13}} = 2.35 \times 10^{-5} = 0.00235\%$

From example 2.1, it is obvious that the amount of uncompensated ions needed to charge the electric field across the membrane is very small. Even for the smallest cells, more than 99.9% of all ions are compensated by ions of the opposite charge.

The principle of space-charge neutrality therefore holds in any volume in the biological system except within the plasma membrane, where the electric field is nonzero. Another way to express this point is Gauss's law, which states that the flux of electric field E through any closed surface (i.e., the integral $\int E d\mathbf{a}$ over the surface) equals 4π times the total charge *enclosed* by the surface.

$$\int \mathbf{E} \cdot d\mathbf{a} = 4\pi \int \rho d\nu = 4\pi q,$$

where $\mathbf{E}$ is the electric field, $\mathbf{a}$ is the oriented area, ρ is the charge density, ν is the volume, and q is the total charge in the enclosed surface. $\mathbf{E} \cdot d\mathbf{a}$ is the scalar product of $\mathbf{E}$ and $d\mathbf{a}$. (For reference, see Purcell 1965.)

Gauss's law says that the electric field at the surface of a given volume, no matter what the shape, as long as it is enclosed by the surface, is proportional to the total charge the volume contains. Take a spherical cell as an example (figure 2.1). The electric field within the plasma membrane (hypothetical surface 1 or HS_1) is nonzero, because there are net negative charges enclosed in surface 1 (example 2.1). The electric field outside the plasma membrane (HS_2), on the other hand, is zero, because the negative charges inside the membrane are compensated by the positive charges outside the membrane; thus the total charge inside surface HS_2 is zero.

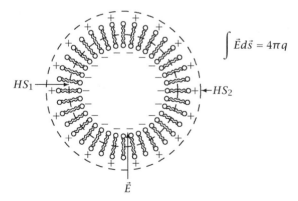

$$\int \vec{E} d\vec{s} = 4\pi q$$

Figure 2.1 Schematic diagram of a spherical cell. The two hypothetical spherical surfaces (HS_1 and HS_2, dashed lines) enclose different amounts of electric charges. According to Gauss's law, electric field E is nonzero (pointing inward) at HS_1.

2.3   The Nernst-Planck equation (NPE)

Under physiological conditions, ion movement across the membrane is influenced by both electrical field and concentration gradients. This is because ion concentrations inside and outside the cell are different, and electric field is nonzero within the plasma membrane due to separation of charges across the membrane. The concentration differences of various ions are caused by active ion transporters (pumps; this is active distribution of ions) and by selective permeabilities of ions of the plasma membrane (Donnan equilibrium or passive distribution of ions). Details on ion distribution will be discussed in section 2.5. Separation of ions across a cell membrane is caused by selective permeability, usually to K^+, of the cell membrane. This allows K^+ ions to diffuse out of the cell, down their concentration gradient ($[K^+]_{in} > [K^+]_{out}$), resulting in net negative charges inside the cell and positive charges outside the cell. Such charge separation results in the electric field across the membrane pointing inward while the membrane is at rest.

The ion flux under the influence of both concentration gradient and electric field can be written by combining the diffusion and drift flux, i.e.,

$$
\begin{aligned}
J &= J_{drift} + J_{diff} \\
&= -\mu z [C] \frac{\partial V}{\partial x} - D \frac{\partial [C]}{\partial x}.
\end{aligned}
$$

By using Einstein's relation, we can express the diffusion coefficient in terms of mobility, and thus simplify the flux equation.

$$
J = -\left(\mu z [C] \frac{\partial V}{\partial x} + \frac{\mu k T}{q} \frac{\partial [C]}{\partial x} \right).
\tag{2.3.5}
$$

Equation 2.3.5 is the Nernst-Planck equation (NPE) of the ion flux form (J is in molecules/sec-cm^2). If one divides J by Avogadro's number, one can obtain the NPE of the molar form. NPE in molar form:

$$
\begin{aligned}
\mathbf{J} &= J/N_A = \frac{-\mu z [C]}{N_A} \frac{\partial V}{\partial x} - \frac{\mu k T}{N_A q} \frac{\partial [C]}{\partial x} \\
&= -\left(u z [C] \frac{\partial V}{\partial x} + u \frac{RT}{F} \frac{\partial [C]}{\partial x} \right).
\end{aligned}
\tag{2.3.6}
$$

Since current is the product of ion flux and the charge it carries, the NPE of the current density form can be obtained by multiplying the molar flux by the total molar charge, zF. NPE in current density form:

$$I = \mathbf{J} \cdot zF = -\left(uz^2F[C]\frac{\partial V}{\partial x} + uzRT\frac{\partial [C]}{\partial x}\right), \tag{2.3.7}$$

where $\mathbf{J}$ is expressed in mol/sec-cm^2; N_A is Avogadro's number (6.02×10^{23}/mol); R is the gas constant (1.98 cal/$^\circ$K-mol); F is Faraday's constant ($96,480$ C/mol); u is μ/N_A : molar mobility (cm^2/V-sec-mol); and I is A/cm^2.

The Nernst-Planck equation describes the ionic current flow driven by electrochemical potentials (concentration gradient and electric field). The negative sign indicates that I flows in the opposite direction as $\frac{\partial V}{\partial x}$ increases and in the opposite (same) direction as $\frac{\partial [C]}{\partial x}$ increases if z is positive (negative). This equation describes the passive behavior of ions in biological systems. Ions flow *down* their concentration gradients and the electric fields. It is the equation that is most widely used for ion flux in neurophysiology. In later sections and chapters, we will apply this equation to many physiological conditions. Several fundamental equations describing electric current flow across the membrane will be derived from this equation.

2.4 The Nernst equation

The NPE gives the explicit expression of ionic current in terms of concentration and electric potential gradients. If one examines the electric current across the cell membrane, it is very important to determine under what condition the net cross-membrane current is zero, i.e., the membrane is at rest. This condition can easily be derived from the NPE by setting the total cross-membrane current to zero, i.e.,

$$I = -\left(uz^2F[C]\frac{\partial V}{\partial x} + uzRT\frac{\partial [C]}{\partial x}\right) = 0.$$

$$\frac{\partial V}{\partial x} = \frac{-RT}{zF}\frac{1}{[C]}\frac{\partial [C]}{\partial x} \rightarrow \int_{x_1}^{x_2}\frac{dV}{dx}dx = -\frac{RT}{zF}\int_{x_1}^{x_2}\frac{d[C]}{[C]dx}dx.$$

Change variables:

$$\int_{V_1}^{V_2}dV = -\frac{RT}{zF}\int_{[C]_1}^{[C]_2}\frac{d[C]}{[C]}.$$

Therefore,

$$V_2 - V_1 = -\frac{RT}{zF}ln\frac{[C]_2}{[C]_1}. \tag{2.4.8}$$

The membrane potential of a cell is defined

$$V_m \overset{\text{def}}{=} V_{\text{in}} - V_{\text{out}}. \tag{2.4.9}$$

The equilibrium potential of ion i, defined as the cross-membrane potential at which membrane current carried by ion i equals zero, is therefore

$$E_i = V_m (I = 0) \overset{\text{def}}{=} V_{\text{in}} - V_{\text{out}} = \frac{RT}{zF} \ln \frac{[C]_{\text{out}}}{[C]_{\text{in}}}. \tag{2.4.10}$$

Equation 2.4.10 is the Nernst equation, which gives an explicit expression of the equilibrium potential of an ion species in terms of its concentrations inside and outside the membrane.

Equation 2.4.10 also implies that when the membrane is at the equilibrium potential of an ion species, the cross-membrane voltage and concentration gradient exert equal and opposite forces that counter each other. Take the K^+ ions as an example. $[K^+]_{\text{in}} > [K^+]_{\text{out}}$ in most cells, and thus K^+ ions tend to flow outward (down their concentration gradient). The equilibrium potential calculated by the Nernst equation gives a negative value (e.g., -75 mV; see table 2.1). This negative $V_m (= V_{\text{in}} - V_{\text{out}})$ results in an inward-pointing electric field and thus drives the K^+ (positively charged) ions to flow inward. The two forces, outward chemical gradient and inward electric field, thus cancel each other and result in zero cross-membrane ionic current.

It is not difficult to show that at $T = 20°C$ and $z = +1$,

$$E_i = 58 \text{ mV } \log_{10} \frac{[C]_{\text{out}}}{[C]_{\text{in}}}.$$

(The detailed derivation is given in the answer to homework problem 2.1.)

For humans or warm-blooded animals, body temperature is about $37°C$.

$$E_i = 62 \text{ mV } \log_{10} \frac{[C]_{\text{out}}}{[C]_{\text{in}}}.$$

Table 2.1 illustrates the concentrations and equilibrium potentials (calculated by using the Nernst equation) of major ions in the frog muscle cells, the squid giant axon, and typical mammalian cells.

Although there are species differences, the concentration distribution of the four major ions in most animal cells follows the same general rules: $[K^+]_{\text{in}} > [K^+]_{\text{out}}$, $[Na^+]_{\text{in}} < [Na^+]_{\text{out}}$, $[Cl^-]_{\text{in}} < [Cl^-]_{\text{out}}$, and $[Ca^{2+}]_{\text{in}} < [Ca^{2+}]_{\text{out}}$. According to the Nernst equation, these ion distributions result in positive E_{Na} and E_{Ca}, and negative E_K and E_{Cl}. The causes of these asymmetrical ion distributions across cell membranes will be discussed in the next section.

Table 2.1 Ion concentrations and equilibrium potentials

	Inside (mM)	Outside (mM)	Equilibrium Potential (NE) $E_i = \frac{RT}{zF} \ln \frac{[C]_{out}}{[C]_{in}}$
Frog muscle (Conway 1957)			$T = 20°C = 293°K$
K^+	124	2.25	$58 \log \frac{2.25}{124} = -101$ mV
Na^+	10.4	109	$58 \log \frac{109}{10.4} = +59$ mV
Cl^-	1.5	77.5	$-58 \log \frac{77.5}{1.5} = -99$ mV
Ca^{2+}	4.9[†]	2.1	$29 \log \frac{2.1}{10^{-4}} = +125$ mV
Squid axon (Hodgkin 1964)			
K^+	400	20	$58 \log \frac{20}{400} = -75$ mV
Na^+	50	440	$58 \log \frac{440}{50} = +55$ mV
Cl^-	40-150	560	$-58 \log \frac{560}{40-150} = -66 - (-33)$ mV
Ca^{2+}	0.4[†]	10	$29 \log \frac{10}{10^{-4}} = +145$ mV
Typical mammalian cell			$T = 37°C = 310°K$
K^+	140	5	$62 \log \frac{5}{140} = -89.7$ mV
Na^+	5-15	145	$62 \log \frac{145}{5-15} = +90.7 - (+61.1)$ mV
Cl^-	4	110	$-62 \log \frac{110}{4} = -89$ mV
Ca^{2+}	1-2[†]	2.5-5	$31 \log \frac{2.5-5}{10^{-4}} = +136 - (+145)$ mV
[†](10^{-4}) free			

2.5 Ion distribution and gradient maintenance

Ions in biological systems are not uniformly distributed. The intracellular concentrations of Na^+, Cl^-, and Ca^{2+} in most animal cells are lower than those in the extracellular space, whereas the extracellular K^+ concentration is lower than that inside the cells (see table 2.1). These ionic concentration gradients across the cell membrane constitute the driving forces (or chemical potentials) for ionic currents flowing through open channels in the membrane. In other words, the ionic concentration gradients act like DC batteries for cross-membrane currents: the larger the gradients, the stronger the currents. The relationship between cross-membrane currents and concentration gradients of individual ions will be discussed in the next chapter.

Most biological membranes are, to varying degrees, permeable to small ions like K^+ and Cl^-. This implies that if there were no maintenance mechanisms, the concentrations of these ions inside and outside the cells would be the same because ions would flow from the high-concentration side to the low-concentration side until there were no driving forces (or concentration gradients). However, all living cells manage to maintain their ionic concentration gradients although their membranes are permeable to the ions. There are two types of maintenance mechanisms, described in the next two sections.

2.5.1   Active transport of ions

There are proteins in the plasma membrane of most animal cells that are capable of pumping ions from one side of the membrane to the other, often against their concentration gradients. The actions of such proteins often consume energy, which in some cases comes from the hydrolysis of ATP molecules, and in other cases from the chemical potential of other ions. It is not the intention of this chapter to discuss ion transport in great detail. Instead, the basic properties of major ion transporters are briefly summarized below:

Na^+-K^+ **pump** This is probably the most important ion transporter in biological membranes. This transporter is driven by the energy derived from the hydrolysis of ATP (it is therefore an ATPase). Three Na^+ are pumped out for every two K^+ pumped in, thus the net exchange of ions is one cation out per pump turnover. This type of pump is called *electrogenic* (discussed in section 2.7.3.3). The pump can be blocked by cyanide, DNP, or ouabain. The Na^+-K^+ pump is ubiquitous in the plasma membrane of virtually all animal cells. It is the primary cause of the Na^+ and K^+ concentration gradients across the plasma membrane: $[Na^+]_{out} > [Na]_{in}$ and $[K^+]_{in} > [K^+]_{out}$.

Na^+-Ca^{2+} **exchanger** These drive 3 Na^+ inward and 1 Ca^{2+} outward. Since Na^+ is driven *down* its concentration gradient, it provides the energy source. Therefore, this pump is not directly driven by ATP but rather by the Na^+ concentration gradient that is maintained by the ATP-driven Na^+-K^+ pumps. The primary function of this exchange is to keep $[Ca^{2+}]_{in}$ low.

Ca^{2+} **pump** This is an ATP-driven pump located in the endoplasmic reticulum and plasma membrane. It requires Mg^{2+} as a cofactor and drives Ca^{2+} into the endoplasmic reticulum and out of the plasma membrane to keep the cytosolic Ca^{2+} low.

Bicarbonate-Cl^- exchanger This exchanger is driven by Na^+ influx, and it pumps HCO_3^- in and Cl^- out of the plasma membrane. This system is inhibited by SITS and DIDS. Its primary function is to keep $[Cl^-]_{in}$ low and the intracellular pH high.

Cl^--Na^+-K^+ cotransporter This is driven by Na^+ influx and transports Na^+ (inward), K^+ (inward), and Cl^- (inward) in a ratio of $1:1:2$. It can be blocked by furosemide and bumetamide.

All the processes listed above are used by living cells to maintain ionic concentration gradients across their plasma membranes. Ions whose concentration gradients are maintained by such active (energy-consuming) processes are said to be *actively distributed*.

2.5.2 Passive distribution of ions and Donnan equilibrium

As well as by active transporters, ion concentration gradients can be maintained by the selective permeabilities of the plasma membrane to various ions. As mentioned in previous sections, most membranes at rest are permeable to K^+ and perhaps Cl^- but are much less permeable to Na^+ and Ca^{2+}. Moreover, many impermeant anions, such as SO_4^{2-} and small charged proteins, are found inside cells. These differences in ion permeability of the cell membrane, as shown below, can result in ion concentration gradients across the cell membrane. Since this type of ion distribution requires no energy, it is called *passive distribution*.

If a cell membrane is permeable to several ion species, and if no active transport is present for these ions, the ions are said to be *passively distributed* and the membrane potential of this cell should be equal to the equilibrium potential (determined by Nernst equation) of each of these ions. i.e.,

$$V_m = \frac{RT}{zF} \ln \frac{[C]_{out}}{[C]_{in}}$$

for all permeable ions.

Let C^{+m} = cation of valence m; A^{-n} = anion of valence n. Then

$$\left[\frac{C^{+m}_{out}}{C^{+m}_{in}}\right]^{\frac{1}{m}} = \left[\frac{A^{-n}_{in}}{A^{-n}_{out}}\right]^{\frac{1}{n}}. \tag{2.5.11}$$

This is the Donnan rule of equilibrium.

Taking the example of the frog muscle, where K^+ and Cl^- are the two permeable ions at rest, then at Donnan equilibrium,

$$\frac{[K^+]_{out}}{[K^+]_{in}} = \frac{[Cl^-]_{in}}{[Cl^-]_{out}}. \tag{2.5.12}$$

In most cells, there are a sizable number of negatively charged molecules (A^-, proteins, etc.) in the cytoplasm that are *not* permeant to the membrane, and because of space-charge neutrality,

$$[K^+]_{in} = [Cl^-]_{in} + [A^-]_{in} \text{ and } [K^+]_{out} = [Cl^-]_{out}.$$

These relations plus the Donnan rule for K^+ and Cl^- yield

$$[K^+]_{in}^2 = [K^+]_{out}^2 + [A^-]_{in}[K^+]_{in}.$$

Therefore,

$$[K^+]_{in} > [K^+]_{out} \text{ and } [Cl^-]_{out} > [Cl^-]_{in}.$$

These results indicate that even without active transporters, the concentration of K^+ inside is higher than that outside of the cell, and the opposite is true for Cl^-. These distributions agree qualitatively with the direction of concentration gradients of K^+ and Cl^- in most animal cells (see table 2.1). The origin of these ion distribution differences is the existence of intracellular impermeable anions. These anions attract more K^+ into the cell and expel more Cl^- out of the cell, in accordance with the principles of space-charge neutrality.

In conclusion, the concentration gradients across the membrane of animal cells are maintained by two processes. In some cells, active transporters are abundant and thus ions are mainly distributed actively across the membrane. In other cells, where fewer transporters exist, passive distribution may take a substantial role in maintaining concentration gradients across the cell membrane.

Example 2.2

Consider a hypothetical two-compartment system separated by a membrane permeable to K^+ and Cl^-, but not to A^-. No active pump is involved.

	I	II	
A^-	100	0	
K^+	150	150	(in mM)
Cl^-	50	150	

a. Is the system in electrochemical equilibrium (ECE)?

b. If not, in what direction will each ion move? What are the final equilibrium levels each ion will reach in I and II?

Answer to example 2.2

a. Each compartment is in space-charge neutrality

$$I: \quad \overset{+}{100} + \overset{+}{50} \;=\; \overset{-}{150},$$
$$II: \quad \overset{+}{150} \;=\; \overset{-}{150},$$

but Cl^- is obviously not in ECE; Cl^- tends to diffuse from II to I.

b. Diffusion tends to move 50 mM of Cl^- from II to I, but then space-charge neutrality will be violated.

$$I: \quad \overset{-}{100} + \overset{-}{50} + \overset{-}{50} \;>\; \overset{+}{150},$$
$$II: \quad \overset{-}{100} \;<\; \overset{+}{150}.$$

Therefore, 50 mM K^+ tends to move with Cl^- from II to I. However, this will result in imbalance of K^+ concentration, and K^+ will diffuse back from I to II, etc. A simple way to calculate the final ECE of this system is to use the Donnan rule:

$$\frac{[K^+]_I}{[K^+]_{II}} = \frac{[Cl^-]_{II}}{[Cl^-]_I}.$$

In order to satisfy space-charge neutrality, equal amounts of K^+ and Cl^- must be moved from compartment II to I.

Let X be the amount of KCl that must be moved from II to I to achieve ECE. Then

$$\frac{150 + X}{150 - X} = \frac{150 - X}{50 + X}.$$

Solving this equation yields $X = 30$, so 30 mM of K^+ and 30 mM of Cl^- will flow from II to I to achieve electrochemical equilibrium. Thus,

	I	II
A^-	100	0
K^+	180	120
Cl^-	80	120

(in mM)

Answer to example 2.2 (continued)

To double-check:

1. Space-charge neutrality:

   $I: \quad \overset{-}{100} + \overset{-}{80} \quad = \quad \overset{+}{180}.$

   $II: \quad \quad \overset{-}{120} \quad = \quad \overset{+}{120}.$

2. ECE: $\frac{[K^+]_I}{[K^+]_{II}} = \frac{180}{120} = 1.5 = \frac{120}{80} = \frac{[Cl^-]_{II}}{[Cl^-]_I} \quad V_I - V_{II} = -58\log 1.5 =$
   -10.2 mV. However, the system is in *osmotic imbalance:* total
   ionic strength in $I = 100 + 180 + 80 = 360 >$ that in $II = 120 +$
   $120 = 240$. Thus, water will flow from II to I.

2.6 Effects of Cl⁻ and K⁺ on membrane voltage

In the previous section, we discussed the passive distribution of ions under equilibrium conditions (Donnan equilibrium). We will now describe
the behavior of membrane potential in response to sudden changes of
extracellular Cl⁻ and K⁺, the two permeable ions of the cell (nonequilib-
rium conditions). Hodgkin and Horowicz (1959) investigated this problem
extensively by using the isolated frog muscle fibers. Extracellular concentrations of Cl⁻ (solid trace at top of figure 2.2) or K⁺ (dashed trace in figure 2.2) were abruptly changed $[Cl^-]_{out} : 120$ mM $\rightarrow 30$ mM; or $[K^+]_{out} :$
2.5 mM $\rightarrow 10$ mM.

The membrane potential of the muscle in response to these changes is
shown in the lower portion of figure 2.2. The potential changes at seven
different instances (a–g) are explained as follows for Cl⁻.

(a) At rest, the cell is permeable to K⁺ and Cl⁻; thus $V_m = E_K = E_{Cl} =$
-98.5 mV. (b) When $[Cl^-]_{out}$ *suddenly* drops from 120 to 30 mM, E_{Cl}
becomes $-58\log \frac{30}{2.4} = -63.5$ mV, but E_K is still -98.5 mV. V_m depolarizes to a value in between $(-77$ mV$)$. (c) However, (b) is not at ECE
since $\frac{[K^+]_{out}}{[K^+]_{in}} = \frac{2.5}{125} = 0.02 \neq \frac{2.4}{30} = \frac{[Cl^-]_{in}}{[Cl^-]_{out}}$. Thus, Cl⁻ and K⁺ diffuse out
of the muscle cell until a new Donnan equilibrium is reached: $[K^+]_{in} \times$
$[Cl^-]_{in} = [K^+]_{out} \times [Cl^-]_{out} = 2.5 \times 30 = 75$. Because $[K^+]_{in} >> [Cl^-]_{in}$
and some H₂O flows out to maintain osmotic balance, $[K^+]_{in}$ is practically
unchanged. (d) $[Cl^-]_{in}$ at $(d) \approx \frac{1}{4}[Cl^-]_{in}$ at $(a) = \frac{1}{4}(2.4$ mM$) = 0.6$ mM,
because $[K^+]_{out} \times [Cl^-]_{out}$ at (a) $= 2.5 \times 120 = 300 = 4 \times [K^+]_{in} \times [Cl^-]_{in}$
at (d), and therefore $E_{Cl} = -58\log \frac{30}{0.6} = -98.5$ mV $= +58\log \frac{2.5}{125} =$

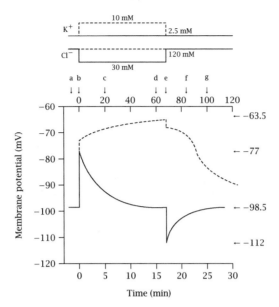

Figure 2.2   Effects of sudden reduction of extracellular Cl⁻ (solid traces) or sudden elevation of extracellular K⁺ (dashed traces) on the membrane potential of the isolated frog muscle. (Adapted from Hodgkin and Horowicz 1959.)

E_K. Therefore, $V_m = -98.5$ mV. (e) When $[Cl^-]_{out}$ returns from 30 mM to 120 mM, $E_{Cl} = -58 \log \frac{120}{0.6} = -134$ mV, and E_K is still at -98.5 mV, so V_m hyperpolarizes (overshoot) to a value in between (-112 mV). (f) The system again is not in ECE $\left(\frac{[K^+]_{out}}{[K^+]_{in}} \neq \frac{[Cl^-]_{in}}{[Cl^-]_{out}} \right)$, until (g) when Cl⁻ and K⁺ ($+ H_2O$) flow in (space-charge neutrality) to achieve $\frac{[Cl^-]_{in}}{[Cl^-]_{out}} = \frac{2.4}{120} = \frac{2.5}{125} = \frac{[K^+]_{in}}{[K^+]_{out}}$, which is the same as (a).

Similarly, in response to the change of extracellular K⁺, the membrane potential changes can be described as follows. (a) At rest, the cell is permeable to K⁺ and Cl⁻, $V_m = E_K = E_{Cl} = -98.5$ mV. (b) When $[K^+]_{out}$ increases from 2.5 to 10 mM, $E_K = 58 \log \frac{10}{104} = -59$ mV, E_{Cl} still $= -98.5$ mV. V_m depolarizes to a value (-73 mV) in between. (c) K⁺, Cl⁻, and H_2O flow into the cell and move V_m from -73 to -65 mV. (d) Both $[K^+]_{in}$ and $[Cl^-]_{in}$ are increased. -65 mV $= 58 \log \frac{10}{132} = -58 \log \frac{120}{9}$. Therefore, $[K^+]_{in} \approx 132$, $[Cl^-]_{in} \approx 9$ mM. (e) At return to $[K^+]_{out} = 2.5$ mM, $E_K = 48 \log \frac{2.5}{132} \approx -100$ mV, but the instantaneous jump at (e) is very small as the K⁺ channel conductance is voltage-dependent (smaller at depolarized voltages); and (f) and (g) K⁺, Cl⁻, and H_2O slowly move out of the cell; the ionic concentrations and V_m return to condition (a).

2.7   Movement of ions across biological membranes

In the previous section, we described the principles of ion movement in biological systems. We derived the Nernst-Planck equations, which dealt with ion movement under the influences of chemical potential and electric field. In this chapter, we shall employ the Nernst-Planck formulation and use it to study the behavior of ion movement within and across biological membranes.

Biological membranes consist of lipid bilayers with protein molecules anchored from either side or, often, anchored through the whole thickness of the membrane. Some of the cross-membrane proteins form pores, or channels, that allow ions to flow through and result in cross-membrane currents. Properties of ion channels are one of the most important topics of modern physiology, and we will spend a great deal of time discussing them in later chapters. In this chapter, we shall take a classical view of ion flux across the membrane. This view does not address individual ion channels in the membrane; rather, it makes several assumptions and provides a simplified description of the behavior of the whole membrane. This classical view of electrodiffusion across biological membranes was first formulated by Goldman (1943) and subsequently developed by Hodgkin and Katz (1949). It is therefore called the Goldman-Hodgkin-Katz model, and because the model assumes constant electric field in the membrane, it is also named the constant field model.

2.7.1   Membrane permeability

Before we describe the constant field model, we must introduce a very important term, the membrane permeability P. P is defined empirically by

$$J = -P\triangle[C], \tag{2.7.13}$$

where J is molar flux (mol/cm^2-sec), and P is membrane permeability to ion i (cm/sec). From sections 2.1 and 2.2,

$$J = -D\frac{d[C]}{dx}.$$

Within the membrane, if one assumes that [C] drops linearly with respect to x, $\frac{d[C]}{dx}$ can be written as $\frac{\triangle[C]}{l} \cdot \beta$, where β is the water-membrane partition coefficient for ion i (dimensionless), and l is the thickness (cm) of the membrane (see figure 2.3).

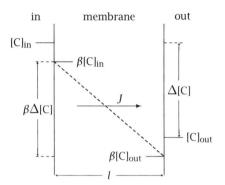

Figure 2.3 Concentration profile of ions through a membrane. The ordinate represents concentration C, and the abscissa represents distance. J is the ionic flux and β is the water-membrane partition coefficient.

$$\frac{d[C]}{dx} = \frac{\triangle[C]\beta}{l},$$

$$D = D^*.$$

Then

$$J = -\frac{\triangle[C]D^*\beta}{l}, \qquad (2.7.14)$$

Thus,

$$P = \frac{D^*\beta}{l}.$$

D^* is the diffusion coefficient for ion i within the membrane (cm^2/sec).

Hence, permeability P is governed by the solubility and diffusion coefficient of ion i in the membrane. By the Einstein relation,

$$D^* = \frac{kT}{q}\mu^* = \frac{u^*RT}{F} \text{ (molar form).}$$

Thus,

$$P = \frac{\beta u^*RT}{lF}, \qquad (2.7.15)$$

where u^* is the molar mobility of ion i within the membrane (cm^2/V-sec-mol). Equation 2.7.15 illustrates that the membrane permeability to a given ion species is proportional to the mobility of the ion in the membrane (u^*), the absolute temperature (T), and the relative solubility of the ion in the aqueous and membrane phases (β). It is inversely proportional to the thickness of the membrane.

2.7.2 The Goldman-Hodgkin-Katz (GHK) model

The Nernst-Planck equation derived in the last chapter describes ion current flow in aqueous media. When ions flow across the membrane, they pass through cross-membrane protein molecules that form aqueous pores that connect the interior (cytoplasm) and the exterior of the cell. In general, ions flowing through open channels may or may not obey the Nernst-Planck equation (NPE). In the case of simple aqueous pores, ions move down their electrochemical gradients and can be described approximately by the NPE. In the case where complex energy barriers or blocking sites are involved within a channel, NPE fails. In this section, we consider the case where NPE holds within the membrane, and the electric field within the membrane is constant. This model is called the Goldman-Hodgkin-Katz (GHK) constant field model and has been widely used to describe the ionic current flow across the cell membrane. As described later in this book, this model can be used to describe certain electrical properties of excitable membranes and certain ionic currents. But it falls short in describing other membrane properties because they do not follow the constant field assumption or the NPE. There are three basic assumptions for the GHK constant field model: (1) Ion movement within the membrane obeys the Nernst-Planck equation; (2) ions move across the membrane independently (without interacting with each other); and (3) the electric field $\mathbf{E}$ in the membrane is constant (i.e., the electric potential drops linearly across the membrane; $E = -\frac{dV}{dx} = -\frac{V}{l}$).

Based on the first assumption, the ionic current *across* the membrane can be described by the NPE:

$$I = -\left(u^* z^2 F[C] \frac{dV}{dx} + u^* zRT \frac{d[C]}{dx} \right).$$

$\frac{dV}{dx} = \frac{V}{l}$, based on assumption 3.

Recall that the negative sign indicates that I flows in the opposite direction as $\frac{dV}{dx}$. Now I is defined to be positive when flowing from "in" to "out"; thus the first term is positive and the second term is negative (see figure 2.4). Therefore,

$$I = u^* z^2 F[C] \frac{V}{l} - u^* zRT \frac{d[C]}{dx}. \tag{2.7.16}$$

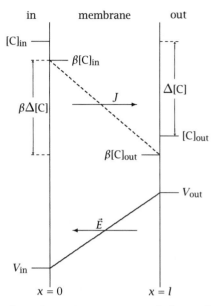

Figure 2.4 Concentration and electrical potential profiles of a membrane for the constant field model. The ordinate represents concentration C (upper portion) and electrical potential V (lower portion). The abscissa represents distance. J is the ionic flux. β is the water-membrane partition coefficient and $\vec{E}$ is the electric field.

Let $y = I - \dfrac{u^* z^2 F[C]V}{l}$.

Then

$$\frac{dy}{dx} = \frac{dI}{dx} - \frac{u^* z^2 FV}{l}\frac{d[C]}{dx},$$

and

$$\frac{dI}{dx} = 0 \quad \text{because } I \text{ is in steady state.}$$

Substituting equation 2.7.16 into the above equation yields

$$y = u^* z^2 F[C]\frac{V}{l} - u^* zRT\frac{d[C]}{dx} - \frac{u^* z^2 F[C]V}{l} = \frac{RTl}{zFV}\frac{dy}{dx},$$

$$\int_{x=0}^{x=l} dx = \frac{RTl}{zFV}\int_{y(C_{x=0})}^{y(C_{x=l})}\frac{dy}{y} \rightarrow l = \frac{RTl}{zFV}\ln\frac{I - \frac{u^* z^2 FV\beta}{l}[C]_{\text{out}}}{I - \frac{u^* z^2 FV\beta}{l}[C]_{\text{in}}}.$$

Note : $[C]_{x=0} = \beta[V]_{in}, [C]_{x=l} = \beta[C]_{out}$. Therefore,

$$
\begin{aligned}
I &= \frac{u^* z^2 F V \beta}{l} \left[\frac{[C]_{out} e^{\frac{-zFV}{RT}} - [C]_{in}}{e^{\frac{-zFV}{RT}} - 1} \right] \\
&= PzF\xi \left(\frac{[C]_{in} - [C]_{out} e^{-\xi}}{1 - e^{-\xi}} \right)
\end{aligned}
\tag{2.7.17}
$$

(Recall that $P = \dfrac{\beta u^* RT}{lF}$ [equation 2.7.15], and define $\xi \overset{\text{def}}{=} \dfrac{zVF}{RT}$.)

Equation 2.7.17 is called the *GHK current equation*, which gives the current-voltage relation (*I-V* relation) of the membrane or ionic current.

Because of the assumption of independence (assumption 2), this equation can be split into two expressions representing the independent, unidirectional flux:

$$
\text{efflux} : I_{out} = PzF\xi \frac{[C]_{in}}{1 - e^{-\xi}},
\tag{2.7.18}
$$

and

$$
\text{influx} : I_{in} = -PzF\xi \frac{[C]_{out} e^{-\xi}}{1 - e^{-\xi}}.
\tag{2.7.19}
$$

Note that the GHK current equation predicts that the membrane current (either total or unidirectional) is a nonlinear function of membrane potential (figure 2.5). The current-voltage relation depends on the ratio, $[C]_{out}/[C]_{in}$. As $\frac{[C]_{out}}{[C]_{in}} = 1$, $I = PzF\xi[C]_{out} = \frac{Pz^2F^2[C]_{out}}{RT}V$, the *I-V* relation is linear. When $\frac{[C]_{out}}{[C]_{in}} < 1$, *I-V* shows outward rectification (slope increases with membrane voltages). When $\frac{[C]_{out}}{[C]_{in}} > 1$, *I-V* shows inward rectification (slope decreases with membrane voltages). The intersecting points of these *I-V* curves with the abscissa (*V*-axis) are the equilibrium potentials of the ion at each concentration ratio: They follow the values calculated from the Nernst equation. This is because the Nernst equation is derived under the condition that the net cross-membrane current equals zero (see section 2.1.3).

Figure 2.6 shows that, according to the assumption of independence, total current can be represented as the sum of inward and outward currents.

The GHK current equation gives the *I-V* relation of ion permeation across the membrane under the assumptions stated above. If we know

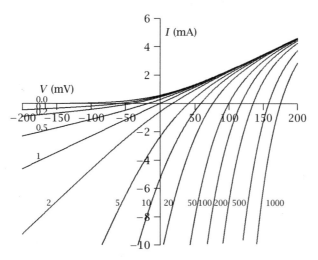

Figure 2.5 Current-voltage relations given by equation 2.7.17 (GHK current equation) for various values of $[\mathrm{C}]_{out}/[\mathrm{C}]_{in}$ (indicated by small numbers near each curve).

the major permeant ions for a cell, under the assumption of no electrogenic pump, the *resting potential* of the cells can be calculated by setting the total current across the membrane equal to zero.

For a cell that is permeable to K^+, Na^+, and Cl^- ions,

$$
\begin{aligned}
I &= I_K + I_{Na} + I_{Cl} \\
&= P_K zF\xi \frac{[K^+]_{in} - [K^+]_{out}e^{-\xi}}{1 - e^{-\xi}} + P_{Na}zF\xi \frac{[Na^+]_{in} - [Na^+]_{out}e^{-\xi}}{1 - e^{-\xi}} \\
&\quad + P_{Cl}zF\xi \frac{[Cl^-]_{in} - [Cl^-]_{out}e^{-\xi}}{1 - e^{-\xi}} \\
&= P_K zF\xi \frac{y - we^{-\xi}}{1 - e^{-\xi}},
\end{aligned}
\tag{2.7.20}
$$

where

$$
y = [K^+]_{in} + \frac{P_{Na}}{P_K}[Na^+]_{in} + \frac{P_{Cl}}{P_K}[Cl^-]_{out},
$$

$$
w = [K^+]_{out} + \frac{P_{Na}}{P_K}[Na^+]_{out} + \frac{P_{Cl}}{P_K}[Cl^-]_{in}.
$$

At steady state $I = 0$, then $y - we^{-\xi} = 0$.

$$
e^{\xi} = \frac{w}{y} = \frac{P_K[K^+]_{out} + P_{Na}[Na^+]_{out} + P_{Cl}[Cl^-]_{in}}{P_K[K^+]_{in} + P_{Na}[Na^+]_{in} + P_{Cl}[Cl^-]_{out}}.
$$

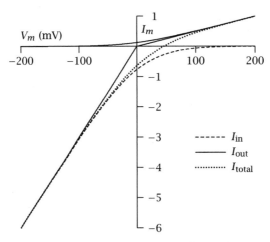

Figure 2.6   Unidirectional current-voltage relations given by Equations 2.7.18 and 2.7.19 for $[C]_{out}/[C]_{in} = 5$. The solid curve is the outward current and the dashed curve is the inward current. The dotted curve is the total current.

Therefore,

$$V = \frac{RT}{F} \ln \frac{P_K[K^+]_{out} + P_{Na}[Na^+]_{out} + P_{Cl}[Cl^-]_{in}}{P_K[K^+]_{in} + P_{Na}[Na^+]_{in} + P_{Cl}[Cl^-]_{out}}. \tag{2.7.21}$$

Equation 2.7.21 is the *GHK voltage equation*, which describes the steady-state membrane potential (the resting potential) of a cell.

The GHK equations give explicit relations between I, V, and $[C]$ and between V_{rest}, P, and $[C]$. These are important measurable parameters; thus, GHK equations have been very powerful and commonly used tools for analyzing excitable membranes.

2.7.3   Applications of GHK equations

Although the GHK equations are derived from simple assumptions in the constant field model, they can be used to describe a wide range of membrane behavior. A number of important membrane properties agree with the GHK equations reasonably well. A few examples are given below.

2.7.3.1   Resting potential   Take the squid giant axon as an example. At rest, the ratio of permeabilities $P_K : P_{Na} : P_{Cl} = 1 : 0.03 : 0.1$, from the GHK voltage equation of the membrane to K^+, Na^+, and Cl^-:

$$V_{rest} = 58 \log \frac{1(10) + 0.03(460) + (0.1)40}{1(400) + 0.03(50) + 0.1(540)} = -70 \text{ mV}.$$

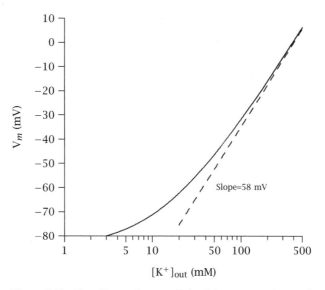

Figure 2.7 The effects of extracellular K^+ concentration on the membrane potential. The solid curve is given by the GHK voltage equations for K^+, Na^+, and Cl^-, and the dashed line is given by the Nernst equation for K^+.

This gives reasonable agreement with experimental results. Additionally, if one varies the external K^+ concentration ($[K^+]_{out}$), the resting potential follows the predictions of the GHK voltage equation very well (solid curve in figure 2.7).

As $[K^+]_{out}$ increases,

$$V_{rest} = \frac{RT}{F} \ln \frac{P_K[K^+]_{out} + P_{Na}[Na^+]_{out} + P_{Cl}[Cl^-]_{in}}{P_K[K^+]_{in} + P_{Na}[Na^+]_{in} + P_{Cl}[Cl^-]_{out}}$$

depends more on K^+, so

$$
\begin{aligned}
V_{rest} &\approx \frac{RT}{F} \ln \frac{P_K[K^+]_{out}}{P_K[K^+]_{in}} \\
&= 58 \text{ mV} \log \frac{[K^+]_{out}}{[K^+]_{in}},
\end{aligned}
$$

which is the Nernst equation (dashed line in figure 2.7). Therefore, at high $[K^+]_{out}$, the solid curve and dashed line approach each other.

But at normal $[K^+]_{out}$ concentration, $[K^+]_{out} = 10$ mM, and K^+ only contributes about one-third of the resting potential. Thus, the GHK voltage equation must be used to describe the resting potential.

2.7.3.2 Action potential During the action potential $P_K : P_{Na} : P_{Cl}$ changes from $1 : 0.03 : 0.1$ to $1 : 15 : 0.1$, so

$$V_m = 58 \log \frac{1(10) + 15(460) + 0.1(40)}{1(400) + 15(50) + 0.1(540)} = +44 \text{ mV}.$$

2.7.3.3 Effects of electrogenic pumps on membrane potential By using the GHK voltage equation, we can estimate, for example, the contribution of the Na^+-K^+ pump on the membrane potential. This is illustrated in the following example.

Example 2.3

I_i :   passive current

I_P :   current generated by pump

r :   the number of Na^+ pumped out for each K^+ pumped in

At steady state, $I_{\text{total}} = I_i + I_P = 0$. Therefore,

$$\left.\begin{array}{r} I_{Na} + I_{NaP} = 0 \\ I_K + I_{KP} = 0 \\ rI_{KP} + I_{NaP} = 0 \end{array}\right\} = rI_K + I_{Na} = 0.$$

Use the GHK current equation:

$$\frac{F\xi}{1 - e^{-\xi}} \left[rP_K \left([K^+]_{\text{in}} - [K^+]_{\text{out}} e^{-\xi} \right) + P_{Na} \left([Na^+]_{\text{in}} - [Na^+]_{\text{out}} e^{-\xi} \right) \right] = 0.$$

Therefore,

$$V = \frac{RT}{F} \ln \frac{rP_K[K^+]_{\text{out}} + P_{Na}[Na^+]_{\text{out}}}{rP_K[K^+]_{\text{in}} + P_{Na}[Na^+]_{\text{in}}}.$$

As mentioned earlier, r for Na^+-K^+ pump is 1.5 (3 Na^+ out for 2 K^+ in). Then for squid axon at rest,

$$V_m^P = 58 \log \frac{1.5 \times 1 \times 10 + 0.03 \times 440}{1.5 \times 1 \times 400 + 0.03 \times 50} = -88 \text{ mV}.$$

If no pump, or electrically neutral pump, $r = 1$. Then,

$$V_m = 58 \log \frac{1 \times 10 + 0.03 \times 440}{1 \times 400 + 0.03 \times 50} = -82 \text{ mV}.$$

Therefore, V_m (Na^+-K^+ pump) − V_m (neutral pump) = $-88 - (-82) = -6$ mV. The Na^+-K^+ pump contributes about -6 mV (7%) of the resting potential of the squid axon. In some mammalian cells, the contribution can be 15% of the resting potential.

2.8 Review of important concepts

1. Electric signals in excitable cells are carried by dissociated Na^+, K^+, Cl^-, and Ca^{2+} ions. Movement of ions is dictated by four physical laws: Fick's law of diffusion, Ohm's law of charge drift, Einstein's relation between diffusion and mobility, and the principle of space-charge neutrality.

2. Ion flux (or current) in biological systems can be described by the Nernst-Planck equation. The equilibrium potential of an ion species across the cell membrane can be expressed in terms of the ratio of ion concentrations inside and outside the cell (Nernst equation).

3. Ionic concentration gradients across the cell membrane are maintained by two processes: active distribution (by active transporters) and passive distribution (by Donnan's rule of equilibrium).

4. For membranes where the constant field (Goldman-Hodgkin-Katz) assumptions hold, the membrane current can be described by the GHK current equation, and the membrane voltage can be described by the GHK voltage equation.

2.9 Homework problems

1. In this chapter we derived the Nernst equation. Show that

$$\frac{RT}{F} \ln \frac{[C]_{\text{out}}}{[C]_{\text{in}}} = 58(\text{mV}) \log_{10} \frac{[C]_{\text{out}}}{[C]_{\text{in}}} \text{ when } T = 20°C,$$

where
R = ideal gas constant = $1.98 \text{ cal}(°K)^{-1} \cdot \text{mol}^{-1}$,
F = Faraday constant = $96,000$ C-equivalent^{-1}.

Hint: 1 cal = 4.2 joules; 1 V = 1 joule/C.

2. In this chapter, we mentioned that the space-charge neutrality is a very good approximation in excitable cells because very few uncompensated changes are needed to produce large electrical potentials across the membrane. If the membrane capacitance is 1 $\mu F/cm^2$ (i.e., 10^{-6} uncompensated coulombs of charge on each side of the 1 cm^2 membrane are needed to produce 1 V across the membrane),

and the concentrations of ions inside and outside the cell are about 0.1 M, calculate the fraction of uncompensated ions on each side of the membrane required to produce 100 mV

  (a) across a 1 cm^2 patch of membrane (fraction of 1 cm^3 cytoplasm);

  (b) in a spherical cell (10 μm radius); and

  (c) in a cylindrical cell (1 μm radius, 100 μm long).

3. A membrane is permeable to H$_2$O, K$^+$, and Cl$^-$ but is not permeable to a large organic ion R$^+$. The initial concentrations of RCl and KCl on the two sides of the membrane are given:

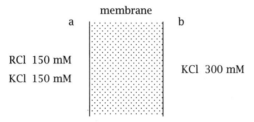

  (a) What are the final concentrations of R$^+$, K$^+$, and Cl$^-$ on each side of the membrane at equilibrium?

  (b) What is V_m at equilibrium?

  (c) Will there be any osmotic pressure? If so, in which direction?

4. Using the radioisotope tracers, the ion fluxes were measured from isolated frog muscle fibers. Results are shown as

$$
\begin{array}{ccc}
\text{(IN)} & & \text{(OUT)} \\
[\text{Na}^+]_{\text{in}} = 9.2 \text{ mM} & \rightleftharpoons & [\text{Na}^+]_{\text{out}} = 120 \text{ mM} \\
& J_{\text{in}}^{Na}=3.5 & \\
\end{array}
$$

$$
\begin{array}{ccc}
& J_{\text{ef}}^{K}=8.8 & \\
[\text{K}^+]_{\text{in}} = 140 \text{ mM} & \rightleftharpoons & [\text{K}^+]_{\text{out}} = 2.5 \text{ mM} \\
& J_{\text{in}}^{K}=5.4 & \\
\end{array}
$$

at $V_m = -90$ mV.

All fluxes were measured in steady state, and they are in 10^{-12} mol-cm^{-2}-sec^{-1}.

    (a) Calculate P_K and P_{Na} from the above data.

    (b) What is the resting potential of this cell if the membrane is permeable to K^+, Na^+, and Cl^-, and Cl^- is passively distributed across the membrane?

    (c) Repeat (b) for $P_K/P_{Na} = 0$ and for $P_K/P_{Na} = 1$.

5. The unicellular organism *Paramecium caudatum* shows a resting potential (RP) and an action potential (AP) that are similar in many respects to corresponding neural potentials. With the cell in "typical pond water," the following measurements were made with an intracellular electrode:

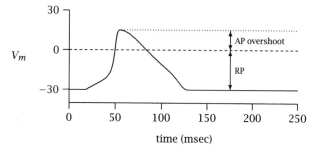

If one varies $[K^+]_{out}$ only, or $[Ca^{2+}]_{out}$ only, one observes the following:

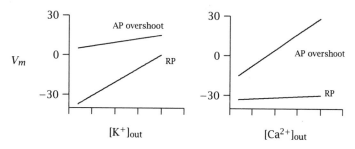

In the following questions, assume that the membrane of *P. caudatum* is normally permeable only to K^+, Ca^{2+}, and water.

    (a) In the resting state, which of these is true? Explain concisely.

      i. $P_K > P_{Ca}$

      ii. $P_K = P_{Ca}$

      iii. $P_K < P_{Ca}$

(b) Which is true during the peak of the AP? Explain concisely.

(c) Compared to the ionic concentrations of "typical pond water," is $[K^+]_{in}$ greater than, equal to, or less than $[K^+]_{out}$? Explain.

(d) Compare also $[Ca^{2+}]_{in}$ with $[Ca^{2+}]_{out}$.

(e) When the posterior end of the organism is mechanically tapped, the membrane transiently hyperpolarizes. What permeability change(s) might be responsible? Explain.

6. (a) *Briefly* state the assumptions for the constant field model.

(b) Sketch approximately the *I-V* relations predicted by the constant field model for various ratios of intracellular and extracellular ion concentrations, i.e., when $\frac{[C]_{in}}{[C]_{out}} = 0, 0.1, 1, 30$, or ∞.

(c) Using the data provided in the figure below, calculate the ratio of P_{Na}/P_K that predicts the resting potential as a function of $[K^+]_{out}$ for the *Myxicola* neuron. Note: $[Na^+]_{out} = 430$ mM, $[Na^+]_{in} = 12$ mM, $[K^+]_{in} = 270$ mM, and $P_{Cl} = 0$.

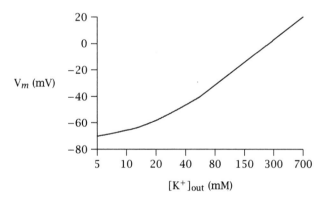

7. The ionic concentrations inside and outside of a neuron are given below:

	Inside (mM)	Outside (mM)
K^+	168	6
Na^+	50	337
Cl^-	41	340

(a) In the presence of K^+ and Na^+ channel blockers, the neuron is permeable only to Cl^-. If Cl^- movement across the cell membrane obeys the assumptions of the constant field model, sketch the current-voltage relation of this neuron. Label the axes and mark the intersecting point (with appropriate value and unit) of this relation with the voltage axis (i.e., when $I = 0$). Is this relation inward rectified or outward rectified? What causes this rectification?

(b) In the absence of K^+ and Na^+ channel blockers, the neuron is permeable to all three ions and the permeability ratio is $P_K : P_{Na} : P_{Cl} = 1 : 0.019 : 0.381$. What is the membrane potential of the neurons under this condition?

8. The intracellular and extracellular concentrations of K^+, Na^+, and Cl^- and anion A^- of a neuron are given below:

	Inside (mM)	Outside (mM)
K^+	150	150
Na^+	10	100
Cl^-	50	250
A^-	110	0

The plasma membrane of the neuron is permeable to K^+ and Cl^- but not to Na^+ and A^-.

(a) Is the neuron in electrochemical equilibrium (ECE)? Is the principle of space-charge neutrality obeyed?

(b) In what direction will each ion move if the neuron is not in ECE? What are the final equilibrium intracellular and extracellular concentrations of each of the four ions listed above?

(c) What are the values of the final equilibrium potentials of K^+, Na^+, and Cl^-? What is the resting potential of this neuron?

9. A large dissociated neuron (with $[K^+]_{in} = 150$ mM, $[Na^+]_{in} = 10$ mM, $[Cl^-]_{in} = 50$ mM, anions $[A^-]_{in} = 110$ mM, $[Ca^{2+}]_{in} = 10^{-4}$ mM) is placed in a chamber containing a small volume (about the same volume as that of the neuron) of culture medium (which consists of 150 mM K^+, 9 mM Na^+, 250 mM Cl^-, and 5 mM Ca^{2+}. The permeability ratio of the plasma membrane of the neuron at rest is $P_K : P_{Na} : P_{Cl} : P_A : P_{Ca} = 1 : 0 : 1 : 0 : 0$.

(a) Is the neuron in electrochemical equilibrium (ECE) immediately after it is placed in the culture medium? Explain concisely.

What are the final equilibrium intracellular and extracellular concentrations of K^+, Na^+, Cl^-, A^-, and Ca^{2+}? What is the resting potential of this neuron in the culture medium after reaching equilibrium?

(b) After reaching equilibrium, immediately after the onset of a sustained stimulus, $P_K : P_{Na} : P_{Cl} : P_A : P_{Ca} = 1 : 10 : 1 : 0 : 0$. Two seconds after the stimulus onset, $P_K : P_{Na} : P_{Cl} : P_A : P_{Ca} = 1 : 10 : 10 : 0 : 0$. Draw the voltage response of this neuron to the sustained stimulus (> 5 sec). Label the membrane voltage at rest, immediately after the stimulus onset, and at steady state with appropriate values and units.

3   Electrical Properties of the Excitable Membrane

3.1   Equivalent circuit representation

Biological membranes exhibit properties similar to those in electric circuits. It is customary in membrane physiology to describe the electrical behavior of biological membranes in terms of electric circuits. The equivalent electric circuit of a typical biological membrane is given in figure 3.1. The membrane capacitance represents the membrane dielectric property as a whole (mainly from the lipid bilayer), and it is independent of local variations of permeation channels. The capacitance per square centimeter (specific membrane capacitance) remains quite constant in most neurons, with a value very close to 1 μF/cm^2. The membrane resistance (or its reciprocal, membrane conductance) represents ion permeation through cross-membrane protein molecules (or channels). Unlike membrane capacitance, membrane conductances (or resistances) in excitable cells are often highly dependent on cross-membrane voltage and time. The voltage- and time-dependent changes in membrane conductance are the primary controllers of the electrical signals in excitable cells; we will focus mainly on membrane (or ionic) conductances in this and later chapters.

From the equivalent circuit representation (figure 3.1) and Kirchhoff's laws, one can write the differential equation for the total current flowing across a patch of membrane, I_m, which is the sum of the capacitive current, I_C, and the ionic current, I_i, flowing through the channels. The *driving force* of I_i is $V_m - E_r$, and the resistance of the channels is R_m.

$$
\begin{aligned}
I_m = I_C + I_i \;\; &= \;\; C_m \frac{dV_m}{dt} + \frac{(V_m - E_r)}{R_m} \\
&= \;\; C_m \frac{dV_m}{dt} + G_m(V_m - E_r).
\end{aligned}
\tag{3.1.1}
$$

where V_m is the membrane potential; I_m is the total membrane current

Biological membrane   Equivalent circuit representation

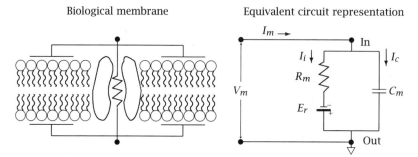

Figure 3.1   Equivalent circuit representation of the biological membrane. Parameters are given in the text.

(A/cm^2); C_m is the specific membrane capacitance (F/cm^2); R_m is the specific membrane resistance (Ω-cm^2); G_m is the specific membrane conductance (S/cm^2) $= \frac{1}{R_m}$; and E_r is the resting potential of the cell. If a single ion were responsible for I_i, then E_r would be replaced with the equilibrium potential or *Nernst* potential, E_i of ion i.

The ionic current in equation 3.1.1 is given as the product of total membrane conductance G_m and its driving force $(V_m - E_r)$. As we will describe later, ionic currents of a cell can usually be broken down into individual ion species, i.e., $I_i = I_{Na} + I_K + I_{Cl} + \cdots +$. Each ion species carries a current that can be written as the product of the conductance to that ion and the difference between the membrane potential and the equilibrium potential, e.g., $I_{Na} = g_{Na}(V_m - E_{Na})$.

The driving force is a useful concept that will be used often throughout this book. It is the difference between the actual membrane potential (at some point in time) and the *reversal potential* for the current. For the case illustrated in figure 3.1 and equation 3.1.1 in which the ionic current is driven by the resting potential, the reversal potential (V_{rev}) is equal to the resting potential, E_r, and is simply the value of membrane potential at which the current reverses polarity or

$$(V_m - E_r) = (V_m - V_{rev}). \tag{3.1.2}$$

If the current is carried by a single ion, then V_{rev} in Equation 3.1.2 becomes the equilibrium potential for that particular ion (E_i). A *large* driving force means a large difference between the membrane potential and the energy source (e.g., E_i) driving the ion through the membrane. If instead the current is being carried by several ions, then V_{rev} will be the reversal potential of the net current. The concept of a reversal potential is also used for synaptic potentials and will be discussed again in chapters 4 and 13.

The equivalent circuit representation is another important concept that will be used frequently in this book. The reader is encouraged to consult with appendix A for a review of circuit theory if the principles governing current flow in electrical circuits are not familiar.

3.2 Membrane conductance

Membrane conductance, G_m, is the reciprocal of membrane resistance ($G_m = \frac{1}{R_m}$). G_m is the sum of all ionic conductances in the cell membrane. In excitable cells, there are many types of ion conductances (ion channels). Some of them open when the cell is at rest, and others open when the cell is excited. Ionic conductances may stay constant with respect to changes in membrane potential, or they may vary with membrane voltage. Ion conductances can be activated by transmembrane voltage and by extracellular or intracellular ligands. The activation can be abrupt (nearly instantaneous), or it can be slow (time dependent). In view of the complexity of ionic conductances in excitable membranes, it is useful to define membrane and ionic conductances systematically. The convention used in this section follows that used by Jack, Noble, and Tsien (1975).

3.2.1 Linear membrane

A linear membrane is one that exhibits a linear relation between membrane ionic current I_i and the transmembrane potential (*I-V* relation is always a straight line):

$$I_i = G_m(V_m - E_i),$$

where G_m is a constant. Two additional definitions need to be introduced here: (1) membrane chord conductance = $G_m = \frac{I_i}{V_m - E_i}$; and (2) membrane slope conductance = $\frac{dI_i}{dV_m}$.

For linear membranes, chord conductance is always equal to the slope conductance:

$$G_m = \frac{I_i}{V_m - E_i} = \frac{dI_i}{dV_m}.$$

It is worthwhile to note that there is no uniform convention in neurophysiology for the usage of symbols for membrane or ionic conductances (refer also to chapter 4 and appendix A for discussion of conventions for symbols). Some authors use capital letter G, whereas others use small letter g. In this book, we use G_m as the membrane conductance and g_i (i can be Na^+, K^+, Cl^-, Ca^{2+}, or leak) as ionic conductances. $\frac{dI}{dV}$ is used as slope conductance.

3.2.2 Nonlinear membrane

A nonlinear membrane is one that shows a nonlinear relationship between membrane ionic current and the transmembrane potential, and/or its membrane conductance varies with time.

Case I Nonlinear membrane; I_i is a nonlinear function of V_m:

$$I_i = F(V),$$
$$\text{chord conductance} = \frac{F(V)}{(V - E_r)},$$
$$\text{slope conductance} = \frac{dI_i}{dV} = F'(V).$$

An example is given in figure 3.2.

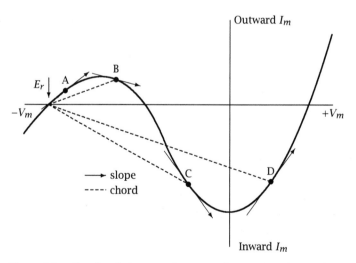

Figure 3.2 Chord and slope conductance of a current voltage relation (solid thick curve) at various points (A–D). E_r: membrane resting potential. Slope of the solid arrows: slope conductance. Slope of the dashed lines: chord conductance. (Adapted from Jack et al. 1975.)

At various points, the chord and slope conductances can be listed below:

	chord(G_m)	slope$\frac{dI_i}{dV}$
A	+	+
B	+	−
C	−	−
D	−	+

Here we can see the difference between the chord and slope conductances of a membrane: Chord conductance provides a measure of the value of ionic current *at* any given potential, whereas slope conductance determines the response to small variations in voltage or current *about* any given point.

The definition of chord conductance is ambiguous for the case shown in figure 3.2: E_r may be chosen for any of the three intersecting points with the abscissa, and thus the values of chord conductance will be different. This ambiguity will be overcome when the *I-V* relations of individual ions are considered (the above *I-V* relation is for more than one ionic current.)

Case II Nonlinear membrane; I_i is a function of both V_m and t: $I_i = f(V, t)$.

In cases where I_i varies with both V_m and time, such as the voltage-clamp record shown in figure 3.3, membrane ionic currents are usually analyzed in the following way. The time dependence of the membrane current is measured only at two selected instances: immediately after the onset of membrane potential change (instantaneous current, marked with $*$ in figure 3.3); and after the current reaches steady-state value (steady-state current, marked with ∞ in figure 3.3). These two measurements of time are taken before ($*$) and after (∞) the time-varying current takes place, and therefore the currents measured at these two instances *do not vary with time*. In this way, I^* and I^∞ become single variable functions (functions only of V_m, but not t), and this simplifies the analysis a great deal. Employing this approach, nonlinear membranes whose I varies with both V_m and t are characterized in the following components.

Instantaneous conductance This is achieved by changing V very quickly, which does not allow enough time for the time-dependent process to occur.

$$I_i^* = f(V, t^*),$$

where t^* is short enough to be considered a constant. This makes I_i^* a function of V only, and thus,

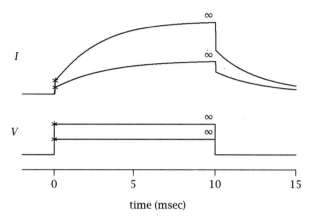

time (msec)

Figure 3.3 Membrane current (I) recorded under voltage-clamp conditions to voltage steps (V). $*$ indicates the instantaneous current, and ∞ indicates the steady-state current.

$$G_m^* = \frac{I_i^*}{V - E_r} = \frac{f(V, t^*)}{V - E_r} \quad \text{instantaneous chord conductance;}$$

$$\frac{dI_i^*}{dV} = f'(V, t^*) \quad \text{instantaneous slope conductance;}$$

and

$$I_i^* = f(V, t^*) \quad \text{instantaneous current-voltage relation.}$$

Steady-state conductance This can be achieved by allowing V to be held long enough for steady-state current to be developed: $I_i^\infty = f(V, t) = \infty$. This again makes I_i *not* a function of time, but a function only of voltage.

$$G_{m\infty} = \frac{I_i^\infty}{V - E_r} = \frac{f(V, \infty)}{V - E_r} \quad \text{steady-state chord conductance;}$$

$$\frac{dI_i^\infty}{dV} = f'(V, \infty) \quad \text{steady-state slope conductance;}$$

and

$$I_i^\infty = f(V, \infty) \quad \text{steady-state current-voltage relation.}$$

Example 3.1

The instantaneous and steady-state I-V relations of a neuron obtained by voltage-clamp experiments are shown below. The time-dependent current follows first-order kinetics with a time constant $\tau = 0.1$ second.

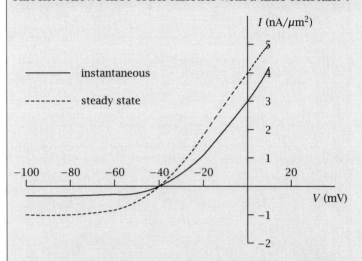

a. Draw the membrane ionic current, with respect to time, after the membrane voltage is stepped from $V_H = -40$ mV to $V_c = 0$ mV, and to $V_c = -80$ mV. Label the current and time axes with appropriate units.

b. Draw the membrane ionic current with respect to time after the membrane voltage is stepped from $V_H = -80$ mV to $V_c = -100$ mV. Label the current and time axes with appropriate units.

c. Plot in approximate scales the instantaneous and steady-state slope conductances (i.e., $\frac{\partial I^*}{\partial V}$ and $\frac{\partial I_\infty}{\partial V}$) vs. membrane voltage V.

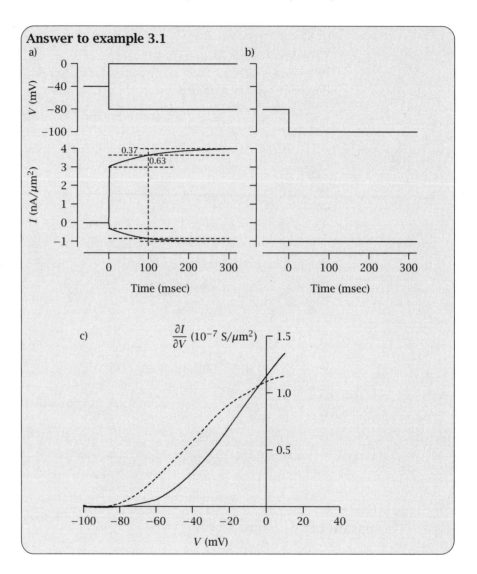

Answer to example 3.1

3.3 Ionic conductances

Ionic current can be written as the product of the membrane conductance to ion i (g_i) and the driving force ($V - E_i$), where E_i is the equilibrium potential of the ion.

$$I_i = g_i(V - E_i), \qquad E_i = \frac{RT}{zF} \ln \frac{[C]_{out}}{[C]_{in}}.$$

All the variables and definitions in section 3.2 may be applied to individual ionic currents I_{Na}, I_K, I_{Cl} and I_{Ca}, etc., so $I_i = f(V, E_i, t)$. However, two restrictions must be introduced if I_i is passive (i.e., if I_i flows down its electrochemical gradient):

1. $I_i = f(V, E_i, t) = 0$ when $V = E_i$, and
2. $I_i = f(V, E_i, t) \begin{subarray}{l} >0 \\ <0 \end{subarray}$ when $V \begin{subarray}{l} > \\ < \end{subarray} E_i$.

These two restrictions make the *I-V* curve of individual ionic current cross the voltage axis ($I_i = 0$) once and only once. Additionally, the chord conductance $g_i = \frac{I_i}{V - E_i}$ is always positive, whereas the slope conductance $\frac{dI_i}{dV}$ can be either positive or negative.

It is not too difficult to understand the conceptual implications of the two restrictions for individual ionic currents. First, each ion species has one and only one equilibrium potential, determined by the concentration ratio inside and outside the cell in accordance with the Nernst equation. This condition allows the *I-V* curve to cross the voltage axis only once at $V = E_i$ (thus, $I_i = g_i(V_m - E_i) = 0$). Second, because ionic current flows down its electrochemical gradient (passive current), the chord conductance g_i has to be positive. This is obvious when one examines the equation $I_i = g_i(V_m - E_i)$. When $V_m > E_i$, it is outward (positive) because the electrochemical gradient favors outward flux, and g_i is therefore positive; when $V_m < E_i$, I_i is inward (negative) because the electrochemical gradient favors inward flux, thus g_i is also positive. Negative chord conductance exists only for *active* ionic currents (those that are actively pumped against the electrochemical gradient of the ion).

3.4 The parallel conductance model

Electrical properties of excitable membrane in many cases resemble those described in electric circuits. For example, the membrane capacitance resembles a capacitor, membrane resistances resemble resistors, and the equilibrium potential of individual ions resembles the electromotive force (emf). It is customary in electrophysiology to describe the electrical properties of biological membrane by electric circuit diagrams (equivalent circuit representation). The equivalent circuit of a biological membrane consisting of K^+, Na^+, and Cl^- conductances is shown in figure 3.4. g_K, g_{Na}, and g_{Cl} are the chord conductances of K^+, Na^+, and Cl^- currents. C_m

is the membrane capacitance, and E_K, E_{Na}, and E_{Cl} are the equilibrium potentials for K$^+$, Na$^+$, and Cl$^-$. V_m is the transmembrane potential, and arrows on g_K and g_{Na} indicate that those two conductances vary with V_m and time. The total membrane current in figure 3.4 can be written as

$$I = I_C + I_K + I_{Na} + I_{Cl},$$

or

$$I = C\frac{dV}{dt} + g_K(V - E_K) + g_{Na}(V - E_{Na}) + g_{Cl}(V - E_{Cl}). \tag{3.4.3}$$

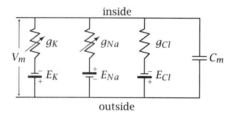

Figure 3.4   Parallel conductance model for ionic currents in a membrane. Parameters are given in the text.

At rest (steady state), $I_{\text{total}} = 0$ and $\frac{dV}{dt} = 0$, thus

$$V = \frac{g_K E_K + g_{Na} E_{Na} + g_{Cl} E_{Cl}}{g_K + g_{Na} + g_{Cl}}. \tag{3.4.4}$$

Equations 3.4.3 and 3.4.4 are very useful relations. 3.4.3 describes the total membrane current through various ionic conductances, and 3.4.4 gives the resting membrane potential in terms of ionic *conductances* and equilibrium potentials. (Recall that the GHK voltage equation gives the resting membrane potential in terms of membrane *permeabilities* to various ions and ion concentrations inside and outside the cell.) We will see and use equations 3.4.3 and 3.4.4 many times later in this book.

Example 3.2

The following concentration values (in mM) have been calculated for the giant cell of the sea snail *Aplysia:*

$[K^+]_{in} = 168$ $\quad [Na^+]_{in} = 50$ $\quad\quad [Cl^-]_{in} = 41$

$[K^+]_{out} = 6$ $\quad [Na^+]_{out} = 337$ $\quad [Cl^-]_{out} = 340$

and at rest, $P_K : P_{Na} : P_{Cl} = 1 : 0.019 : 0.381$.

a. What is V_{rest} as predicted by the GHK equation?

b. What would be the effect of a tenfold increase in the external K^+ concentration on the resting membrane potential?

c. The resting membrane conductances have been measured in this cell to be: $g_K = 0.57$ μS; $g_{Na} = 0.11$ μS; and $g_{Cl} = 0.32$ μS. What is the resting potential of this cell predicted by the parallel-conductance model?

Answer to example 3.2

$$
\begin{aligned}
\text{a.} \quad V_{rest} &= \frac{RT}{F} \ln \frac{P_K[K^+]_{out} + P_{Na}[Na^+]_{out} + P_{Cl}[Cl^-]_{in}}{P_K[K^+]_{in} + P_{Na}[Na^+]_{in} + P_{Cl}[Cl^-]_{out}} \\
&= 58 \log \frac{6 + (0.019)(337) + (0.381)(41)}{168 + (0.019)(50) + (0.381)(340)} \\
&= -59.6 \text{ mV}.
\end{aligned}
$$

$$
\begin{aligned}
\text{b.} \quad V_{rest} &= 58 \log \frac{60 + (0.019)(37) + (0.381)(41)}{168 + (0.019)(50) + (0.381)(340)} \\
&= -32.5 \text{ mV}.
\end{aligned}
$$

The cell is depolarized by 27.1 mV.

$$
\begin{aligned}
\text{c.} \quad E_K &= 58 \log \frac{6}{168} = -83.9 \text{ mV} \\
E_{Na} &= 58 \log \frac{337}{50} = +48 \text{ mV} \\
E_{Cl} &= -58 \log \frac{340}{41} = -53.3 \text{ mV}.
\end{aligned}
$$

$$
\begin{aligned}
V_{rest} &= \frac{g_K E_K + g_{Na} E_{Na} + g_{Cl} E_{Cl}}{g_K + g_{Na} + g_{Cl}} \\
&= \frac{(0.57)(-83.9) + (0.11)(48.1) + (0.32)(-53.3)}{(0.57 + 0.11 + 0.32)} \\
&= -59.6 \text{ mV}.
\end{aligned}
$$

3.5 Current-voltage relations

The current-voltage (*I-V*) relation is the most commonly used tool for analyzing membrane conductances. Ionic current density is plotted as a function of membrane voltage, and the slope of this relation gives the slope conductance of the membrane current. An example of *I-V* relations is given in figure 2.5, where membrane current density is plotted against membrane potential, based on the GHK constant field assumptions. In a given cell, the *I-V* relation can be obtained by the voltage-clamp method (see appendix A), which allows measurements of membrane current at any given "clamped" voltage. The *I-V* relations obtained with this method are usually plotted with current as the ordinate and voltage as the abscissa. Alternatively, *I-V* relations may also be obtained by measuring membrane voltages while injecting constant current pulses into the cell, a method known as current clamping (see appendix A). Membrane voltages obtained with this method are usually plotted as the ordinates and the currents as the abscissas. Such plots are sometimes called voltage-current (*V-I*) relations.

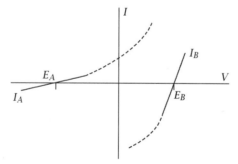

Figure 3.5 *I-V* relations of ions A and B with equilibrium potentials E_A and E_B. The solid lines show the linear portion and the dashed lines show the nonlinear portion of the *I-V* relations.

The current-voltage relations of most biological membranes are nonlinear over the entire physiological voltage range (−100 mV to +100 mV). Nevertheless, some *I-V* relations are approximately linear over a finite voltage range. In other words, the membrane conductance, or ion conductance, within a certain voltage range is ohmic (G_m is constant) as illustrated by I_A and I_B in figure 3.5.

I_A and I_B are linear with constant slopes within certain voltage ranges.

The slope is the conductance of the membrane (or a population of ion channels; i.e., $G_m = \frac{I_m}{V-E}$ = slope conductance = chord conductance = constant. Further analysis of linear membrane properties is given in chapter 4. Nonlinear membrane properties are discussed in chapters 5-7.

3.6 Review of important concepts

1. Electric properties of biological membrane can be characterized by parameters of electric circuits. The dielectric property of the membrane bilayer can be represented by electric capacitance, and the ion channels can be represented by electric conductances.

2. A linear membrane exhibits constant membrane conductance (or resistance), whereas nonlinear membranes exhibit conductances that vary with respect to membrane voltage and/or time.

3. Membrane voltage of a cell is determined by the equilibrium potentials and the relative conductances of permeant ions (parallel conductance model).

3.7 Homework problems

1. According to the constant field model,

$$I = PzF\xi \frac{[C]_{in}e^{\xi}-[C]_{out}}{e^{\xi}-1}, \text{ and}$$
$$P = \frac{\beta u RT}{aF}, \quad \xi. = \frac{zVF}{RT}.$$

   Show that the membrane slope conductance

$$\frac{dI}{dV} = \frac{PF^2z^2\xi}{RT} \frac{[C]_{out}}{(e^{\xi}-1)}$$

   for very small current (i.e., $I \rightarrow 0$).

2. The relative conductances for K^+, Na^+, and chloride of the plasma membrane of an invertebrate photoreceptor are given below:

   In darkness: $g_K : g_{Na} : g_{Cl} = 1 : 0.005 : 0.1$.

   Under constant light: $g_K : g_{Na} : g_{Cl} = 1 : 20 : 0.1$.

   The equilibrium potentials of these ions are
   $E_K = -90$ mV, $E_{Na} = +50$ mV, and $E_{Cl} = -50$ mV.

(a) What are the resting potentials of this photoreceptor in darkness and under constant light?

(b) The photoreceptor receives a synaptic input from a feedback interneuron in the retina, and this interneuron, when stimulated, alters the chloride conductance of the photoreceptor. Stimulation of the interneuron in darkness results in a sustained depolarization of about 10 mV in the photoreceptor. Does the interneuron stimulation cause an increase or a decrease in g_{Cl}? Why? If the input resistance of the photoreceptor is ohmic with a value of 10^8 Ω, what is the value of the change in g_{Cl}?

(c) From the information described in (b), and knowing that g_{Cl} is constant between -100 mV and $+50$ mV, calculate the amplitude of the voltage response of the photoreceptor to the same interneuron stimulation under constant light.

3. The current-voltage (I-V) relations of ionic currents (ions a and b) at resting state are given in the figure as A and B, and those at excitation state are given as A' and B'. (A and A' for ion a; B and B' for ion b.)

(a) What are the ionic conductances (g_a and g_b) at rest and at excitation states? Give the values of g_a and g_b in appropriate units.

(b) What is the resting potential of the cell?

(c) What is the peak membrane potential during excitation?

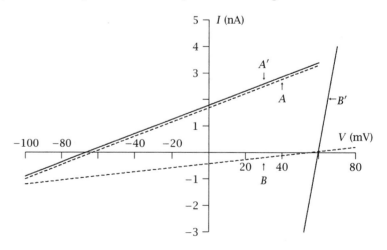

4. The instantaneous and steady-state *I-V* relations of a neuron obtained from voltage-clamp experiments are shown below:

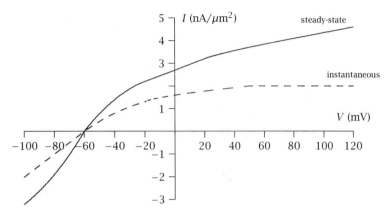

The time-dependent current follows first-order kinetics with the time constant $\tau = 0.1$ sec.

(a) Draw the membrane current with respect to time after the membrane voltage is stepped from $V_H = -60$ mV to $V_c = 0$ mV and to $V_c = -80$ mV. Label the current and time axes with the appropriate units.

(b) Repeat (a) after the membrane voltage is stepped from $V_H = +50$ mV to $V_c = +100$ mV.

4   Functional Properties of Dendrites

4.1   Introduction

In the previous chapter the basic concepts for the electrical properties of a linear membrane were presented. In this chapter we will build upon those concepts to present what are sometimes called the *cable properties* of neurons as an aid to understanding the function of neuronal dendrites. As we will see in this chapter, it is important to consider the electrical properties of neurons as they relate to specific geometries. For example, the electrical properties of neurons shaped like spheres are quite different from those of neurons with branched dendritic trees. Where did the term *cable properties* come from? One of the simplest geometrical shapes approximated by some parts of a neuron is a cylinder. The cylinder has a conductive core surrounded by an outer shell or membrane that has different electrical properties from its core. Many of the simplifications we will make will be to allow parts of a neuron, such as its axon or pieces of its dendrites, to be represented by such cylinders. The mathematics of cylinders in which current flows down the center and across the sides (also called *core conductors* or electrical cables) has been around for many years, dating back in some cases to the first transatlantic cable used to transmit telegraphy. Any discussion of the cable properties of neurons therefore borrows heavily from previous work on electrical cables.

4.2   Significance of electrotonic properties of neurons

In general, neurons in the CNS have extensive dendritic trees that receive thousands of excitatory, inhibitory, and neuromodulatory synaptic inputs. Fast inhibitory inputs (e.g., those using y-aminobutyric acid, or

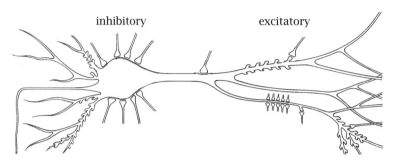

Figure 4.1   Diagram of a CNS neuron with synaptic inputs. (Adapted from L.L. Iversen. Copyright © 1979 by Scientific American, Inc. All rights reserved.)

GABA, as a transmitter) have traditionally been thought to occur mostly on the somatic membranes of neurons, but this notion is an oversimplification, because GABA inputs exist on dendrites as well.

Figure 4.1 illustrates in schematic form a neuron in the CNS. Inhibitory synapses are depicted as terminating on the cell body or soma while excitatory synapses are depicted as being distributed throughout the dendritic tree. Most excitatory synapses terminate on dendritic spines. Note that this diagram is not to scale. Synapses are actually much smaller in relationship to the illustrated size of the soma and dendrites. The dynamic interaction of excitatory and inhibitory inputs to both the somatic and dendritic membranes determines whether the neuron will be sufficiently depolarized to generate one or more action potentials. What role do the dendrites play in this dynamic process called *synaptic integration*?[1] The goal of this chapter is to provide a theoretical framework for understanding the basics of dendritic integration of synaptic inputs on a quantitative level. For example, what are the functional differences between a proximal and a distal synapse? How do synapses on the same or different dendritic branches interact? How might spines alter synaptic inputs? These are some of the questions that can be answered once we have a theory to describe the electrical properties of dendrites.

The theory we will develop and use to address these and other questions is *core conductor* or *linear cable theory*. To apply this theory to real neurons (see, for example, figure 4.2) we will have to make a number of critical assumptions, several of which are unlikely to be valid for all sit-

---

[1]The term *integrate* to describe dendritic processing of multiple synaptic inputs is used in keeping with the definition "to form, coordinate, or blend into a functioning or unified whole" rather than in a mathematical sense.

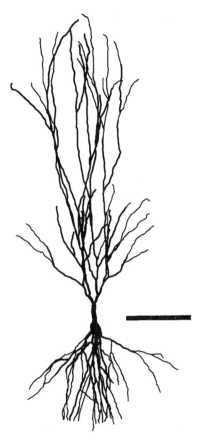

Figure 4.2 Example of a horseradish peroxidase (HRP) stained hippocampal CA3 neuron from a rat. Scale bar is 250 μm. (Kindly provided by M. O'Boyle, R. Gonzales, and B. Claiborne.)

uations, but which nevertheless will provide us with a foundation upon which to build a more complete understanding of the functional properties of dendrites.

It has been shown that animals raised in a sensorially enriched environment have CNS neurons with more extensive dendritic trees. Moreover, neurons from patients with diseases such as Down's syndrome, epilepsy, and certain senile dementias have significantly different dendritic trees than normal subjects. There also are interesting developmental changes in dendrites. For example, figure 4.3 illustrates dentate granule cells from a rat at different stages of development. What are the functional consequences of these changes in dendritic structure?

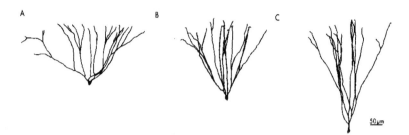

Figure 4.3 Developmental changes in dendritic structure of rat dentate granule cells. (A) from a young animal; (B) from an intermediate-aged animal; (C) from an adult. (Adapted from Rihn and Claiborne 1990.)

There are prominent theories in which it has been proposed that learning is associated with growth of dendrites, growth of or changes in shapes of spines, or growth of or changes in shapes of synapses. Whether or not these theories are correct is not known. Without a theoretical framework within which to explore the functional consequences of, for example, growth of dendrites or changes in shapes of spines, we would not be able to go much beyond correlating such changes with learning. Ideally, we would want to know whether the change in structure could be causative for the change in the observed behavior. Furthermore, as we will see, an understanding of the electrical properties of dendrites is critical for evaluating the errors associated with the electrophysiological measurements of synaptic function.

As mentioned above, many excitatory synapses terminate on dendritic spines. Given an assumption that memory is stored as a change in synaptic weights, changes in spine shape would be a possible substrate for such a process. There are a number of theories for the function of dendritic spines:

1. Chemical and/or electrical isolation (compartmentation)

2. Attenuation of electrical signal

3. Amplification of electrical signal

4. Parking place for synapse (an increase in the surface area of dendrites)

5. Impedance matching

6. Modulation

After we have developed cable theory in this chapter and discussed synaptic transmission in chapters 11–14, we will use that information in chapters 13 and 15 to evaluate some of these possible functions for spines.

Some theoretical framework is needed to address these issues. This framework is supplied by linear cable or core conductor theory and is used to investigate the passive electrotonic spread of electrical signals in dendrites (also called *electrotonus*). *Electrotonic* is a rather arcane term that is used to describe passive electrical signals, that is, signals (current or voltage) that are not influenced by the voltage-dependent properties of the membrane. It will become obvious that this theory is too simple to explain the complexities of dendrites, but at least it is a good starting point. In particular, many dendrites have voltage-gated channels at different locations, and these channels will influence the integrative properties of the neuron in important ways. Before we can deal with these complexities, however, we must understand the basics.

Another important reason it is necessary to understand linear cable theory, as alluded to above, is for evaluating the electrical measurements one may make of synaptic function. Typically, the way in which measurements are made is to record from the soma of a neuron with an electrode while stimulating a synaptic input located somewhere in the dendrites. This situation is depicted in figure 4.4. What we measure in the soma will obviously be a distorted view of what takes place at the site of synaptic input. How it will be distorted will depend, in part, on the passive properties of the dendrites. This issue will be addressed in some detail in this chapter and again in chapter 13.

We will develop linear cable theory and its application to neurons in stages. First, we will discuss cable properties in general and derive equations that will be applied to a number of different physical situations. Second, we will apply the theory to infinite and semi-infinite cables. This situation is particularly applicable to long axons. Third, we will apply the theory to finite cables and then to finite cables with lumped soma. This situation represents the application of the theory to dendrites, and we will show how a complex dendritic tree can be reduced to a simpler cable structure. After deriving the appropriate equations for each situation, we will apply them to investigate how cable properties influence synaptic inputs.

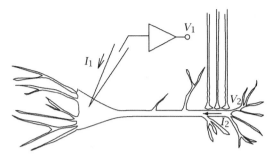

Figure 4.4   Diagram of a typical situation in which a recording electrode is in the soma of a neuron and the synaptic inputs are remotely located on the dendrites. Current, I_1, can be passed and voltage, V_1, recorded by the microelectrode, while current, I_2, is injected by the synapses and resulting voltage, V_2, is generated in the dendrites. (After Carnevale and Johnston 1982.)

4.3   Isopotential cell (sphere)

For the first stage of our development and application of linear cable theory to neurons, we will consider the case of a spherical cell in which the membrane potential is uniform at all points of the sphere. The cell is thus considered to be *isopotential*. The next assumption we will make is that the membrane resistance (R_m) is constant and independent of voltage. The cell is therefore *ohmic* (see appendix A). We will also let the resting potential be zero so that all potentials are referenced to the resting potential.

Referring to figure 4.5, current injected into such an isopotential cell will distribute uniformly across the surface of the sphere. The current flowing across a unit area of the membrane will be the sum of the capacity current and the resistive current, or

$$I_m = C_m \frac{dV_m}{dt} + \frac{V_m}{R_m}, \tag{4.3.1}$$

where I_m is uniform everywhere across the surface.

By convention, outward current is represented in the upward direction in diagrams such as figure 4.5, and inward current in the downward direction. Outwardly applied current from an external circuit will cause a depolarization (more positive) of the membrane potential while inwardly applied current will make the membrane potential more negative and hyperpolarize the neuron. These polarities can be understood easily by remembering the polarity of the potential drop across a resistor (i.e., mem-

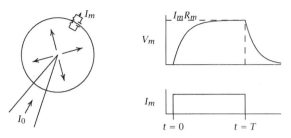

Figure 4.5 Injection of current into an isopotential cell. A step of injected current and the resulting membrane potential are illustrated on the right (see text for details).

brane resistance) for a given direction of current flow (see appendix A). If the current is not externally applied but arises from changes in membrane conductance, then the situation is a little different. Inward ionic current will cause a depolarization and outward ionic current a hyperpolarization. In this situation the inward current does not cause the depolarization directly the way the externally applied current does in figure 4.5, but merely deposits positive charge to the inside of the membrane capacitance. This will be discussed more fully in chapters 5, 6, and 7.

For the case of a finite step of current where I_0 goes from zero to a steady value at time 0 and then back again at time T,

$$V_m = I_m R_m (1 - e^{-t/\tau_m}) \qquad\qquad 0 < t < T, \qquad\qquad (4.3.2)$$

where

$$\tau_m = R_m C_m. \qquad\qquad (4.3.3)$$

This is simply the solution of the first-order differential equation for current flow across the membrane (equation 4.3.1). The derivation of this equation is given in appendix A. For the turn off of the current step (or the relaxation of the potential), we get a slightly different equation

$$V_m = I_m R_m e^{-t/\tau_m} \qquad\qquad T < t. \qquad\qquad (4.3.4)$$

For the onset of the current pulse, the change in membrane potential is described by a single exponential equation where the time at which V_m reaches $(1 - e^{-1})$, about 63% of its final value, is called τ_m, the membrane time constant. At the end of the current pulse, the time (measured from the end of the pulse) at which V_m has decayed to 37% of its value just before the end of the pulse is also equal to τ_m. If I_m is applied for a long time (i.e., $t \to \infty$), then

$$V_m(\infty) = I_m R_m. \qquad\qquad (4.3.5)$$

This is called the steady-state value of membrane potential during a step of current. As stated above, for an isopotential cell the injected current distributes uniformly across the surface (i.e., Kirchhoff's current law) so we need to derive a relationship for I_m in terms of I_0. For a spherical cell with I_0 injected at the center, I_m will be related to the surface area of the sphere by

$$I_m = I_0/4\pi a^2, \tag{4.3.6}$$

where a is the radius of sphere, and I_m has units of current/unit-area. The *input resistance* of a cell (any cell) is defined as

$$R_{\text{input}} = R_N \stackrel{\text{def}}{=} \frac{V_m(\infty)}{I_0}. \tag{4.3.7}$$

This is an important definition that will be used throughout this book. The definition of R_N does not depend on the geometry of the neuron, nor on where the measurement is made, although, obviously, the value of R_N does. To obtain an expression for the input resistance of a sphere, we need to combine equations 4.3.6 and 4.3.7:

$$R_N = \frac{I_m R_m}{I_m 4\pi a^2} = \frac{R_m}{4\pi a^2}. \tag{4.3.8}$$

This equation states that the input resistance of a spherical cell varies as the reciprocal of the surface area. That is, bigger cells have lower input resistances for any given R_m, and vice versa. In other words, adding resistors in parallel (i.e., adding more surface area) decreases the total resistance. This is an important concept to keep in mind and, in general, is true regardless of the geometry of the neuron. With nonisopotential neurons, however, the relationship between input resistance and surface area is more complex than that for a sphere.

4.4 Nonisopotential cell (cylinder)

With this section we begin the more complex (but also the more interesting) development of linear cable theory as applied to nonisopotential neurons, or parts thereof. The simplest and most useful geometrical structure to deal with is a cylinder (see figure 4.6). The parts of a neuron that most closely resemble a cylinder would be an axon or sections of dendrites (this will be discussed later).

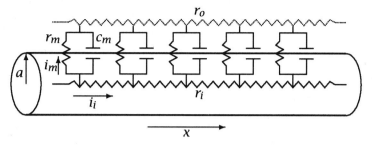

Figure 4.6 Diagram for current flow in a uniform cylinder such as an axon or segment of dendrite.

The equivalent electrical circuit is shown superimposed on the cylinder in figure 4.6. In addition to the parameters of membrane resistance and capacitance, we have added the resistance of the cytoplasm, r_i, the resistance of the extracellular space, r_o, and distance along the cylinder, x. (The reason for the switch to small letter r will be explained in section 4.4.1.) For most of this book, the extracellular space will be considered isopotential for the sake of simplicity (i.e., $r_o = 0$). Including r_o might be important for axons packed tightly together in a nerve trunk or for neurons in the CNS where current flow in the extracellular space may be significant. We will nevertheless ignore r_o for the time being but will try to point out where in the equations it would appear if it were included.

Current injected into the cylinder will, by Kirchhoff's current law, flow both across the membrane (i_m) and along the inside of the cylinder or cable (i_i). Before we can derive the equations governing the membrane potential along a cable, we need to make some key assumptions.

Assumptions:

1. The membrane parameters are assumed to be linear and uniform throughout. That is, r_m, r_i, and c_m are constants, they are the same in all parts of the neuron, and they are not dependent on the membrane potential (i.e., passive).

2. We assume that current flow is along a single spatial dimension, x, the distance along the cable. Radial current is therefore assumed to be 0.

3. As mentioned above, we assume for convenience that the extracellular resistance, r_o, is 0.

Referring again to figure 4.6, for current injected at some point along the cylinder, V_m is a function of time *and* distance from the site of injection. Along the x-axis

$$\frac{\partial V_m(x,t)}{\partial x} = -r_i i_i. \tag{4.4.9}$$

This is just Ohm's law (the decrease in V_m with distance is equal to the current times the resistance). The negative sign denotes that voltage decreases with increasing x (negative slope). Some of i_i, however, "leaks" out across the membrane (analogous to a leaky hose) through r_m and c_m, so that i_i is not constant with distance. The decrease in i_i with distance is equal to the amount of current that flows across the membrane or i_m. This follows from Kirchhoff's current law and is stated mathematically as

$$\frac{\partial i_i}{\partial x} = -i_m. \tag{4.4.10}$$

Combining equations 4.4.9 and 4.4.10, we obtain

$$\frac{\partial^2 V_m}{\partial x^2} = -r_i \frac{\partial i_i}{\partial x} = r_i i_m. \tag{4.4.11}$$

Recalling from chapter 3,

$$i_m = i_C + i_i = c_m \frac{\partial V_m}{\partial t} + \frac{V_m}{r_m}, \tag{4.4.12}$$

we can now combine equations 4.4.11 and 4.4.12 into

$$\frac{1}{r_i} \frac{\partial^2 V_m}{\partial x^2} = c_m \frac{\partial V_m}{\partial t} + \frac{V_m}{r_m}. \tag{4.4.13}$$

These fairly simple steps have led us to what has been called the *cable equation*. Equation 4.4.13 was derived for a one-dimensional cylinder (cable) and is extremely important. We will be using solutions to this equation to derive equations that describe a number of specific situations (infinite cable, finite cable, and finite cable with lumped soma). There are a number of other forms of the *cable equation*

$$\lambda^2 \frac{\partial^2 V_m}{\partial x^2} = \tau_m \frac{\partial V_m}{\partial t} + V_m, \tag{4.4.14}$$

where

$$\lambda = \sqrt{\frac{r_m}{r_i}} = \sqrt{\frac{aR_m}{2R_i}}. \tag{4.4.15}$$

λ is called the *space* or *length constant*. The physical meaning of λ will be described later. If we needed to include the resistance of the extracellular space, λ would be altered as follows:

$$\lambda = \sqrt{\frac{r_m}{r_i + r_o}},$$

where r_o is in Ω per unit length of cable (e.g., Ω/cm). No other equations would be changed. As mentioned in the previous section, $\tau_m = r_m c_m = R_m C_m$ and is called the *membrane time constant*.

4.4.1 Units and definitions

For historical reasons, and for convenience, we will make a distinction here and elsewhere between parameters that are geometry specific (that is, dependent on a cylinder) and those that are not. The parameters that are defined only for a cylinder will be designated by small letters (r_i, r_m, and c_m) while parameters that are independent of any specific geometry will be designated by capital letters (R_i, R_m, and C_m). Except for this one deviation for cylinder-specific parameters, we will continue to abide by the convention for the use of lowercase vs. uppercase letters mentioned in chapter 3 and appendix A. For example, lowercase g_{Na} would be used for a conductance that varies as a function of voltage and/or time, while G_r would be used for a resting conductance that is constant. Also, capital letters are used for peak values. The definitions for the cylinder-dependent parameters are given below.

1. r_i = axial resistance (Ω/cm)
2. r_m = membrane resistance (Ω-cm)
3. c_m = membrane capacitance (F/cm)

All of the above are for a unit length of cable and have meaning only with respect to a uniform cylinder. Think of r_i as an infinitely thin disk of the cytoplasm with the same radius as the inside of the cylinder. Obtaining the total resistance of the cytoplasm for a given length of the cylinder would be like stacking together these thin disks until you reach length l, and would be given by $l \cdot r_i$.

In the case of r_m and c_m, each can be thought of as an infinitely thin ring of membrane, again with the same radius as the cylinder. The total membrane resistance for a given length of cable would be obtained by stacking together these thin rings of membrane and would be equal to r_m/l. Obtaining the total membrane capacitance would be a similar stack of these thin rings of membrane but they would be equal to $l \cdot c_m$. The difference between the calculations for total membrane resistance and for membrane capacitance results from the different electrical definitions of resistance and capacitance (see appendix A). Increasing the membrane area by increasing the length of the cylinder is the same as adding resistors in parallel. Total resistance decreases as you add resistors in parallel, and hence you divide r_m by l. In the case of capacitance, adding capacitors in parallel increases the total capacitance, hence you multiply c_m by l.

The specific resistivity and capacitance of the membrane are independent of geometry and are defined as follows.

1. R_i = specific intracellular resistivity (Ω-cm)

2. R_m = specific membrane resistivity (Ω-cm^2)

3. C_m = specific membrane capacitance (F/cm^2)

The cytoplasmic or intracellular resistivity R_i must be multiplied by the length and divided by the cross-sectional area to obtain total resistance (that is, $R_{\text{total}} = R_i \cdot l/A$, as discussed in appendix A). This parameter is in its most general form and can be converted to resistance only if one knows the geometry of the conductor.

The membrane resistivity, R_m, and membrane capacitance, C_m, are the values for a unit patch (area) of the membrane. They are converted to total resistance and total capacitance by either dividing or multiplying, respectively, by the surface area.

The general forms of the membrane parameters are related to the cable specific parameters by the following equations:

$$
\begin{aligned}
R_i &= \pi a^2 r_i, \\
R_m &= 2\pi a r_m, \\
C_m &= c_m/2\pi a.
\end{aligned}
$$

4.4.2 Solutions of cable equations

The cable equation can be put into its most general and useful form by normalizing by the length constant and time constant. Let $X = x/\lambda$ and $T = t/\tau_m$. Then equation 4.4.14 becomes

$$
\frac{\partial^2 V_m}{\partial X^2} - \frac{\partial V_m}{\partial T} - V_m = 0, \tag{4.4.16}
$$

where X and T are dimensionless.

It is desirable to solve the cable equation to get V_m as a function of time and space in the cable. This can be done for a number of different cases using a variety of methods. We will not attempt to derive a solution in this chapter but instead will use general solutions derived by others (see Jack, Noble, and Tsien (1975), Hodgkin and Rushton (1946), and Rall (1977)) and add the appropriate parameters that allow each solution to be applied to a particular experimental situation. For example, different equations are obtained when dealing with steady-state vs. transient voltage changes. Also, different geometries, for example finite-length vs. infinitely long cables, require different boundary conditions that result in different equations. We will therefore formulate the applicable equations for each of the following specific situations and then use the equations in a number of (we hope) interesting examples.

4.4.2.1 Infinite cable, current step The first experimental situation we will deal with is that of a step of current injected into an infinitely long cable (figure 4.7).

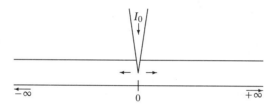

Figure 4.7   Current injected into an infinite cable at $x = 0$. Note that the injected current flows in both directions.

A general solution to the cable equation is

$$V_m(T,X) = \frac{r_i I_0 \lambda}{4} \left[e^{-X} \mathrm{erfc}\left(\frac{X}{2\sqrt{T}} - \sqrt{T} \right) \right. \tag{4.4.17}$$
$$\left. - e^{X} \mathrm{erfc}\left(\frac{X}{2\sqrt{T}} + \sqrt{T} \right) \right],$$

where $\mathrm{erfc}(x)$ = complementary error function (and erf(x) = error function). This is a formidable looking equation in which some of the terms (e.g., erfc) may be unfamiliar. The error function should not be intimidating, because it has some very simple properties:

$$\mathrm{erfc}(x) = 1 - \mathrm{erf}(x) = 1 - \frac{2}{\sqrt{\pi}} \int_0^x e^{-y^2} dy = \frac{2}{\sqrt{\pi}} \int_x^\infty e^{-y^2} dy,$$

where $\mathrm{erf}(0) = 0$, $\mathrm{erf}(\infty) = 1$, and $\mathrm{erf}(-x) = -\mathrm{erf}(x)$.

The error function is just a convenient equation that simplifies the solutions for many differential equations. In the next two sections we will modify equation 4.4.17 so that it describes two special cases of interest to neurophysiologists: steady-state voltage distribution along the cable after a step of current, and the change in voltage as a function of time for a step of current.

Steady-state solution ($T \rightarrow \infty$)   Using the above definitions for the erfc, as T gets large in equation 4.4.17, $\text{erfc}\left(\frac{X}{2\sqrt{T}} + \sqrt{T}\right)$ becomes $1 - \text{erf}(\infty) = 0$, while $\text{erfc}\left(\frac{X}{2\sqrt{T}} - \sqrt{T}\right)$ becomes $1 - \text{erf}(-\infty) = 2$. Equation 4.4.17 then reduces to a rather simple exponential equation

$$V_m(\infty, X) \;=\; \frac{r_i I_0 \lambda}{2} e^{-X} \tag{4.4.18}$$

or

$$V_m(\infty, x) \;=\; \frac{r_i I_0 \lambda}{2} e^{-x/\lambda}. \tag{4.4.19}$$

This equation says that the distribution of membrane potential (in the steady state) from the site of current injection decays as a single exponential with distance (see figure 4.8). The distance at which the potential has decayed to $1/e$ (~0.37) of the value at $x = 0$ is the space constant or λ. λ will be used frequently in the coming sections, so its definition needs to be clear. It is defined only in terms of an infinite (or semi-infinite) cable and reflects the steady-state properties of the cable. For example, if the diameter of the cable increased, λ would increase, if the membrane resistivity increased, λ would increase, and so forth.

The best analogy to use here is that of a leaky hose. Think of a hose with small holes in it (a sprinkler or soaker hose) that extends as far as you can see, but with one end attached to a faucet. It is easy to imagine that as you get farther from the faucet, there is less water pressure inside the hose to force water through the little holes. If you add more holes or increase the size of the holes (decrease R_m), then more water leaks out and the fall-off in pressure along the length of the hose increases. If the diameter of the hose is increased (but keeping the same number of holes per unit area or same R_m), then there will be proportionally less water leaking out and the decrease in pressure with distance decreases. This leaky hose analogy is very useful, and we will resort to it frequently.

In addition to the description of membrane potential along the cable, we will also want expressions for input resistance of the infinite (and semi-infinite) cable. Remember that input resistance is just the steady-state

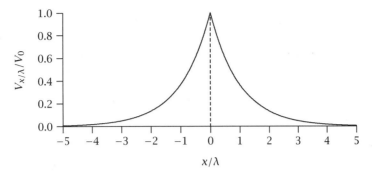

Figure 4.8 Decay of potential along infinite cable. The site of current injection is at $x = 0$.

membrane potential, $V_m(\infty)$, evaluated at $x = 0$, divided by the injected current, or, from equation 4.4.19

$$R_N = \frac{r_i\lambda}{2} = \frac{r_i\sqrt{r_m/r_i}}{2} = \frac{\sqrt{r_m r_i}}{2}$$

$$= \frac{1}{2}\left(\frac{R_m}{2\pi a} \cdot \frac{R_i}{\pi a^2}\right)^{1/2}.$$

$$R_N(\text{infinite cable}) = \frac{1}{2}\left(\frac{R_m R_i}{2\pi^2 a^3}\right)^{1/2} = \frac{\sqrt{R_m R_i/2}}{2\pi a^{3/2}}. \tag{4.4.20}$$

This equation for the input resistance of an infinite cable is an important equation that we will use again in developing the Rall model for a neuron. Note the term in the denominator on the right-hand side of the equation, that is, the radius of the cable raised to the 3/2 power. The input conductance of an infinite cable, G_N ($= 1/R_N$), will therefore be directly proportional to the radius of the cable raised to the 3/2 power ($G_N \propto a^{3/2}$), or

$$G_N(\text{infinite cable}) = \frac{2\pi a^{3/2}}{\sqrt{R_m R_i/2}}. \tag{4.4.21}$$

Equation 4.4.19 describes the decay of membrane potential from the site of current injection ($X = 0$) to $+\infty$ or to $-\infty$. If we are dealing with a semi-infinite cable, we will be interested only in the potential distribution from $X = 0$ to $+\infty$, which will be described by equation 4.4.19 except for one important difference—the amplitude of the membrane potential at $X = 0$ will be larger, twice as large to be exact. Think of the input resistance from $X = 0$ to $+\infty$ as a single resistor to ground and the input resistance from $X = 0$ to $-\infty$ as another, identical resistor to ground.

Each of these resistors will have a value equivalent to the input resistance of a semi-infinite cable, while the input resistance of an infinite cable will be equal to the parallel combination of these two resistors. The total resistance of two identical resistors in parallel is half of the value of either resistor by itself. This fact leads to the equations for the relationships between input resistance (and conductance) of semi-infinite (semi) and infinite cables (inf):

$$R_{N(semi)} \quad = \quad 2 \cdot R_{N(inf)} = \frac{(R_m R_i / 2)^{1/2}}{\pi a^{3/2}}. \tag{4.4.22}$$

$$G_{N(semi)} \quad = \quad \frac{G_{N(inf)}}{2} = \frac{\pi a^{3/2}}{\sqrt{R_m R_i / 2}}. \tag{4.4.23}$$

Transient solution at x, X=0 The other special case for an infinite cable that we want to derive is the so-called transient solution of the cable equation. This solution will describe the change in membrane potential with respect to time following the injection of a step of current at $X = 0$. The simplest transient solution of interest is when we let $X = 0$. From equation 4.4.17 and the definitions of the error function, when $X = 0$

$$V_m(T,0) = \frac{r_i I_0 \lambda}{2} \text{erf}(\sqrt{T}), \tag{4.4.24}$$

(remember $\text{erfc}(-\sqrt{T}) = 1 + \text{erf}(\sqrt{T})$ and $\text{erfc}(\sqrt{T}) = 1 - \text{erf}(\sqrt{T})$).

Equation 4.4.24 describes the change in membrane potential as a function of time at the site of current injection for an infinite cable. This equation, however, also describes the response for a semi-infinite cable except for a factor of 2 (i.e., $V_m(T,0)_{semi} = 2 \cdot V_m(T,0)_{inf}$). Because most of us have little appreciation for the error function, it is best to calculate and graph equation 4.4.24 and compare the result to that obtained for the response of a spherical cell to a current step (equation 4.3.2). This comparison is made in figure 4.9; note that the amplitudes are normalized. The result is somewhat nonintuitive, that is, the membrane potential of the infinite cable changes faster toward its steady-state value than that of the isopotential sphere. The easiest way to explain this result is that in the infinite cable, part of the injected current flows down the cable to charge distant sites of the cable. The time constants governing the charging of sites distant from the site of current injection are faster than the membrane time constant and contribute to the early phase of the voltage by making it change faster than that for the sphere.

Equation 4.4.24 was an extremely important result of Rall's early work, as it demonstrated why early estimates of membrane time constants from

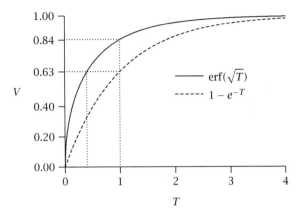

Figure 4.9 Comparison of charging curves described by an error function and a single exponential function. The dotted line from $T = 1$ depicts the time at which the exponential function has reached 0.63 of its final value and the error function 0.84. Note that if one chooses a point of 0.63 on the error function, a time of less than one time constant (i.e., $T \simeq 0.4$) would be obtained (see text for further explanation). (After Jack et al. 1975.)

neurons were too low. Many investigators had assumed that the membrane potential response to a current step (also called the *charging curve*) would be a single exponential and estimated the membrane time constant, τ_m, from the time to reach 63% of the steady-state value. For a charging curve similar to that for an infinite (or semi-infinite) cable, this would result in an estimated value for τ_m that was about half of the actual value (refer to figure 4.9). At $T = 1$ the charging curve for an infinite cable has already reached 84% of its final value while at 63%, $T \simeq 0.4$.

Transient solution at different X The transient solution of the cable equation for an infinite cable at $X = 0$ given above is extremely useful because it describes the charging curve at the site of current injection, the usual site of potential measurement. It would also be useful, however, to know what the transient change in membrane potential would be at sites different from that of current injection, even if we are unable actually to measure the membrane potential at those sites. To determine this we must use the entire solution given in equation 4.4.17 and reproduced below:

$$V_m(t,x) = \frac{r_i I_0 \lambda}{4} \left[e^{-x/\lambda} \mathrm{erfc}\left(\frac{x/\lambda}{2\sqrt{t/\tau_m}} - \sqrt{t/\tau_m} \right) \right.$$
$$\left. - e^{x/\lambda} \mathrm{erfc}\left(\frac{x/\lambda}{2\sqrt{t/\tau_m}} + \sqrt{t/\tau_m} \right) \right]. \qquad (4.4.25)$$

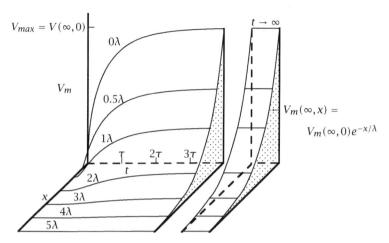

Figure 4.10   Solution of the cable equation as a function of time and distance for a step of current injected at $x = 0$ into a semi-infinite cable. The length of the cable extends along the x axis.

Equation 4.4.17 was calculated for different times and distances and is plotted in three dimensions in figure 4.10. The origin represents the site of current injection at $t = 0$. (The cable is coming out of the page, and note that the plots for an infinite and semi-infinite cable are the same, the only difference being that the V_{max} for each differs by a factor of 2.) There are several aspects of figure 4.10 that should be noted. First, at large t, the distribution of potential along the cable is simply the steady-state solution of equation 4.4.17 seen in the previous section. The curve is a single exponential described by the equation indicated in the figure. Second, at intermediate values of t, between 0 and the steady state, the decay of potential with distance along the cable is much greater than at the steady state. This is an important concept to which we will return later. Third, at $x = 0$, the charging curve is identical to that plotted for equation 4.4.24 in figure 4.9. And fourth, at values of x increasing from 0, the charging curves show slower rising phases and reach a smaller steady-state value.

Conduction velocity of a passive, decremental wave of membrane potential   It is sometimes useful to know how long it takes for a passive change in membrane potential to occur at different sites along a cable. We can derive a "conduction velocity" for such a passive wave from equation 4.4.17. This conduction velocity can then be compared to that for an actively propagating action potential (see chapters 6 and 7). The conduction velocity can be defined as the time to conduct half of the maximum steady-state value at X. The first step is to obtain an equation that is normalized with respect to the steady-state values at different X. This is achieved by dividing $V_m(T, X)$ by $V_m(\infty, X)$ or

$$\frac{V_m(T, X)}{V_m(\infty, X)} = \frac{\frac{r_i I_0 \lambda}{4}\left[e^{-X}\text{erfc}\left(\frac{X}{2\sqrt{T}} - \sqrt{T}\right) - e^X \text{erfc}\left(\frac{X}{2\sqrt{T}} + \sqrt{T}\right)\right]}{\frac{r_i I_0 \lambda}{2}e^{-X}}$$

$$= \frac{1}{2}\left[\text{erfc}\left(\frac{X}{2\sqrt{T}} - \sqrt{T}\right) - e^{2X}\text{erfc}\left(\frac{X}{2\sqrt{T}} + \sqrt{T}\right)\right].$$

As stated above, the conduction velocity of the decremental wave (passive spread) can be estimated by calculating the time needed to conduct the voltage to the $1/2$ maximum value at X. So when

$$\frac{V_m(T, X)}{V_m(\infty, X)} = \frac{1}{2} = \frac{1}{2}\left[\text{erfc}\left(\frac{X}{2\sqrt{T}} - \sqrt{T}\right) - e^{2X}\text{erfc}\left(\frac{X}{2\sqrt{T}} + \sqrt{T}\right)\right],$$

$$X = 2T - 0.5, \tag{4.4.26}$$

or

$$x = \frac{2\lambda}{\tau_m}t - 0.5\lambda. \tag{4.4.27}$$

Remember from physics that velocity is simply $\theta = dx/dt$, so conduction velocity of the wave is

$$\theta = \frac{dx}{dt} = \frac{2\lambda}{\tau_m}. \tag{4.4.28}$$

Figure 4.11 plots in two dimensions the charging curves at different distances from the site of current injection. These are extracted from figure 4.10. When the curves are plotted in this way, it becomes obvious that the rising phase of the curves is slower at increasing distance from the origin. Also, if one measures the time to reach half of the final value of each curve, one finds that it also increases with increasing distance. Figure 4.12 makes this measurement, and the comparison of the measurements among the curves, somewhat easier. In this figure we have normalized the curves by plotting the ratio $V(T, X)/V(T, \infty)$ against T. The 0.5 values can readily be determined from these curves. These 0.5 values are the points that satisfy equation 4.4.26, from which we derived the equation for conduction velocity (equation 4.4.28). From figure 4.12

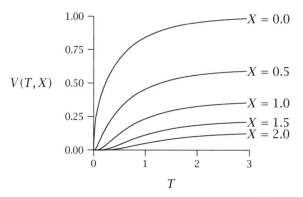

Figure 4.11   Charging curves at different distances along the cable from the site of current injection. (After Jack et al. 1975.)

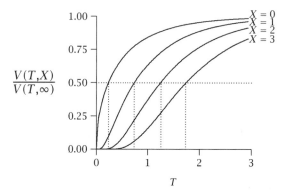

Figure 4.12   Charging curves at different distances along the cable from the site of current injection but normalized in amplitude. One can determine the conduction velocity of a decremental wave from this figure by looking at the time it takes for each curve to reach 0.5 of its final amplitude (indicated by dotted lines). (After Jack et al. 1975.)

we can see that it takes about 1 time constant for the passive wave to travel 1.5 space constants.

How does fiber size influence conduction velocity of passive spread? This can be derived quite easily by expanding equation 4.4.28 as follows:

$$\theta = \frac{2\lambda}{\tau_m} = \frac{2\sqrt{\frac{R_m a}{2 r_i}}}{R_m C_m} = \left(\frac{2a}{R_m R_i C_m{}^2}\right)^{1/2}. \tag{4.4.29}$$

Thus $\theta \propto \sqrt{a}$, the fiber radius, assuming that R_m, R_i, and C_m do not vary with a.

Example 4.1

To get a feeling for passive conduction velocity, let's work through a simple example. Consider two infinite cables with diameters of 1 and 10 μm but otherwise identical parameters (R_m = 20,000 Ω-cm^2, R_i = 100 Ω-cm, and C_m = 1 μF/cm^2). Using these values, τ_m = 20 msec, $\lambda_{1\mu m}$ = 7.1 cm, and $\lambda_{10\mu m}$ = 22.4 cm. So,

$$\theta_{1\mu} = 2\lambda_{1\mu}/\tau_m = 0.7 \text{ cm/msec},$$
$$\theta_{10\mu} = 2\lambda_{10\mu}/\tau_m = 2.2 \text{ cm/msec}.$$

This means that for a 1 μm cable, a passive wave will propagate at the rate of less than 1 cm/msec, vs. about 2 cm/msec for the larger-diameter cable. To calculate distance along the cable traveled by the wave in a given time one must use equation 4.4.26. Remember, however, that this propagation is highly decremental. Unlike the action potential (chapter 6), the amplitude of this passive wave will decrease greatly with distance along the cable.

4.4.2.2 Finite cable, current step Although the derivation for the infinite cable allowed us to introduce important concepts such as space constant, transient and steady-state distributions of membrane potential along a cable, and conduction velocity of a passive wave, the infinite-cable solution is of less interest experimentally. This is because there are few anatomical structures other than axons that approximate an infinite cable. The far more useful solution (which is also more complex mathematically) is that for the finite length cable. That solution will be derived in this section.

Steady-state solution The steady-state solution will be derived first. Under steady-state conditions, we can let $\frac{\partial V_m}{\partial T} = 0$ so that the cable equation can be reduced to

$$V_m = \frac{\partial^2 V_m}{\partial X^2}.$$

A general solution for this second-order differential equation is

$$V_m(\infty, X) = A_1 e^X + A_2 e^{-X}. \tag{4.4.30}$$

A cable can usually be considered finite if the length of the cable, l, is less than about twice the space constant, or $0 < l < 2\lambda$. For such a cable, there is an important boundary condition that must be considered—the termination at the ends of the cable. The end at the origin ($x = 0$) will be

Figure 4.13   Diagram of a finite cable used for derivations.

terminated with a zero-conductance membrane so that no current flows
in the negative direction. (In a later section we will add a soma to this
point, but for now assume that all the current injected at $x = 0$ will flow
in the positive x direction.) At the far end of the cable there can be either
of two types of termination, a sealed end or an open end.

The sealed end can be thought of as simply a disk of membrane that
caps or seals the end. Because of the high resistivity of the membrane
and the small surface area at the end of the cable, the resistance of this
cap or seal can be considered infinite so that no current flows across the
end of the cable. In electrical terms this would be called an *open circuit*
condition. In contrast, the open end type of termination would occur
if the cable were cut and nothing sealed the end. In this condition the
extracellular solution would be in direct contact with the cytoplasm, and
current would flow freely out the end of the cable. This would be called
the *short circuit* condition in electrical parlance. As we will see, the type
of termination chosen will have a dramatic effect on the distribution of
membrane potential along the cable.

Before we proceed, two terms should be defined, one of which was introduced previously. These are X and L. X is defined as before ($X = x/\lambda$)
and will be called the *electrotonic distance* along the cable. L is defined as
$L = l/\lambda$ and will be called the *electrotonic length* of the cable. Note that at
the end of the cable $X = L$.

In addition to these terms we also need to introduce two mathematical
functions that may not be familiar to all readers. These are hyperbolic
cosine, cosh, and hyperbolic sine, sinh. cosh and sinh are defined below
and are plotted in figure 4.14.

$$\cosh(X) \quad = \quad \frac{e^X + e^{-X}}{2}. \tag{4.4.31}$$

$$\sinh(X) \quad = \quad \frac{e^X - e^{-X}}{2}. \tag{4.4.32}$$

Using these definitions, we can derive several useful relations:

$$\sinh(-X) \quad = \quad -\sinh(X),$$

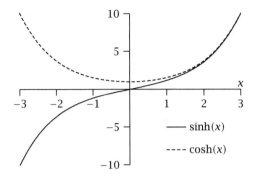

Figure 4.14 Plot of hyperbolic sine and cosine functions.

$$\cosh(-X) = \cosh(X),$$

$$\frac{d}{dX}\sinh(X) = \cosh(X),$$

$$\frac{d}{dX}\cosh(X) = \sinh(X),$$

and from figure 4.14, $\cosh(0) = 1$ and $\sinh(0) = 0$.

The general solution of the cable equation can be rewritten using the hyperbolic functions as

$$V_m(\infty, X) = B_1\cosh(X) + B_2\sinh(X),$$

or

$$V_m(\infty, X) = C_1\cosh(L - X) + C_2\sinh(L - X). \qquad (4.4.33)$$

From the above equation at $X = L$,

$$V_m(\infty, L) = V_L = C_1\cosh(0) + C_2\sinh(0) = C_1.$$

Let $B_L = C_2/V_L$ and substitute back into equation 4.4.33, so that

$$V_m(\infty, X) = V_L[\cosh(L - X) + B_L\sinh(L - X)]. \qquad (4.4.34)$$

B_L is the boundary condition for different end terminations. At $X = 0$,

$$V_m(\infty, 0) = V_0 = V_L[\cosh(L) + B_L\sinh(L)],$$

or

$$V_L = V_0/[\cosh(L) + B_L\sinh(L)].$$

Substituting this back into equation 4.4.34, we obtain

$$V_m(\infty, X) = V_0 \frac{\cosh(L - X) + B_L \sinh(L - X)}{\cosh(L) + B_L \sinh(L)}, \tag{4.4.35}$$

which is the steady-state solution of the cable equation for a finite-length cable. We are now in a position to consider the effects of the two types of end terminations. B_L is equivalent to the ratio of G_L/G_∞, where G_L is the conductance of the terminal membrane and G_∞ is the conductance of a semi-infinite cable. For the two types of terminations, G_L will be 0 for the sealed end and ∞ for the open end. We will consider three cases (only two of which are new):

1. If $B_L = 1$ (that is, $G_L = G_\infty$), then

$$V_m(\infty, X) = V_0 e^{-X},$$

which is the same as for a semi-infinite cable seen in section 4.4.2.1.

2. If $B_L = 0$ ($G_L \ll G_\infty$), then

$$V_m(\infty, X) = V_0 \frac{\cosh(L - X)}{\cosh(L)}. \tag{4.4.36}$$

This condition occurs when the conductance of the terminal membrane is small (large resistance) compared to the input conductance of a semi-infinite cable with the same properties. This is the open circuit or sealed-end condition.

3. If $B_L = \infty$ ($G_L \gg G_\infty$), then

$$V_m(\infty, X) = V_0 \frac{\sinh(L - X)}{\sinh(L)}. \tag{4.4.37}$$

This condition occurs when the conductance of the terminal membrane is large (small resistance) compared to the input conductance of a semi-infinite cable. This is called the short-circuit or open-end condition.

An interesting side issue worth mentioning here is that a voltage clamp applied to one end of a cable is mathematically equivalent to a short circuit at that end. The use of a voltage clamp will thus fundamentally alter the electrotonic structure of a neuron, and the implications of this are dealt with in Rall (1969).

The equations for a finite cable with open or sealed ends at electrotonic lengths of 0.5, 1, and 2 are plotted in figure 4.15 along with the equation

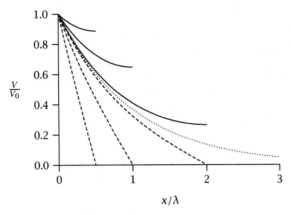

Figure 4.15 Comparison of voltage decays along finite cables of different electrotonic lengths and with different end terminations. Current is injected at $x = 0$. The solid lines are for finite cables with sealed ends, the dashed lines are for finite cables with open ends, and the dotted line is for a semi-infinite cable. (After Rall 1959.)

for a semi-infinite cable. Current is injected into one end of the cable ($X = 0$) and the steady-state membrane potential (normalized to the value at $X = 0$) is plotted as a function of distance along the cable. For the sealed-end condition, note that there is less decay of potential with distance along the finite cable compared to the infinite cable. This is an extremely important concept to understand. Look, for example, at the cable with electrotonic length of 1. At a distance of 1 λ, the potential is greater than that at the same distance along a semi-infinite cable. Remember that the definition of λ as the distance at which the potential has decayed to $1/e$ of its initial value holds only for an infinite (or semi-infinite) cable.

The open-end condition may be easier to understand. The decay of potential along the length of the cable is much greater for an open-end than for a sealed-end cable and also much greater than for a semi-infinite cable.

The leaky hose analogy may again be useful here. Think of the expected pressure of two identical hoses, both extending to infinity. Now take one and cut it at a length about 6 feet from the faucet. First, leave the end open and try to think about which hose has more pressure at a point 5 feet from the faucet. Since you probably can imagine much of the water gushing out of the open end, it should be easy to conclude that the pressure will be less along the hose with the open end (greater loss of pressure with distance) than the infinitely long hose. Now take the hose with the cut end and put a cap on it. Which hose now has more pressure 5 feet from the faucet?

The water that would have flowed down the length of the infinitely long hose now is blocked and instead can contribute to the pressure along a shorter length. Therefore, the shorter hose will have greater pressure (less loss of pressure with distance) than the long hose. There are more precise mathematical descriptions of why the finite, sealed-end cable has less attenuation of potential with distance, having to do with reflections, but those are beyond the scope of this book.

Transient solution   The derivation of the transient solution for a finite cable is formidable, but it does eventually reduce to a fairly simple infinite series that should be relatively easy to understand and remember. We go back to the original cable equation

$$\frac{\partial^2 V_m}{\partial X^2} - \frac{\partial V_m}{\partial T} - V_m = 0.$$

A general solution of this second-order partial differential equation is

$$V_m(T, X) = [A\sin(\alpha X) + B\cos(\alpha X)]e^{-(1+\alpha^2)T}. \tag{4.4.38}$$

As before, we must establish the boundary conditions. The first condition is simply that the change in potential with respect to distance at each end is zero, or

$$\frac{\partial V_m}{\partial X}(T, L) = \frac{\partial V_m}{\partial X}(T, 0) = 0.$$

To use these boundary conditions, we must take the first derivative with respect to X of equation 4.4.38 and plug in our boundary conditions to this derivative, or

$$\frac{\partial V_m}{\partial X} = [\alpha A\cos(\alpha X) - \alpha B\sin(\alpha X)]e^{-(1+\alpha^2)T}.$$

At $X = 0$, $\dfrac{\partial V_m}{\partial X} = 0$, so $A = 0$.

At $X = L$, $\alpha B\sin(\alpha L) = 0$, so $\alpha_n = n\pi/L$, $\tag{4.4.39}$

where n represents a series of positive integers from 0 to ∞, and

$$V_m(T, X) = \sum_{n=0}^{\infty} B_n\cos(n\pi X/L)e^{-[1+(n\pi/L)^2]T}. \tag{4.4.40}$$

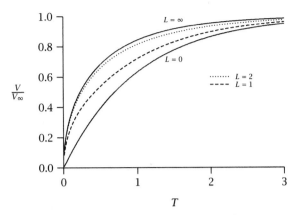

Figure 4.16 Comparison of normalized charging curves for finite cables of different electrotonic lengths. A step of current is injected, and the voltage is measured, at $x = 0$. (After Jack et al. 1975.)

The above equation represents an infinite series of exponential terms that can be further simplified by first expanding

$$V_m(T, X) = B_0 e^{-T} + B_1 \cos(\alpha_1 X) e^{-(1+\alpha_1^2)T} +$$
$$B_2 \cos(\alpha_2 X) e^{-(1+\alpha_2^2)T} + \cdots +$$
$$B_n \cos(\alpha_n X) e^{-(1+\alpha_n^2)T},$$

where $\alpha_n = n\pi/L$ and $1 + \alpha_n^2 = \tau_0/\tau_n$, so $L = n\pi[(\tau_0/\tau_n) - 1]^{-1/2}$. Simplifying,

$$V_m(t, X) = B_0 e^{-t/\tau_0} + B_1 \cos(\alpha_1 X) e^{-t/\tau_1} + \qquad (4.4.41)$$
$$B_2 \cos(\alpha_2 X) e^{-t/\tau_2} + \cdots + B_n \cos(\alpha_n X) e^{-t\tau_n}.$$

This can also be written as

$$V_m(t, x) = C_0 e^{-t/\tau_0} + C_1 e^{-t/\tau_1} + \qquad (4.4.42)$$
$$C_2 e^{-t/\tau_2} + \cdots + C_n e^{-t\tau_n}.$$

This is the complete solution for a finite-length cable and is important to remember. This solution will also describe a semi-infinite cable, which is just a special case of the finite cable where $L \to \infty$. The charging curve for a step of current injected into a finite cable at $X = 0$ is, by equation 4.4.42, the sum of an infinite number of exponential terms. Plots of charging curves for cables of different lengths are illustrated in figure 4.16. The first term of the equation, $C_0 e^{-t/\tau_0}$, will represent the membrane time constant (that is, $\tau_0 = \tau_m$) if the cable has uniform membrane properties.

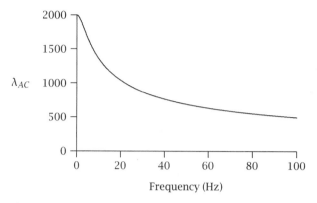

Figure 4.17   Example of space constant (in μm) as a function of frequency for a representative cable. Zero frequency would represent λ_{DC}.

Note again that the charging of a finite-length cable is faster than that for an isopotential cell.

Nonsteady-state voltage attenuation   Before we leave the general discussion of finite cables, however, an additional point should be mentioned. The steady-state decay of potential along finite and infinite cables was illustrated in figure 4.15. It is important to emphasize that the decay of potential along a cable will be much different (much greater) if the applied current is changing with time. We can illustrate this best by showing an equation for the so-called AC length constant. The length constant we derived previously was for the steady-state (DC) condition only. If the injected current changes with time (AC) a totally different length constant is obtained. This is given below and illustrated in figure 4.17.

$$\lambda_{AC} = \lambda_{DC} \sqrt{\frac{2}{1 + \sqrt{1 + (2\pi f \tau_m)^2}}}, \qquad (4.4.43)$$

where f is frequency in Hz.

The equation for steady-state voltage attenuation (equation 4.4.36) derived previously must also be modified for applied currents or voltages that are changing with time. For alternating sine wave voltage attenuation, equation 4.4.36 becomes

$$V_m(\infty, X) = V_0 \sqrt{\frac{\cosh(2aY) + \cos(2bY)}{\cosh(2aL) + \cos(2bL)}}, \qquad (4.4.44)$$

where V_0 is the peak amplitude of the sine wave voltage change at the site of current injection, $Y = L - X$ (where L and X are calculated using

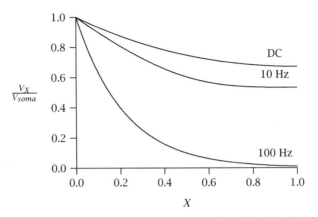

Figure 4.18   Voltage attenuation along a finite-length cable ($L = 1$) for current injections (DC to 100 Hz) at $X = 0$ (i.e., soma) ($R_m = 50,000\ \Omega$-cm^2).

λ_{DC}), and a and b are the real and imaginary parts of $\sqrt{1 + j\omega\tau_m}$, where $j = \sqrt{-1}$ and ω = frequency in radians. Equation 4.4.44 can be solved using

$$r = \sqrt{1 + \omega^2\tau_m^2},$$
$$\theta = \arctan\omega(\tau_m),$$
$$a = \sqrt{r}\cos(\theta/2),$$
$$b = \sqrt{r}\sin(\theta/2).$$

Using this relation, figure 4.15 can be replotted for a sealed-end cable of $L = 1$ at a variety of different frequencies. This is illustrated in figure 4.18. Note that the decay of potential along a cable is much greater for nonsteady-state conditions. This was also noted and discussed in reference to figure 4.10 (see also Spruston et al. 1993).

Example 4.2

Assume that you have a finite-length, sealed-end cable of $l = 1$ cm and $L = 1$. If sufficient current is injected to reach 10 mV at one end, what will the potential be at $X = 0.6$ and $X = L$? How does this compare to that at $X = 1$ along a semi-infinite cable?

For a finite cable at $X = 0.6$,

$$
\begin{aligned}
V_m(\infty, X) &= V_0 \frac{\cosh(L - X)}{\cosh(L)} \\
&= 10 \text{ mV} \frac{\cosh(0.4)}{\cosh(1)} \\
&= 10 \text{ mV} \frac{1.1}{1.54} \\
&= 7.1 \text{ mV}.
\end{aligned}
$$

For a finite cable at $X = L$,

$$
\begin{aligned}
V_m(\infty, X) &= V_0 \frac{\cosh(L - X)}{\cosh(L)} \\
&= 10 \text{ mV} \frac{\cosh(0)}{\cosh(1)} \\
&= 10 \text{ mV} \frac{1}{1.54} \\
&= 6.5 \text{ mV}.
\end{aligned}
$$

For a semi-infinite cable at $X = 0.6$,

$$
\begin{aligned}
V_m(\infty, X) &= 10 \text{ mV } e^{-X} \\
&= 10 \text{ mV } e^{-0.6} \\
&= 5.5 \text{ mV}.
\end{aligned}
$$

For a semi-infinite cable at $X = 1$,

$$
V_m(\infty, X) = 10 \text{ mV } e^{-X} = 10 \text{ mV } e^{-1} = 3.7 \text{ mV}.
$$

If the current is injected in the form of a 50 Hz sine wave, what will the length constant be for this cable?

$$
\lambda_{DC} = l/L = 1 \text{ cm}
$$

$$
\lambda_{AC} = \lambda_{DC} \sqrt{\frac{2}{1 + \sqrt{1 + (2\pi f \tau_m)^2}}} = \lambda_{DC}(0.35) = 0.35 \text{ cm}.
$$

4.5 Rall model for neurons

In the previous sections we presented the equivalent circuit representation of a membrane and then derived expressions for the passive properties of spherical and cylindrical structures (cables). For cables we demonstrated the differences between infinite, semi-infinite, and finite-length cables. In this section we will use the descriptions for the semi-infinite and finite-length cables to derive the Rall model of a neuron. This model provides a theoretical framework for understanding the spread of current and voltage in complicated dendritic trees.

4.5.1 Derivation of the model

Before we can derive the Rall model, we must once again list the underlying assumptions.

1. Uniform R_i, R_m, C_m. The membrane properties are independent of voltage (i.e., are passive), and the electrical properties are identical for all of the membrane comprising the soma and dendrites. (These are key assumptions that are almost certainly not true in the strictest sense. The assumptions are probably reasonable when looking at different parts of the same axon, but it seems highly unlikely that identical channel types and channel densities exist in all parts of the soma and dendrites of a neuron. This would mean that some amount of nonuniformity, in particular of R_m, is likely to exist. Furthermore, it is likely that R_m is not strictly passive but displays some amount of voltage dependency. Nevertheless, by using this assumption and deriving the Rall model we can obtain a "simplest case" from which one can explore the effects of nonuniform and nonpassive electrical properties in different parts of a neuron.)

2. $R_o = 0$. Again, this assumption is for convenience only, as R_o can be incorporated into the expressions if desired.

3. Soma isopotential. We will assume that the soma can be represented as an isopotential sphere.

4. All dendrites terminate at the same electrotonic length. This is another important assumption that is unlikely to be completely true. The assumption allows all dendritic branches to be collapsed into a single cable. If this were not the case, then individual dendritic branches would have to be treated separately.

5. Radial current = 0.

4.5.1.1 Equivalent (semi-infinite) cylinder An imaginary neuron with
a branched dendritic tree is illustrated in schematic form in figure 4.19.
X_1, X_2, and X_3 are branch points encountered when traversing along one
pathway from the soma out to the end of the dendrites; d_0, d_{11}, d_{12}, etc.,
are the branches at these intersections. The branches attached to X_3 (as
well as the final branch points in all of the other trees) are assumed to
extend to infinity and thus are semi-infinite cables. One of Rall's major
contributions was to show how a complicated tree structure such as this
could be reduced to a single equivalent cable. This is illustrated below.

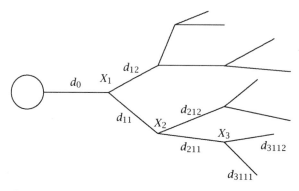

Figure 4.19   Schematic of a neuron with a branched dendritic tree. d is the diameter of
the respective branches and X_1, X_2, and X_3 are three representative branch points.

In a previous section we derived an equation for the input conductance
of a semi-infinite cable. Equation 4.4.23 is repeated here in slightly different form:

$$G_N = (\pi/2)(R_m R_i)^{-1/2}(d)^{3/2}. \tag{4.5.45}$$

This equation can be simplified if we let $K = (\pi/2)(R_m R_i)^{-1/2}$. Equation 4.5.45 then reduces to $G_N = K(d)^{3/2}$. Notice that the input conductance of a semi-infinite cable is proportional to the diameter of the cable
raised to the $3/2$ power.

To simplify the branched dendritic tree, we start at the last branch point,
X_3. If the cable d_{3111} were detached from X_3 with a sealed end, its input
conductance would be

$$G_N(d_{3111}) = K(d_{3111})^{3/2}.$$

A similar equation exists for cable d_{3112}. Since total conductance is just the sum of all parallel conductances, the attachment of the two cables d_{3111} and d_{3112} to each other would yield a total input conductance of

$$G_N(X_3) = G_N(d_{3111}) + G_N(d_{3112}) = K[(d_{3111})^{3/2} + (d_{3112})^{3/2}].$$

If instead of a branch point at X_3, cable d_{211} were extended to infinity and detached at the same spot (with a sealed end), its input conductance would be

$$G_N(d_{211}) = K(d_{211})^{3/2}.$$

Rall's insight was to recognize that if

$$(d_{211})^{3/2} = (d_{3111})^{3/2} + (d_{3112})^{3/2},$$

then mathematically (and electrically) having the branch point X_3 with branches d_{3111} and d_{3112} is *exactly equivalent* to extending branch d_{211} to infinity.[2] This is illustrated in figure 4.20.

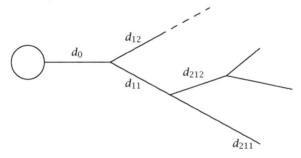

Figure 4.20   Schematic of the dendritic tree after eliminating branch point X_3.

If we had done the same simplification at the branch point along d_{212}, then at X_2 there would be two semi-infinite cables, d_{211} and d_{212}, attached to branch d_{11}. As before, if

$$(d_{11})^{3/2} = (d_{211})^{3/2} + (d_{212})^{3/2},$$

then

$$G_N(X_2) = G_N(d_{211}) + G_N(d_{212}) = K(d_{11})^{3/2},$$

---

[2]In electrical terms this equivalency also means that there is *impedance matching* at all branch points. Without impedance matching there would be reflections of electrical signals at branch points.

and having branch point X_2 with semi-infinite cables d_{211} and d_{212} is exactly equivalent to extending cable d_{11} to infinity.

If we assume that the same simplifications can be made at all the branch points extending from cable d_{12} to the end of the dendrites, then at branch point X_1 there would be two semi-infinite cables attached, d_{11} and d_{12}. Again, if

$$(d_0)^{3/2} = (d_{11})^{3/2} + (d_{12})^{3/2},$$

then

$$G_N(X_1) = G_N(d_{11}) + G_N(d_{12}) = K(d_0)^{3/2},$$

and branch point X_0 and branches d_{11} and d_{12} can be eliminated by extending branch d_0 to infinity. Our neuron with branched dendrites is now exactly equivalent to a single semi-infinite cable with diameter of d_0 attached to the soma. This is illustrated in figure 4.21.

Figure 4.21   Spherical soma and equivalent, semi-infinite cable after simplification of the dendritic tree according to the 3/2 power rule.

For purposes of illustration, we used only two branches at each branch point. We could have had any number of branches at each branch point as long as the following relationship was true:

$$d_P{}^{3/2} = \sum d_D{}^{3/2}, \tag{4.5.46}$$

where d_P is the *parent* dendrite and d_D is the *daughter* dendrite. Equation 4.5.46 is the so-called *3/2 power rule*. If at all branch points equation 4.5.46 holds, then the entire dendritic tree can be reduced to an equivalent semi-infinite cylinder. This is a very powerful simplification of an otherwise complex geometrical structure.

If the neuron has more then one dendritic tree, then, as long as each tree obeys the 3/2 power rule, the neuron can be reduced to an isopotential soma with multiple equivalent cylinders attached. The multiple equivalent cylinders can also be reduced to a single equivalent cylinder with diameter equal to $(\sum d_j{}^{3/2})^{2/3}$, where d_j represents the diameter of each of the equivalent cylinders.

It is natural to wonder whether the 3/2 power rule actually holds for neuronal dendrites. Rall stated that he did not propose this as a "law of nature," but only as a convenient idealization. Nevertheless, a number of studies have suggested that the branching in dendrites of spinal motor neurons, cortical neurons, and hippocampal neurons closely approximates the 3/2 power rule.

4.5.1.2 Equivalent (finite) cylinder Dendrites are usually better represented by finite-length cables, that is, l is usually $< 2 \lambda$ for dendrites. Thus, a more useful implementation of the Rall model reduces a branching dendritic structure to an equivalent finite cylinder. The derivation of the equivalent finite cylinder follows the derivation given above for a semi-infinite cylinder, except that the equation for the input conductance of a finite-length cylinder must be used instead of that for a semi-infinite cylinder. Furthermore, we must make the key assumption that *all dendrites end at the same electrotonic length, L,* and that the tips of the dendrites are sealed.[3]

The equation for the input conductance of a finite cable is given below.[4]

$$R_N = (2/\pi)(R_m R_i)^{1/2}(d)^{-3/2} \coth(L), \qquad (4.5.47)$$

or

$$G_N = (\pi/2)(R_m R_i)^{-1/2}(d)^{3/2} \tanh(L). \qquad (4.5.48)$$

As before, we see that the input conductance of a finite-length cable is proportional to the diameter raised to the 3/2 power. Equation 4.5.48 can be simplified, however, by using the relations $\lambda = (r_m/r_i)^{1/2}, R_m = \pi d r_m$, and $R_i = \pi(d/2)^2 r_i$, so that

$$G_N = \frac{1}{\lambda r_i} \tanh(L). \qquad (4.5.49)$$

These equations for input conductance, the 3/2 power rule of the previous section, and the assumption stated above that the ends of all dendrites end at the same L, can be used to reduce the tree of figure 4.19 to a single equivalent cylinder of diameter d_0 and electrotonic length L in the same step-by-step manner as presented in the previous section. Furthermore, we showed previously that $L = l/\lambda$ so that for the tree of figure 4.19

$$L = \frac{l_0}{\lambda_0} + \frac{l_{11}}{\lambda_{11}} + \frac{l_{211}}{\lambda_{211}} + \frac{l_{3111}}{\lambda_{3111}},$$

---

[3]Actually, the dendrites could all end in either the sealed- or open-end condition, but the sealed-end condition is the only one that makes any practical sense for real dendrites.

[4]For interested readers, the derivation of this equation is given in section 4.8 at the end of this chapter.

where the l's represent the length of each of the segments. Note that the total length of the different branches can vary. The requirement is only that the total electrotonic length (L) be the same at the end of all the branches. In the more general case

$$L = \frac{l_0}{\lambda_0} + \frac{l_{11}}{\lambda_{11}} + \frac{l_{211}}{\lambda_{211}} + \cdots + \tag{4.5.50}$$

out to one end.

4.5.1.3 Finite cylinder with lumped soma Some neurons (e.g., hippocampal dentate granule cells) can be reasonably represented by a finite cable with lumped soma. A better representation for some other neurons may be a lumped soma with several finite-length equivalent cylinders. If the multiple cylinders are of the same L, however, then they can be collapsed into a single cylinder with diameter of $(\sum d_j^{3/2})^{2/3}$, as given in the previous section.

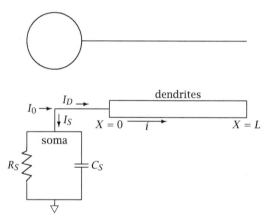

Figure 4.22 Diagram of finite cable with lumped soma along with the circuit elements for the soma.

The electrical circuit representation for the finite cylinder attached to a lumped soma is illustrated in figure 4.22. This is also called the *ball-and-stick* model, for obvious reasons. As we will spend some time discussing this model, we will need to introduce a new term called the *dendritic to somatic conductance ratio*, or ρ. ρ is defined as the input conductance of the dendrites (or the equivalent cylinder of the dendrites) divided by the input conductance of the soma

$$\rho = \frac{G_D}{G_S}.$$

Stated another way, when a current is injected into the soma, ρ is the proportion of the current that flows into the dendritic tree relative to that which flows across the somatic cell membrane. The total input conductance of the neuron is simply

$$G_N = G_D + G_S,$$

or

$$G_N = G_S(1 + \rho).$$

Example 4.3

A dendritic tree of a neuron has the diameter and length of each branch indicated in the figure below:

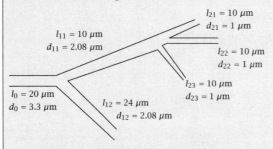

$l_{21} = 10\ \mu m$
$d_{21} = 1\ \mu m$

$l_{11} = 10\ \mu m$
$d_{11} = 2.08\ \mu m$

$l_{22} = 10\ \mu m$
$d_{22} = 1\ \mu m$

$l_{23} = 10\ \mu m$
$d_{23} = 1\ \mu m$

$l_0 = 20\ \mu m$
$d_0 = 3.3\ \mu m$

$l_{12} = 24\ \mu m$
$d_{12} = 2.08\ \mu m$

a. Is the 3/2 power law obeyed?

b. What is the electrotonic length of the equivalent cylinder?

c. If all four terminals at the right hand side are sealed, what is the input conductance of this dendritic tree? (Let $R_m = 2000$ Ω-cm^2 and $R_i = 60$ Ω-cm.)

Answer to example 4.3

a. $(d_{21})^{3/2} + (d_{22})^{3/2} + (d_{23})^{3/2} = 1 + 1 + 1 = 3,$

$(d_{11})^{3/2} = (2.08)^{3/2} = 3,$

$(d_{11})^{3/2} + (d_{12})^{3/2} = 3 + 3 = 6,$

$(d_0)^{3/2} = (3.3)^{3/2} = 6.$

Therefore,

$(d_{11})^{3/2} = (d_{21})^{3/2} + (d_{22})^{3/2} + (d_{23})^{3/2},$

$(d_0)^{3/2} = (d_{11})^{3/2} + (d_{12})^{3/2}.$

Yes, the 3/2 power law is obeyed.

Answer to example 4.3 (continued)

b.
$$\lambda_0 = \sqrt{\frac{d_0 R_m}{4R_i}} = \sqrt{\frac{3.3\ \mu\text{m} \cdot 2000\ \Omega \times 10^8\ \mu\text{m}^2}{4 \times 60\ \Omega \times 10^4\ \mu\text{m}}},$$
$$= \sqrt{27.5 \times 10^4\ \mu\text{m}^2},$$
$$= 5.24 \times 10^2\ \mu\text{m}.$$

$$\frac{\lambda_{11}}{\lambda_0} = \left(\frac{d_{11}}{d_0}\right)^{1/2} = \left(\frac{2.08}{3.3}\right)^{1/2} = 0.79.$$

$$\frac{\lambda_{21}}{\lambda_0} = \left(\frac{d_{21}}{d_0}\right)^{1/2} = \left(\frac{1}{3.3}\right)^{1/2} = 0.55.$$

$$L = \frac{l_0}{\lambda_0} + \frac{l_{11}}{\lambda_{11}} + \frac{l_{21}}{\lambda_{21}},$$
$$= \frac{20}{\lambda_0} + \frac{10}{0.79\lambda_0} + \frac{10}{0.55\lambda_0},$$
$$= \frac{50.84\ \mu\text{m}}{\lambda_0},$$
$$= \frac{50.84\ \mu\text{m}}{5.24 \times 10^2\ \mu\text{m}},$$
$$= 0.1.$$

c. $\lambda = 5.24 \times 10^2\ \mu\text{m}.$

$$r_i = \frac{60\ \Omega\text{-cm}}{\pi a^2} = 7 \times 10^8\ \Omega/\text{cm}.$$

$$G_0 = \frac{1}{\lambda r_i} \tanh(L),$$
$$= \frac{\tanh(L)}{(5.24 \times 10^{-2}\ \text{cm})(7 \times 10^8\ \Omega/\text{cm})},$$
$$= 2.7 \times 10^{-9}\ \text{S}.$$

4.5.2 Experimental determination of L, ρ, and τ_m

The ball-and-stick model is usually described by the three parameters, L, ρ, and τ_m, as well as the input conductance, G_N. More information is needed to uniquely determine other parameters, for example R_i, a, and l, but these are not necessary to construct the model. A number of methods have been used to determine L, ρ, τ_m, and G_N for a neuron. We

will present one here in some detail and then give references at the end
of the chapter for some of the others.

The typical experimental situation is one in which a recording is made
from a neuron with either a sharp microelectrode or a whole-cell patch
electrode. One approach for determining passive properties of a neuron is
to inject a long ($\gg \tau_0$) step of current and record the change in membrane
potential with respect to time, i.e., the charging curve.

In a previous section, the transient response of a finite cable to a current
step was given as (equation 4.4.40)

$$V_m(T, X) = \sum_{n=0}^{\infty} B_n \cos(n\pi X/L)e^{-[1+(n\pi/L)^2]T}.$$

This equation must be modified for the addition of a lumped soma at
$X = 0$.[5]

The solution for the ball-and-stick model is

$$V_m(T, X) = V(\infty, X) - \sum_{n=0}^{\infty} B_n \cos[\alpha_n(L-X)]e^{-(1+\alpha_n^2)T}, \tag{4.5.51}$$

where

$$B_n \cos(\alpha_n L) = \frac{2Bo\ \tau_n/\tau_0}{1 + (\alpha_n L)^2/(k^2 + k)}, \tag{4.5.52}$$

$$V(\infty, X) = IR_N \frac{\cosh(L-X)}{\cosh(L)}, \tag{4.5.53}$$

and

$$k = -\alpha_n L \cot(\alpha_n L) = \rho\ L \coth(L). \tag{4.5.54}$$

Note that the α_n's are not as simple as they were for the finite cable presented previously, because they are now dependent on both L and ρ.

$V_m(T, X)$ is an infinite series of exponentials that can be rewritten as

$$V_m(t, X) = C_0 e^{-t/\tau_0} + C_1 e^{-t/\tau_1} + C_2 e^{-t/\tau_2} + \cdots + C_n e^{-t/\tau_n}, \tag{4.5.55}$$

which was obtained by letting $C_n = B_n \cos[\alpha_n(L-X)]$ and $\tau_n/\tau_0 = 1/(1 + \alpha_n^2)$. Also note that

$$\tau_0 > \tau_1 > \tau_2 > \ldots > \tau_n,$$

---

[5]Details of the derivation are given in section 4.9 at the end of this chapter.

and that $\tau_m = \tau_0$ in the ideal case of uniform R_m and C_m. As with most such series, only the first few terms of equation 4.5.55 are usually necessary to describe V_m adequately. For illustrative purposes, we will assume that only the first two terms are relevant and experimentally measurable. The membrane potential response to a step of current at $X = 0$ is then

$$V_m(t) = C_0 e^{-t/\tau_0} + C_1 e^{-t/\tau_1}.$$

The above equation means that the membrane potential response to a step of current consists of two exponential terms. There are a number of computer programs that will fit multiple exponentials to an arbitrary curve. Let's assume, however, that we do not have access to a computer. There is a fairly simple and conceptually revealing method for extracting exponential terms from a waveform. It is called peeling. If we take the logarithm of the above equation, we obtain

$$\ln(V_m) = \ln(C_0 e^{-t/\tau_0} + C_1 e^{-t/\tau_1}).$$

As t gets large, the last term on the right side approaches zero before the first term does (remember, $\tau_0 > \tau_1$), so at large t we are left with

$$\ln(V_m) = \ln(C_0) - (1/\tau_0) \cdot t.$$

This equation is in the form of $y = b + mx$, the equation for a straight line. Therefore, a plot of the charging curve on semilog graph paper (plot of $\ln(V_m)$ vs. t) will yield a straight line as t gets large. The slope of this line will be $-(1/\tau_0)$ and the intercept will be $\ln(C_0)$. If we then extrapolate this line back to $t = 0$, subtract it from the rest of the charging curve (i.e., $C_0 e^{-t/\tau_0}$ is subtracted from V_m), and plot this difference, we end up with a curve described by the following equation:

$$\ln(V_m) = \ln(C_1) - (1/\tau_1) \cdot t.$$

This procedure can be repeated if there are more than two exponential terms that need to be extracted from the charging curve. The peeling of the first term of the charging curve and plotting the difference is illustrated in figure 4.23.

Although this method of "peeling" exponentials need seldom be used anymore, it is still a useful exercise as it illustrates the differences in the various exponential terms as well as the meaning of the coefficients, C_n's. As mentioned previously and as shown by the peeling procedures, the higher-order time constants in the series get progressively smaller (show

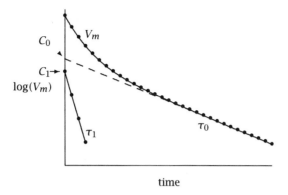

Figure 4.23 Example of the method for peeling exponentials from a multiexponential charging curve. The log of the membrane potential in response to a step of current is plotted vs. time from a nonisopotential neuron (see text for further explanation).

faster decay). The coefficients, on the other hand, are essentially amplitudes or weights for each of the exponential terms. The value of C_n determines how much each exponential term (regardless of its time constant) contributes to the charging curve. In general, the C_n's for $n \geq 2$ are quite small compared to C_0 and C_1. In many cases even C_1 is small compared to C_0 so that the charging curve is dominated by a single exponential, τ_0.

Another approach for determining the τ_n's and C_n's makes use of a brief ($< \tau_0/20$) current pulse instead of a long pulse. With the use of a brief pulse, the ratio of C_1 to C_0 (or C_2 to C_0) is greater than with the long pulse and thus there is greater weighting of the higher order exponential terms in the charging curve, making them easier to measure. The exponentials and their coefficients can be determined exactly as described above, but now the coefficients determined from the short pulse method (C_{Sn}'s) are related to the coefficients from the long pulse method (C_{Ln}'s) by (Durand et al. 1983)

$$C_{Sn} = C_{Ln}(1 - e^{-w/\tau_n})$$
(4.5.56)

where w is the pulse width. One of the problems with this method is that any leakage around the recording electrode can cause spurious results. It is therefore best to use the brief pulse method in conjunction with whole-cell patch recordings.

Regardless of whether one uses the brief or long pulse method, if at least two exponentials can be extracted from the charging curve, either by peeling or by some other procedure, then the time constants and their coefficients can be used to determine the parameters of the ball-and-stick model for the neuron under study. As seen in figure 4.23, C_{L0} (or C_{S0}) and C_{L1} (or C_{S1}) are the intercepts and τ_0 and τ_1 can be derived from the slopes of the lines. From equation 4.5.52, we get the following

$$C_{L0} = B_0,$$

$$C_{L1} = B_1 \cos(\alpha_1 L),$$

or

$$C_{L1} = \frac{2 C_{L0} \, \tau_1/\tau_0}{1 + (\alpha_1 L)^2/(k^2 + k)}.$$

Using the relation for k (equation 4.5.54),

$$(\alpha_1 L)^2 = k^2/[\cot(\alpha_1 L)]^2 \quad \text{and} \quad k = -\alpha_1 L \cot(\alpha_1 L)$$

so

$$\left| \frac{C_{L1}}{(2C_{L0}\tau_1/\tau_0) - C_{L1}} \right| = \cot(\alpha_1 L)[\cot(\alpha_1 L) - 1/(\alpha_1 L)]. \qquad (4.5.57)$$

The above is a transcendental equation in that it has an infinite number of roots or solutions. Given measurements for $C_{L1}, C_{L0}, \tau_1, \tau_0$, one can usually use equation 4.5.57 to solve for L. The procedure is to first guess at L[6] and then iteratively calculate the right side of the equation, improving on the estimate of L each time, until the equation converges on the correct solution. Once L is calculated, ρ can be obtained from the following:[7]

$$\rho = \frac{-\alpha_1 \cot(\alpha_1 L)}{\coth(L)}, \qquad (4.5.58)$$

where

$$\alpha_1 = \left[\frac{\tau_0}{\tau_1} - 1 \right]^{1/2}.$$

Assuming that the neuron under investigation can be represented by an equivalent cable with lumped soma, the fairly simple procedures of injecting a step of current and measuring the resulting charging curve can give us R_N (from the steady-state V_m/I) and, as shown above, ρ, L, and τ_m.

---

[6]A good starting guess can be obtained from the solution for L of a finite cable or

$$L = \pi \left[\frac{1}{(\tau_0/\tau_1) - 1} \right]^{1/2}.$$

[7]see section 4.9 for derivation

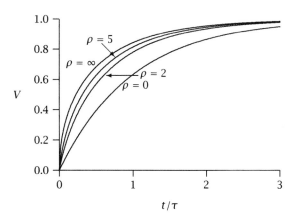

Figure 4.24 Comparison of charging curves from neurons represented by an isopotential soma with a finite cable but with different values for ρ. (After Jack et al. 1975.)

How does the charging curve vary as a function of ρ? We illustrated previously the charging curves for different L's, and in figure 4.24 we illustrate charging curves for different ρ's. The limits are at $\rho = 0$, which would represent somatic dominance or a large soma attached to a thin equivalent cylinder, and at $\rho = \infty$, which would represent dendritic dominance or a small soma attached to a large diameter equivalent cylinder. Note that a $\rho \geq 5$ is barely distinguishable from $\rho = \infty$. ρ thus affects the shape of the charging curve appreciably only when it is between 0 and 5.

Early estimates of R_N, ρ, L, and τ_0 for hippocampal neurons (for description of CA3, CA1, and granule cells, refer to chapter 14) were about 20,000 Ω-cm^2, 1.5, 0.9, and 20 msec, respectively. Unfortunately, only the granule cells are well fit by a ball-and-stick model so that many of the assumptions for determining L and ρ do not hold. Also, the measurements were made using sharp microelectrodes, which are known to introduce a leak into the soma and make the assumption of uniform R_m invalid (that is, τ_0 will not be equal to τ_m and the value of ρ is artifactually low). Nevertheless, what is clear from these and later estimates is that R_N and τ_m are large and L is small ($L < 1$) for all cell types even though they have quite different morphologies. This has led to the idea that cortical neurons are *electrotonically compact*. Although this idea will be discussed more fully below and in chapter 13 (also see Spruston et al. 1993), it is worth emphasizing that "compactness," as used here with respect to small L and ρ, is only relevant for the steady state (DC) signals. As we saw in the section on finite cables, λ can become very small (or large L) when dealing with non-steady-state (AC) signals. Since most of the electrical signals of interest in neurophysiology are changing with time, the term electrotonically compact is a bit of a misnomer, because it applies to conditions that are unusual for neurons.

Newer techniques have been used to reevaluate the passive membrane properties of hippocampal neurons. Using perforated-patch recording methods, R_N and τ_0 are now believed to be about 135, 104, and 446 MΩ and 66, 28, and 43 msec for CA3, CA1, and dentate granule cells, respectively (Spruston and Johnston 1992). These τ_0 values suggest $R_m = 66,000$, 28,000, and 43,000 Ω-cm^2 for the three cell types. Even these measurements of R_N and τ_0, however, may have been influenced by spontaneous synaptic input, and the values could be as much as 20% higher. These and other studies have also made clear that several of the key assumptions of the Rall model, such as passive R_m and all dendrites ending at the same L, may not hold. An understanding of the passive properties of neurons is still a useful first step, however, for investigating properties of dendrites and how they modify (or integrate) synaptic inputs. These topics are discussed in the next two sections and in chapter 13.

4.5.3 Application to synaptic inputs

As mentioned earlier, the main reason for studying the electrotonic properties of neurons is to help us understand the function of dendritic synaptic inputs. Although the mechanisms of synaptic transmission will be dealt with in some detail in later chapters, we will discuss here some of the ways in which the electrotonic properties of dendrites affect synaptic integration.

In the early 1950s, when the field of neurophysiology was in its infancy, scientists believed that the distal dendrites of cortical and spinal motor neurons were not functionally connected to the rest of the neuron. In other words, they believed that the synaptic inputs to the distal dendrites would have little or no effect on the firing properties of the neuron—they were just too far away. The function of the distal synaptic inputs was obviously not known, and many scientists believed that the synaptic potentials recorded in spinal motor neurons must arise from the more proximal inputs.

Rall's development of his neuron model fundamentally changed the way neurophysiologists thought about dendritic trees. He estimated that the electrotonic length of spinal motor neurons was on the order of 1.5 and that distal synaptic inputs would indeed produce measurable signals in

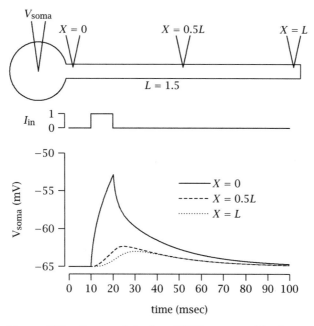

Figure 4.25 Shapes of simulated EPSPs measured in the soma resulting from a current step injected at three different locations on a ball-and-stick model. The model was implemented as a compartmental model (see chapter 13).

the soma.[8] Even though Rall demonstrated that distal synaptic inputs could produce measurable signals in the soma, the synaptic potentials were nonetheless attenuated and distorted in shape by the electrical properties of the dendrites.

The first example we will use to demonstrate this fact is a ball-and-stick model of a neuron with the injection of brief steps of current into the soma, into a middle region of the dendrites, and into the distal dendrites. The changes in membrane potential measured in the soma in response to each of these current steps are illustrated and compared in figure 4.25. Assuming that a synaptic input injects a brief step of current into the dendrites, the results are quite interesting and informative. First, as expected, the amplitude of the synaptic potential, or, in the nomenclature of neurophysiology, *excitatory postsynaptic potential* (EPSP), measured in the soma in response to distal input is smaller than that from either the middle or somatic inputs. In other words, because of the electrical properties of the

---

[8]Rall also showed that the charging curve resulting from a step of current to the soma was not a single exponential, as was first believed, and that by peeling, a better (higher) estimate of τ_m could be obtained.

dendrites, distal synaptic signals are attenuated upon reaching the soma. Second, the rise time and peak amplitude of the inputs is progressively slowed and delayed for inputs at increasing distance from the soma. This slowing arises from the distributed resistance and capacitance of the dendritic membrane. And third, the final decay time of all the inputs is the same. That is, the potential developed from a brief synaptic current injection cannot dissipate or decay slower than the membrane time constant.

One can argue (as we will show later) that synaptic inputs do not inject a step of current. (In fact, they produce a locally brief conductance change that may or may not inject current, but this will be explained later.) A better representation for the shape of the current injected by a synapse is given by the so-called *alpha function*. The equation is given below and graphically illustrated in figure 4.26.

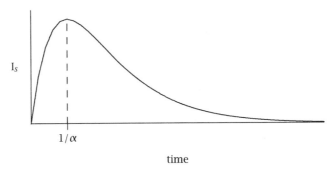

Figure 4.26   The alpha function, $I_s = I_0(t/\alpha)e^{-\alpha t}$, is illustrated. The time to peak is equal to $1/\alpha$.

$$I_s = I_0(t/\alpha)e^{-\alpha t}, \qquad\qquad\qquad (4.5.59)$$

where I_0 is the maximum current (it could be either positive or negative, depending on whether it is being used to represent outward or inward current) and $1/\alpha$ is the time to peak (in $\sec^{-1}$) of the current. The value chosen for α determines the shape (rise time and decay time) of the synaptic current. Using this better representation for a synaptic input, a similar experiment to that above, where synaptic inputs are placed at three different locations on the model, is illustrated in figure 4.27.

The results are identical, that is, the EPSPs are smaller and slower for inputs at increasing electrotonic distances from the soma. In fact, Rall has shown that there is a quantitative relationship between electrotonic distance and shape of the synaptic potential measured in the soma. This is best illustrated by another example. Using the same ball-and-stick model,

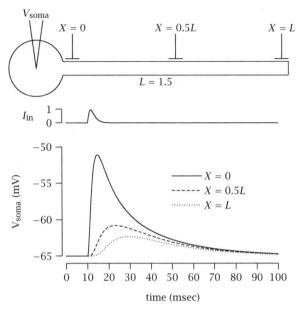

Figure 4.27 Shapes of simulated EPSPs measured in the soma. The synaptic inputs were represented by a conductance change in the form of an alpha function ($I_s = G_s(t/\alpha)e^{-\alpha t}(V_m - E_s)$). I_{in} illustrates the shape of the current injected by the synapse at each of the indicated locations in the dendrites. Same compartmental model as in figure 4.25.

figure 4.28 plots the rise time (measured as the time from 10%–90% of peak) vs. the half width (measured as the width of the EPSP at 50% of peak amplitude) for different α functions injected at different electrotonic distances from the soma. This is called a shape index plot. As we saw in the previous two figures, the shapes of the EPSPs change in a predictable fashion for inputs at different electrotonic distances. In fact, one could, in theory, use such a relationship to estimate the electrotonic distance of a synapse by first measuring the shape of the EPSP and then finding its value on a representative shape index plot for that neuron. This method was used by Rall and others for localizing inputs to spinal motor neurons and by Johnston and Brown for estimating the electrotonic distance of mossy fiber synapses on CA3 pyramidal neurons in hippocampus (figure 4.28).

Few if any dendrites actually have passive membranes as the above analyses assume. Considerable interest is currently focused on identifying and localizing voltage- and ligand-gated ion channels in dendrites and then, along with linear cable theory, developing principles of dendritic integration.

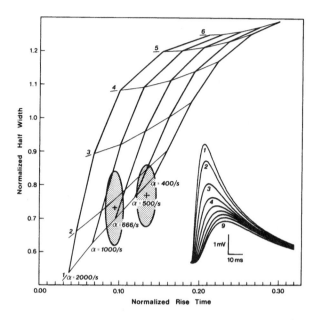

Figure 4.28 Shape index plot for mossy fiber EPSPs in a nine-compartment model of a CA3 pyramidal neuron (from Johnston and Brown 1983). $1/\alpha$ represents the time to peak of the synaptic current (varied from 0.5–2.5 msec), and the numbers represent the compartments into which the model synapse was placed. The ovals represent the means $\pm 2 \times$ the SEMs of rise times and half widths of the EPSPs recorded from two different neurons. The conclusion from this analysis is that the mossy fiber synapses from these two cells were at electrotonic distances equivalent to those of compartments 1 and 2 in this model (see also chapter 13). The inset illustrates simulated EPSPs measured from the soma compartment (i.e., # 1) when the model synapse was placed in each of the indicated compartments (for α=500/sec).

4.6 Two-port network analysis of electrotonic structure

As mentioned previously, the Rall model requires assumptions of spatial uniformity of membrane properties, voltage independence of R_m, the 3/2 power rule for branch points, and all dendrites terminating with the same L. The latter is a particularly restrictive assumption that is unlikely to be true except for a few types of neurons. Without the ability to reduce dendritic trees to an equivalent cylinder, even multiple equivalent cylinders, very little can be determined about the electrotonic structure of a neuron beyond the measurement of R_N and τ_m. Since the main reason for wanting to know something about electrotonic structure is for analyzing dendritic synaptic inputs, an alternative approach was developed

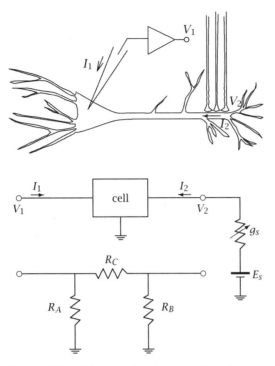

Figure 4.29 Schematic for the reduction of a neuron with synaptic inputs into a two-port network. One port represents the soma and the other the site of synaptic input. The neuron between the ports can be represented by a three-resistor network called a π network, shown in the lower part of the figure. (Adapted from Carnevale and Johnston 1982.)

in the early 1980s and has recently received increased attention. It is the so-called two-port network analysis of synaptic inputs.

The typical recording situation is depicted again in figure 4.29. One is usually not as interested in the electrotonic structure of the neuron per se, as in how the dendrites distort the synaptic signal. This situation can be reduced to two ports or contact points for inputs and outputs. In this case, one port is the site of synaptic input and the other is the site of recording. The rest of the neuron, at least for the moment, is not of interest. We have an input and an output, the synaptic and recording sites, respectively; or, conversely, when we inject current into the recording site it becomes the input while the synaptic site becomes the output. In any event, for this particular synaptic input, we have reduced the neuron to a two-port network. No assumptions need be made about branching structures of the dendrites or spatial uniformity of membrane properties.

The only key assumption remaining is that the membrane properties must be passive.

For the steady-state case, the entire neuron can be reduced to three resistors connected in what is called a π network. Using some simple methods of circuit analysis (see appendix A), we can derive several useful parameters called *coupling coefficients* (or *attenuation factors*). For example, if one injects steady current into the soma, there will be a potential gradient between the soma and the synaptic site, described by

$$V_2 = \frac{R_B}{R_B + R_C} V_1.$$
(4.6.60)

We will introduce the first coupling coefficient, k_{12}, by rearranging the equation

$$k_{12} = \frac{V_2}{V_1} = \frac{R_B}{R_B + R_C}.$$
(4.6.61)

This coupling coefficient describes the steady-state *voltage attenuation* from the soma to a synaptic site. A similar coupling coefficient can be derived for the opposite case, the attenuation of potential from the synapse to the soma in response to current injected into the synaptic port. In this case,

$$k_{21} = \frac{V_1}{V_2} = \frac{R_A}{R_A + R_C}.$$
(4.6.62)

A very important principle can be seen immediately, that is, $k_{12} \neq k_{21}$. In other words, voltage attenuation in dendrites is in most cases not symmetrical. The attenuation of potential from the soma to the synapse is not the same as the attenuation of potential from the synapse to the soma, and, in general, $k_{12} > k_{21}$.

There are several other useful relations that can be derived from this simple two-port network. We will simply present them here and direct the interested reader to the original publications.

Principle 1: In general, $k_{12} \neq k_{21}$, and, in most cases, the voltage attenuation from soma to synapse is less than from synapse to soma (i.e., $k_{12} > k_{21}$).

Principle 2: Charge transfer from synapse to soma is equal to the voltage attenuation from soma to synapse ($q_{\text{soma}}/q_{\text{synapse}} = k_{12}$).

Principle 3: With a voltage clamp to the soma, steady-state current attenuation from synapse to soma is equal to the voltage attenuation from soma to synapse ($i_{\text{soma}}/i_{\text{synapse}} = k_{12}$).

This relatively simple idea of using a two-port description of a neuron and deriving coupling coefficients has recently been extended in a clever way (Brown et al. 1992; Tsai et al. 1994; Zador 1992). The two-port description was generalized from the somatocentric view outlined above, where one port was always the soma and the other was an arbitrary site of synaptic input, to the more general case in which the two ports are any two sites on the neuron. In other words, the two ports could be two different dendritic sites as well as the soma and one dendritic site. The coupling coefficients derived above are now more descriptively called *attenuation factors* (A's) so that

$$A_{12} = \frac{V_1}{V_2} = \frac{1}{k_{12}} \tag{4.6.63}$$

or, more generally,

$$A_{ij} = \frac{1}{k_{ij}}, \tag{4.6.64}$$

where the A_{ij} would represent the attenuation of potential from site i to j for current injected at site i.

Zador, Brown, and colleagues recognized that the logarithm of the attenuation ratio is equal to the electrotonic distance X_{ij} for an infinite cylinder or

$$X_{ij} = \ln |A_{ij}| \tag{4.6.65}$$

for an infinite cable.

What makes this formulation so appealing is that the ln(attenuation ratios) for different parts of a neuron are additive, regardless of the geometry of the neuron. One can thus construct the electrotonic structure of a dendritic tree by adding together the ln(attenuation ratios) for different parts of a neuron. This can be done based on the power of the two-port network description derived above and without any assumptions other than that of passive membrane properties.

Examples of the resulting *morphoelectrotonic transforms* are illustrated in figure 4.30. In this figure a reconstructed CA1 pyramidal neuron is illustrated on the left in normal units of distance (μm). This same neuron is redrawn four times on the right with a scale according to the natural logarithm of the attenuation ratios (1 ln unit). The small neuron diagram in the upper middle represents the ln of the attenuation for the steady-state (0 Hz) voltage decay from the soma *out* the dendrites. The diagram in the upper right represents the steady-state voltage decay from different points in the dendrites *in* toward the soma. The figure in the upper right is larger because the attenuation of potential from the dendrites to the soma is much greater than the attenuation from the soma out the dendrites. The other two figures represent similar measures except that they were done for a 40 Hz sine wave. Obviously, there is much greater attenuation (in either direction) at the higher frequency.

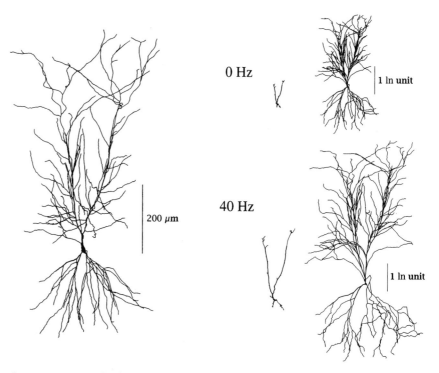

Figure 4.30 Morphoelectrotonic transforms. A reconstructed CA1 neuron is shown on the left, and the morphoelectrotonic transforms for the decay of potential from the soma out the dendrites and from the dendrites toward the soma are illustrated on the right. The neuron diagrams on the right are drawn according to a scale that represents the ln of the voltage attenuation ratios (1 ln unit represents voltage attenuation of $1/e$ from the site of current injection). The top middle transform represents the ln(attenuation) of a DC signal applied to the soma as it decays to different sites in the dendrites. The top right transform represents the ln(attenuation) of a DC signal applied to different sites in the dendrites as it decays toward the soma. The two transforms at the bottom right represent the same measurement as those above but resulting from the application of a 40 Hz sine wave signal instead of 0 Hz. (Kindly provided by Tsai, Carnevale, Claiborne, and Brown.)

4.7 Summary of important concepts

1. Equivalent circuit representation of neurons.
2. Formulation of the cable equation.
3. Passive properties of neurons (r_i, r_m, c_m, R_i, R_m, C_m, τ_m, λ, l, L, ρ, R_N).
4. Differences in transient and steady-state attenuation properties between infinite, semi-infinite, and finite cables.
5. Rall model.
6. How to determine L, ρ, and τ_m.
7. Application of cable theory to synaptic inputs and asymmetrical voltage attenuation in dendrites.

4.8 Derivation of the input conductance of a finite-length cable

Figure 4.31 Diagram of finite cable with current injected into one end. For derivation of input conductance of a finite-length cable.

$$I_0 \;=\; i_i = -(1/r_i)\partial V_m/\partial x = -(1/\lambda r_i)\partial V_m/\partial X, \quad \text{for } T \to \infty.$$

$$R_N \;=\; V_m/I_0 = (V_0/I_0)\cosh(L-X)/\cosh(L)$$

$$=\; -[V_0\lambda r_i/(\partial V_m/\partial X)]\cosh(L-X)/\cosh(L).$$

$$\frac{\partial V_m}{\partial X} \;=\; \frac{V_0\cosh(L)\cdot d/dX\cosh(L-X)}{\cosh(L)^2} - \frac{\cosh(L-X)\cdot d/dX\cosh L}{\cosh(L)^2}$$

$$=\; -\frac{V_0\sinh(L-X)}{\cosh(L)}.$$

At $X = 0$,

$$R_N \;=\; \lambda r_i\frac{1}{\tanh(L)} = \lambda r_i\coth(L),$$

$$G_N \;=\; \frac{1}{\lambda r_i}\tanh(L).$$

4.9   Derivation of potential distribution along finite cable attached to lumped soma

For current step applied at $X = 0$ (refer to figure 4.22),

$$I_0 = I_S + I_D,$$

$$I_S = G_S(V_m + \partial V_m / \partial T),$$

$$I_D = - [1/(\lambda r_i)] [\partial V_m / \partial X]_{X=0}.$$

New boundary condition at $X = 0$ gives

$$\frac{\partial V_m}{\partial X}\bigg]_{X=0} = \lambda r_i [-I_0 + G_S(V_m + \partial V_m / \partial T)].$$

Remember:
$$\rho = G_D/G_S = \frac{\tanh(L)}{\lambda r_i G_S} \text{ at } I_0 = 0.$$

$$\rho \frac{\partial V_m}{\partial X}\bigg]_{X=0} = (V_m + \partial V_m / \partial T) \tanh(L).$$

A general solution for cable equation is

$$V_m(T, X) = [A \sin \alpha(L - X) + B \cos \alpha(L - X)] e^{-(1+\alpha^2)T}.$$

As before,

$$\frac{\partial V_m}{\partial X}\bigg]_{X=L} = 0$$

requires

$$A = 0.$$

At $X = 0$,

$$\rho \alpha B \sin(\alpha L) e^{-(1+\alpha^2)T} = \tanh(L) e^{-(1+\alpha^2)T} [B \cos(\alpha L) - (1 + \alpha^2) B \cos(\alpha L)],$$

$$\rho \alpha \sin(\alpha L) = -\alpha^2 \cos(\alpha L) \tanh(L),$$

or

$$\alpha L \cot(\alpha L) = -\rho L / \tanh(L) = -k.$$

The roots of the equation are α_n, and, as before,

$$\tau_n = \frac{\tau_0}{1 + (\alpha_n)^2}.$$

The solution is now

$$V_m(T, X) = \sum_{n=0}^{\infty} B_n \cos [\alpha_n (L - X)] e^{-(1 + \alpha_n^2)T},$$

where

$$B_n \cos (\alpha_n L) = \frac{2Bo \; \tau_n / \tau_0}{1 + (\alpha_n L)^2 / (k^2 + k)}$$

at $X = 0$ for current step.

4.10 Homework problems

1. In the text the voltage response in an infinite linear cable to a prolonged current step was shown to be

$$V_m(t, x) = \frac{r_i I_0 \lambda}{4} \left[e^{-x/\lambda} \text{erfc} \left(\frac{x/\lambda}{2\sqrt{t/\tau_m}} - \sqrt{t/\tau_m} \right) \right.$$
$$\left. - e^{x/\lambda} \text{erfc} \left(\frac{x/\lambda}{2\sqrt{t/\tau_m}} + \sqrt{t/\tau_m} \right) \right],$$

where

$$\text{erfc}(x) = 1 - \text{erf}(x) = 1 - \frac{2}{\sqrt{\pi}} \int_0^x e^{-y^2} dy, \; \lambda = \left(\frac{aR_m}{2R_i} \right)^{1/2},$$

and

$$\tau_m = R_m C_m.$$

(a) Show that

$$
\begin{aligned}
\text{erfc}(\infty) &= 0, \\
\text{erfc}(0) &= 1, \\
\text{erfc}(-\infty) &= 2, \\
\text{erf}(-x) &= -\text{erf}(x).
\end{aligned}
$$

(b) The input resistance R_N of a cable is defined as $R_N = V_m(\infty, 0)/I_0$. What is the input resistance of an infinite cable if the radius is 25 μm, the membrane resistance is 2000 Ω-cm^2, and the axoplasm resistivity is 60 Ω-cm?

(c) What is the steady-state voltage response measured at $x = 0$, $x = \lambda$, $x = 10\lambda$, to a prolonged current step of 10 nA injected at $x = 0$? Use the values of a, R_m, and R_i in (b).

(d) What is the transient-state voltage response measured at $x = 0$ and $t = \tau_m/2$ to a prolonged current step of 10 nA injected at $x = 0$, $t = 0$? Use the values of a, R_m, and R_i in (b).

2. Show that

$$V_m = A_1 e^X + A_2 e^{-X},$$
$$V_m = A_1 \cosh(X) + A_2 \sinh(X), \text{ and}$$
$$V_m = A_1 \cosh(L - X) + A_2 \sinh(L - X)$$

are all solutions to the steady-state cable equation $V_m = \frac{\partial^2 V_m}{\partial X^2}$.

3. Assume a semi-infinite cable of diameter 1 μm with $R_m = 5000$ Ω-cm^2 and $R_i = 75$ Ω-cm.

(a) Calculate the input conductance, G_∞, for this cable.

(b) If the end of this cable is sealed with a disk of membrane with the same properties as the cable, calculate the resistance, G_L, of the sealed end.

(c) What is the value of B_L for this finite cable? Would you call it an open-circuit or short-circuit termination?

4. Derive the cable equation using the circuit diagram shown. Give the equation in as many forms as you can and define all parameters.

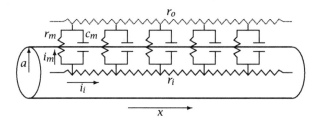

5. For the cables shown below,

$d_1 = d_2 = d_3 = 1$ cm,

$R_{m_1} = R_{m_2} = R_{m_3} = 5000$ Ω-cm^2, and

$R_{a_1} = R_{a_2} = R_{a_3} = 100$ Ω-cm.

Assume all terminators can be considered open circuit (sealed end).

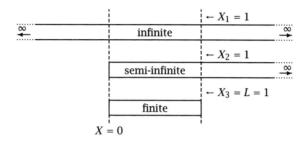

  (a) Calculate the input resistance at $X = 0$ for each of the three cables.

  (b) If 10 mV is imposed at $X = 0$, calculate the voltage at $X = 1$ for each of the three cables.

6. Two infinitely long axons, A and B, are shown below. Both have Na$^+$ channels located only at the regions indicated by the dots, while the rest of the axonal membranes can be considered passive. A prolonged action potential (duration $> 3 \times \tau_m$) is generated in A and B at the left. The resting potential of A and B is -70 mV, and the peak of the prolonged AP is $+30$ mV. The next row of Na$^+$ channels on the right is located 2λ from the ones on the left.

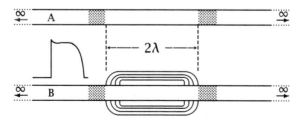

  (a) Assuming threshold is -50 mV, will the prolonged AP in A propagate to the right and produce another AP? Why or why not?

(b) In B, extra membrane tightly surrounds the space between the Na$^+$ channels. This extra membrane effectively increases the membrane resistance and decreases the membrane capacitance by a factor of 4. Will the prolonged AP in B propagate to the right and produce another AP? Why or why not?

(c) How would your answers to (a) and (b) change if the diameter of A and B were doubled?

7. At each of the three cables shown below, a Na$^+$-dependent EPSP ($E_{Na} = +30$ mV) is located at one end of the cable. You are recording with a microelectrode at the other end and are able to make any reasonable type of measurements you wish with the microelectrode while activating the synapse. What will be the approximate *measured* reversal potential for each cable? Assume a resting potential of -70 mV and sealed ends for the cables.

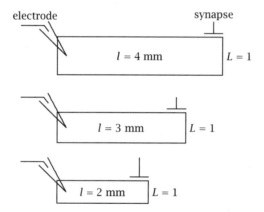

8. You have two unmyelinated axons, identical except that their diameters are 5 μm and 10 μm. (Assume that they are infinite in length.)

(a) What are the relative conduction velocities (passive wave) of the two cables?

(b) What are the relative input resistances of the two cables?

(c) What are the relative space constants of the two cables?

(d) What are the relative time constants of the two cables?

Show all calculations.

9. You are recording from a spherical neuron with $R_m = 25,000$ Ω-cm^2 and $a = 10$ μm;

(a) What is R_N?

(b) Plot (showing at least 4 time points) the response of this cell to a 10 pA step of current. (This should be a detailed plot, not just a rough sketch.)

(c) If we now attach to this cell a semi-infinite cable with $R_i = 100$ Ω-cm and $a = 1$ μm,

   i. What is ρ?

   ii. What is the expected voltage response of this cell with attached cable to the same step of current as before? (Make two plots, one with an absolute amplitude scale that includes this response and the one calculated above, and a second plot with a normalized amplitude scale that again shows both responses. The time courses of each curve can now be less precise than that expected above, but be specific as to the steady-state values of each and their relative shapes.)

(d) If you added a neurotransmitter to the bath (e.g., GABA) in which the receptors for this neurotransmitter were located only on the soma, and this addition to the bath reduced R_m by a factor of 2, what is the new input resistance of this neuron with attached cable? What is ρ?

10. Answer *true* or *false* for each of the following:

(a) Distal EPSPs have slower rise times than proximal EPSPs.

(b) Distal EPSPs have slower decay times than proximal EPSPs.

(c) Proximal EPSPs have longer half widths than distal EPSPs.

(d) Proximal and distal EPSPs have the same time to peak.

(e) An EPSP in the soma would affect an EPSP in the distal dendrites the same as a distal EPSP would affect an EPSP in the soma.

11. Given a finite-length cable of $l=1$ cm, $\lambda =0.5$ cm, and $\tau_m =100$ msec;

(a) What is the electrotonic length of the cable?

(b) What is the electrotonic distance for a synapse 0.5 cm from the origin?

(c) What is the ratio of V_L/V_0 for a DC signal (where V_L is at the end of the cable and V_0 is at the beginning of the cable)?

(d) What is the length constant for a 10 Hz signal? How does the voltage attenuation of cables differ for DC vs. signals that change with time?

12. For the model neuron shown below (not drawn to scale) assume that the 3/2 power law is obeyed and that $R_m = 10,000$ Ω-cm^2 and $R_i = 200$ Ω-cm, $a_1 = a_2 = 1$ μm, and $a = 20$ μm for the soma. Assume also that all ends are sealed and that $\lambda_1 = \lambda_2 = 500$ μm, $\lambda_{11} = \lambda_{12} = \lambda_{21} = \lambda_{22} = 400$ μm; $l_1 = 300$ μm, $l_2 = 200$ μm, and $l_{11} = l_{12} = l_{21} = l_{22} = 100$ μm.

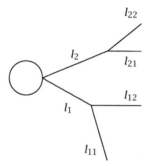

Calculate the following:

(a) L for each of the branches

(b) To what extent can this neuron be reduced to a Rall-type model?

(c) The total input resistance of the neuron

(d) ρ

(e) If the initial resting potential is –70 mV and one applies a steady 1 nA hyperpolarizing current, what will the final (steady-state) potential be if a 2 nA step of depolarizing current is superimposed on the hyperpolarizing current? Assume that the neuron is passive.

13. A motor neuron has a spherical soma 60 μm in diameter. Five dendritic trees identical to that described in example 4.3 and a long axon (which can be considered a semi-infinite cable) of 2 μm in diameter are attached to the soma (let $R_m = 2000$ Ω-cm^2 and $R_i = 60$ Ω-cm).

(a) What is the soma conductance?

(b) What is the total input conductance of the 5 dendritic trees?

(c) What is the input conductance of the axon?

(d) What is the input conductance (G_N) of the whole neuron? Use the values of R_m = 2000 Ω-cm², R_i = 60 Ω-cm.

(e) What is the steady-state membrane potential recorded from the soma when a constant current step of 10 nA is applied at the center? The resting potential of the cell before current injection is –70 mV.

14. You are recording from the soma of a hippocampal neuron and you measure the voltage responses to a step of current shown below.

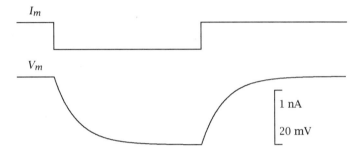

(a) What is the total input conductance of this neuron?

(b) If you know by some other means that the input resistance of the *soma* is 100 MΩ, what is the input conductance of the dendrites? What is ρ?

(c) If you know that the neuron can be represented by an equivalent cable model with lumped soma (i.e., a ball-and-stick) with l = 0.01 cm, a = 0.0005 cm, and L = 1, calculate the passive membrane properties of the cell ($\lambda, r_i, r_m, R_m, \tau_m$, and the radius of the soma, r). If necessary, you can assume C_m = 1 μF/cm² (most people consider this value for C_m a constant for all biological membranes), and a sealed end for the cable.

(d) What would be the voltage at the end of the dendrites if the soma were depolarized by 50 mV? (Assume a sealed end for the equivalent cable.)

(e) What would be the voltage at the same point if the end of the equivalent cable were terminated by a semi-infinite cable with the same membrane parameters?

15. You are recording intracellularly from a neuron and wish to determine its electrotonic structure. You are able to make all the necessary electrophysiological measurements and want to assume that the neuron and its dendrites can be represented by an equivalent finite-length cable with lumped soma. List the most important assumptions necessary to allow you to represent this neuron by such a model. Each assumption should be written in two sentences or less.

16. Three cables are attached together as shown below.

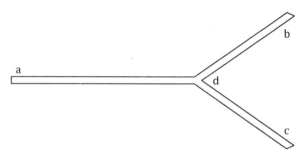

1 mA of steady current is injected at the intersection, d. What are the *steady-state* voltages at d and at one of the ends (a, b, or c)? Make any assumptions you need to make, but state them explicitly. Use the following values: l_{da}= 10 cm, λ_{da}= 5 cm, r_{da} = 42 Ω/cm; l_{db}= 2 cm, λ_{db}= 2 cm, r_{db} = 1660 Ω/cm; l_{dc}= 2 cm, λ_{dc}= 3 cm, r_{dc} = 327 Ω/cm. Note that the diagram is not drawn to scale.

17. (a) State the conditions under which a dendritic tree can be reduced to Rall's equivalent cylinder.

    (b) Plot in the approximate scale the steady-state voltage response $V(\infty, X)$ against X to a current step injected at $X = 0$ in a finite cable when the end of the cable is sealed, infinite, or grounded.

18. In the figure for problem 16 (which is not drawn to scale) you are given a branched cable with the following properties: l_{ad}=4 cm, λ_{ad}=5 cm, $r_{i_{ad}}$=42 Ω/cm, a_{ad}=0.75 cm, l_{db}=3.3 cm, λ_{db}=3.6 cm, a_{db}=0.39 cm, l_{dc}=4 cm, λ_{dc}=4.3 cm, and a_{dc}=0.55 cm. You are asked to determine:

    (a) the steady-state input resistance as measured at a.

    (b) the steady-state voltages at a, b, c, and d given a 1 mA steady current injected into a.

You can assume that all ends are sealed. Justify your methods of analysis.

19. In plots of $\log V_m$ vs. t (using semilog graph paper), where V_m is the membrane voltage response to a step of current, the "peeled" straight lines are used to calculate time constants. Derive an equation for the time constant as a function of graphical measurements V_1, V_2, t_1, t_2.

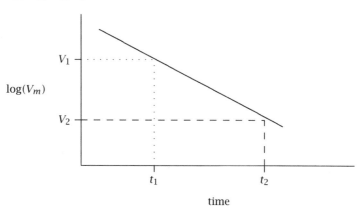

20. The drawing below illustrates four different theoretical conditions (not drawn to scale): an infinite cable, a semi-infinite cable, a finite cable ($L=1$), and a spherical cell. Each has the following properties: $R_m = 10,000$ Ω-cm^2, $R_i = 100$ Ω-cm, $a = 1$ μm for the three cables, and $a = 30$ μm for the sphere. Assume sealed ends where relevant and that the cells are purely passive.

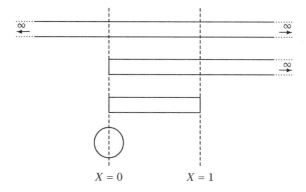

(a) If a 0.1 nA current is injected at $X = 0$, draw on a piece of graph paper, with as much accuracy as you can (i.e., provide numbers where possible), the voltage responses with respect to time for each of the four conditions. Draw each response separately and also normalize and combine the four responses on a single graph.

(b) Calculate the conduction velocity of a passive wave for each of the three cables and determine the conduction time from $X = 0$ to $X = 1$.

(c) If an excitatory synaptic input is applied at $X = 1$ on each of the three cables, draw on a piece of graph paper the *relative* sizes and decay time courses of the three responses as measured at $X = 0$. Assume that the amplitude and time course of the EPSP at $X = 1$ for each of the three cables is identical. Be precise in your description of the decay phase of each of the EPSPs.

21.

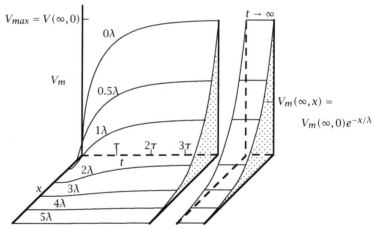

(a) Using the figure above and your knowledge of cable theory, answer the following, *true* or *false*.

    i. $V_m(\tau, \lambda) < V_m(\infty, \lambda)$

    ii. $\frac{V_m(\infty,\lambda)}{V_m(\infty,2\lambda)} > \frac{V_m(\tau,\lambda)}{V_m(\tau,2\lambda)}$

    iii. $\frac{V_m(\tau,\lambda)}{V_m(3\tau,\lambda)} < \frac{V_m(\tau,2\lambda)}{V_m(3\tau,2\lambda)}$

    iv. $V_m(\tau/10, 0) > V_m(\tau/5, 2\lambda)$

(b) Using the figure below and your knowledge of cable theory, answer the following, *true* or *false*.

    i. $V_m(X = 1)_{\text{sealed end}} > V_m(1)_{\text{inf cable}} > V_m(1)_{\text{open end}}$

    ii. $V_m(1)_{\text{open end}} > V_m(1)_{\text{inf cable}}$

   iii. $V_m(1)_{\text{open end, L=2}} > V_m(2)_{\text{sealed end}}$

   iv. $V_m(L)_{\text{open end}} = 0$

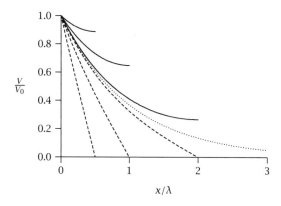

22. Referring to the figure below (not drawn to scale), let $d_1 = 1\ \mu m$, $d_2 = 2\ \mu m$, $d_3 = 3\ \mu m$, $d_4 = 4\ \mu m$, $d_{11} = 0.5\ \mu m$, and $d_{111} = 0.3\ \mu m$. Assume that *all* of the assumptions necessary for the Rall model hold.

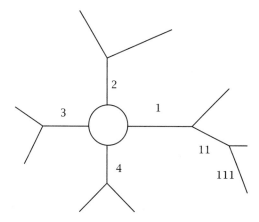

(a) If you wanted to represent all of the 4 trees (i.e., 1, 2, 3, and 4) by a single equivalent cable attached to the soma, what would its diameter have to be?

(b) If there is a 30% attenuation of the potential from the soma to the end of the dendrites, what is the electrotonic length of this neuron?

(c) If we detached the soma, connected all of the somatic ends of the 4 cables together, and sealed those ends, what would the voltage attenuation be from the distal ends of the dendrites to the soma ends?

(d) If we then reattached the soma as before, would the attenuation of potential from the distal ends of the dendrites to the soma be greater than, less than, or the same as the 30% in (b)?

23. In the figure below (not drawn to scale) the input conductance of the soma is 10^{-9} S, $R_i = 200$ Ω-cm, and (in μm) $l_A = 250$, $l_{A_1} = l_{A_2} = 80$, $\lambda_A = 500$, $\lambda_{A_1} = \lambda_{A_2} = 400$, $d_{A_1} = d_{A_2} = 1$; $l_B = 600$, $l_{B_1} = l_{B_2} = 600$, $\lambda_B = 400$, $\lambda_{B_1} = \lambda_{B_2} = 300$, $d_B = 1$. Assume sealed ends, the 3/2 power law holds at all branch points, and $E_r = -70$ mV.

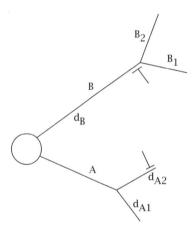

(a) Calculate L for each of the dendritic trees A and B.

(b) Calculate the total input resistance, R_N, for this neuron.

(c) If $E_s = +10$ mV for each of the synapses, what is V_{rev} for each? Calculate V_{rev} for B using two different equations.

(d) If the alpha function used to simulate each synapse is the same, describe qualitatively the relative magnitudes (i.e., A/B) of the amplitudes, rise times, half widths, and decay time constants for the inputs as measured in the soma. Assume that they are activated individually.

5   Nonlinear Properties of Excitable Membranes

5.1   Introduction

In chapters 3 and 4, we considered the situation in which the membrane conductance is constant, that is, the current-voltage relation is linear. In most excitable cells, however, the voltage range over which the membrane may be assumed linear is very restricted. In this chapter, we shall describe nonlinear properties of excitable membrane and derive models and equations applicable to nonlinear membrane conductances.

5.2   Membrane rectification

As mentioned in chapter 3, membrane nonlinearity can come from two sources: voltage dependence and time dependence. When the membrane conductance varies with voltage, the *I-V* relation becomes nonlinear, and the membrane currents (or ionic currents) are said to be rectified. Membrane rectifications are defined in the following ways: (1) outward rectification, a membrane allowing outward current to flow more easily than inward current; and (2) inward rectification, a membrane allowing inward current to flow more easily than outward current. *I-V* relations A and B in figure 5.1 elucidate outward and inward rectifications. For outward rectification (curve A), the slope conductance increases as V is more depolarized, whereas for inward rectification (curve B), the slope conductance decreases as V is more depolarized.

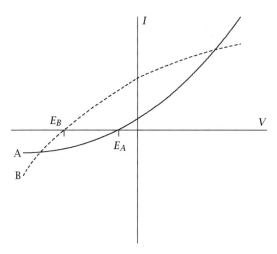

Figure 5.1   Inward and outward rectification. (A) Outward rectification; the slope conductance increases as V increases. (B) Inward rectification; the slope conductance decreases as V increases.

5.3   Models for membrane rectification

Since most biological membranes have nonlinear I-V relations, it is extremely important to understand the mechanisms underlying membrane rectification and time-varying nonlinearities in excitable cells. In this section, we will introduce three models that elucidate how voltage and time dependence of membrane currents may occur. Although these three models will by no means explain all types of membrane nonlinearities, they nevertheless are representative and commonly used in many biological systems. Additionally, these models can be expanded into more complex models that can explain more complicated schemes of membrane nonlinearities. For instance, the *single-energy barrier model* can be expanded into multiple-energy barrier models that are suitable for explaining certain complex ion permeabilities. The *gate model* may be expanded into multiple-state transition schemes that are used to describe the kinetics of single-channel activities in excitable cells. The three models presented in this section, therefore, not only are extremely useful themselves but also serve as foundations for establishing more complex models for analysis of membrane nonlinearities.

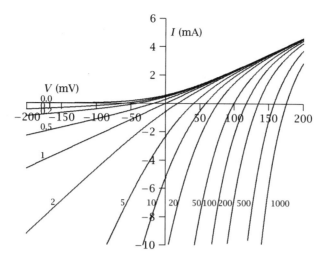

Figure 5.2 Current-voltage relations given by the GHK current equation (equation 2.2.5) for various values of $[C]_{out}/[C]_{in}$. Rectification occurs when concentrations at the two sides of the membrane are unequal ($[C]_{out}/[C]_{in} \neq 1$). Direction of rectification follows the direction of diffusion of cations (see explanation in the text).

5.3.1 Constant field (GHK) model

The constant field model, which is described in chapter 2, provides non-linear I-V relations described by the GHK current equation. Refer to figure 5.2 and recall that

$$I = P\frac{z^2F^2}{RT}V\left(\frac{[C]_{in} - [C]_{out}e^{\frac{-zVF}{RT}}}{1 - e^{\frac{-zVF}{RT}}}\right). \tag{5.3.1}$$

The GHK model is derived with the Nernst-Planck electrodiffusion formulation and the assumptions of constant field and independence (see chapter 2). This model provides I-V curves for cation currents that inward rectify when $\frac{[C]_{out}}{[C]_{in}} > 1$ (when E_r is positive), and outward rectify when $\frac{[C]_{out}}{[C]_{in}} < 1$ (when E_r is negative) (see figure 5.2). It is very important to understand conceptually why the I-V curves rectify this way. Based on the constant field model, ions diffuse down their electrochemical gradients, and the potential drop in the membrane is linear. When $\frac{[C]_{out}}{[C]_{in}} > 1$, the ion concentration outside is higher than inside, and it is easier for the ions to flow from outside to inside, down their concentration gradient. Thus, the current is inward rectified. Conversely, when $\frac{[C]_{out}}{[C]_{in}} < 1$, the ion concentration is higher inside, and thus it is easier for the ions to flow from inside to outside, and the current is outward rectified. The constant field model has been widely used to describe membrane rectification in many biological systems. It gives a satisfactory description for chloride conductance in the skeletal muscle and for the instantaneous I-V relation of the Na^+ and K^+ channels in myelinated nerves. However, there are many membrane conductances that cannot be described by the constant field model. For example, those I-V relations that exhibit inward rectification when E_r is negative, or exhibit outward rectification when E_r is positive, do not fit the I-V relations described by the GHK model. These nonlinearities fail to obey the constant field assumptions and have historically been called *anomalous rectifications*.

5.3.2 Energy barrier model (Eyring rate theory)

An alternative model for nonlinear I-V relations of membrane currents other than the Nernst-Planck formulation (i.e., constant field model) is the thermodynamic approach in which the rate coefficients for chemical reactions are described in terms of energy barriers that must be overcome by reactants. This formulation is often called the energy barrier model. For ionic currents, a simple application of this model is to assume that each ion flowing from one side of the membrane to the other side must overcome an energy barrier (thus it is called the single energy barrier model; more complex models may involve multiple energy barriers).

Based on the law of mass action for chemical reactions, the flux of a reactant is proportional to the concentration of the reactant, and the proportionality constant is named rate coefficient k. Thus, for ion flux through a single energy barrier (figure 5.3), the inward and outward flux can be written as

$$J_{\text{inward}} = k_1 \beta [C]_{\text{out}}, \qquad (5.3.2)$$

$$J_{\text{outward}} = k_2 \beta [C]_{\text{in}}. \qquad (5.3.3)$$

where J is the unidirectional ion flux, k_1 and k_2 are rate coefficients, $[C]_{\text{out}}$ and $[C]_{\text{in}}$ are concentrations (or activities), and β is the partition coefficient of water-membrane for the ion. At thermodynamic equilibrium, the rate coefficients are related to the standard free energy of activation (ΔG_o) crossing the membrane by Boltzmann's constant (see Reif 1965).

$$k_1 = A e^{-\Delta G_0/RT}. \qquad (5.3.4)$$

$$k_2 = A e^{-\Delta G_0/RT}. \qquad (5.3.5)$$

In equations 5.3.4 and 5.3.5 the standard free energies from both directions are assumed to be the same, so k_1 is equal to k_2. When an electric

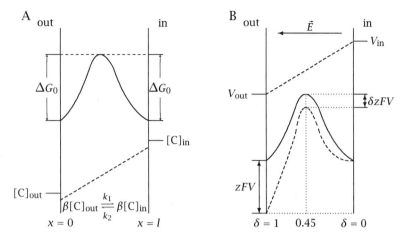

Figure 5.3 (A) Free energy and concentration profiles and (B) membrane potential and free energy profiles of the cell membrane. Ions move across the membrane by "hopping" over a symmetrical barrier of free energy ΔG_0 in the absence of membrane potential difference ($V_{in} - V_{out} = 0$, A). The energy barrier becomes asymmetrical (dashed barrier in B) when a potential difference is present across the membrane (B). See text for detailed explanations.

field is applied to the membrane (figure 5.3), the energy barrier for permeable ions is influenced by a factor of δzFV, where δ is a factor of asymmetry that gives the fractional influence of V on ΔG. $\delta = 1$ if the energy barrier is on the outside margin of the membrane; and $\delta = 0$ if the energy barrier is on the inside margin of the membrane, so $0 < \delta < 1$. The free energy of activation under such conditions is no longer symmetrical, and thus $k_1 \neq k_2$. Under the influence of V the rate coefficient becomes

$$k_1 = Ae^{-(\Delta G_0 + (1-\delta)zFV)/RT} = k_0 e^{-(1-\delta)zFV/RT}, \tag{5.3.6}$$

$$k_2 = Ae^{-(\Delta G_0 - \delta zFV)/RT} = k_0 e^{\delta zFV/RT}, \tag{5.3.7}$$

where

$$k_0 = Ae^{-\Delta G_0/RT}.$$

The net current flow across the membrane in this situation (combining equations 5.3.6 and 5.3.7 with equations 5.3.2 and 5.3.3) is

$$
\begin{aligned}
I &= zF(J_{outward} - J_{inward}) \\
&= zF\beta k_0 \left[[C]_{in} e^{\delta zFV/RT} - [C]_{out} e^{-(1-\delta)zFV/RT} \right].
\end{aligned}
\tag{5.3.8}
$$

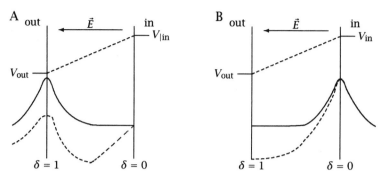

Figure 5.4 Membrane potential and free energy profiles of the membrane with rate-limiting barriers at the outside (A) and inside (B) margins of the membrane. The solid and dashed energy profiles are the barriers in the absence and presence of transmembrane voltage, respectively.

When $\delta = 0$, i.e., the rate-limiting barrier is at the inside margin,

$$I = zF\beta k_0 \left[[C]_{in} - [C]_{out} e^{-zFV/RT} \right], \tag{5.3.9}$$

and the I-V relation is inward rectified (dotted curve in figure 5.5).

When $\delta = 1$, the rate-limiting barrier is at the outside margin and the I-V relation is outward rectified (dashed curve in figure 5.5).

$$I = zF\beta k_0 \left[[C]_{in} e^{z\delta FV/RT} - [C]_{out} \right]. \tag{5.3.10}$$

For any fixed value of δ (e.g., $\delta = 0.5$), the I-V relations depend on the concentration ratio $[C]_{out}/[C]_{in}$ and whether $[C]_{out}$ or $[C]_{in}$ are varying (figure 5.6).

The reason for membrane rectification in this model is obvious: When the rate-limiting barrier is at the outside margin ($\delta = 1$), it is more difficult for ions to get into the membrane from outside than from inside, resulting in outward rectification (easier flow from in to out). The converse is true when the rate-limiting barrier is at the inside margin ($\delta = 0$).

The single energy barrier model has wider applications than the constant field model. Unlike the constant field model, which can only describe outward rectification with negative reversal potentials (E_r) and inward rectification with positive reversal potentials, the energy barrier model can be used to formulate either inward or outward rectified I-V relations with positive or negative reversal potentials.

In biological systems, it is unrealistic to expect that all ions encounter only one energy barrier when moving across a cell membrane that is about 100 Å thick. The single energy barrier model is used to analyze situations when one of the energy barriers is much greater than the other(s). For

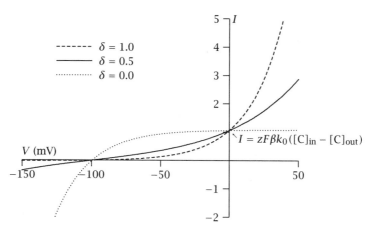

Figure 5.5 Current-voltage relations given by single energy barrier model (equation 5.3.8) for $\delta = 1.0$, $\delta = 0.5$, and $\delta = 0$. $[C]_{out}/[C]_{in} = \frac{1}{53}$.

multiple energy barriers of nearly equal heights, multiple barrier models involving more complex mathematical equations should be employed (Woodbury 1965, 1971). The detailed mathematical descriptions of such models are beyond the scope of this book. Rather, an example of a multiple energy barrier model will be described qualitatively below.

Hille (1975a) used a four-barrier model to represent the diffusion path of Na^+ ions in an open Na^+ channel (figure 5.7A). The relative energy levels required to pass each barrier are labeled in units of RT at the top of the barriers and at the bottom of each well. The possible molecular structures responsible for the energy barriers and wells are given in figure 5.7B. The barrier between A and B (8.5 RT) is the diffusion barrier for Na^+ to arrive from the extracellular space to one side of the COO^- group. The energy well B represents the electrostatic attraction of the COO^- group onto Na^+ ($-1.0\ RT$). The energy barrier C (11.5 RT) represents the energy required to pass through the narrow "selectivity filter" region formed between the COO^- and an oxygen group. Two additional energy barriers (each 9.5 RT) must be passed between the selectivity filter and the interior of the cell.

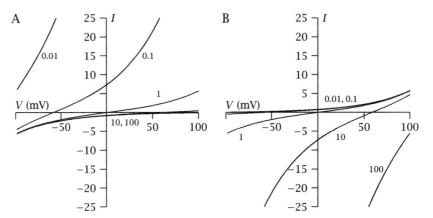

Figure 5.6 Current-voltage relations given by single energy barrier model (equation 5.3.8) for $[C]_{out}/[C]_{in}$ = 0.01, 0.1, 1, 10, and 100, with δ = 0.5 and k_0 = 1. The $[C]_{out}/[C]_{in}$ ratio is changed by (A) fixing $[C]_{out}$ and varying $[C]_{in}$; and (B) fixing $[C]_{in}$ and varying $[C]_{out}$.

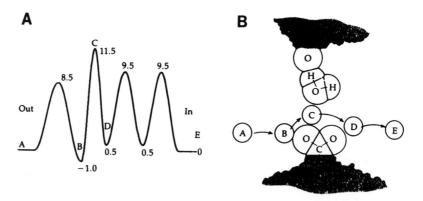

Figure 5.7 Energy barrier profile (A) and binding sites (B) in Na^+ channels. The energy levels relative to bulk solutions are labeled in multiples of RT with values appropriate for Na^+ ions in (A). (From Hille 1975a by copyright permission of the Rockefeller University Press.)

Example 5.1

The current-voltage relation for ionic current I_i in a neuron is shown below.

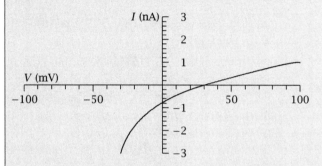

a. What are the slope and chord conductances at $V = -10$ mV and at $V = +100$ mV?

b. What is the ratio of intracellular and extracellular concentrations of the ion i? Is I_i inward or outward rectified?

c. Can the *I-V* relation shown above be explained by the constant field model? If so, what factor is responsible for the rectification?

d. Can the *I-V* relation shown above be explained by the single energy barrier model? If so, what factor is responsible for the rectification?

Answer to example 5.1

a.   At $V = -10$ mV, $g_s =$ 1.1 nA/20 mV =   55 nS,
     $g_c =$ 1.2 nA/30 mV =   40 nS.
   At $V = +100$ mV, $g_s =$ 0.4 nA/100 mV =   4 nS,
     $g_c =$ 1.0 nA/70 mV =   14 nS.

b. $E_i = \dfrac{RT}{zF} \ln \dfrac{[\text{C}]_{\text{out}}}{[\text{C}]_{\text{in}}} = +30$ mV (from figure where $I = 0$),

$$\frac{[\text{C}]_{\text{in}}}{[\text{C}]_{\text{out}}} = \left[\text{antilog}\frac{30}{58} \right]^{-1} = [3.29]^{-1} = 0.3 \quad (z = +1).$$

I_i is inward rectified because slope conductance decreases as V becomes more positive.

c. Yes, it can be explained by the constant field model. The concentration gradient inside and outside the cell is responsible for the rectification: When $[C_i]_{out} > [C_i]_{in}$, it is easier for ion i to flow from outside to inside, and thus it is inward rectified; when $[C_i]_{out} < [C_i]_{in}$, the converse is true.

d. Yes, it can be explained by the single energy barrier model. The position of the energy barrier within the membrane is responsible for the rectification. When the energy barrier is close to the inside of the membrane, it is easier for ions to get into the membrane from the outside; thus the membrane is inward rectified. The converse is true if the barrier is close to the outside of the membrane.

5.3.3 The gate model (Hodgkin and Huxley's model)

The constant field and energy barrier models described in previous sections provide possible mechanisms for voltage dependence (nonlinear I-V relations) of membrane conductance. These models cannot, however, deal with the second source of membrane nonlinearity, that is, time dependence of ion conductance. In order to describe both voltage and time dependence of ion conductances in the squid axon, Hodgkin and Huxley (1952) developed the *gate* model, which proposed that ion currents are flowing through transmembrane channel proteins that form aqueous pores through which ions can diffuse down their concentration gradients. These pores have "gates" that are controlled by voltage-sensitive gating charges or gating particles (see figure 5.8A). The movement of gating particles within the membrane can be described by the single energy barrier model (figure 5.8B). Note that the difference between the gate model and the single energy barrier model described in the previous section is that the energy barrier in the gate model is for the gating particles restricted within the channel protein, whereas that in the single energy barrier model is for the cross-membrane ions. The basic assumptions for the gate model are given below:

1. Ionic channels in the membrane undergo conformational change in response to variation in electric field. This conformation change will cause the channel to move from open to closed states or vice versa:
   $$\text{open} \underset{\alpha(V)}{\overset{\beta(V)}{\rightleftharpoons}} \text{closed}, \text{ where } \alpha(V) \text{ and } \beta(V) \text{ are rate coefficients.}$$

2. The reaction between the open and closed states is a first-order reaction.

Based on these assumptions, the gate model provides descriptions of both voltage and time dependence of ionic channels (figure 5.8).

5.3.3.1 Voltage dependence From figure 5.8, the gate is controlled by membrane-bound charged gating particles: y is the probability of a gating particle in permissive (*a* or open) state, and $(1 - y)$ is the probability of the particle in nonpermissive (*b* or closed) state. As in the single energy barrier model, the energy barrier between the two states (and thus the probabilities of the gating particle staying at each of the two states) is influenced by the transmembrane voltage. In figure 5.8B, membrane depolarization (increase in V_{in}) will lower the energy barrier for the gating particle moving from the nonpermissive state to the permissive state, and thus increase the probability that the particle will stay in the permissive (open) state. Membrane hyperpolarization will do the opposite. Channels that have such gating particle arrangements are activated (opened) by membrane depolarization. Channels with opposite gating particle arrangements, that is, permissive state at the inside margin and nonpermissive state at the outside margin of the membrane, are activated by membrane hyperpolarization.

Mathematically, the voltage dependence of the rate coefficients $\alpha(V)$ and $\beta(V)$ can be derived by similar procedures as the k_1 and k_2 in the energy barrier model.

$$
\begin{aligned}
\alpha(V) &= Ae^{-(\Delta G_0 - \delta zFV)/RT} \\
&= \alpha_0 e^{\delta zFV/RT}. \\
\beta(V) &= Ae^{-(\Delta G_0 + (1-\delta)zFV)/RT} \\
&= \beta_0 e^{-(1-\delta)zFV/RT}.
\end{aligned}
$$

For first-order kinetics, which we will derive in the next section, the steady-state probability for permissive state $y_\infty = \frac{\alpha}{\alpha+\beta}$. Thus,

$$
\begin{aligned}
y_\infty(V) &= \frac{\alpha(V)}{\alpha(V) + \beta(V)} = \frac{\alpha_0 e^{\delta zFV/RT}}{\alpha_0 e^{\delta zFV/RT} + \beta_0 e^{-(1-\delta)zFV/RT}} \\
&= \frac{\alpha_0}{\alpha_0 + \beta_0 e^{-zFV/RT}}. \\
y_\infty(V) &= \frac{1}{1 + e^{-zFV/RT}} \qquad \text{if } \alpha_0 = \beta_0
\end{aligned}
$$

$y_\infty(V)$ is sometimes called the activation function of a channel.

A

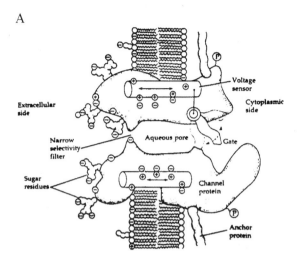

B

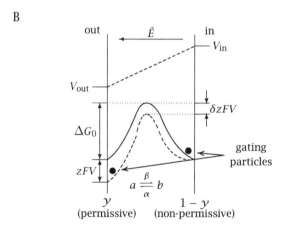

Figure 5.8   (A) Schematic diagram of a voltage-gated channel (from Hille 1992) and (B) the energy profile for gating particles in the voltage sensors. The energy barrier for the gating particle is similar to that for the single energy barrier model for the *permeant ions* (figure 5.3). y is the probability of the gating particle in the permissive state (a) and $1-y$ is the probability of the particle in the nonpermissive state (b). α and β are rate coefficients for the two-state transition.

Note that the rate coefficients $\alpha(V)$ and $\beta(V)$ are dependent on δ, the factor of asymmetry, but $y_\infty(V)$ does not depend on δ. The values of $\alpha(V)$, $\beta(V)$, and $y_\infty(V)$ are plotted in figures 5.9 and 5.10.

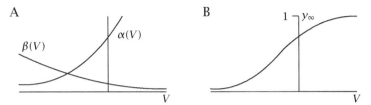

Figure 5.9 Voltage dependence of the rate coefficients α and β (A) and the steady-state probability y_∞ (B) for voltage-gated channels that are activated by membrane depolarization (or inactivated by hyperpolarization).

The plots in figure 5.9 represent the case shown in the energy barrier model in figure 5.8B, in which the channel is opened or *activated* (has a high probability of being opened) *by depolarization* (you can also say inactivated by hyperpolarization).

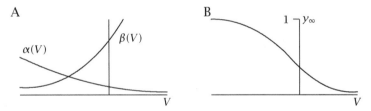

Figure 5.10 Voltage dependence of the rate coefficients α and β (A) and the steady-state probability y_∞ (B) for voltage-gated channels that are activated by membrane hyperpolarization (or inactivated by depolarization).

The plots in figure 5.10 represent the case when the positions of a and b states in the energy barrier diagram are reversed, that is, the channel is opened or *activated by hyperpolarization* (or inactivated by depolarization).

5.3.3.2 Time dependence The gate model assumes a first-order reaction between open and closed states for gating particles. Thus, the probability of the particle in the open state, y, can be described by

$$a \underset{\alpha}{\overset{\beta}{\rightleftharpoons}} b$$
$$y \qquad (1-y)$$

First-order kinetics yields:

$$\frac{dy}{dt} = \alpha(1 - y) - \beta y.$$

At steady state,

$$\frac{dy}{dt} = 0 = \alpha(1 - y_\infty) - \beta y_\infty.$$

Therefore,

$$y_\infty = \frac{\alpha}{\alpha + \beta}.$$

Substituting this into the first-order differential equation,

$$
\begin{aligned}
\frac{dy}{dt} &= \alpha(1 - y) - \beta y = \alpha - (\alpha + \beta)y = y_\infty(\alpha + \beta) - (\alpha + \beta)y \\
&= (\alpha + \beta)(y_\infty - y).
\end{aligned}
$$

Therefore,

$$\int \frac{dy(-1)}{y_\infty - y} = (-1)\int (\alpha + \beta)dt \rightarrow \ln(y_\infty - y) = -(\alpha + \beta)t + C,$$

and

$$y = y_\infty - Ae^{-(\alpha+\beta)t}.$$

Apply boundary condition $y_0 = y(t = 0)$, then

$$y_0 = y_\infty - A \cdot 1 \rightarrow A = y_\infty - y_0.$$

Therefore,

$$y(t) = y_\infty - \left[(y_\infty - y_0)\, e^{-(\alpha+\beta)t}\right]$$

or

$$y(t) = y_0 + \left[(y_\infty - y_0)\left(1 - e^{-(\alpha+\beta)t}\right)\right].$$

If P independent gating particles are involved in gating a channel, then the channel will follow the time course (figure 5.11)

$$Y(t) = [y(t)]^P = \left[y_\infty - (y_\infty - y_0)e^{-(\alpha+\beta)t}\right]^P.$$

The three models described in this chapter elucidate how membrane rectification and time-varying nonlinearities may occur. These models have also been used as foundations of more complicated models. It is

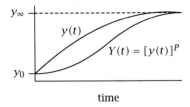

Figure 5.11 Time course of the probability of channel activation: $y(t)$ for channels activated by one gating particle, and $Y(t)$ for channels activated by P identical gating particles.

important to note, however, that several other mechanisms may be involved in mediating membrane rectification. One well-studied example is the inward rectification made by intracellular magnesium in voltage-gated inward rectifier K$^+$ channels (see chapter 7). Readers who are interested in detailed discussions of mechanisms underlying nonlinear currents should consult Hille (1992).

Example 5.2

The membrane current of a neuron to a voltage-clamp step shown in the figure below consists of a leakage current $I_L(V)$ and a time-dependent current $I_y(V,t)$. $I_y(V,t)$ is a voltage-gated current following first-order kinetics, that is,

$$I_y(V,t) = \bar{I}_y(v)y(V,t).$$

$$dy/dt = (y_\infty(V) - y(V,t))/\tau_y,$$

where $y(V,t)$ is the gating variable, $y_\infty(V)$ is the steady-state value of $y(V,t)$, and τ_y is the time constant of $y(V,t)$. The activation curve of $I_y(V,t)$ is shown to the right in the figure below.

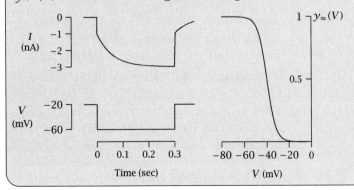

a. Is $I_y(V, t)$ activated by depolarization or hyperpolarization? Explain in one sentence. Is $I_y(V, t)$ an inward or outward current?

b. What is the approximate value of τ_y?

c. What is the amplitude of leakage current when the membrane voltage is stepped from -20 mV to -60 mV?

d. What is the amplitude of the fully activated $I_y(V, t)$, i.e., $\bar{I}_y(V)$?

e. What is the steady-state value of $I_y(V, t)$ when the membrane is stepped from -20 mV to -40 mV?

Answer to example 5.2

a. I_y is activated by hyperpolarization because the y_∞ curve $\to 1$ at hyperpolarized voltage. I_y is *inward* because an inward current is induced by hyperpolarizing voltage pulse.

b. $\tau \approx 0.1$ sec.

c. When $V_H(-20) \to V_c(-60 \text{ mV})$, $I_L = -1$ nA because no I_y is activated at -20 mV; therefore, the jump of current at $t = 0$ is due to leakage.

d. $\bar{I}_y(V) = 2$ nA because $V_H(-20) \to V_c(-60)$ makes $y_\infty = 0 \to y_\infty = 1$.

e. When $V_H(-20) \to V_c(-40) \Rightarrow y_\infty = 0 \to y_\infty = 0.5$, $I_y(-40)\infty = 1$ nA.

5.4 Review of important concepts

1. The current-voltage relation (*I-V* relation) of a voltage-dependent membrane is nonlinear, and the membrane may exhibit inward, outward, or both forms of rectification.

2. Membrane rectification may be mediated by several mechanisms. Membrane rectification mediated by ionic concentration gradients can be explained by the constant field (GHK) model. The energy barrier model attributes membrane rectification to the relative locations of energy barriers within the membrane. The gate (Hodgkin and Huxley) model attributes membrane rectification to voltage-dependent gating of ion channels.

3. The gate (Hodgkin and Huxley) model can be used to describe not only the voltage dependence but also the time dependence (kinetics) of membrane conductances.

5.5 Homework problems

1. (a) Describe concisely the driving forces of ion flux in excitable cells. What factors determine the direction of ion flux?

    (b) Sketch the I-V relations of Na^+ and K^+ conductances of a cell ($[K^+]_{out}/[K^+]_{in} = 0.1$, $[Na^+]_{out}/[Na^+]_{in} = 10$) if both ions follow the assumptions of the constant field model. You are not required to make a quantitative plot; just sketch the general shape (inward or outward rectification) and give the value (in mV) of the intersections of the I-V curves with the voltage axis. Explain why the two I-V curves rectify in the direction you sketch.

2. The figure below shows a family of curves (a–i) generated by the computer according to the "energy barrier model."

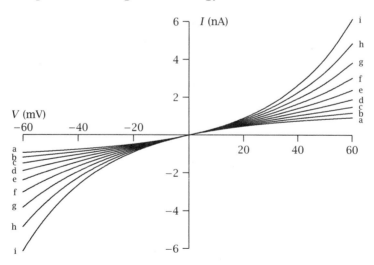

    (a) What is the concentration ratio $[C^+]_{out}/[C^+]_{in}$ of the ion involved?

    (b) Assign appropriate values of δ for each curve. Draw the position of the energy barrier for a, e, and i in a membrane.

(c) Write the expression of I as a function of V and δ for this family of curves.

3. The I-V relations of an ionic current measured at different ratios of ionic concentrations are shown below (curves a–k).

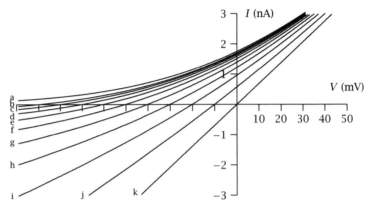

Given $[C^+]_{in}/[C^+]_{out} = a$ and $a_B = 0.028$,

(a) What is the most likely ion that carries this current? Justify your answer in one sentence.

(b) What are the values of a_A, a_C, a_F, a_J, and a_K?

(c) What model(s) can be used to explain the rectification shown in this figure? Give a brief intuitive account of the causes of such membrane rectification.

4. The I-V relations A–E of the ionic conductances (for ions a–e, respectively) in a cell membrane are plotted below.

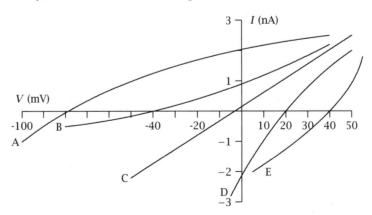

(a) Assign *one* of the following choices to each *I-V* relation (A-E): constant field model; single energy-barrier model; either; neither.

(b) What is the value of $\frac{[C_a]_{out}}{[C_a]_{in}}$ for ion a if the valence of this ion is -1?

(c) What is the valence of ion e if $\frac{[C_e]_{out}}{[C_e]_{in}} = 24$?

(d) Under a given condition, the membrane becomes permeable only to ions a, b, and d, and the permeability ratio for these ions is $P_a : P_b : P_d = 1 : 10 : 0.5$. If $[C_a]_{in} = 10$ mM, $[C_b]_{in} = [C_d]_{in} = 100$ mM, $z_a = -1$, and $z_b = z_d = +1$, what is the resting potential of this cell?

5. The membrane constant of a neuron to a voltage-clamp step shown on the left in the figure below consists of a leak current $I_L(V)$ and a time- and voltage-dependent current $I_y(V,t) = \bar{I}_y y(V,t)$ that follows first-order kinetics. The activation curve of $y_\infty(V)$ is given on the right. $I_y(V,t)$ is carried by an ion whose equilibrium potential is -70 mV.

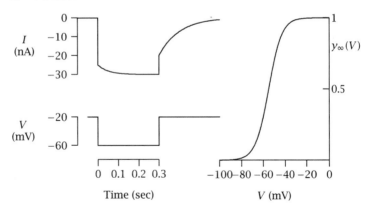

(a) Is $I_y(V,t)$ activated by depolarization or hyperpolarization? Is $I_y(V,t)$ an inward or outward current? Explain your answers in two sentences.

(b) What is the value of fully activated $g_y(V,t)$, i.e., $\bar{g}_y$)? What is the approximate value of τ_y?

(c) Under physiological conditions (no voltage clamping), the resting potential of the neuron is -60 mV. The neuron is spherical with a diameter of 20 μm, the input resistance is 10^8 Ω, and the membrane capacitance (C_m) is 1 μF/cm^2. A 0.5 sec current step of +0.4 nA is injected into the center of the neuron. Draw the voltage response of the neuron to this current step on graph paper. Give the amplitude and time course of each component of this voltage response.

6. Briefly describe the basic assumptions of the single energy barrier model for ion permeation. How does transmembrane voltage affect ion permeation? What factor is responsible for inward or outward rectification of ion currents? What physical entity in the channel constitutes the energy barrier?

7. The voltage-clamp records of a voltage- and time-dependent current $I_y(V,t)$ are given in the figure below (all other currents are eliminated).

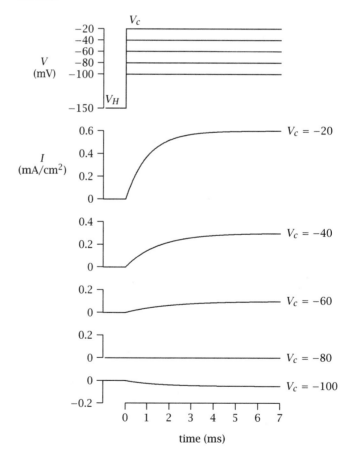

$I_y(V,t)$ can be written as $I_y(V,t) = y(V,t)\overline{g}_y(V - E_y)$,
$\overline{g}_y = 10$ mS/cm^2, and $y(V,t)$ follows first-order kinetics, i.e.,

$$
\begin{array}{ccc}
C & \underset{\alpha}{\overset{\beta}{\rightleftharpoons}} & O \\
1 - y & & y
\end{array}
$$

and

$$\frac{dy}{dt} = \alpha(1 - y) + \beta y.$$

    (a) Estimate the values of τ and I_∞ at each V_c. What is the value of E_y?

    (b) Plot y_∞, α, and β as functions of V.

8. The plasma membrane of a sensory neuron (S_1) at rest is permeable only to Na$^+$ and K$^+$, and the I-V relations for these ions are shown as dashed lines in the figure below. Immediately after a sustained stimulus, the I-V relations for Na$^+$ and K$^+$ become the solid lines, and these I-V relations are maintained throughout the entire stimulation.

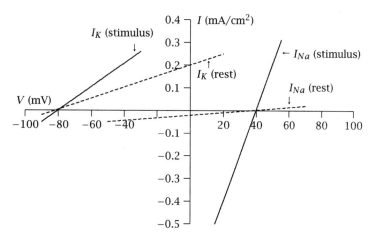

    (a) What is the resting potential of this sensory neuron (S_1)? What is the membrane potential of this sensory neuron (S_1) during the sustained stimulus? Draw the voltage response of S_1 to the sustained stimulus.

Another sensory neuron (S_2) has the exact same membrane properties as S_1 except that it has a population of voltage- and time-dependent Ca^{2+} channels in its plasma membrane ($I_{Ca^{2+}}(V,t)$). The activation curve of $I_{Ca^{2+}}$ is given in the top of the figure below, and $I_{Ca^{2+}}$ follows first-order kinetics with a time constant of 1 sec. The fully activated I-V relation of $I_{Ca^{2+}}$ (when $y_{Ca^{2+}}(V, \infty) = 1$) is given as a dotted line in the plot at the bottom of the figure.

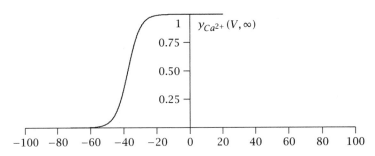

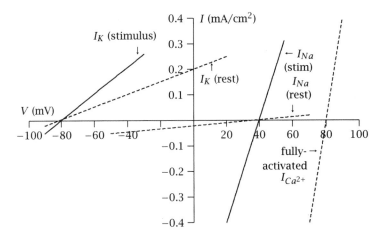

(b) What is the resting potential of S_2? What is the membrane potential of S_2 immediately after the onset of the stimulus (same stimulus as that for S_1)? What is the membrane potential of S_2 10 sec after the onset of the sustained stimulus? Draw the voltage response of S_2 to the sustained stimulus.

6   Hodgkin and Huxley's Analysis of the Squid Giant Axon

6.1   Introduction

Electric signals in excitable cells are transmitted from one part of the cell to another in two ways: the passive spread of graded potentials and the propagation of all-or-none action potentials. Graded potentials are normally observed in interneurons that transmit signals over short distances, whereas action potentials are used by neurons that bear far-reaching processes (axons), carrying signals over long distances.

Graded potentials are carried by ions that diffuse down their electrochemical gradients, and the magnitude of the signal varies with the ionic currents. Action potentials, on the other hand, are carried by voltage- and time-dependent conductances that generate transient all-or-none potential changes (spikes or nerve impulses) that propagate from one part of the cell to another. The purpose of this chapter and part of the next chapter is to describe the basic mechanisms underlying the generation and propagation of action potentials. The best-studied preparation for action potentials is the squid giant axon, and the analytical tool is the gate model of Hodgkin and Huxley.

6.2   Voltage-clamp experiments of the squid axon

6.2.1   Voltage-clamp systems and reasons for voltage clamping

In order to analyze the nonlinear properties of ion conductances underlying action potentials, Hodgkin and Huxley performed a series of voltage-clamp experiments on the squid giant axon. Voltage-clamp experiments usually involve inserting two electrodes (in the case of the squid axon, two silver wires) into the axon, one for recording the transmembrane voltage and the other for passing current into the axon to keep the transmembrane voltage constant (or clamped). The basic circuitry of the voltage-clamp experiment is shown in figure 6.1 and is discussed in appendix A.

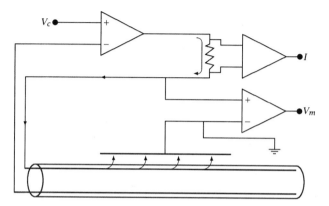

Figure 6.1   Schematic diagram of the two-wire voltage-clamp experiments on the squid axon. One wire is used for monitoring the membrane potential and the other for passing current. The voltage clamp amplifier injects or withdraws charges from the interior of the squid axon in order to hold the membrane voltage constant (voltage is clamped at the command voltage, V_c).

The reason for voltage-clamping the axon is threefold: (1) By keeping the voltage constant, one can eliminate the capacitive current, that is, $I_C = C\frac{dV}{dt} = 0$; (2) by keeping the voltage constant, one can measure the time-dependent characteristics of ion conductances without the influence of voltage-dependent parameters; and (3) by inserting two silver wire electrodes into the axon, one can space-clamp it so that the whole length of the axon is isopotential (silver wires short-circuit the interior of the axon).

6.2.2   Voltage-clamp records

Under voltage-clamp conditions, the current responses to voltage steps reflect the change of ion currents flowing across the membrane. Figure 6.2 shows the current records of the squid axon when the voltage is stepped from a holding voltage (V_H) to a command voltage (V_c) of various levels. When V_c is above -30 mV, a transient inward current followed by a sustained outward current is observed. The amplitude of the early inward current first increases and then decreases as V_c becomes more positive, whereas that of the late outward current increases monotonically with V_c.

So that the current record can be examined closely, current traces in response to $V_c = 0$ mV and $V_c = -120$ mV are given in the left part of figure 6.3. I_C is the capacitive current elicited by the transitions of volt-

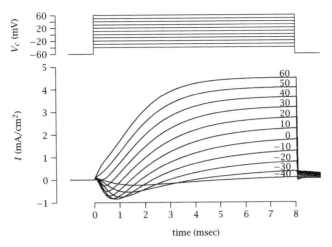

Figure 6.2 Currents measured with voltage clamp of squid axon. Membrane potential was held at −60 mV and then stepped (at 0 msec) to various potentials (shown at the right of each trace) for 8 msec before stepping back to −60 mV.

age from V_H to V_c. The membrane current consists of an early transient inward current and a late steady outward current. The *I-V* relations of the early and late currents are shown in the right part of figure 6.3.

It can be shown experimentally that the two currents are mediated by two separate conductances. If one removes extracellular Na^+ (which eliminates the driving force of the inward Na^+ flux) or adds tetrodotoxin (TTX, which blocks Na^+ channels), one can eliminate the early transient inward current mediated by Na^+ channels. On the other hand, if one removes intracellular K^+ or adds tetraethylammonium (TEA, which blocks K^+ channels), one can eliminate the late outward current mediated by K^+ channels. The separation of the two currents is illustrated in figure 6.4.

6.2.3 Instantaneous current-voltage relation

The voltage-clamp analysis of the squid axon discussed in the previous section shows that I_{Na} and I_K can be measured independently. The next step is to determine g_{Na} and g_K. How is $I_{Na}(I_K)$ related to $g_{Na}(g_K)$? Do they follow Ohm's law (i.e., $I_{Na} = g_{Na}(V - E_{Na})$)? In order to obtain answers to these questions, Hodgkin and Huxley performed the following experiments. Two voltage pulses were applied to the axon. V_1 activates the early (Na^+) current I_1 (• in figure 6.5). While the Na^+ conductance was turned on at $V_1 = -29$ mV, the potential was suddenly stepped to V_2, and the instantaneous current I_2 (◦ in figure 6.5) was recorded. By measuring

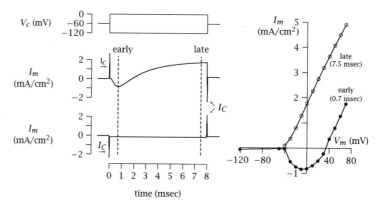

Figure 6.3 Early and late currents of a squid axon when the voltage is stepped from −60 mV to 0 mV or −120 mV (A); and the current-voltage relations of the early and late currents (B).

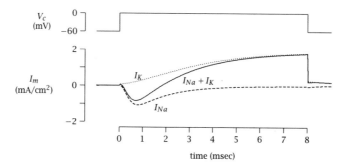

Figure 6.4 Separation of membrane current (solid trace) into Na^+ (dashed trace) and K^+ (dotted trace) currents. I_K is obtained in the presence of TTX or when $[Na^+]_{out} = 0$; I_{Na} is obtained in the presence of TEA. The voltage is stepped from −60 mV to 0 mV for 8 msec.

$I_2 - I_1$ as a function of $V_2 - V_1$, one can estimate the *I-V* relation of the Na^+ channels without time-dependent influence on the current.

For voltage pulse V_1, $I_{Na_1}(V_1, t_1) = g_{Na}(V_1, t_1)(V_1 - E_{Na})$. Then a sudden second voltage pulse V_2, $I_{Na_2}(V_2, t_1^*) = g_{Na}(V_2, t_1^*)(V_2 - E_{Na}) = g_{Na}(V_1, t_1)(V_2 - E_{Na})$ $(g_{Na}(V_1, t_1) = g_{Na}(V_2, t_1^*)$ because g_{Na} does *not* have enough time to change). Therefore, $I_2 - I_1 = g_{Na}(V_1, t_1)(V_2 - V_1)$. $g_{Na}(V_1, t_1)$ is a constant at a fixed voltage V_1 and fixed time t_1. Thus, Ohm's law predicts that $(I_2 - I_1)$ and $(V_2 - V_1)$ are linearly related. Experimental results, shown in figure 6.5, indicated that this is true for the squid axon. I_2 vs. V_2 is a straight line, which implies that $(I_2 - I_1)$ vs. $(V_2 - V_1)$ is a straight line because $I_1(V_1)$ and V_1 are constants. Note that I_1 and I_2 intersect at two points. The left intersecting point is obviously V_1, the

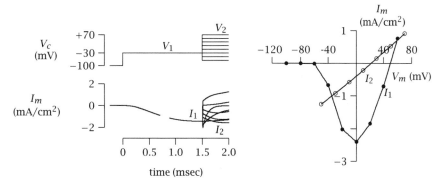

Figure 6.5   Instantaneous current-voltage relation (B) obtained with voltage clamp for the early inward channel. Closed circles indicate normal peak inward current for various depolarizations (A). Open circles indicate variation of I_2 with V_2 as shown in inset on right. $I_2 - I_1$, instantaneous step of current produced by voltage step $V_1 - V_2$. Duration of first pulse = 1.5 msec. (After Hodgkin and Huxley 1952b.)

first voltage pulse; the right intersecting point is E_{Na}, because $I_1(V) = g(V, t_1)(V - E_{Na})$ (• in figure 6.5), and $I_2(V, V_1, t_1^*) = g(V_1 t_1^*)(V - E_{Na})$ (∘ in figure 6.5). $I_1(V) = I_2(V, V_1, t_1^*)$ only under two conditions: (1) $V = V_1$, thus $g(V, t_1) = g(V_1, t_1) = g(V_1, t_1^*)$ (left intersecting point); or (2) $V = E_{Na}$, thus $V - E_{Na} = 0$, therefore, $I_1(V) = I_2(V, V_1, t_1) = 0$ (right intersecting point).

The right intersecting point of I_1 and I_2 is E_{Na}, but it does not lie on the V-axis ($I = 0$). This indicates that I_1 is not totally mediated by Na$^+$, because if it were so, I_1 should be 0 when $V = E_{Na}$. The explanation for the nonzero I_1 at $V = E_{Na}$ is that at $t = t_1 = 1.53$ msec, I_K is nonzero although its value is far from the steady-state maximum level (8 msec). $I_1(= I_{Na} + I_K)$ is therefore shifted toward more positive values in the I-V plot.

Similar experiments were performed for the late (K$^+$) channels, and the results are shown in figure 6.6. V_1 is about +20 mV (right intersecting point of I_1 and I_2), and E_K is about −80 mV (left intersecting point). The instantaneous K$^+$ current is also linear.

The above experiments show that $I_{Na}(I_K)$ and $g_{Na}(g_K)$ are related by Ohm's law in the squid axon. Thus by measuring I_{Na} and I_K experimentally, Hodgkin and Huxley were able to determine g_{Na} and g_K by simply dividing the currents by their driving forces, $(V - E_{Na})$ and $(V - E_K)$.

It is worth noting, however, that not all excitable cells exhibit linear instantaneous I-V relations for Na$^+$ and K$^+$ channels; for example, nodes of Ranvier do not.

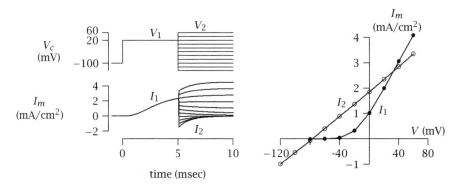

Figure 6.6 Instantaneous current-voltage relation obtained with voltage clamp for the late outward current. Similar protocols are used for the late current as for the early current (see figure 6.5).

6.2.4 g_{Na} and g_K of the squid axon

From the analysis described in the last section, it can be seen that the instantaneous Na$^+$ and K$^+$ conductances in the squid axon are linear. This allowed Hodgkin and Huxley to obtain $g_{Na}(V, t)$ and $g_K(V, t)$ by dividing $I_{Na}(V, t)$ and $I_K(V, t)$ (from their voltage-clamp data) by $(V - E_{Na})$ and $(V - E_K)$, respectively. The procedures and results are shown in figure 6.7.

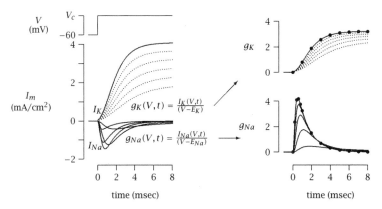

Figure 6.7 Time course of g_K (dashed traces) and g_{Na} (solid traces) at various voltages (V_c) obtained from I_K and I_{Na} traces, according to Ohm's law.

6.3 Hodgkin and Huxley's model

Using the experimental results of $g_{Na}(V,t)$ and $g_K(V,t)$ described in the previous section and the gate model in the last chapter, Hodgkin and Huxley proposed a landmark model that quantitatively described the behavior of Na$^+$ and K$^+$ channels, nerve excitation, and conduction. First, they used the parallel conductance model to describe the major ionic conductances in the squid axon.

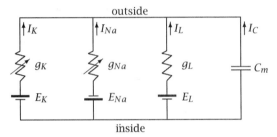

Figure 6.8 Parallel conductance model for the squid axon. g_K and g_{Na} are voltage and time dependent, and g_L is constant. The total membrane current is described by equation 6.3.1.

$$I_m = C_m \frac{dV}{dt} + I_K + I_{Na} + I_L,$$

where I_L = leak current, which is carried mainly by Cl$^-$ and other ions, since all currents obey Ohm's law. Thus,

$$\begin{aligned} I_m &= C_m \frac{dV}{dt} + g_K(V,t)(V - E_K) + g_{Na}(V,t)(V - E_{Na}) \\ &\quad + g_L(V - E_L). \end{aligned} \tag{6.3.1}$$

Hodgkin and Huxley proposed that the Na$^+$ and K$^+$ conductances were controlled by gating particles, and thus g_K and g_{Na} can be written as products of gating variables and maximum conductances:

$$g_K(V,t) = Y_K(V,t)\overline{g}_K,$$

and

$$g_{Na}(V,t) = Y_{Na}(V,t)\overline{g}_{Na},$$

where Y_K and Y_{Na} are gating variables between 0 and 1, and $\overline{g}_K$ and $\overline{g}_{Na}$ are maximum conductances.

From the time course of the *measured* g_{Na} and g_K (see the previous section), Hodgkin and Huxley found that Y_K and Y_{Na} do *not* follow simple exponentials (thus not a single $y(t)$; see gate model in chapter 5). Instead, they follow power functions of the exponential. Thus Hodgkin and Huxley proposed:

$$g_K(V,t) = Y_K(V,t)\overline{g}_K = n^4\overline{g}_K,$$

and

$$g_{Na}(V,t) = Y_{Na}(V,t)\overline{g}_{Na} = m^3 h\overline{g}_{Na},$$

where n, m, and h are the gating variables ($y(t)$, see chapter 5) in the gate model and follow first-order kinetics (exponential time course). Recalling the kinetics of the gating variable $y(t)$, one can write $n(t)$, $m(t)$, and $h(t)$ the same way:

$$\frac{dn}{dt} = \alpha_n(1-n) - \beta_n n, \quad n_\infty = \frac{\alpha_n}{\alpha_n + \beta_n}, \quad \tau_n = \frac{1}{\alpha_n + \beta_n}. \tag{6.3.2}$$

$$\frac{dm}{dt} = \alpha_m(1-m) - \beta_m m, \quad m_\infty = \frac{\alpha_m}{\alpha_m + \beta_m}, \quad \tau_m = \frac{1}{\alpha_m + \beta_m}. \tag{6.3.3}$$

$$\frac{dh}{dt} = \alpha_h(1-h) - \beta_h h, \quad h_\infty = \frac{\alpha_h}{\alpha_h + \beta_h}, \quad \tau_h = \frac{1}{\alpha_h + \beta_h}. \tag{6.3.4}$$

Equations 6.3.2, 6.3.3, and 6.3.4 yield the following solutions: (see gate model in the last chapter for $y(t)$):

$$n(t) = n_0 - \left[(n_0 - n_\infty)\left(1 - e^{-t/\tau_n}\right)\right],$$

$$m(t) = m_0 - \left[(m_0 - m_\infty)\left(1 - e^{-t/\tau_m}\right)\right],$$

and

$$h(t) = h_0 - \left[(h_0 - h_\infty)\left(1 - e^{-t/\tau_h}\right)\right],$$

or

$$h(t) = h_\infty + \left[(h_0 - h_\infty)e^{-t/\tau_h}\right].$$

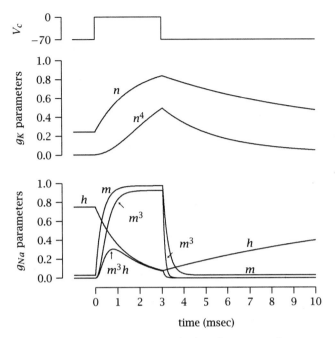

Figure 6.9 Time courses of n, n^4, m, m^3, h, and m^3h following a depolarizing voltage step (from -70 mV to 0 mV; duration of the step is 3 msec). n and m follow the $(1 - e^{-t/\tau})$ time course (activated by depolarization), whereas h follows the $e^{-t/\tau}$ time course (inactivated by depolarization).

Note that n and m are in $(1 - e^{-t/\tau})$ form and h is in the form $e^{-t/\tau}$ because of the difference in boundary conditions. That is, $h_0 > h_\infty$, whereas $n_\infty > n_0, m_\infty > m_0$. n and m are *activated by depolarization*, whereas h is *inactivated by depolarization*. One can then substitute the solutions of $n(t)$, $m(t)$, and $h(t)$ into $g_K(t)$ and $g_{Na}(t)$:

$$g_K(t) \;=\; \overline{g}_K n^4 = \overline{g}_K \left[n_0 - (n_0 - n_\infty)(1 - e^{-t/\tau_n}) \right]^4. \tag{6.3.5}$$

$$
\begin{aligned}
g_{Na}(t) \;&=\; \overline{g}_{Na} m^3 h \\
&=\; \overline{g}_{Na} \left[m_0 - (m_0 - m_\infty)(1 - e^{-t/\tau_m}) \right]^3 \left[h_\infty + (h_0 - h_\infty)e^{-t/\tau_h} \right] \\
&=\; \overline{g}_{Na} m_\infty^3 h_0 \left(1 - e^{-t/\tau_m} \right)^3 e^{-t/\tau_h}, \tag{6.3.6}
\end{aligned}
$$

because m_0 and h_∞ are neglectably small. The time course of g_K and g_{Na} described by equations 6.3.5 and 6.3.6 fitted the experimental data very well (see figure 6.7; smooth lines are equations 6.3.5 and 6.3.6, and circles are data points). The gate model also provides a quantitative description of the voltage dependence of g_K and g_{Na} as follows.

$$\alpha_n(V) = \alpha_n^0 e^{yzFV/RT} \qquad n_\infty(V) = \frac{\alpha_n(V)}{\alpha_n(V)+\beta_n(V)}$$
$$\beta_n(V) = \beta_n^0 e^{-(1-y)zFV/RT} \qquad \tau_n = \frac{1}{\alpha_n(V)+\beta_n(V)}$$

$$\alpha_m(V) = \alpha_m^0 e^{yzFV/RT} \qquad m_\infty(V) = \frac{\alpha_m(V)}{\alpha_m(V)+\beta_m(V)}$$
$$\beta_m(V) = \beta_m^0 e^{-(1-y)zFV/RT} \qquad \tau_m = \frac{1}{\alpha_m(V)+\beta_m(V)}$$

$$\alpha_h(V) = \alpha_h^0 e^{-(1-y)zFV/RT} \qquad h_\infty(V) = \frac{\alpha_h(V)}{\alpha_h(V)+\beta_h(V)}$$
$$\beta_h(V) = \beta_h^0 e^{yzFV/RT} \qquad \tau_h = \frac{1}{\alpha_h(V)+\beta_h(V)}$$

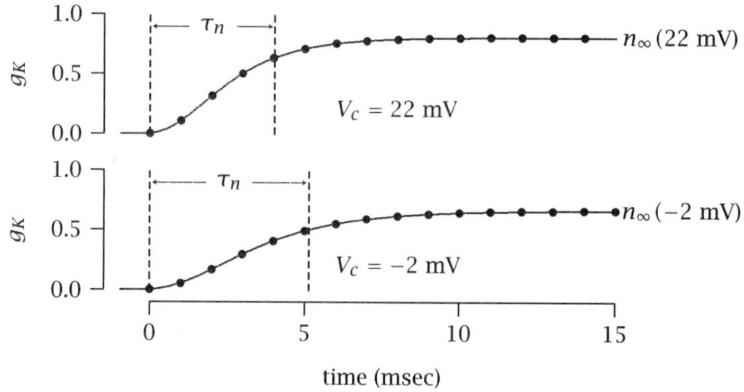

Figure 6.10 Measurement of τ_n and n_∞ from g_K traces at two voltages (−2 mV and 22 mV).

Hodgkin and Huxley determined all of the above parameters from their experimental data in the following way. They first measured τ_n, τ_m, τ_h, n_∞, m_∞, and h_∞ from the time records of g_K and g_{Na} at various voltages (an example is given in figure 6.10 for measuring τ_n and n_∞ at two voltage levels), and then they calculated α_n, β_n, α_m, β_m, α_h, and β_h by the following relationships:

$$\alpha_n = n_\infty/\tau_n \qquad \beta_n = (1 - n_\infty/\tau_n),$$
$$\alpha_m = m_\infty/\tau_m \qquad \beta_m = (1 - m_\infty)/\tau_m,$$
$$\alpha_h = h_\infty/\tau_h \qquad \beta_h = (1 - h_\infty)/\tau_h.$$

Hodgkin and Huxley plotted the values of α and β against transmembrane voltage and found that they can be fitted by the following empirical equations:

$$\alpha_n(V) = 0.01(-V + 10)/\left[e^{\frac{-V+10}{10}} - 1\right], \tag{6.3.7}$$

$$\beta_n(V) = 0.125 e^{\frac{-V}{80}}, \tag{6.3.8}$$

$$\alpha_m(V) = 0.1(-V + 25) / \left(e^{\frac{-V+25}{10}} - 1\right), \tag{6.3.9}$$

$$\beta_m(V) = 4 e^{\frac{-V}{18}}, \tag{6.3.10}$$

$$\alpha_h(V) = 0.07 e^{\frac{-V}{20}}, \tag{6.3.11}$$

$$\beta_h(V) = 1 / \left(e^{\frac{-V+30}{10}} + 1\right). \tag{6.3.12}$$

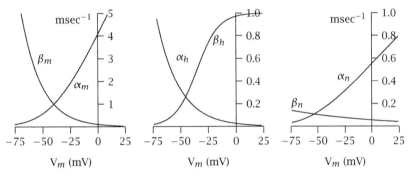

Figure 6.11　Voltage dependence of the rate coefficients of the Hodgkin and Huxley model.

These plots are shown in figure 6.11. It is important to note that the α's and β's in figure 6.11 follow the function $\alpha(V)$ and $\beta(V)$ predicted by the gate model (figures 5.9 and 5.10). α_n and α_m increase with membrane depolarization because n and m particles are activated by membrane depolarization. α_h decreases with membrane depolarization because the h particle is inactivated by depolarization.

Using the equations given above, the values of n_∞, m_∞, h_∞, τ_n, τ_m, and τ_h are calculated and plotted in figure 6.12. The m_∞, h_∞, and n_∞ also follow the y_∞ function predicted by the gate model (figures 5.9 and 5.10). The $n_\infty(V)$, m_∞, and h_∞ are called the steady-state *activation curves*. They give the voltage range and slope of activation for voltage-gated channels.

Equations 6.3.2–6.3.12, shown in the last section, are called the Hodgkin and Huxley equations. By putting them in equation 6.3.1, one obtains

$$I_m = C_m \frac{dV}{dt} + \overline{g}_K n^4 (V - E_K) + \overline{g}_{Na} m^3 h (V - E_{Na}) + g_L (V - E_L).$$

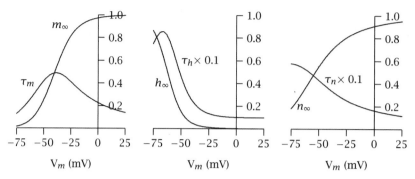

Figure 6.12   Steady-state activation curves (n_∞, m_∞, and h_∞) and the voltage dependence of the time constants of the Hodgkin and Huxley model.

Using numerical methods, Hodgkin and Huxley solved these equations and obtained remarkable fits between the recorded and the calculated action potentials (figure 6.13, left). Moreover, the calculated voltage-clamp records (figure 6.13, right) fit the experimental data very well.

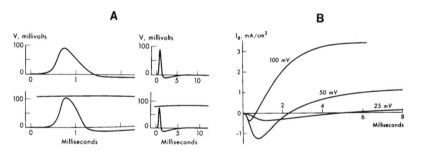

Figure 6.13   (A) Calculations, above, and experimental records, below, for propagating impulse on fast time scale, left, and slow scale, right; and (B) the ionic membrane currents after the indicated potential increase as calculated from the Hodgkin-Huxley equations. (From Cole 1968.)

Hodgkin and Huxley's model is certainly a triumph of classical biophysics in answering fundamental biological questions. Not only does it give quantitative accounts of Na^+ and K^+ fluxes, voltage- and time-dependent conductance changes, and the waveforms of action potentials, but also, as we will see next, it accounts for the conduction of action potentials along nerve fibers.

6.4 Nonpropagating and propagating action potentials

6.4.1 Hodgkin and Huxley equations for nonpropagating and propagating action potentials

For cells that are space-clamped and for which the membrane can be excited uniformly, cable properties are not involved, and the action potential is *nonpropagating:*

$$I_m = C_m \frac{dV}{dt} + I_K + I_{Na} + I_L \quad \text{(nonpropagating)}.$$

However, under physiological conditions, neurons are not voltage- or space-clamped, and action potentials initiated at one point will propagate along the axon. To describe action potential propagation, one should first derive the cable equation that illustrates how ions diffuse along the axons.

The equivalent circuit of a cable is given in figure 6.14 (see also chapter 4). V_m is now a function of time *and* distance.

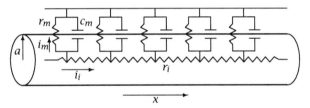

Figure 6.14   Schematic diagram illustrating current flow along a cylindrical axon (see figure 4.7).

Along the x-axis, $\dfrac{\partial V_m(x,t)}{\partial x} = -r_i i_i.$ \hfill (6.4.13)

Some of i_i, however, leaks out across the membrane through r_m and c_m, so i_i is not constant with distance.

$$\frac{\partial i_i}{\partial x} = -i_m. \qquad (6.4.14)$$

Combining equations 6.4.13 and 6.4.14,

$$\frac{\partial^2 V_m}{\partial x^2} = -r_i \frac{\partial i_i}{\partial x} = r_i i_m.$$

Thus,

$$i_m = \frac{1}{r_i} \frac{\partial^2 V_m}{\partial x^2}. \qquad\qquad (6.4.15)$$

From linear cable theory (chapter 4), the membrane current I_m along a cable is given by

$$I_m = \frac{a}{2R_i} \frac{\partial^2 V}{\partial x^2}.$$

Combining equation 6.4.1 with the equation for current in the parallel conductance model, we have

$$I_m = \frac{a}{2R_i} \frac{\partial^2 V}{\partial x^2} = C_m \frac{\partial V}{\partial t} + I_K + I_{Na} + I_L.$$

This is a second-order partial differential equation, which is very difficult to solve. It is known, however, that action potentials propagate with a constant speed (at least in axons with constant diameter), so one can use the wave equation

$$\frac{\partial^2 V}{\partial X^2} = \frac{1}{\theta^2} \frac{\partial^2 V}{\partial t^2},$$

where θ = conduction velocity (cm/sec). This would simplify the propagating Hodgkin and Huxley equation to

$$\frac{a}{2R_i \theta^2} \frac{d^2 V}{dt^2} = C_m \frac{dV}{dt} + I_K + I_{Na} + I_L.$$

This is a second-order ordinary differential equation, which is relatively easy to solve. From this wave equation, one can obtain

$$\theta = \sqrt{Ka/2R_iC_m} \propto \sqrt{a}.$$

K is a constant which is experimentally estimated to be 10.47 (msec^{-1}). Hodgkin and Huxley calculated conduction velocity and obtained

$$\theta = \sqrt{Ka/2R_iC_m} = 18.8 \text{ m/sec},$$

while the experimentally measured value of conduction velocity in the squid axon is

$$\theta = 21.2 \text{ m/sec}.$$

The Hodgkin and Huxley equations therefore give a very good fit to the experimental data.

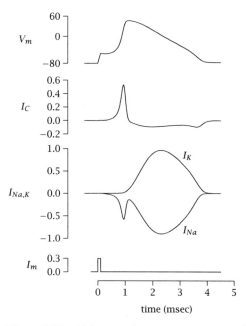

Figure 6.15 Voltage and current responses calculated for a nonpropagating action potential initiated by a brief square pulse current. The top diagram shows the potential change as a function of time. The second diagram shows the capacity current, I_C, given by $C_m dV/dt$. The initial, nearly square, wave is attributable to the applied current. The third diagram shows the K^+ and Na^+ components of the ionic current. Note that $I_m = I_C + I_K + I_{Na}$. (After Jack et al. 1975.)

6.4.2 Variations in voltage and currents during nonpropagating and propagating action potentials

Under certain conditions, the membrane of a cell or a portion of a cell can be excited uniformly and the cable complications can be eliminated. This can be achieved experimentally by space-clamping the cell (e.g., by inserting a long metal wire along the axon) or in cases where the cell is short enough for the membrane to be uniformly polarized during the action potential. In such circumstances, the relation between ion current flow across the membrane and the membrane potential is quite simple: All ionic current (I_i) is used to charge the local membrane capacitor, and none flows as local circuit (cable) current (figure 6.15). I_m is the stimulus current that initiated the action potential by depolarizing the cell above the threshold. I_m is zero during the action potential, thus

$$I_m = I_C + I_i = C_m \frac{dV}{dt} + I_i = 0; \text{ hence}$$

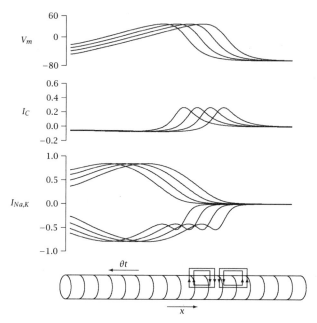

Figure 6.16   Calculated voltage changes and currents during an action potential propagated along an axon. The traces in each family of curves are separated by 0.1 msec. The top curve shows the membrane potential. Below are shown the changes in I_C, I_K and I_{Na}. The wave is propagating from right to left and, as noted in the text, the abscissa may also be regarded as time since $x = \theta t$. (After Jack et al. 1975.)

$$I_i = -I_C = -C_m \frac{dV}{dt} \text{ for nonpropagating action potentials.}$$

In the case of propagating action potentials, cable properties have to be considered. Figure 6.16 shows that an action potential is conducting from right to left. Upon arrival, the action potential depolarizes the local membrane, which causes an increase of g_{Na} and results in a net inward ionic current. This current enters the cell and diffuses laterally forward (left) and backward (right) along the axon and forms local circuit loops. The forward current depolarizes the local membrane and causes an increase of g_{Na}, which results in more inward current at that location. This process continues, causing the action potential to propagate in the forward direction. The backward current also depolarizes the local membrane, but since the action potential has just passed that location, the threshold is high (g_{Na} is low and g_K is high), and the action potential is not generated. The ionic current I_i in the propagating action potential not only charges the local membrane capacitor but also flows longitudinally and results in conduction of the action potential from one part of a cell to another.

6.5 Noble's model for nerve excitation: simplified *I-V* relations

In order to illustrate the characteristics of nerve and muscle excitation without going into too much mathematical detail, Noble (1966) made simplifying assumptions and provided a semiquantitative account of excitable membrane. This model is extremely useful for an intuitive understanding of membrane excitation. Assumptions: (1) m is significantly faster than n and h, so m will reach its steady-state value *almost* instantaneously at each potential. Since it is not, strictly speaking, instantaneous, we call it the *momentary I-V relation* for I_{Na}. (2) The instantaneous K$^+$ *I-V* relation is linear, as shown in squid axon (but not necessarily true in other cells). The momentary *I-V* relation of excitable membrane can be written as $I_i = I_{Na} + I_K$, which can be plotted as in figure 6.17.

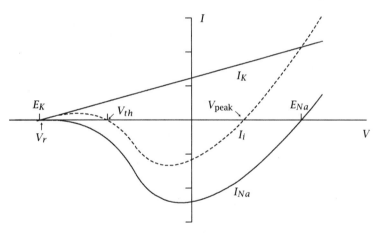

Figure 6.17   Diagram illustrating the form of simplified (momentary) current-voltage relation, $I_i = I_{Na} + I_K$, obtained by allowing fast Na$^+$ activation reaction (m) to be in a steady state while slower reactions (h and n) are held constant.

When the squid axon is at rest, the membrane is permeable primarily to K$^+$; thus the resting potential V_r ($I = 0$, the leftmost intersecting point) is very close to E_K. The slope conductance at this intersecting point is positive. This positive slope conductance makes V_r a *stable* point, because if a positive perturbation in V_m occurs, the *I-V* relation gives rise to an increment of positive (outward) current, which hyperpolarizes the cell and brings V_m back to V_r. For a negative V_m perturbation, the *I-V* relation gives rise to a negative (inward) current, which depolarizes the cell and brings V_m back to V_r.

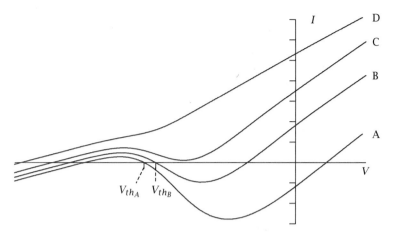

Figure 6.18   Diagram illustrating changes in momentary current-voltage relations with time on depolarization.

When the membrane potential reaches threshold V_{th} ($I = 0$, the middle intersecting point), the slope conductance is negative. This makes the *I-V* relation *unstable* because positive voltage perturbation results in negative (inward) current, which further depolarizes the cell and brings the potential further away from V_{th}. Negative voltage perturbation results in positive (outward) current, which further hyperpolarizes the cell. Therefore, when the membrane potential reaches V_{th}, it either depolarizes to generate an action potential, or hyperpolarizes back to V_r. Membrane potential never stably stays at V_{th}.

The rightmost intersecting point V_{peak} is the peak potential level an action potential can reach (when $g_{Na} >> g_K$). The slope conductance at this point is positive, and therefore the voltage should be stable in principle. However, the stability at V_{peak} is only transient because depolarization activates n and h, which increases I_K and decreases I_{Na}. This results in an outward shift of the *I-V* relation, and the time course of this shift is shown in figure 6.18 (A $\rightarrow$ B $\rightarrow$ C $\rightarrow$ D). Consequently, V_{peak} will become more negative and eventually disappear (when $I_K > I_{Na}$ at all potentials; C and D). Action potentials cannot be generated at instances C and D, and the membrane potential will spontaneously repolarize back to V_r.

By using these simplified *I-V* relations, one can explain the following important characteristics of actions potentials:

Threshold: Depolarizing current is needed to bring the membrane from a positive slope conductance region (E_r) to a negative slope conductance region for regenerative membrane potential change (dep $\rightarrow$ I_{inward} $\rightarrow$ dep

$\rightarrow I_{\text{inward}} \ldots$), which sets off the all-or-none action potential. Depolarization below threshold will result in repolarization back to V_r.

Accommodation: (Slow stimulus current is less effective than a rapid rising current to elicit excitation, and if the current is slow enough, excitation may not occur at all.) When the membrane is depolarized rapidly, the I-V relation will follow curve A because n and h do not have time to change appreciably, and thus the threshold will be at V_{th_A}. If the depolarization occurs slowly, then curve B will be followed because n increases and h falls, and the threshold will be at V_{th_B}, a higher value. If the depolarization occurs even more slowly, the I-V relation will follow curve C and there will be no threshold, and excitation will not occur. *Anode break* (an action potential observed at the offset of a hyperpolarizing current step) can be explained similarly.

Refractory period: During the action potential, the system moves from A to C and D, so that at the end of the action potential, the membrane is inexcitable (this is the absolute refractory period). After the action potential, the system moves back to B (relative refractory period—threshold high; V_{th} is more positive) and then to A (normal excitability).

6.6 Gating current

Hodgkin and Huxley (1952d) pointed out in their gate model that every voltage-dependent step must have an associated charge movement. Take the Na$^+$ channel as an example. Hodgkin and Huxley assumed that the Na$^+$ conductance is proportional to the probability of some gating particles that are near the *outside* of the membrane, that is,

$$g_{Na} = \overline{g}_{Na} y_{\text{out}}, \tag{6.6.16}$$

where

$$y_{\text{in}} + y_{\text{out}} = 1. \tag{6.6.17}$$

The gating particle can be in either of the two positions:
y_{in} = probability of gating particle on the inside, and
y_{out} = probability of gating particle on the outside.

Assuming the gating particles move *independently* in the membrane, then the probabilities of their being at any given state are proportional to $e^{-\xi/kT}$ where ξ is the energy of the particle in the state (Boltzmann's distribution). Thus,

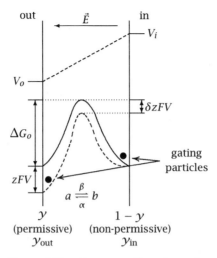

Figure 6.19   Energy profiles of a gating particle in the membrane under the influence of the electric field. y_{out} is the probability of the particle near the outside margin of the membrane and y_{in} is the probability of the particle near the inside margin. (See also legend for figure 5.9.)

$$y_{in} \propto e^{-\xi_{in}/kT}, \quad y_{out} \propto e^{-\xi_{out}/kT},$$

when a transmembrane voltage (V) is developed (see figure 6.19).

$$y_{in} \propto e^{-\xi_{in}/kT} \qquad y_{out} \propto e^{-(\xi_{out}-zeV)/kT}.$$

Therefore,

$$\frac{y_{out}}{y_{in}} = e^{-(\xi_{out}-\xi_{in})/kT+zeV/kT} = e^{(w+zeV)/kT},$$

where $w = -(\xi_{out} - \xi_{in})$ and is the work required for a particle to move across the membrane when $V = 0$, z = number of charges on the gating particle, e = elementary electric charge, V = membrane potential, k = Boltzmann's constant, and T = absolute temperature.

   Combining equations 6.6.16 and 6.6.17, we get

$$g_{Na} = \overline{g}_{Na} \frac{e^{(w+zeV)/kT}}{1 - e^{(w+zeV)/kT}}. \tag{6.6.18}$$

When V is large and negative, $w << zeV$ and $e^{(w+zeV)/kT} << 1$.

$$g_{Na} \cong \overline{g}_{Na} e^{zeV/kT} = \overline{g}_{Na} e^{zV/25(\text{mV})}. \tag{6.6.19}$$

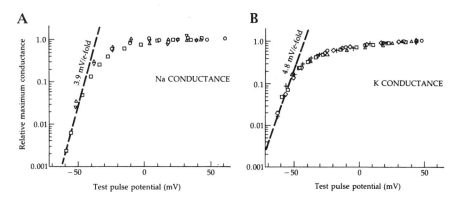

Figure 6.20 Peak g_{Na} (A) and steady-state g_K (B) are measured during depolarizing voltage steps under voltage clamp. Symbols are measurements from several squid giant axons, normalized to 1.0 at large depolarizations, and plotted on a logarithmic scale against the potential of the test pulse. Dashed lines show limiting equivalent voltage sensitivities of 3.9 mV per *e*-fold increase of g_{Na} and 4.8 mV per *e*-fold increase of g_K for small depolarizations. (From Hille 1992, adapted from Hodgkin and Huxley 1952a.)

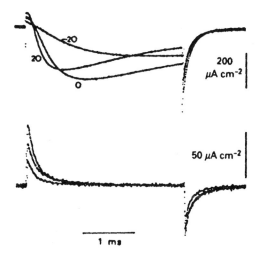

Figure 6.21 Na^+ ionic and gating currents in squid axon, produced during depolarizations under voltage clamp. The upper traces show currents recorded in an artificial sea water with only one fifth of the normal Na^+ concentration, for depolarizations from -70 mV to -20, 0, and $+20$ mV. The initial brief outward current is gating current, followed by the much larger inward Na^+ ionic current. The lower set shows the gating currents alone, after blockage of the Na^+ ionic currents with TTX. K^+ currents were eliminated by using K^+-free solutions for both internal and external media. (From Aidley 1989, adapted from Bezanilla 1986.)

Experimentally, Hodgkin and Huxley showed that for large negative potentials, $g_{Na} << \overline{g}_{Na}$ (figure 6.20A), and the semi-log plot shows that g_{Na} increases by e-fold for every 4 mV (3.9 mV) increase in membrane voltage. Thus,

$$g_{Na} \propto e^{V/4(mV)}. \qquad (6.6.20)$$

Comparing equations 6.6.19 and 6.6.20 yields that $z = 6$. Therefore, in order to open one Na^+ channel, 6 charges must move across the membrane or 3 dipoles must be reversed.

The movement of this gating charge is *outward* for *depolarizing* voltage pulse, and it gives rise to a small *outward current*. However, this outward *gating current* is masked by the inward Na^+ current and the outward capacitive current. In order to remove these currents, Armstrong and Bezanilla (1974) used Na^+-free solutions + TTX (blocks Na^+ channels) + Cs^+ internally to block K^+ channels. In addition, to eliminate the linear capacitive current, they averaged the responses to an equal number of positive and negative steps of potential from a negative holding level. The resultant average current is shown in figures 6.21 and 6.22A.

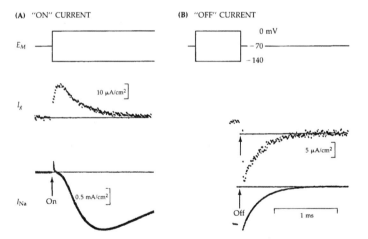

Figure 6.22 Gating current (I_g) and I_{Na} recorded by adding responses to symmetrical positive and negative pulses applied to the squid giant axon. I_g was measured in Na^+-free solutions with TTX to block Na^+ channels and internal Cs^+ to block K^+ channels. Since I_g is small, 50 traces had to be averaged in the recording computer to reduce the noise. I_{Na} is measured in normal artificial sea water without TTX. (A) Depolarization from rest elicits an outward "on" I_g that precedes opening of Na^+ channels. (B) Repolarization elicits an inward "off" I_g coinciding with closing of channels (a different axon). (From Hille 1992, adapted from Armstrong and Bezanilla 1974 by copyright permission of the Rockefeller University Press.)

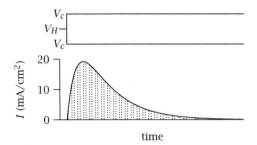

Figure 6.23 Integral of the gating current (shaded area) gives the net gating charge transfer during the voltage step.

Gating current is not blocked by TTX, which blocks I_{Na}. Both I_g and I_{Na} are blocked by $ZnCl_2$, predepolarization, or glutaraldehyde.

The net charge transfer in gating during a single voltage clamp step is given by $Q = zeD$, where z = number of charges on one gating particle, e = elementary electric charge, and D = density of gates $(1/\mu m^2)$.

The integral of the time course of the gating current (shaded area, figure 6.23) gives the net charge transformed during a single voltage-clamp step. Experimental results showed $Q = 1882e/\mu m^2$. Kinetic and pharmacological studies attribute almost all mobile gating charges in axons to gating of Na^+ channels. From earlier analysis, 6 charges were thought to be required to open 1 Na^+ gate. Therefore,

$$D = Q/ze = \frac{1882e/\mu m^2}{6e} = 314 \text{ gates}/\mu m^2.$$

Experimentally, Hodgkin and Huxley estimated that $g_{Na} = 1200$ pS/μm^2. Thus, the single-channel conductance for Na^+ channels

$$\gamma = \frac{\overline{g}_{Na}}{D} = \frac{1200 \text{ pS}/\mu m^2}{314 \text{ gates}/\mu m^2} = 3.82 \text{ pS}.$$

The Na^+ gating current shown on the previous page was obtained with short voltage pulses (~ 0.75 msec), which do not elicit much inactivation of the Na^+ channels. Therefore, the total charge transfer at the offset (B, time integral of the inward gating current) is approximately equal to that at the onset, that is, $Q_{on} = Q_{off}$, showing that activation is quick and reversible.

For longer voltage-clamp pulses, the total charge transfer at the offset is less than that at the onset (figure 6.24). After a 10 msec voltage pulse, Q_{off} may be only 30% of Q_{on}, showing that 70% of the gating particles are

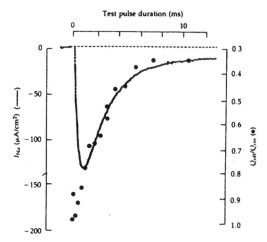

Figure 6.24   Comparison of the time course of inactivation of I_{Na} (solid line) with the immobilization of gating charge (circles) in the squid axon. Gating-charge movement is determined by integrating the rapid "on" and "off" I_g for test pulses of different durations. The fraction of charge returning quickly at the "off" step decreases with increasing pulse length (but not offset of right scale) in parallel with inactivation of Na^+ channels. (From Hille 1992, adapted from Armstrong and Bezanilla 1977 by copyright permission of the Rockefeller University Press.)

immobilized. Immobilization occurs with the same time course as Na^+ channel inactivation (see figure 6.24).

The gating current of K^+ channels is much more difficult to detect than the gating current of Na^+ channels. This is because of the lower density of K^+ channels and their slower activation. Gilly and Armstrong (1980) discovered a prominent slow phase of the gating current that is insensitive to the local anesthetic dibucaine (which reduces Na^+ gating current) in the squid axon. White and Bezanilla (1985) found that the maximum charge transfer associated with the K^+ gating current is about $490e/\mu m^2$ (figure 6.25), and according to the Hodgkin and Huxley measurement of g_K (figure 6.20B), $g_K \propto e^{V/5(\text{mV})}$. Similar to equation 6.6.19, $g_K = \overline{g}_K e^{zV/25(\text{mV})}$ and $z = 5$; therefore 5 charges must be moved to open one K^+ channel.

K^+ channel density, D, can be estimated by

$$D = \frac{Q}{ze} = \frac{490e/\mu m^2}{5e} = 98 \text{ channels}/\mu m^2.$$

This is a high estimate of D: others assume that z for K^+ channel gating is $7-13\ e$, which makes $D = 36 - 70$ channels$/\mu m^2$.

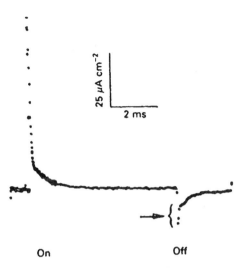

25 μA cm^{-2}

2 ms

On   Off

Figure 6.25   K$^+$ gating current in a perfused squid axon. The internal solution contained impermeant organic cations to eliminate the K$^+$ ionic current, and the external solution contained Tris nitrate, TTX, and dibucaine to eliminate the Na$^+$ ionic current and reduce the Na$^+$ gating current. The gating currents were produced by a 6 msec depolarization from −110 to 0 mV, followed by a return to −60 mV. The "on" response consists of a very brief Na$^+$ gating current followed by a more slowly changing K$^+$ gating current. In the "off" response these two components are even more distinct, and the Na$^+$ gating current (arrow) is smaller. (From Aidley 1989, adapted from White and Bezanilla 1985 by copyright permission of the Rockefeller University Press.)

It is difficult to study the kinetics and charge mobilization of K$^+$ gating current in the squid axon because it is often masked by the Na$^+$ gating current (figure 6.25). This problem was overcome recently by studying the gating current of the *Shaker* K$^+$ channel expressed in *Xenopus* oocytes, which do not contain voltage-gated Na$^+$ channels (Bezanilla et al. 1991). Figure 6.26 shows the gating currents recorded in normal *Shaker* K$^+$ channels that exhibit inactivation (A) and mutant *Shaker* K$^+$ channels that do not exhibit inactivation (B). The gating charge transfer (time integral of the gating current) at the voltage step offset (Q_{off}) for the normal *Shaker* K$^+$ channel (A) is much less than the gating charge transfer at the voltage step onset (Q_{on}). For the mutant channels in which inactivation is removed (B, similar to the K$^+$ channels in the squid axon), $Q_{off} = Q_{on}$ although the time course of the gating current at the voltage offset is slower than that at the voltage onset.

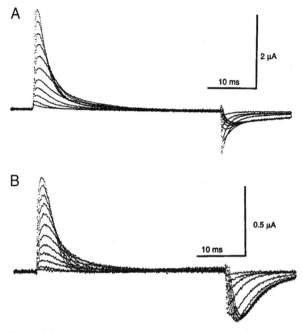

Figure 6.26   Gating currents recorded from normal *Shaker* K$^+$ channels (A) and from the non-inactivating mutant *Shaker* K$^+$ channels (B) expressed in *Xenopus* oocytes. The gating currents were activated by depolarizing voltage-clamp steps ranging from -70 mV to 30 mV in 10 mV steps. (From Bezanilla et al. 1991. Used by permission of AAAS, copyright © 1991 AAAS.)

Example 6.1

The Na$^+$ ionic current I_{Na} and Na$^+$ gating current I_g in the squid axon produced by a depolarizing voltage step (from -70 mV to $+20$ mV) under voltage clamp are shown in the figure below. The lower trace shows currents recorded in an artificial sea water with only one fifth of the normal Na$^+$ concentration, for depolarizations from -70 mV to $+20$ mV. The initial brief outward current is gating current, followed by the much larger inward Na$^+$ ionic current. The upper trace shows the gating currents alone, after blocking the Na$^+$ ionic currents with TTX and eliminating capacitive currents. K$^+$ currents were eliminated by using K$^+$-free solutions for both internal and external media.

Example 6.1 (continued)

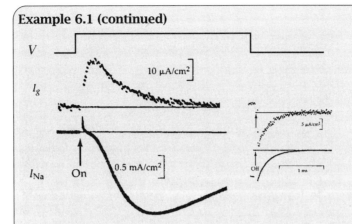

a. Explain why I_g is outward at the pulse onset and inward at the pulse offset. Why is the amplitude of the I_g at the offset smaller than that at the onset?

b. Why is there a transient inward I_{Na} at the offset of the voltage pulse?

c. Under what condition do you expect to observe symmetrical I_g at the onset and offset of the voltage pulse?

Answer to example 6.1

a. I_g is outward at pulse onset because the depolarization results in an increase of an outward electric field, which pushes the gating particles outward. At pulse offset, the electric field changes in the opposite direction, and thus I_g is inward. I_g is smaller at the offset because a substantial fraction of Na^+ channels is in the inactivated state after the voltage pulse is maintained for about 3.25 msec.

b. The inward transient I_{Na} at offset is caused by the instantaneous voltage drop (from -20 to -70 mV), which results in a much larger driving force ($V - E_{Na}$), and the time-dependent closure of Na^+ channels. (Na^+ channels do not have enough time to close immediately after the pulse offset.)

c. I_g will be symmetrical if the voltage pulses are shorter than 1 msec, which is not long enough for significant inactivation to take place.

6.7   Review of important concepts

1. The Na^+ and K^+ conductances mediating the action potentials in the squid axon were studied by Hodgkin and Huxley (1952a–d) by using the voltage-clamp technique and the gate model.

2. The instantaneous *I-V* relations of the Na^+ and K^+ channels in the squid axon are linear. Voltage-clamp results suggest that the Na^+ channel is activated by three gating particles (*m*) and one inactivating particle (*h*), and the K^+ channel is activated by four gating particles (*n*).

3. The voltage- and time-dependent parameters of the Na^+ and K^+ conductances were quantitatively described by the Hodgkin and Huxley equations. The simulated action potential based on these equations is in excellent agreement with the recorded action potential. The propagation and conduction velocity of action potentials can also be described by these equations.

4. Mechanisms mediating various dynamic features of action potentials such as threshold, refractory period, and accommodation can be described by changes of *I-V* relations of the Na^+ and K^+ currents in the cell.

5. The gating currents of the Na^+ and K^+ channels were recorded by eliminating all ionic currents and the capacitive current. From the charge transfer of the gating currents, one can estimate the number of gating particles per charge, channel density, and single-channel conductance in the membrane.

6.8 Homework problems

1. Ionic currents involved in the action potential of a cardiac muscle fiber have been studied by the voltage-clamp technique. When membrane potential is stepped from its resting value of -77 mV to -50 mV, an initial inward current is seen, which is carried by Na^+.

   (a) Assume internal Na^+ concentration is normally 30 mM, and external Na^+ concentration is normally 150 mM. Draw to approximate scale the initial current traces for a step to -50 mV when external Na^+ concentration is normal; and when external Na^+ concentration is reduced to 30 mM, to 10 mM, and to 1 mM by replacement of Na^+ with an impermeant cation.

   (b) If the peak inward Na^+ current with normal Na^+ concentration is 1 mA/cm^2, calculate the peak Na^+ current in each of the other cases.

   (c) External Na^+ concentration is adjusted so that the initial inward current during a voltage clamp step to -50 mV is abolished. However, when the membrane potential is stepped from -77 mV to -20 mV, a longer-lasting inward current is recorded. Can this be due to the opening of further Na^+ channels at this membrane potential? Explain in one sentence. Assuming that internal and external K^+ and Cl^- concentrations are comparable to those of frog muscle, could it be due to the opening of K^+ channels or of Cl^- channels? Explain each answer.

   (d) The suggestion has been made that Ca^{2+} carries this current. If external Ca^{2+} concentration is 2.5 mM and internal Ca^{2+} concentration is less than 10^{-2} mM, in what range is E_{Ca}? Would the Ca^{2+} current at a membrane potential of -20 mV be in the right direction to account for the observed current? Justify your answer.

2. After a particular step depolarization in Hodgkin and Huxley's squid axon, the parameter n follows the curve

   $$n = 0.891 - 0.376e^{-t/1.7\text{(msec)}}.$$

   t is in msec, and $\bar{g}_K$ is known to be 24.3 mS/cm^2.

   Plot g_K as a function of time, using 1 msec steps for 10 msec. What is the steady-state g_K (i.e., $g_{K\infty}$)?

3. After the same depolarization as in problem 2, the parameters m and h follow the curves $m = 0.963 \left(1 - e^{-t/0.252(\text{msec})}\right)$, and $h = 0.605 e^{-t/0.84(\text{msec})}$. t is in msec, and $\bar{g}_{Na} = 70.7$ mS/cm^2.

   Plot g_{Na} as a function of time using 0.5 msec steps for 5 msec. What is the largest value of g_{Na} reached?

4. Using the figure below (from Hodgkin and Huxley 1952d), calculate the steady-state K$^+$ current ($I_{K\infty}$) after the axon is stepped from resting potential (−60 mV) to 0 mV (in the Hodgkin and Huxley paper, from 0 to −60 mV). $\bar{g}_K$ is given to be 36 mS/cm^2, and the equilibrium potential of K$^+$ is −72 mV.

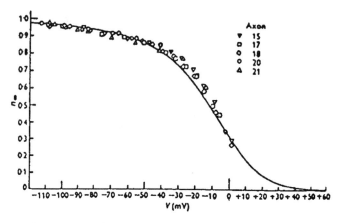

5. An alga *Chara globularis* is known to generate positive-going action potentials. The major ions in both its cytoplasma and the pond water it lives in are Na$^+$, K$^+$, and Cl$^-$, and their concentrations are as follows:

	Cytoplasma (mM)	Pond water (mM)
Na$^+$	57	0.031
K$^+$	65	0.046
Cl$^-$	112	0.040

The resting potential of the cell is −182 mV, and the peak amplitude of the action potential is +198 mV.

   (a) What is (are) the primary permeable ion(s) for this cell at rest?

   (b) What is (are) the primary permeable ion(s) for this cell during the peak of an action potential?

(c) In a pump that pumps both Na^+ and Cl^- into the cell at a ratio of $1 : 1$, what is the contribution of this pump to the resting potential of the cell?

(d) The voltage-clamp analysis of this cell reveals the following results:

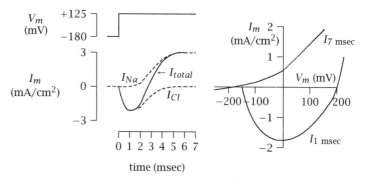

What are the values of g_{Cl} and g_{Na} at $V_m = -50$ mV and at $V_m = +150$ mV?

6. Currents measured with voltage clamp of squid axon are shown in the figure below. Membrane held at $V_H = -60$ mV was stepped to potentials V_c shown in the right-hand side of each current trace.

(a) What caused the small sustained inward current measured after the membrane was stepped to $V_c = -125$ mV? Give your answer in one sentence.

(b) The peak amplitude of the early inward current increased when V_c was varied from -30 mV to -5 mV. Why?

(c) The peak amplitude of the early inward current decreased when V_c was varied from -5 mV to $+57$ mV. Why?

(d) Give the approximate threshold (V_{th}) of this axon. Justify your answer in one sentence.

(e) If the early inward current is carried by Na^+, what is the approximate equilibrium potential for Na^+ in this axon? Explain.

(f) If the late outward current is carried by K^+ ions and E_K is known to be closed to -65 mV, why is there no reversal of the late outward current in these measurements?

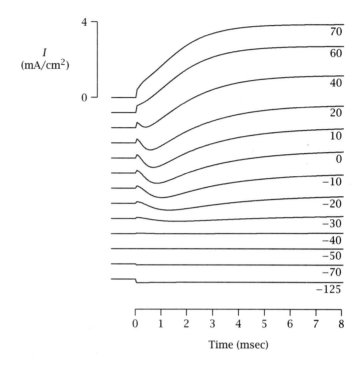

7. When a normal, healthy squid axon is voltage-clamped in artificial sea water, one obtains the following membrane current record in response to a step change in membrane potential from $V_m = -70$ mV to $V_m = 0$ mV.

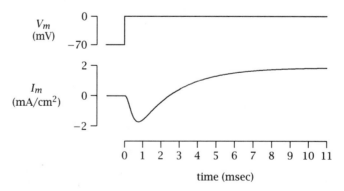

Draw similar plots of I_m vs. t (when V_m is stepped from -70 mV to 0 mV) when the recordings are made under each of the following experimental conditions. For each of your plots, explain in one or two sentences how and why your graph differs from that drawn above.

(a) TTX is added to the bath surrounding the axon.

(b) TEA is added to the interior of the axon.

(c) $[Na^+]_{out}$ is adjusted so that $[Na^+]_{out} = [Na^+]_{in}$.

(d) $[K^+]_{out}$ is adjusted so that $[K^+]_{out} = [K^+]_{in}$.

(c) Ouabain, a specific inhibitor of the Na^+-K^+ pump, is added to the bath five minutes before the experiment.

8. Using the normal data in the figure below, plot the current at 0.5 and 4.5 msec after the start of the stimulus vs. membrane potential. On the same graph, plot the corresponding currents in Na^+-free solution. Does the replacement of external Na^+ affect the late outward current?

   For a depolarization to −4 mV, measure and tabulate the current in the figure below in normal and Na^+-free solutions vs. time, using 0.25 msec intervals. From these data, plot I_K and I_{Na} vs. time on the same graph. Assume $E_{Na} = +60$ mV and $E_K = -70$ mV. Calculate and plot g_{Na} and g_K (in mS/cm^2) vs. time.

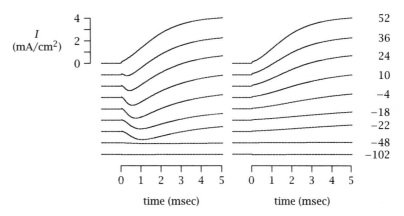

9. The momentary *I-V* relations of I_{Na}, I_K, and $I_i = I_{Na} + I_K$ of the squid giant axon are given below:

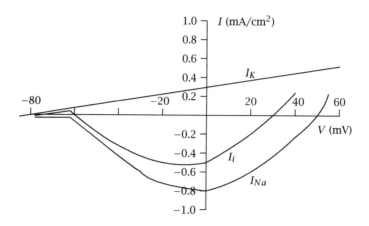

(a) What are the conductance ratios g_K/g_{Na} of this axon
   i. at rest,
   ii. at threshold of action potential, and
   iii. at the peak of an action potential?

(b) If I_{Na} shown in the above figure does not change with time (i.e., h particles are inactivated), but g_K becomes 5 times larger at time T, draw the I-V relation of I_K, I_{Na}, and I_i at time T on graph paper. What are the resting potential, threshold potential, and peak voltage of an action potential of the axon at time T?

10. Records of I_K obtained by voltage- and space-clamping of the squid giant axon in the presence of TTX are given below. The holding potential is -80 mV and the V_c are 25 mV, 60 mV, 85 mV, and 100 mV. $\bar{g}_K = 83$ mS/cm^2, and $E_K = -80$ mV.

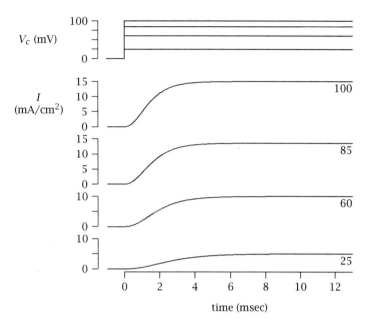

(a) From the above data and what you have learned about Hodgkin and Huxley's experiments, estimate the values of $g_{K\infty}$ at each V_c.

(b) What are the values of α_n and β_n at each V_c?

11. In an excitable cell, the membrane conductance is mediated by calcium and chloride channels only. Using Ca^{2+} and Cl^- channel blockers, one can measure Ca^{2+} and Cl^- currents separately. The current traces and I-V relations obtained under voltage-clamp conditions are shown below.

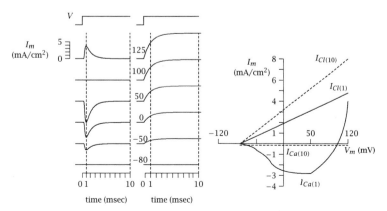

(a) Draw the I-V relation of the total membrane current $(I_{Ca^{2+}} + I_{Cl^-})$ at $t = 1$ msec and at $t = 10$ msec in the left figure. Estimate the threshold voltage (V_{th}) and the peak action potential voltage (V_{peak}) of this cell.

(b) What are the intracellular concentrations of Ca^{2+} and Cl^- in this cell if the extracellular concentrations of calcium and chloride are 5 mM and 50 mM, respectively?

12. The gating current I_g obtained by averaging equal numbers of positive and negative voltage-clamp steps of equal amplitude (in the presence of agents that block all ionic currents) is shown below.

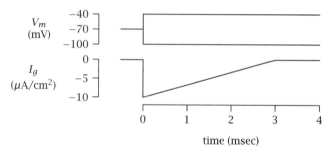

This gating current is known to be associated with opening K^+ channels in the cell, and 5 gating particles (each carries charge $e = 1$ electronic charge) are required to open one K^+ channel.

(a) What is the density (in channels/cm^2) of K^+ channels in this cell membrane?

(b) If the peak K^+ conductance $\bar{g}_K$ of this cell is 0.2 S/cm^2, what is the single-channel conductance of the K^+ channels?

(c) Is the K^+ current activated by depolarization or by hyperpolarization? Explain in one sentence.

13. The gating current I_g obtained by averaging voltage-clamp currents elicited by equal numbers of positive and negative voltage steps of equal amplitude (in the presence of agents that block all ion currents) is as shown:

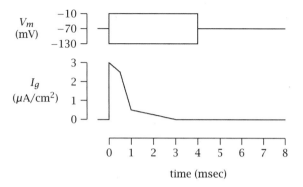

This gating current is known to be associated with opening Na$^+$ channels in a snail neuron, and 2 gating particles (each carries charge $e = 1$ electronic charge; $e = 1.6 \times 10^{-19}$ C) are required to open one Na$^+$ channel. In the absence of Na$^+$ channel blocker, the activation curve of the Na$^+$ channel is shown below:

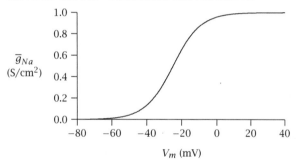

(a) What is the density (channel/cm^2) of Na$^+$ channels in the membrane of the snail neuron?

(b) What is the single-channel conductance of the Na$^+$ channels in the snail neuron?

(c) Assuming the Na$^+$ channels do not inactivate, redraw figure A on the graph paper, and add the gating current trace after the offset of the voltage-clamp steps (from 4 msec to 8 msec along the time scale).

14. The I-V relation of the Na$^+$ channels in the node of Ranvier is given in figure A, and the current records of the two-pulse voltage-clamp experiment carried out in this preparation are given in figure B. Figures A and B are obtained while other conductances (such as g_K) are totally blocked.

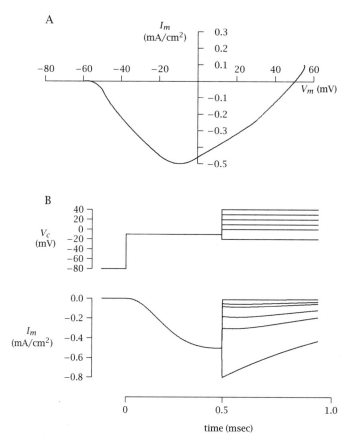

(a) Redraw figure A on graph paper, and plot the instantaneous I-V relation of the Na^+ channels on the same graph, based on the records given in figure B. Mark E_{Na} on your graph.

(b) Is the instantaneous g_{Na} ohmic? If not, what model(s) can be used to explain the nonlinearity?

(c) Repeat (a) if figures A and B are obtained while g_K is not blocked. g_K (at $t = 0.5$ msec) = 1 mS/cm^2 and g_K is approximately linear. $E_K = -80$ mV.

15. The time course of immobilization of "off" gating charges of Na^+ channels is given in figure A, and macroscopic Na^+ current I_{Na} and the Na^+ gating current (I_g) evoked by a brief depolarizing pulse are given in figures B and C, respectively.

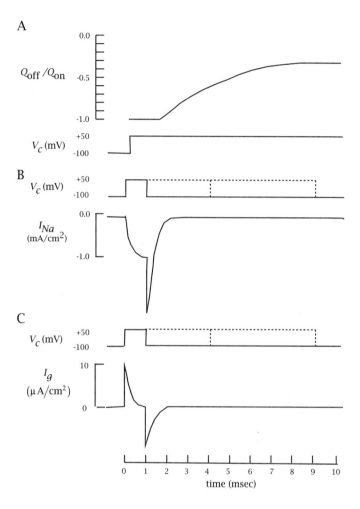

(a) Draw all components of I_{Na} and I_g with appropriate values (on figures B and C) for voltage pulses of the same amplitude ($V_H = -100$ mV, $V_c = 50$ mV) but longer durations: 4 msec and 9 msec. Label explicitly the values of $I_{Na^{off}}$ and the approximate values of $I_{g^{off}}$ with appropriate units.

(b) What is the major difference in the current waveforms between Hodgkin and Huxley's prediction and the actual experimental results? What does this difference imply in evaluating the Hodgkin and Huxley model?

7 Functional Diversity of Voltage-Gated Conductances

7.1 Introduction

In the last two chapters we presented the classical description of the nerve membrane derived from the work of Hodgkin and Huxley and based largely on experiments from squid axon. Although the type of analysis used by Hodgkin and Huxley for characterizing voltage-gated ionic currents is extremely useful, their representation of the nerve membrane in terms of two voltage-gated conductances, Na^+ and K^+, is inadequate for describing excitable membrane in other parts of a neuron, for example, the cell body, dendrites, and presynaptic terminal. In this chapter we will present an abbreviated survey of the major classes of voltage-gated conductances. We will also show how the quantitative model of Hodgkin and Huxley, the very general gate model presented in chapter 5, can be modified and used to describe many types of voltage-gated conductances. The different ionic conductances will be distinguished based on their ion selectivity, current-voltage relationships, activation/inactivation properties, and sensitivity to pharmacological agents.

It is worth mentioning here what we are *not* going to do in this chapter, namely, discuss the *molecular* diversity of ion channels. With modern techniques of gene cloning, we now know that within each family of ion channel (both voltage-gated and ligand gated), there is a great molecular diversity. That is, there are many different genes that express each of the subunits that comprise the channel. The functions of these different genes are, in many cases, not known, and this is an extremely active and rapidly changing area of research.

This chapter will instead emphasize the *functional* diversity of membrane currents with a view toward the whole cell rather the single channel. In fact, it is the functional diversity of conductances at the whole-cell level that has enabled molecular physiologists to classify channels into broad functional classes, for example, delayed-rectifier or A-type K^+ channels. As we will see, some of this classification is based on pharmacology and on the responses of different membrane currents to neurotransmitter agonists and antagonists.

Although this chapter is meant to be somewhat of a survey, it will require that you be able to use quantitative information (e.g., *I-V* curves and activation/inactivation parameters) to understand and predict how a particular ionic current might affect the electrophysiological functioning of a neuron. For the most part, we will restrict the discussion to voltage-gated ionic currents measured in neurons and, in particular, mammalian cortical neurons.

7.2   Cellular distribution of ion channels

Two important questions that form the basis of much research by neuroscientists are: What types of ion channels are present in a particular neuron, and where are they located? The types of ion channels and their distribution largely determine the electrophysiological behavior of a neuron. Unfortunately, the distribution of voltage-gated ion channels throughout a cortical neuron is currently unknown. A few general principles, however, can be gleaned from the present state of knowledge. First, axons are electrophysiologically simple compared to other parts of the neuron. The main role of the axon is to conduct a nondecrementing action potential (AP) from one point to another. The major types of ion channels present in axons produce a fast inward Na^+ current and a delayed outward K^+ current. Given that Hodgkin and Huxley did their experiments on the squid axon, it is not surprising that they described only two types of voltage-gated currents, Na^+ and K^+. In myelinated axons, there are primarily only fast Na^+ channels (at high density) and K^+ (and maybe Cl^-) leak channels at individual nodes of Ranvier. Repolarization of the AP occurs through Na^+ inactivation and current flow through K^+ selective leak channels rather than from the activation of a delayed-rectifier type K^+ current.

Second, the number of different types of ion channels in the soma is large. Functionally distinct Na^+ currents, many different types of K^+ currents, Ca^{2+} currents, and a few Cl^- and nonspecific cation currents have been measured in the cell bodies of cortical neurons.

Third, not much is known about dendrites and presynaptic terminals

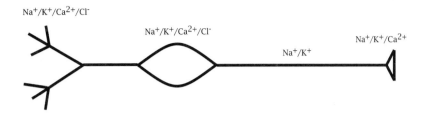

Figure 7.1 Heterogeneous distribution of ion channel types in soma, dendrites, and axon.

other than that they are probably at least as complex as the cell body in terms of their complement of different types of ion channels. For example, the synapse is the site of neurotransmitter release, and thus the message passed to the next neuron is dependent in part on the types of voltage-gated ion channels present in the presynaptic terminal. We know that there are at least one or two types of Ca^{2+} and K^+ channels in the terminal, but the extent of the channel diversity in the terminal is not yet known. This is also true for dendrites, and, as we saw in chapter 4, the types of channels present in dendrites can have important effects on the ways in which a neuron responds to the thousands of synapses distributed throughout the dendritic tree. Ca^{2+} and K^+ channels are located in most dendritic trees, but it is not clear which types. Also, some dendritic trees appear to have Na^+ channels and are capable of propagating all-or-none action potentials while others are not. Thus, dendritic trees are clearly not passive but contain (at least) several different types of voltage-gated ion channels. In conclusion, there is a well-recognized heterogeneous distribution of different types of ion channels in a neuron, at least in terms of the axon vs. the rest of the cell. It would be surprising if there weren't also a heterogeneous distribution of both types and densities of ion channels in soma, dendrites, and synaptic terminals. The distribution of these channel types will have important functional implications for dendritic intregration of synaptic inputs as well as for neurotransmitter release at the synapse.

7.3 Propagation of the action potential in myelinated axons

In chapter 6 we discussed the local circuit theory for the propagation of the action potential and derived equations for the conduction velocity of the action potential in unmyelinated axons. A simplified diagram of the local circuits for the propagation of an action potential in unmyelinated axons is shown in figure 7.2. In the case of myelinated axons, conduction velocity depends critically on the myelination. For unmyelinated axons recall from chapter 6 that conduction velocity is proportional to the square root of the fiber diameter, or $\theta \propto \sqrt{d}$.

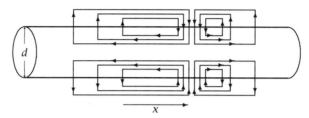

Figure 7.2 Local-circuit currents (rectangular arrows) established along an unmyelinated axon when an action potential is initiated. The direction of propagation is indicated by the arrow at the bottom, and the site of the action potential is at the location of inward current.

In figure 7.3 a myelinated axon is schematized to illustrate the parameters d (diameter of axon), D (diameter of axon + myelin), l (length of internodal region), and n (length of node). Because the myelin acts as an insulator, it reduces transmembrane resistive and capacitive currents in the internodal region.[1] An action potential generated at one node therefore causes a local-circuit current to flow only from one node to the next until threshold is reached at the next node. This causes the action potential to jump literally between nodes and is called *saltatory* conduction (saltare means "to leap" in Latin). Given this different mode of propagation in myelinated axons, what parameters determine conduction velocity?

If we assume that d/D = constant (this is approximately true, although the constant (around 0.6) may be different for peripheral and central fibers), then from previous discussions of conduction velocity, θ will depend on d and thus on D. Because the action potential propagates only from node to node, velocity must also be dependent on the distance between nodes or l. In order for propagation to occur from node to node, the local-circuit current must be sufficient to reach threshold at each node. This means that the distance between nodes must not be too large. The maximum length of the internodal region in order for threshold to be

---

[1]Each layer of myelin adds resistance and capacitance in series with the transmembrane resistance and capacitance of the axon. Recall from appendix A that resistors in series sum while capacitors in series sum as reciprocals. This leads to an effective increase in transmembrane resistance but a *decrease* in transmembrane capacitance in the internodal region.

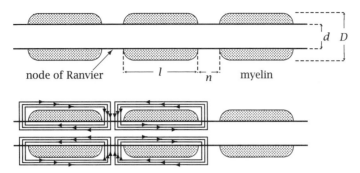

Figure 7.3 Schematic diagram of a myelinated axon (above) and the local-circuit currents (below) during the propagation of an action potential (see text for explanation of symbols).

reached from an action potential in an adjacent node will depend on the diameter D or $D \propto l$. This makes sense because with a larger D comes a larger d as well as more layers of myelin—each of which reduces the loss of current between nodes. l can therefore be longer and still allow for propagation of the action potential from one node to the next. Experimentally this has been shown to be approximately true, that is, a plot of D vs. l for many different fibers is close to a straight line.

For myelinated axons the time for an action potential to travel a given distance is directly proportional to l. For example, think of a 1 cm length of unmyelinated axon. Take that same 1 cm length of axon and break it up into 10 pieces and put 1 cm of an internodal region in between each of those 10 segments (or nodes). The time it takes an action potential to travel from one end of this fictitious myelinated axon is roughly the same as that for the intact unmyelinated axon (the length of the unmyelinated portion has not changed), except that now the entire axon is about 10 times longer. Therefore the conduction velocity increases by a factor of 10 (the ratio of internodal length to node length, or l/n). Since $D \propto l$, then $\theta \propto D$ for myelinated axons. More thorough treatments of this subject can be found in Rushton (1951) and in Jack, Noble, and Tsien (1975).

7.4 Properties of different membrane currents

Please refer to table 7.1 (page 208) while reading this section.

In the sections that follow (adapted from Brown et al. 1990), we will describe the different types of voltage-gated ionic currents in nerve membranes. The different types will be distinguished on the basis of ion selectivity, electrophysiological properties—primarily from voltage-clamp (or whole-cell) measurements—and pharmacology (blockers). It should be stressed that within each class or family of channel (for example, Na^+ channels or A-type K^+ channels), there may be large numbers of genes expressing different channels (different *at least* on the basis of sequence). The functions of the different channels within each family are not fully understood, and thus it is possible that each of the classes will eventually be further subdivided based on channel function, location of a channel within a neuron, location of a channel within the brain, or developmental sequence of channel expression.

7.4.1 Sodium currents $I_{Na(\text{fast})}$ and $I_{Na(\text{slow})}$

Two functionally distinct Na^+ currents have been recorded in cortical neurons. Both currents are sensitive to TTX, but they differ in their inactivation properties. They are called $I_{Na(\text{fast})}$ and $I_{Na(\text{slow})}$. $I_{Na(\text{fast})}$ is essentially the Na^+ current, described in previous chapters, that is responsible for the action potential. $I_{Na(\text{slow})}$, on the other hand, is a non- or slowly inactivating Na^+ current that plays a role in repetitive firing and subthreshold behavior of the cell. $I_{Na(\text{slow})}$ may represent a separate type of channel or a separate gating mode of the same channel. It is also possible to have a subthreshold or noninactivating Na^+ current from $I_{Na(\text{fast})}$ if the m_∞ and h_∞ curves are shifted toward each other along the voltage axis leading to a significant degree of overlap of the two curves. $I_{Na(\text{slow})}$ could therefore be called a *window current* and is illustrated in figure 7.4.

The idea behind a window current is that within the voltage region of the overlap, both the m_∞ and h_∞ parameters have nonzero values. Remember that m_∞ is just the value of the activation variable after a long period of time. If one steps the membrane potential into this region of overlap of the two curves, for example, to -50 mV in figure 7.4, the value of m_∞ is about 0.2 and that of h_∞ is also about 0.2. Using the equation for Na^+ conductance presented in the last chapter,

$$g_{Na} = m^3 h \cdot \overline{g}_{Na},$$

then

$$g_{Na} = (0.2)^3 (0.2) \cdot \overline{g}_{Na} \simeq 0.002 \overline{g}_{Na},$$

a small but finite conductance. A similar calculation at -40 mV gives a steady-state g_{Na} of about $0.006\,\overline{g}_{Na}$, while at -70 mV it gives about 0.0001.

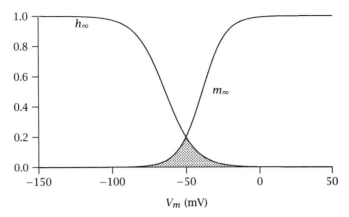

Figure 7.4   Overlap of m_∞ and h_∞ curves (shaded area) produces a "window" current (see text for explanation).

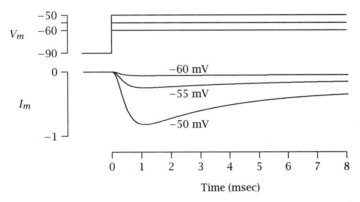

Figure 7.5   Voltage-clamp measurements of $I_{Na(slow)}$. Note that there is very little inactivation at -55 and -60 mV. V_m is in mV, and I_m is in nA.

Negative to about -70 mV the steady-state activation parameter is too small to give much conductance, while at potentials above about -30 mV the steady-state inactivation parameter is too small. Within the region of overlap, however, a significant amount of steady-state Na^+ conductance could be present to produce this window current.

A voltage-clamp experiment in which a slow Na^+ current is revealed is illustrated in figure 7.5. Voltage steps to membrane potentials in the range of -60 to -50 mV elicit inward currents that do not fully inactivate. Steps to more positive potentials, however, produce inward currents that do inactivate completely, as was seen for the fast Na^+ current in the last chapter. Another type of voltage-clamp experiment in which a voltage ramp is used in place of a step is shown in figure 7.6. The use of a ramp command can be helpful: It is possible to construct a complete *I-V* curve from one ramp command. The slope of the ramp, however, is critical because the resulting current from any particular slope will depend on the activation and inactivation kinetics of the currents. In the case of a slow Na$^+$ current, a reasonably slow ramp command (i.e., 10 mV/sec) will reveal a region of negative slope in the resulting *I-V* curve. This area of negative slope simply means that a slowly inactivating inward current is present in this voltage region that will tend to depolarize the neuron. If a slower ramp (i.e., 1 mV/sec) were given to the neuron, then the ramp would only elicit an outward K$^+$ current, because the Na$^+$ current would be mostly inactivated by the slow ramp. A fast ramp, however, would activate the fast Na$^+$ current just as would a step command. This is illustrated in figure 7.6.

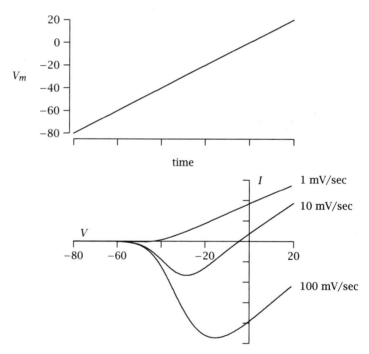

Figure 7.6   Ramp voltage clamp of a neuron with $I_{Na(slow)}$, $I_{Na(fast)}$, and $I_{K(DR)}$. A slow ramp command (1 mV/sec) will activate only outward current (i.e., $I_{K(DR)}$). An intermediate ramp (10 mV/sec) will reveal a region of negative slope due to the slow inactivation of $I_{Na(slow)}$ in this voltage range, while a faster ramp (100 mV/sec) will activate $I_{Na(fast)}$ producing a much larger inward current (see text for further explanation). *V* is in mV, *I* is in nA.

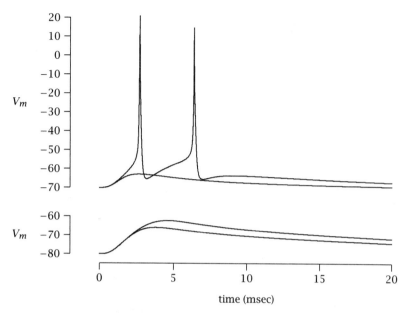

Figure 7.7 Amplification of EPSPs by $I_{Na(slow)}$. EPSPs are shown with and without $I_{Na(slow)}$ present in the neuron. The bottom set of traces are with the neuron hyperpolarized from rest. With $I_{Na(slow)}$ present the EPSP appears larger. At the normal resting potential (upper traces) the presence of $I_{Na(slow)}$ allows the EPSP to trigger APs. V_m is in mV.

What would the function of a slow Na^+ current be? Because the slow Na^+ current is activated near resting potentials, it would tend to amplify small depolarizations. For example, a small EPSP might activate this current and produce a larger depolarization than would be obtained with the EPSP alone. An example of this is shown in figure 7.7. Another possible function of this current might be to sustain repetitive firing in a neuron or to help initiate a burst of action potentials.

As mentioned in the introduction, the location of voltage-dependent Na^+ currents in a neuron is not known precisely. Na^+ currents are certainly present in the axon and soma of most neurons. Some neurons also have Na^+ currents in their presynaptic terminals and in at least parts of their dendritic trees. For example, no Na^+ currents appear to be present in cerebellar Purkinje cell dendrites, while in hippocampal pyramidal neurons, Na^+ currents have been found in proximal areas of their dendritic trees. Na^+ channels are also phosphorylated by several protein kinases including protein kinase C and the cAMP-dependent kinase, PKA.

7.4.2 Calcium currents

Ca^{2+} currents have some similarities and some differences from Na^+ currents. They are similar in that both are activated by depolarization, both are inward currents at normal membrane potentials, and both have some degree of inactivation, depending on the type of Ca^{2+} or Na^+ current being measured. They differ in that Ca^{2+} channels are selective for Ca^{2+} while Na^+ channels are selective for Na^+. Also, the circuit model for ionic current used for describing Na^+ (and K^+) currents does not hold for Ca^{2+} currents. The general equation for ionic current

$$I_i = g_i(V_m - E_i)$$

will not adequately represent Ca^{2+} currents. The internal Ca^{2+} concentration of neurons is exceedingly low, on the order of 10^{-8} M. The predicted reversal potential of I_{Ca} using the chord conductance model given above and the Nernst equation for Ca^{2+} would therefore be about +150 mV. Because of the low internal concentration of Ca^{2+}, however, there is no real reversal potential for Ca^{2+} currents (figure 7.8). Most reversal potential measurements of Ca^{2+} currents are actually contaminated by outward K^+ currents. One can think of the lack of a reversal potential as being due to the fact that there are insufficient numbers of intracellular free Ca^{2+} ions to carry the current in the outward direction. This is not an entirely accurate description, as it also depends on the mechanisms for ion permeation through the channel. Nevertheless, the steady-state Ca^{2+} current is better represented by the GHK current equation (see chapter 2), which predicts the rectification of the *I-V* curve at depolarized potentials:

$$I_{Ca}(V) = P_{Ca}\frac{4F^2}{RT}V\left(\frac{[Ca^{2+}]_{\text{in}}\exp^{(2VF/RT)} - [Ca^{2+}]_{\text{out}}}{\exp^{(2VF/RT)} - 1}\right), \qquad (7.4.1)$$

where P_{Ca} is the Ca^{2+} permeability.

Several measurement difficulties result because of this rectification. First, extrapolation of the *I-V* curve will grossly underestimate the Ca^{2+} equilibrium potential. Second, the concept of a chord conductance is not very meaningful for Ca^{2+} currents. Third, any outward current measured at extreme positive potentials is probably due to contamination from K^+ currents.

At least four functionally and pharmacologically distinct, broad classes of Ca^{2+} currents have been observed in neurons. These are the L-, T-, N-, and P-type currents, and they will be described below.

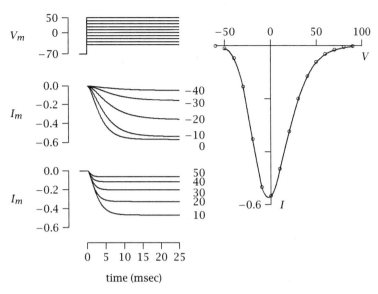

Figure 7.8 *I-V* curve (right) from whole-cell measurements of L-type Ca^{2+} currents (left). V_m is in mV, and I_m is in nA. The numbers to the right of each current trace are the command potentials (in mV) for that trace.

7.4.2.1 High-threshold calcium current $I_{Ca(L)}$**.** The high-threshold (L for Long-lasting) Ca^{2+} current was for many years the main Ca^{2+} current measured from neurons and heart muscle. Although Llinás in 1981 suggested that there might be another Ca^{2+} current operating near rest, it wasn't until 1984 that the work of Carbone and Lux (1984) firmly established the existence of another, low-threshold, Ca^{2+} current. The high-threshold Ca^{2+} current has a number of interesting properties. As the name implies, it is activated at potentials quite depolarized from rest (half maximal activation at around -15 mV); it shows inactivation that is dependent on $[Ca^{2+}]_{in}$ but not voltage; and it is modulated by several different protein kinases. It is blocked by many divalent cations, especially by low concentrations of Cd^{2+}. It is also blocked by dihydropyridines such as nimodipine, but the block is voltage dependent in that the block is greater at more depolarized membrane potentials. The pharmacological tools used for separating the L channel from other Ca^{2+} channels, at least in vertebrate neurons, are the dihydropyridine agonist Bay K8644 and the dihydropyridine antagonists nimodipine or nifedipine. Bay K8644 greatly increases channel open time, resulting in a big increase in total Ca^{2+} current (see figure 7.9). The L channel contributes to Ca^{2+} spikes in neurons, and may also be involved in Ca^{2+} signaling in dendrites. Its single chan-

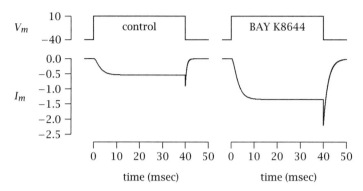

Figure 7.9 Effects of Bay K8644 on high-threshold Ca^{2+} currents. Note the increase in the current with Bay K8644. V_m is in mV, and I_m is in nA.

nel conductance is about 25 pS in isotonic Ba^{2+}, which is about $3 \times$ that measured in normal extracellular $[Ca^{2+}]$.

Using a Hodgkin-Huxley type formulation (see chapter 6) and equation 7.4.1 above, the L-type Ca^{2+} current can be reasonably represented by[2]

$$I_{Ca(L)} = m^2 h_{Ca} \cdot I_{Ca(L)}(V), \tag{7.4.2}$$

where m is the activation variable that is dependent on voltage and time and h_{Ca} is the inactivation variable that is dependent on $[Ca^{2+}]_{in}$ but not voltage or time.

7.4.2.2 Low-threshold calcium current $I_{Ca(T)}$. The low-threshold (T for Transient) type Ca^{2+} current is activated at potentials near rest (half maximal activation around −40 mV). It shows strong voltage-dependent (but not Ca^{2+}-dependent) inactivation at depolarized potentials and thus is also called a transient current. Its activation and inactivation parameters are similar to the fast Na^+ current; it is modulated by muscarinic receptors. The low-threshold Ca^{2+} current appears to be important for spontaneous burst firing of neurons and for subthreshold activity, because it exhibits only partial inactivation near the resting potential. By holding the membrane potential negative to rest and stepping to positive potentials, one can separate the low-threshold from the high-threshold Ca^{2+} current (see figure 7.10). The low-threshold current is insensitive to dihydropyridines and currently known toxins, but is sensitive to Ni^{2+} at relatively

---

[2]We will depart somewhat from the nomenclature of chapter 6 and use the symbols m and h as the activation and inactivation state variables, respectively, for all of the different ionic currents. Their meaning for each of the currents will be explained as needed.

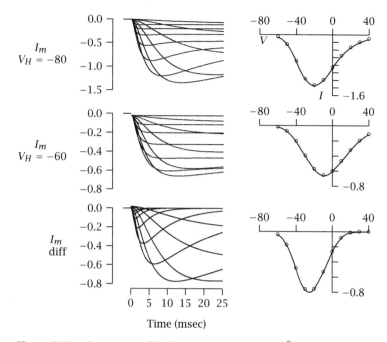

Figure 7.10 Separation of high- and low-threshold Ca^{2+} currents. Voltage steps from a negative holding potential ($V_H = -80$ mV) reveals an inward current that partly inactivates with time. The inactivating portion of this current (the low-threshold Ca^{2+} current) is obtained (lower left panel) by taking the difference between the currents measured from holding potentials of -80 mV and -60 mV. The *I-V* curves represent the measurements made at holding potentials of -80 mV (top right) and -60 mV (middle right), and the difference between the measurements made at -80 and -60 mV (bottom right). V_m is in mV, and I_m is in nA.

low concentrations (10–50 μM). It can be reasonably represented by

$$I_{Ca(T)} = m^2 h_V \cdot I_{Ca(T)}(V), \tag{7.4.3}$$

where m is as above and h_V is dependent on voltage and time but not $[Ca^{2+}]_{in}$. The single channel conductance for the T-type channel is 8–10 pS using either Ba^{2+} or Ca^{2+} as the charge carrier.

The differences in the inactivation properties of low- and high-threshold Ca^{2+} currents are illustrated in figure 7.11.

7.4.2.3 High-threshold Ca^{2+} current $I_{Ca(N)}$. Another Ca^{2+} current, which has several properties that are intermediate between the L and T currents, has been described in a number of different neurons. It is called the N-type current, for Neither L nor T, and has a single channel conductance of 12–16 pS in isotonic Ba^{2+} (about 2.5 $\times$ higher than in

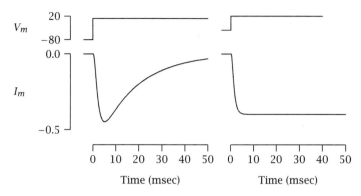

Figure 7.11   Differences in the inactivation properties of low- (left) and high-threshold (right) Ca^{2+} currents. Note that the lack of inactivation shown for the high-threshold, or L-type, Ca^{2+} current occurs only under conditions where the $[Ca^{2+}]_{in}$ is well buffered. Otherwise the high-threshold Ca^{2+} current will display Ca^{2+}-dependent inactivation. V_m is in mV, and I_m is in nA.

normal Ca^{2+}). The N-type current is activated at potentials intermediate between those for the T and L channels (half maximal at around -25 mV), and it shows slower voltage-dependent inactivation than the T current. $I_{Ca(N)}$ also shows some Ca^{2+}-dependent inactivation and is blocked by ω-conotoxin-GVIA, but it is insensitive to dihydropyridines. It is modulated by many neurotransmitters. The function of this channel is unknown, but it may be partly responsible for transmitter release at nerve terminals. It can be represented by

$$I_{Ca(N)} = m^2 h_V h_{Ca} \cdot I_{Ca(N)}(V), \tag{7.4.4}$$

where the m, h_V, and h_{Ca} variables are as defined for the previous two equations.

7.4.2.4   P-type calcium current $I_{Ca(P)}$ Another type of high-threshold Ca^{2+} current that was first described in cerebellar Purkinje neurons is called the P channel (for Purkinje). At the present time less is known about the P-type Ca^{2+} current than the other three types. The P-type current has a single-channel conductance that is similar to the N-type current, but it does not show much inactivation and its main distinguishing feature now appears to be its pharmacological profile. The P-type current is insensitive to dihydropyridines and ω-conotoxin-GVIA, but is blocked by a Funnel web spider venom, ω-Agatoxin-IVA. Evidence from the squid giant synapse and hippocampus suggests that the P-type channel might also be involved in neurotransmission in some cells.

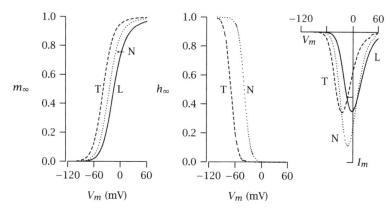

Figure 7.12 *I-V* curves and m_∞ and h_∞ for the L-, T-, and N-type Ca^{2+} currents. The P-type current has activation and inactivation properties similar to that of L.

The cellular distribution of the L, T, N, and P channels is not known. Certainly, some type(s) of Ca^{2+} channels exist in presynaptic terminals. One possibility is that N and P channels are responsible for fast transmitter release (e.g., glutamate and GABA) and L channels are responsible for slower release of neurotransmitters such as catecholamines and neuropeptides. There are also other classes of Ca^{2+} channels, some of which are just beginning to be described (e.g., Q-type and R-type), whose cellular distribution is not yet known. Ca^{2+} spikes have been recorded from isolated dendrites and therefore some type(s) of Ca^{2+} channels exist there as well. It remains to be determined whether transmitter release and/or Ca^{2+} signaling in dendrites can be accounted for by L-, T-, N-, and P-type channels or if there are unique channels in these structures. A comparison of the activation and inactivation properties and *I-V* curves of the L-, T-, and N-type currents is illustrated in figure 7.12.

7.4.3 Potassium currents

The greatest diversity of voltage-gated channels selective for one type of ion may be that for K^+ channels. There are at least four types of voltage-gated K^+ currents in neurons, two types of Ca^{2+}- and voltage-gated K^+ currents, one type of hyperpolarization-gated K^+ current, and at least three "other" types. Functionally, K^+ currents can be thought of as falling into three broad classes: those that contribute to the resting potential; those that are activated at subthreshold potentials (low-threshold K^+ currents); and those that are activated by action potentials and contribute to afterpotentials and repetitive firing (high-threshold K^+ currents). Al-

though the grouping of these many types of K⁺ currents is somewhat
arbitrary, we will follow the usual practice of discussing first the voltage-
gated currents, then the Ca^{2+}- and voltage-gated currents, and then the
other types of K⁺ currents.

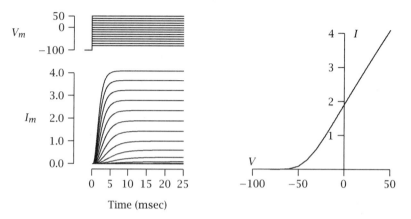

Figure 7.13   Activation of $I_{K(DR)}$ using step commands. A representative *I-V* curve is
shown on the right. V_m is in mV, and I_m is in nA.

7.4.3.1 Voltage-gated K⁺ currents

Delayed-rectifier current $I_{K(DR)}$ This delayed rectifier current is similar to that described previously by Hodgkin and Huxley. In hippocampal
pyramidal neurons, however, it has been reported to have relatively slow
activation, with a time to peak of some 50–100 msec and even slower inactivation. Such a slow activation would make it ill suited to participate
in the repolarization of the AP. Data from dentate granule cells, however,
suggest activation time constants of around 5 msec (figure 7.13), more
in line with the current being involved with AP repolarization. Nevertheless, other K⁺ currents may also take part in the AP. $I_{K(DR)}$ is sensitive to
high concentrations of tetraethylammonium (TEA) and may show a slow
voltage-dependent inactivation. An equation that can describe $I_{K(DR)}$ in
cortical neurons is

$$I_{K(DR)} = m^3 h \overline{g}_{K(DR)} (V_m - E_K),   \tag{7.4.5}$$

where m and h depend on voltage and time.

Transient (A) current $I_{K(A)}$ This transient K⁺ current is prominent in
cortical neurons. It activates within 5–10 msec, inactivates with a time

constant of 20–30 msec, and can usually be separated from $I_{K(DR)}$ by subtraction. It is insensitive to TEA, but is blocked by 4-aminopyridine (4-AP) at $> 100\mu M$ concentrations. This current likely participates in spike repolarization and contributes to the resting potential (note the overlap of the m_∞ and h_∞ curves near rest in figure 7.14). Because this current activates rapidly and then inactivates, it prevents neurons from responding to fast depolarizations (see figure 7.22). An equation to describe $I_{K(A)}$ is

$$I_{K(A)} = mh\overline{g}_{K(A)}(V_m - E_K),\tag{7.4.6}$$

where m and h depend on voltage and time.

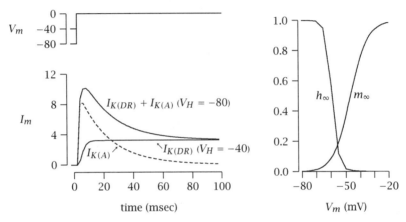

Figure 7.14 The properties of $I_{K(A)}$ and the separation of $I_{K(A)}$ from $I_{K(DR)}$ using different holding potentials are indicated. The activation and inactivation curves for $I_{K(A)}$ are shown on the right. V_m is in mV, and I_m is in nA. (Adapted from Connor and Stevens 1971b.)

Slowly inactivating "delay" current $I_{K(D)}$ This current is similar to the A-type current described above in that it activates rapidly and inactivates completely. It differs from the A current because 1) it shows slower inactivation (several sec); 2) its activation and inactivation curves are 15–20 mV more negative, and thus it is mostly inactivated at rest; and 3) it is more sensitive to 4-AP ($< 100\ \mu M$) (see figure 7.15). It is also sensitive to dendrotoxin (DTX). Only a small portion of this current is likely to contribute to the resting potential. Both of the transient K$^+$ currents ($I_{K(A)}$ and $I_{K(D)}$) participate in spike repolarization. It has been suggested that an $I_{K(D)}$-like current is present in presynaptic terminals because in many preparations transmitter release is sensitive to very low concentrations of 4-AP ($\sim 10\ \mu M$).

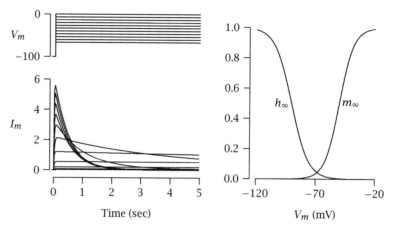

Figure 7.15   Properties of $I_{K(D)}$. Note the slower inactivation compared to that of $I_{K(A)}$. V_m is in mV, and I_m is in nA.

Muscarine-sensitive current or M current $I_{K(M)}$ The M current is a true voltage-gated K^+ current, although it was first discovered because of its sensitivity to agonists for muscarinic cholinergic receptors. $I_{K(M)}$ is a non-inactivating, subthreshold K^+ current that is blocked by activation of muscarinic receptors. It is slowly activated at potentials depolarized to rest (τ_{act} about 50 msec). It contributes to spike accommodation, repetitive firing, and a medium duration after hyperpolarization (see figure 7.22). It is blocked by Ba^{2+} and may be sensitive to several other neurotransmitters through direct G-protein coupling (see figures 7.16 and 7.17). An equation to describe $I_{K(M)}$ is

$$I_{K(M)} = m\overline{g}_{K(M)}(V_m - E_K), \tag{7.4.7}$$

where m is dependent on voltage and time.

Voltage-gated (hyperpolarization) currents   There are a number of membrane currents that are activated by hyperpolarization and undergo no inactivation. These include a K^+-selective current ($I_{K(IR)}$, or inward rectifier), a nonselective monovalent cation current I_Q (also called I_f and I_h in some preparations), and a Cl^--selective current $I_{Cl(V)}$. The inward rectifier is so named because it passes K^+ ions in the inward direction better than in the outward direction. It has first-order activation kinetics, and its voltage range for activation shifts as a function of $[K^+]_{out}$ (see figure 7.18). The voltage dependency of these hyperpolarization activated currents is due to a block by Mg^{2+} ions from the inside of the neuron,

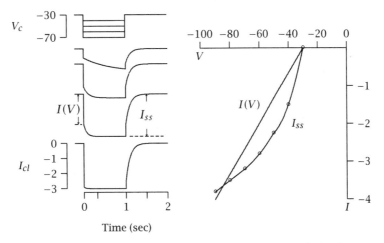

Figure 7.16 Properties of $I_{K(M)}$. This current is measured by holding at positive potentials ($\sim$–30 mV) to fully activate $I_{K(M)}$ and then stepping negative. $I_{K(M)}$ deactivates during each command, causing a relaxation of the clamp current (I_{cl}). The kinetics of $I_{K(M)}$ can be determined in this fashion. Also, when $I(V)$ is plotted vs. the different step commands (V_c), the total membrane conductance when $I_{K(M)}$ is activated can be determined. This consists of the leakage conductance plus that associated with $I_{K(M)}$. I_{ss} represents the total membrane current when $I_{K(M)}$ has been turned off by the steps. The difference between $I(V)$ and I_{ss} is a measure of $I_{K(M)}$, and their intersection is the reversal potential for $I_{K(M)}$. V is in mV, and I is in nA. (After Brown and Adams 1980.)

which is removed with hyperpolarization. The function of these currents is not clear, although they may contribute partly to the resting potential and also $I_{Cl(V)}$ has been proposed to affect the passive properties of dendrites. It is important to recognize that because of these conductances, hyperpolarization of a neuron will frequently result in a *decrease* in the input resistance and thus a rectification of the *V-I* and *I-V* relationships for that neuron (see figure 7.19).

7.4.3.2 Calcium-gated K$^+$ currents There are at least two important Ca^{2+}-gated K$^+$ currents for which information is available. We will discuss them below. There are also at least two Ca^{2+}-gated "other" currents consisting of a Ca^{2+}-gated Cl$^-$ current $I_{Cl(Ca)}$ and a Ca^{2+}-gated nonselective cation current. These will not be discussed further. One problem with determining the time course of these Ca^{2+}-gated currents is that their decays depend on the time course of changes in intracellular [Ca^{2+}], which is not uniform in different parts of a neuron, and on the location of the channels with respect to the sites of Ca^{2+} entry. For example, a change in intracellular [Ca^{2+}] will decay slower in the soma than in the dendrites because of a larger surface to volume ratio, and therefore the time course of a given

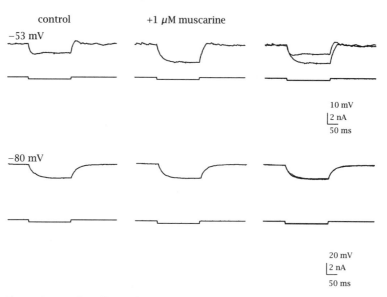

control   +1 μM muscarine
−53 mV

10 mV
2 nA
50 ms

−80 mV

20 mV
2 nA
50 ms

Figure 7.17 The effects of muscarine on membrane properties measured under current-clamp conditions. At potentials depolarized to rest (top sets of traces, upper voltage, lower current), a hyperpolarizing current step turns off $I_{K(M)}$, resulting in a depolarization during the step. This is blocked by muscarine. With prior hyperpolarization (bottom sets of traces, upper voltage, lower current), however, there is no effect of $I_{K(M)}$ (or of muscarine) during the step because $I_{K(M)}$ is already deactivated. The traces with and without muscarine are superimposed on the right. (From Williams and Johnston 1990.)

Ca^{2+}-gated current might also be longer in the soma. The characteristics of a Ca^{2+}-gated current observed from the soma depends critically on the site of origin of the current.

Fast, calcium-gated potassium current $I_{K(C)}$ This is a large current that is activated rapidly (within 1–2 msec) by the combination of Ca^{2+} *and* depolarization (see figure 7.20). It deactivates in 50–150 msec, depending on the membrane potential, and is very sensitive to TEA. In some neurons it is also blocked by charybdotoxin, a peptide from scorpion venom. These channels are likely to be very close to Ca^{2+} channels and probably participate in spike repolarization. They also play a role in repetitive firing. An equation to describe $I_{K(C)}$ is

$$I_{K(C)} = m_{(V+Ca)}\overline{g}_{K(C)}(V_m - E_K), \tag{7.4.8}$$

where the activation variable, $m_{(V+Ca)}$, is dependent on voltage, time, and $[Ca^{2+}]_{in}$.

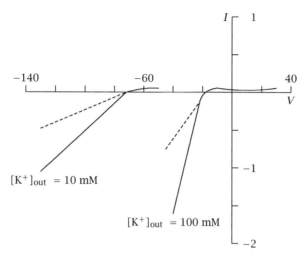

Figure 7.18 Properties of $I_{K(IR)}$ at different concentrations of extracellular K$^+$. The dashed line represents the membrane I-V curve in the absence of $I_{K(IR)}$. V is in mV, and I is in nA. (After Hagiwara et al. 1976.)

After-hyperpolarization current $I_{K(AHP)}$ $I_{K(AHP)}$ is a smaller current that is activated more slowly following Ca^{2+} entry than is $I_{K(C)}$. The $I_{K(AHP)}$ channels are therefore probably more remote from Ca^{2+} channels than are the $I_{K(C)}$ channels. $I_{K(AHP)}$ is also relatively voltage insensitive and is not blocked by TEA. $I_{K(AHP)}$ generates a long, slow hyperpolarization follow-ing a single action potential, but this slow (s)AHP is even more prominent after a train of APs (see figures 7.21 and 7.22). This current plays an im-portant role in repetitive firing and spike accommodation, is modulated by a number of neurotransmitters, and is blocked in some neurons by the bee venom apamin. One suggestion for the slow activation kinetics of this current is that the channels might be located at some distance from the site of Ca^{2+} entry. In contrast, the channels responsible for $I_{K(C)}$ are likely to be quite close to Ca^{2+} channels. $I_{K(AHP)}$ can be described as

$$I_{K(AHP)} = m_{Ca}\bar{g}_{K(AHP)}(V_m - E_K),\qquad(7.4.9)$$

where m_{Ca} is dependent on [Ca^{2+}]$_{in}$ only.

7.4.3.3 Other K$^+$ currents The most prominent and important of the other currents is, of course, the leak current. The leak current is de-fined as a linear, voltage-independent current. As mentioned earlier, K$^+$ leak channels are particularly important for spike repolarization at the nodes of Ranvier. There may also be other leak channels that are selec-

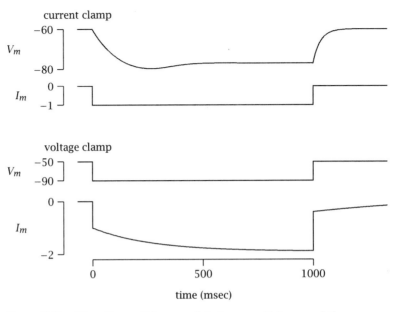

Figure 7.19   The effects of $I_{K(IR)}$ and I_Q in current clamp and the measurement of a hyperpolarization current in voltage clamp. A hyperpolarizing step of current activates $I_{K(IR)}$ and I_Q and decreases the effective input resistance in a time-dependent manner. This results in the "sag" or "droop" of the membrane response. In voltage clamp a hyperpolarizing step command elicits a time-dependent increase in an inward current due to $I_{K(IR)}$ and I_Q. V_m is in mV, and I_m is in nA.

tive for other ions, for example, Cl⁻. The actual leak current in neurons is probably much smaller than previously assumed, because much of the measured leak current was artificially induced by microelectrodes.

An ATP sensitive K⁺ current may play a role in cortical neurons. This is a channel that is normally closed, but opens when the internal ATP concentration falls, for example during periods of anoxia.

A Na⁺ activated K⁺ current has been described by Schwindt, Spain, and Crill (1989). This current, however, may be at least partly a Ca²⁺-gated K⁺ current, because internal Na⁺ is known to cause the release of intracellular Ca²⁺.

Stretch-activated channels have been observed in a variety of preparations. Although they may not be important for cortical neurons, they probably form the basis of mechanotransduction in sensory receptors.

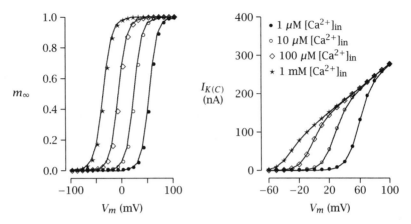

Figure 7.20 Voltage and Ca^{2+} dependence of $I_{K(C)}$. The activation curves at different internal Ca^{2+} concentrations are shown on the left with the resulting I-V curves indicated on the right.

7.4.4 Nonselective cation currents

There are a number of other currents, some activated by hyperpolarization, some by depolarization, and some by increases in intracellular $[Ca^{2+}]_{in}$, that are characterized by their nonselectivity to cations. The properties and functions of these currents are not well understood.

7.5 Functions of different membrane currents

The role of each of the major types of K^+ currents in an action potential is illustrated in figure 7.22. Note that the A- and D-type K^+ currents prevent the neuron from being rapidly depolarized, while A-, D-, C-, and DR-type currents participate in repolarization of the action potential. C-, M-, and AHP-type currents are involved in afterhyperpolarizations and thus affect the repetitive firing properties of a neuron.

Examples of the voltage-clamp responses to a step depolarization resulting from each of the major currents discussed in this chapter are schematized in figure 7.23. Note that no Ca^{2+}-dependent inactivation and only one voltage level is illustrated. The relative differences in activation kinetics of the different K^+ currents are also illustrated. For example, $I_{K(A)}$ activates rapidly and would thus tend to filter out high-frequency changes in membrane potential in the depolarizing direction, while $I_{K(M)}$ activates slowly and would thus tend to filter out low-frequency signals.

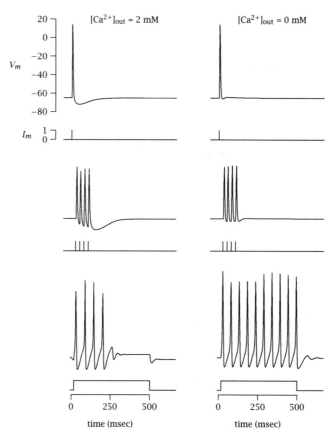

Figure 7.21   Properties of $I_{K(AHP)}$. $I_{K(AHP)}$ affects the firing frequency during a current step (I_m, bottom traces) and produces a slow hyperpolarization after a train of action potentials. $I_{K(AHP)}$ disappears in the absence of extracellular Ca^{2+}. V_m is in mV, and I_m is in nA.

The repetitive firing properties of a cortical neuron are illustrated in figure 7.22. The roles of the different K^+ currents in the firing properties are indicated. Refer also to figure 7.21.

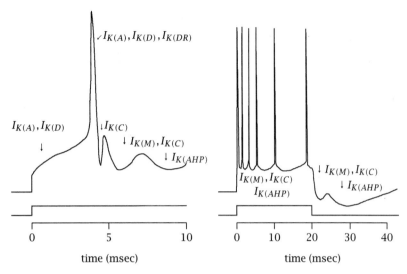

Figure 7.22   Effects of the various K$^+$ currents on different phases of an action potential in a cortical neuron. The current step is at the bottom. (Adapted from Storm 1990.)

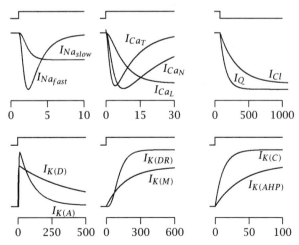

Figure 7.23   Comparison of the time courses of a number of different voltage and Ca^{2+}-dependent currents in hippocampal neurons. The voltage commands used to elicit each current are indicated at the top of each set of traces. (Adapted from Storm 1990.)

Table 7.1 Voltage-gated ionic currents in cortical neurons

Current	Symbol	Ion	V_{th}	Inactivation	Blocked by	Modulation	Function
1. Voltage-gated (depolarization)							
Na$^+$ currents							
Fast	$I_{Na(fast)}$	Na$^+$	−50	Fast	TTX		spike
Slow	$I_{Na(slow)}$	Na$^+$	−65	Slow	TTX		prepotential
Ca^{2+} currents							
High-threshold	$I_{Ca(L)}$	Ca^{2+}	−15	Slow Ca^{2+}-dep	Cd^{2+} DHP	NE (+) ACh (−) ACh (+)	spike
Low-threshold	$I_{Ca(T)}$	Ca^{2+}	−40	Fast V-dep	Ni^{2+}		burst firing
High-threshold	$I_{Ca(N)}$	Ca^{2+}	−25	Medium V & Ca^{2+}-dep	Cd^{2+} ωCTX-GVIA	NE (+,−) Aden. (−) Others (−)	spike (?) presyn. (?)
High-threshold	$I_{Ca(P)}$	Ca^{2+}	−20	Slow	ωAga-IVA		presyn. (?)
K$^+$ currents							
Delayed rectifier	$I_{K(DR)}$	K$^+$	−40	Slow	TEA (10 mM) 4-AP (> 0.1 mM)	ACh (−)	spike repolar.
Transient	$I_{K(A)}$	K$^+$	−60	Fast	4-AP (< 0.1 mM)		
Delay current	$I_{K(D)}$	K$^+$	−75	Slow	4-AP (< 0.1 mM)	DTX	delayed firing, spike repolar.
M current	$I_{K(M)}$	K$^+$	−65	None	Ba^{2+}	ACh (−) 5-HT (−) Somato. (+)	spike train accommod. mAHP

Current	Symbol	Ion	V_{th}	Inactivation	Blocked by	Modulation	Function
2. Voltage-gated (hyperpolarization)							
Slow inward rectifier	I_Q	Na + K	-80	None	Cs$^+$, THA		rest V_m
Fast inward rectifier	$I_{K(IR)}$	K$^+$	-60	Slow	Cs$^+$, Ba^{2+}	G_o (+)	
Time-depend. Cl$^-$ currents	$I_{Cl(V)}$	Cl$^-$ Cl$^-$	-20 -60	None None	Cd^{2+} Cd^{2+}	PBs	dendrites (?)
3. Ca^{2+}-gated							
Fast K$^+$ current	$I_{K(C)}$	K$^+$	-40	None	TEA (1 mM)		spike repolar. *f&m*AHP
Slow K$^+$ current	$I_{K(AHP)}$	K$^+$	None	None	Ba^{2+}	ACh (−) NE (−) 5-HT (−) Hist. (−)	spike train accommod. *s*AHP
Cl$^-$ current	$I_{Cl(Ca)}$	Cl$^-$					
Cation current		Na + K				ACh (+)	AHP (?)
4. Other currents							
Leak (?)	$I_{K(L)}$	K$^+$	None	None	Ba^{2+}	ACh (−)	rest V_m
Cl$^-$	I_{Cl}	Cl$^-$					
Anoxic	$I_{K(ATP)}$	K$^+$					hyperpol.
Na$^+$ Act. K$^+$	$I_{K(Na)}$	K$^+$					
Stretch		Na + K					mechanorec.

7.6 Summary of important concepts

1. Heterogeneous distribution of ion channels.
2. Saltatory conduction.
3. Window current.
4. General description of membrane currents in terms of Hodgkin-Huxley parameters.
5. General features of different membrane conductances.
6. Function of different currents in firing behavior of neuron.

7.7 Homework problems

1. You are recording from a giant axon that contains only one type of voltage-gated ion channel in its membrane (a Na^+ channel). Its resting potential is -100 mV, and the maximum Na^+ conductance (if all channels were open) is 100 nS. The Na^+ equilibrium potential is $+50$ mV.

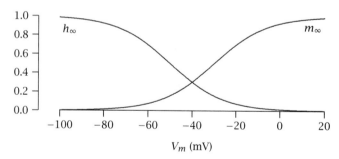

   (a) Using the h_∞ and m_∞ curves provided, calculate and then plot the steady-state *I-V* curve for this axon.

   (b) Describe the types of experiments that might have been used to determine the h_∞ and m_∞ curves.

   (c) What effects would this current have on the properties of the axon in current clamp?

   (d) Plot the new *I-V* curve that would result if the h_∞ and m_∞ curves were each shifted by 20 mV (h_∞ in the hyperpolarizing direction; m_∞ in the depolarizing direction).

2. You are recording from a neuron (assume it is isopotential) with an input resistance (at its resting potential of –80 mV) of 100 MΩ, and you inject a step of 0.5 nA depolarizing current. Plot the approximate membrane potential of this neuron vs. time assuming that the only voltage-gated current is an A-type K$^+$ current. Make a similar plot of the same neuron but with the A-type current blocked with 4-AP. Compare the two graphs.

3. What would you predict the shape of the action potential to be in a hippocampal neuron if you could specifically block $I_{K(C)}$? Draw a picture to illustrate your answer.

4. What would you predict would happen to the repetitive firing ability of a neuron if you could block $I_{K(AHP)}$? Draw a picture to illustrate your answer.

5. (a) Describe the experiments that you would perform, in as much detail as necessary, to distinguish a low-threshold from a high-threshold Ca^{2+} current.

   (b) Given the attached I-V curve for a neuron with a single type of Ca^{2+} channel, calculate the chord conductance at +40 mV, assuming a reversal potential of +80 mV. Is there anything wrong with doing this for a Ca^{2+} channel? Explain.

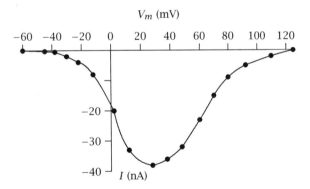

6. The graphs shown below illustrate activation and inactivation curves for three different types of voltage-gated K$^+$ conductances. Assume that E_K = –100 mV and that the conductances follow first-order kinetics for activation and inactivation.

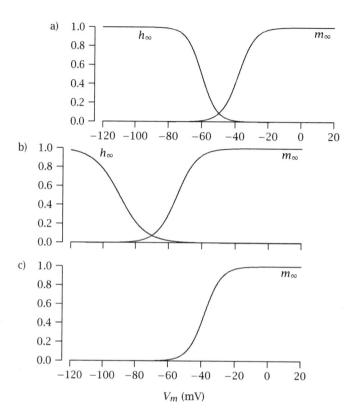

(a) If $\bar{g}_K$ = 100 nS, plot the steady-state I-V curves for the three conductances.

(b) If the time constant for activation is much faster than that for inactivation, draw the peak I-V curve for the three conductances.

(c) Which of the known K$^+$ conductances matches most closely the three conductances shown here?

(d) A completely different voltage-gated conductance happens to have an identical activation curve as that shown in (c). This is a Cl$^-$ conductance. Draw the I-V curve for this conductance, assuming E_{Cl} = -40 mV and $\bar{g}_{Cl}$ = 50 nS.

7. You are voltage clamping an isopotential neuron in which the total membrane current is given by the following equation:

$$I_m = g_L(V_m - E_L) + \bar{g}_x mh(V_m - E_x).$$

(a) Outline the experiments you would do to determine g_L, E_L, $\bar{g}_x$, E_x, m_∞, h_∞. Be specific and provide as much detail as necessary.

(b) If $E_L = -70$ mV, $g_L = 10$ nS, and $\bar{g}_x$ is the muscarine-sensitive K^+ conductance with $\bar{g}_x = 20$ nS and $E_x = -100$ mV, draw the expected current responses for step commands from -100 mV to 0 mV for 1 sec and back to -100 mV, and then in a separate experiment from 0 mV to -50 mV for 1 sec and back to 0 mV. Be precise in your drawings by indicating the amplitudes of the expected current responses and their approximate time courses. Assume that m_∞ is 1 at 0 mV and 0.5 at -50 mV. Also, assume first-order kinetics with a time constant of 100 msec.

(c) Show the results from an identical experiment to that above but with an ED_{50} concentration of muscarine in the bath.

8. (a) Using the figure for problem 1, plot the steady-state I-V curve for an ionic current (I_x) in which $\bar{g}_x = 50$ nS and $E_x = +50$ mV.

(b) If, under voltage clamp, one steps from -80 mV to -40 mV, plot the instantaneous I-V curve (using tail currents) one would obtain upon stepping from -40 mV back to different potentials.

(c) Again, using the figure for problem 1, plot another steady-state I-V curve for an ionic current (I_y) in which $\bar{g}_y = 20$ nS and $E_y = -100$ mV.

(d) Repeat (b) but for I_y.

Assume that I_x and I_y follow first-order kinetics.

9. Discuss briefly the functional differences between $I_{Na(\text{fast})}$ and $I_{Na(\text{slow})}$. Include in your answer how each is measured and how each might influence the firing behavior of a neuron.

10. Equations for three different Ca^{2+} currents were given in the text as

$$I_{Ca_1} = m^2 h_v h_{Ca} I_{Ca_1}(V),$$
$$I_{Ca_2} = m^2 h_v I_{Ca_2}(V),$$
$$I_{Ca_3} = m^2 h_{Ca} I_{Ca_3}(V).$$

Discuss briefly each term in each of the above equations. How are each of the currents likely to be different from each other? Based on your knowledge of Ca^{2+} currents, what are the common names of the above Ca^{2+} currents and what are their likely functions in a cell?

11. Given the activation curve shown and its equation of

$$I_x = m\overline{g}_x(V_m - E_x),$$

what kind of membrane current is this likely to be? Give as many examples of real currents with this sort of activation curve as you can. What are some possible functions for a current such as this? Assume a resting potential of −60 mV.

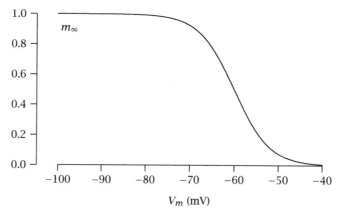

12. Which membrane currents participate in the different phases of the waveform illustrated here?

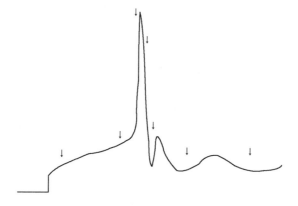

8   Molecular Structure and Unitary Currents of Ion Channels

8.1   Introduction

In previous chapters we described the properties of ion conductances responsible for membrane permeabilities at resting states and during excitation. Ion conductances are mediated by populations of individual channels, and the gross membrane currents reflect the behaviors of individual single channels in the plasma membrane. Therefore, it is extremely important to study the properties of single ion channels in excitable membranes. Ion channels are transmembrane macromolecular pores that allow ions to diffuse down their electrochemical gradients. It has historically been difficult to study the structure and function of single ion channels because of their complex molecular structure and the tiny amount of current flowing across each channel. Two technical breakthroughs in recent years have greatly enhanced the speed of research into single ion channels: the recombinant DNA and patch-clamp techniques. The former helps to determine the molecular structure of single ion channels, whereas the latter enables the direct measurement of ion current flowing through single open channels.

It is beyond the scope of this book to provide a complete account of the structure and function of all ion channels known to date. Readers interested in this topic should consult Hille (1992). In this chapter, we will briefly describe the structure of two representative channels, the Na^+ channel (voltage-gated) and the nicotinic acetylcholine channel (ligand-gated), so that readers can acquire some general ideas of the molecular configurations of channel molecules. Additionally, we will give examples of typical single-channel currents (unitary current) carried by both voltage-gated (Na^+, K^+, and Ca^{2+}) and ligand-gated (ACh- and GABA-gated) channels. The kinetics and functional characteristics of the unitary currents will be described briefly.

8.2   Molecular structure of ion channels

As described earlier, there are two types of ion channels. One is gated by transmembrane voltage (voltage-gated channels) and the other is gated by intracellular or extracellular ligands (ligand-gated channels). A single channel is a protein molecule that resides in the lipid bilayer and forms an aqueous pore that allows ions to pass through. A typical example of the voltage-gated channels is the Na^+ channel. In the electric eel, the Na^+ channel protein is a single large polypeptide with a molecular weight of about 260 kD. In the rat brain, the Na^+ channel consists of three subunits: a large α subunit similar to the 260 kD peptide in the electric eel, and two smaller subunits, β_1 and β_2, with molecular weights of 39 kD and 37 kD respectively (figure 8.1A). The voltage-gated Na^+ channel is formed by the α subunit, whereas the β subunits are believed to be involved in regulating channel kinetics and stability. The α subunit is a 1820 (electric eel) and a 2005 to 2009 (rat brain) amino acid protein containing four homologous domains, each of which has six (S1–S6) transmembrane segments (figure 8.1B). S1 and S3 have some negatively charged residues, whereas S5 and S6 are nonpolar. The S2 segments contain negative charges and are believed to line the walls of the pore permeable to cations (figure 8.1C). S4 is highly conserved and has an amphipathic structure in which every third amino acid is either a lysine or an arginine, the most positively charged amino acid residues. This segment is believed to be the voltage sensor of the Na^+ channel. Catterall (1986, 1992) proposed a *sliding helix* model for the S4 segment in response to changes in transmembrane voltage (figure 8.1D). This model suggests that the S4 segments respond to membrane depolarization by rotating 60° and moving outward by about 5 Å. Since each positively charged residue in S4 is about 5 Å apart from another positively charged residue, the rotating movement results in translocation of charge equivalent to moving one charge across the whole membrane. Catterall also suggests that the S4 segments of the III and IV domains move two steps (10 Å) outward, whereas S4 of the I and II domains moves one step. The total charge movement is six charges per channel, a number that agrees perfectly with the conclusion from the gating current data (chapter 6).

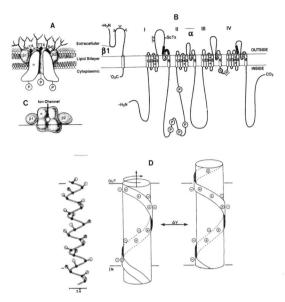

Figure 8.1 Subunit structure of brain Na$^+$ channel. (A) View of a cross section of a hypothetical Na$^+$ channel consisting of a single transmembrane α-subunit of 260 kD in association with a β1-subunit of 36 kD and a β2-subunit of 33 kD. The β1-subunit is associated noncovalently, whereas the β2-subunit is linked through disulfide bonds. All three subunits are heavily glycosylated on their extracellular surfaces, and α-subunit has receptor sites for the α-scorpion toxins (ScTx) and tetrodotoxin (TTX). The intracellular surface of the α-subunit is phosphorylated by multiple protein kinases (P). (B) Primary structures of α- and β1-subunits of a Na$^+$ channel illustrated as transmembrane folding diagrams. The bold line represents polypeptide chains of α- and β1-subunits; the length of each segment is approximately proportional to its true length in rat brain Na$^+$ channel. Cylinders represent probable transmembrane α helices. Other segments that are probably membrane associated are drawn as loops in extended conformation like the remainder of the sequence. Sites of experimentally demonstrated glycosylation, ψ; cAMP-dependent phosphorylation, P in a circle; protein kinase C phosphorylation, P in a diamond; amino acid residues required for tetrodotoxin binding (small circles with +, −, or open fields depict positively charged [Lys^{1422}], negatively charged, or neutral [Ala^{1714}] residues, respectively); amino acid residues that form inactivation particles, h in a circle. (C) View of Na$^+$ channel from extracellular side illustrating formation of transmembrane pore in center of α-subunit. (D) Sliding helix model of voltage-dependent activation. Left: ball-and-stick, 3-dimensional representation of S4 helix of domain IV of Na$^+$ channel. Darkened circles represent α-carbon of each amino acid residue. Open circles specify amino acids in single letter code and show direction of projection of side chain away from core of helix. Positively charged amino acids are indicated in bold letters. Right: movement of S4 helix in response to depolarization [change in voltage (ΔV)]. Transmembrane helix S4 is represented as a cylinder with a spiral ribbon of positive charge. At resting membrane potential (left), all positively charged amino acid residues are paired with fixed negative charges on other transmembrane segments of channel, and the S4 segment is maintained in that position by the force of negative internal membrane potential. Depolarization reduces the force, holding positive charges in their inward position. The S4 helix is then proposed to undergo a spiral motion through a rotation of ∼ 60° and outward displacement of ∼ 5 Å. This movement leaves an unpaired negative charge on the inner surface of membrane and reveals an unpaired positive charge on the external surface of membrane to give a net gating charge transfer (Δ Q) of +1. (From Catterall 1992.)

A typical example of a ligand-gated channel is the nicotinic acetylcholine receptor channel (ACh channel). It is a 2305 amino acid protein containing five subunits. The probable arrangement of the subunits in the membrane is shown in figure 8.2.

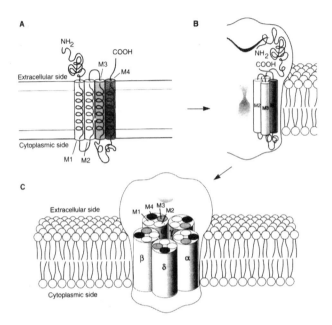

Figure 8.2   Subunit arrangement of the ACh receptor channel. (A) The four membrane-spanning components of one subunit (labeled M1–M4). (B) A hypothetical folding arrangement for one subunit in the channel with the M2 segment facing the channel. (C) A hypothetical arrangement of the five subunits forming an aqueous channel, with the M2 segment always on the inside forming the lining of the channel. (From Kandel et al. 1991.)

The five subunits of the ACh channel are made of four different poly-peptides—α, β, γ, and δ—each of which is a glycoprotein of about 55 kD that transverses the membrane. The ACh channel consists of two α sub-units and one each of β, γ, and δ subunits, with a total molecular weight of about 275,000. The five subunits are arranged symmetrically about a central transmembrane pore. The diameter of the pore is about 65 Å on the extracellular side and about 20 Å on the cytoplasmic side. Ion selectivity experiments suggest that a portion of this ion passage pore is as narrow as 5–10 Å in diameter.

The binding sites of ACh are believed to be located in the N-terminal domain (near cysteine 192) of the α subunits. Two ACh molecules (one for each α subunit) must bind to each ACh receptor to fully open the channel.

8.3   Patch-clamp records of single-channel currents

In order to record the current flow through a single ion channel, Neher and Sakmann (1976) developed the patch-clamp technique, which involves sealing a small area of membrane with polished glass electrodes and allowing voltage clamping of that small sealed area of membrane so that current flow through individual channels can be measured directly. If the sealed area is small enough and the channel density is low enough, one can obtain a sealed patch of membrane that contains only one active ion channel, and the opening and closing of this channel can be directly monitored by measuring the unitary current. The circuitry of the patch-clamp recording system is given in appendix A. The various configurations of patch clamping are given in figure 8.3. The four recording configurations are: cell-attached, whole-cell recording, outside-out patch, and inside-out patch. The upper left diagram is the configuration of a pipette in simple mechanical contact with a cell, as used for single-channel recording (Neher et al. 1978). With slight suction, the seal between the membrane and the pipette increases in resistance by two to three orders of magnitude, forming a cell-attached patch. This leads to two different cell-free recording configurations (the outside-out and inside-out patches). Voltage-clamp currents from whole cells can be recorded after disruption of the patch membrane (Hamill et al. 1981).

The perforated-patch technique (figure 8.4) uses nystatin, a pore-forming antibiotic, in the patch electrode to provide electrical access to the interior of the cell. The nystatin pore allows permeation only of small monovalent cations, with the advantage that intracellular modulatory molecules are not washed out. Specifically, washout of Ca^{2+} currents is a problem common in whole-cell recordings in which the patch of membrane under the electrode is ruptured and mixing of intracellular and electrode solutions occurs. The use of nystatin eliminates this problem and allows long-duration recording of Ca^{2+} currents in small cells.

Unitary currents and ensemble averages of Na^+, K^+, and Ca^{2+} currents are shown in figures 8.5 and 8.6, respectively. Each of these channels opens and closes abruptly and randomly, with higher opening probability when the membrane is depolarized. Na^+ channels open briefly with short delays after the onset of the depolarizing voltage step, and the openings are clustered early in each trace. The ensemble average of many of these traces gives a smooth, transient inward current that resembles the macro-

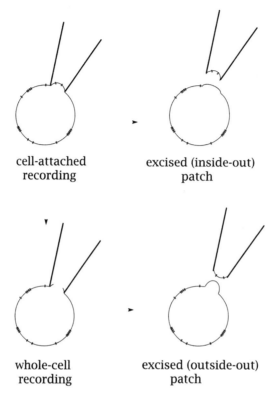

cell-attached   excised (inside-out)
recording   patch

whole-cell   excised (outside-out)
recording   patch

Figure 8.3   Configurations of patch clamping.

scopic (whole-cell) current described by Hodgkin and Huxley (see chapter 6). The K^+ channels open with longer delay and durations, and the openings are maintained throughout the duration of the voltage step. The ensemble average gives a smooth, sustained outward current that resembles the Hodgkin and Huxley whole-cell K^+ current. The unitary T-type Ca^{2+} channels exhibit short early openings that give transient inward ensemble average currents. The unitary L-type Ca^{2+} channels exhibit openings throughout the duration of the voltage step, and the ensemble average is more sustained.

Unitary currents of nicotinic ACh channels and GABA channels are given in figure 8.7. These channels also open and close abruptly in a random fashion, with higher opening probabilities when ligands are present. A common characteristic of these ligand-gated channels is that the opening probability can be increased not only by the natural ligand (ACh or GABA) but also by a number of analogues of the natural ligands. These analogues

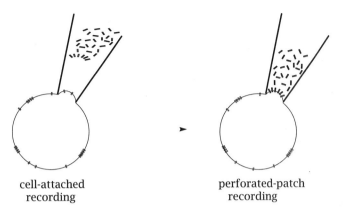

cell-attached
recording

perforated-patch
recording

Figure 8.4   Perforated-patch configuration.

open the same channels, but they usually give different opening kinetics (including opening frequency and duration, bursting behavior, etc.).

From the examples described above and overwhelming evidence elsewhere, we know that all ion channels open and close in a random and abrupt fashion. In other words, the channels vary between conductive and nonconductive states stochastically, with opening (or closing) probabilities controlled by transmembrane voltage or by ligand binding. Mathematically, this type of behavior is nondeterministic, and it cannot be described by exact and explicit mathematical equations. Stochastic analysis is needed to deal with such behavior.

The behavior of the sum of many single channels of the same type, or the ensemble average, on the other hand, can be approximated by explicit mathematical expression. The sum of many single-channel currents gives rise to the macroscopic current (sometimes called whole-cell current if it is the sum of all channels in a given cell). It is worth noting that the ensemble averages of the unitary Na^+ and K^+ currents given in the examples above exhibit similar waveforms as those of the I_{Na} and I_K obtained by Hodgkin and Huxley from the whole squid axon. Therefore, it is possible to explain macroscopic currents in terms of unitary currents. Additionally, analyzing the behavior of single-channel currents can provide detailed molecular mechanisms of channel gating, channel structure-function relationships, and gating kinetics that are beyond the information capacity of macroscopic current records.

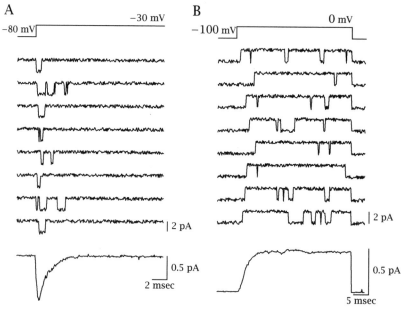

Figure 8.5   Unitary currents (upper 8 traces) and the ensemble average of unitary currents (lowest trace) of a Na^+ (A) and K^+ (B) channel. The membrane voltage is stepped from $V_H = -80$ mV to $V_c = -30$ mV for (A) and from $V_H = -100$ mV to 0 mV for (B). Records in (A) are simulated after data from Horn and Vandenberg (1984), and records in (B) are simulated with the reaction scheme shown in figure 10.1.

Example 8.1

The figure below shows the single-channel current of the ACh-gated channel in the muscle end plate. The voltage across the patch of membrane was clamped at various voltages (V_c).

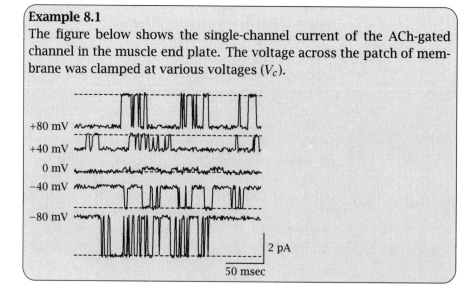

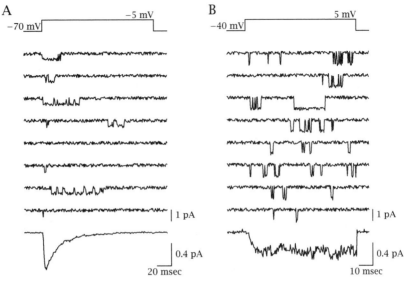

A   −70 mV   −5 mV

B   −40 mV   5 mV

1 pA

0.4 pA

20 msec

1 pA

0.4 pA

10 msec

Figure 8.6   Unitary currents (upper 8 traces) and the ensemble average of unitary currents (lowest trace) of a T-type (A) and L-type (B) calcium channel. The membrane voltage is stepped from $V_H = -70$ mV to $V_c = -5$ mV for (A) and from $V_H = -40$ mV to $V_c = 5$ mV for (B). Records are simulated after data from Fisher et al. 1990.

Example 8.1 (continued)

a. Plot the current-voltage relation of this channel.

b. What is the equilibrium potential of the ionic current flowing through this channel?

c. What are the slope and chord conductances of this channel at $V_m = -60$ mV?

d. If this channel is permeable to K^+ and Na^+ ions only, and $I_K = -80$ mV and $E_{Na} = +50$ mV, what is the relative conductance (i.e., g_K/g_{Na}) of this channel when it is open?

A

B

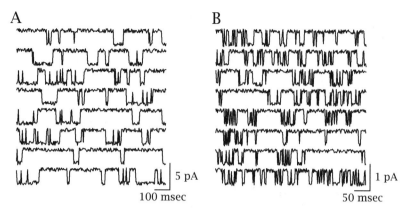

5 pA

100 msec

1 pA

50 msec

Figure 8.7 Unitary currents from an ACh-gated channel (A) and a GABA-gated channel (B). Records in (A) are simulated after data from Colquhoun and Sakmann 1985, and records in (B) are simulated after data from Gray and Johnston 1985.

Answer to example 8.1

a. See figure below.

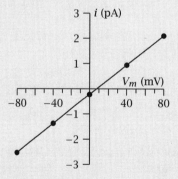

b. From the figure above, the equilibrium potential $E \approx +7.5$ mV.

c. Since the I-V relation is linear, the slope conductance = chord conductance, which is represented by

$$\frac{1.52 \text{ pA}}{52 \text{ mV}} = 0.029 \times 10^{-12} \times 10^3 \text{ S} = 29 \text{ pS}.$$

d. $E = 7.5 = \dfrac{(g_K/g_{Na})(-80) + (+50)}{1 + (g_K/g_{Na})}.$

Therefore, $g_K/g_{Na} = 0.486.$

8.4 Review of important concepts

1. The molecular structures of both voltage-gated and ligand-gated channels are determined by recombinant DNA techniques. Subunits, voltage sensors, transmembrane helices, configurations of aqueous pores, and ligand binding domains of individual channels are identified.

2. Unitary (single-channel) currents are recorded with the patch-clamp technique developed by Neher and Sakmann (1976). Unitary currents of both voltage- and ligand-gated channels show that ion channels open and close randomly and abruptly. The opening and closing probabilities are controlled by transmembrane voltage or by ligand binding.

9 Stochastic Analysis of Single-Channel Function

9.1 Introduction

In the previous chapter we described the behavior of single ion channels and concluded that individual channels, gated either by voltage or by ligands, open and close stochastically. One cannot, for example, predict exactly when a channel is going to make a transition (open → closed or closed → open). Instead, one can only give the *probability* of the occurrence of a transition within a certain period of time. Mathematically, this type of behavior is called nondeterministic, or random (stochastic) data.

Experimental data can be classified into two types: (1) Deterministic data are those that can be described by an explicit mathematical relationship. An example is a rigid body of mass m suspended from a fixed point by a linear spring of spring constant k, released from a position A away from the equilibrium point at $t = 0$. The exact location of the rigid body at any time $t \geq 0$ can be described by $x(t) = A \cos \sqrt{\frac{k}{m}} t$. (2) Nondeterministic data are those that cannot be described by any explicit mathematical relationship because each observation of the phenomenon will be unique. In other words, any given observation will represent only one of many possible results that might have occurred. These data are called *random data* or *stochastic data.* An example is the voltage output of thermal noise generators. Simultaneous recordings from two identical generators give two different time history records (see figure 9.1).

In order to analyze single-channel data that are nondeterministic, new mathematical tools unfamiliar to most physiologists have to be introduced. Stochastic analysis has been used by physicists and engineers for many years, and its application to neurophysiology has proven to be essential and very fruitful. The objectives of applying stochastic analysis to ion channels are at least fourfold: (1) to provide a quantitative, probabilis-

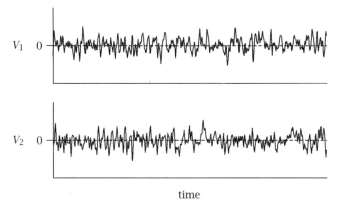

Figure 9.1   Voltage output of two identical thermal noise generators.

tic description of single-channel behaviors that cannot be formulated by deterministic equations; (2) to provide quantitative models describing the molecular mechanisms underlying channel gating and transition kinetics of conformation states of individual channels; (3) to establish a quantitative relationship between single-channel function and the molecular structures responsible for the function; and (4) to establish quantitative relationships between macroscopically measured parameters (e.g., from whole-cell currents) and the microscopic parameters (e.g., from single-channel models), so that ion conductances of whole cells can be explained by molecular events.

9.2   Basic descriptive properties of random data

In this section, we briefly describe the basic mathematical tools for analyzing random data. We follow the conventions used by Bendat and Piersol (1986). Readers who are familiar with these tools can skip this section and go to section 9.3.

Four types of statistical functions are generally used to describe the properties of random data.

9.2.1   Mean square values

1. **Mean square value:** $\Psi_x{}^2$, which furnishes the *general intensity* of any random data, is the average of the square values of the time history of the random process $x(t)$, or

$$\Psi_x{}^2 = \lim_{T \to \infty} \frac{1}{T} \int_0^T x^2(t) dt,$$

where T is the duration of the sample. $|\Psi_x|$ is called the *root mean square* or *rms* value.

2. **Mean value:** μ_x, which describes the *static* or *time-invariant* component of the data, is the average of all values over the sampling time:

$$\mu_x = \lim_{T \to \infty} \frac{1}{T} \int_0^T x(t) dt.$$

3. **Variance:** $\sigma_x{}^2$, which describes the *dynamic* or *fluctuating* component of the data, is the mean square value about the mean value:

$$\sigma_x{}^2 = \lim_{T \to \infty} \frac{1}{T} \int_0^T [x(t) - \mu_x]^2 dt.$$

$|\sigma_x|$ is called the *standard deviation*. Note that

$$\sigma_x{}^2 = \lim_{T \to \infty} \frac{1}{T} \int_0^T \left[x^2(t) - 2x(t)\mu_x + \mu_x{}^2 \right] = \Psi_x{}^2 - 2\mu_x{}^2 + \mu_x{}^2 = \Psi_x{}^2 - \mu_x{}^2.$$

9.2.2 Probability functions

1. **Probability density function** (pdf) $p(x), f(x)$: The pdf furnishes information about data in the *amplitude domain*. The pdf of random data describes the probability that the data will assume a value within some defined range Δx at any instant of time, or

$$p(x) = \lim_{\Delta x \to 0} \frac{\text{Prob } [x < x(t) < x + \Delta x]}{\Delta x}.$$

A sample time history record $x(t)$ is given in figure 9.2.

Then

$$\text{Prob}[x < x(t) < x + \Delta x] = \lim_{T \to \infty} \frac{T_x}{T};$$

thus,

$$p(x) = \lim_{\Delta x \to 0} \left[\lim_{T \to \infty} \frac{T_x}{T} \right].$$

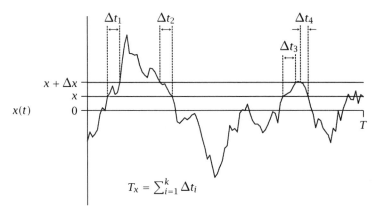

Figure 9.2   Probability measurement of a time history record $x(t)$.

Therefore, $p(x)$ gives the fractional time that event $x(t)$ spends between x and $x + \Delta x$ when $\Delta x \to 0$.

2. Probability distribution function or cumulative distribution function (cdf) $P(x)$, $F(x)$: The cdf describes the probability that the data will assume a value between ∞ and x, or

$$F(x) = P(x) = \text{Prob}[x(t) \le x].$$

Recall the definition of pdf—it is obvious that

$$F(x) = P(x) = \int_{-\infty}^{x} p(\xi)d\xi \ \text{ and } \ \frac{dP(x)}{dx} = p(x), \ \frac{dF(x)}{dt} = f(x).$$

Therefore, $P(x)$ equals the integral of $p(x)$ from $-\infty$ to x (or the total area under $p(x)$ from $-\infty$ to x).

The relationship between $f(x)$ and $F(x)$ is given in figure 9.3. The median M, and the lower and upper quartiles O_1 and O_3 are such that $F(M) = \frac{1}{2}$, $F(O_1) = \frac{1}{4}$, $F(O_3) = \frac{3}{4}$, and, for example, the 57th percentile is the value x for which $F(x) = 0.57$. From the definitions, it is easy to see the following:

$P(a) \le P(b)$ if $a \le b$,

$P(-\infty) = 0$,

$P(\infty) = 1$, and

$$\int_{-\infty}^{\infty} p(x)dx = 1.$$

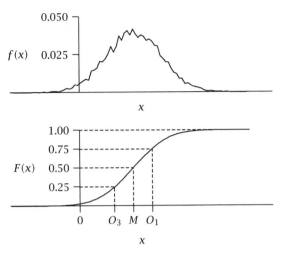

Figure 9.3 Probability density function (pdf), $f(x)$ and cumulative distribution function (cdf), $F(x)$.

The *expected value* (average value) of any real single-valued continuous function $g(x)$ of the random variable $x(t)$ is given by

$$E[g(x(t))] = \int_{-\infty}^{\infty} g(x)p(x)dx.$$

In other words, $E[g(x)]$ is the sum of the function $g(x)$ weighted by its pdf $p(x)$ over all values of x. Therefore,

the mean value $\mu_x = E[x(t)] = \int_{-\infty}^{\infty} xp(x)dx,$

the mean square value $\Psi_x^2 = E[x^2(t)] = \int_{-\infty}^{\infty} x^2 p(x)dx,$ and

the variance $\sigma_x^2 = E\left[(x(t) - \mu_x)^2\right] = \int_{-\infty}^{\infty} (x - \mu_x)^2 p(x)dx.$

Example 9.1

Consider a random variable X with an *exponential distribution*, i.e., pdf is

$$f(x) = \begin{cases} \lambda e^{-\lambda x} & \text{if } x \ge 0 \\ 0 & \text{if } x < 0 \end{cases}$$

where λ is a positive constant.

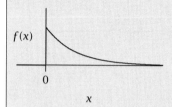

a. Verify that $f(x)$ satisfies the condition for a pdf.

b. Show that $\mu_x = \frac{1}{\lambda}$ and $\sigma_x{}^2 = \frac{1}{\lambda^2}$.

c. Find the median.

d. Show that the cdf is $F(x) = 1 - e^{-\lambda x}$ (for $x \ge 0$).

Answer to example 9.1

a. $f(x) \ge 0$ and $\displaystyle\int_0^\infty \lambda e^{-\lambda x} dx = \left[-e^{-\lambda x}\right]_0^\infty = (0) - (-1) = 1.$

b. To find the mean and variance, we use integration by parts. We also need to use the results that $xe^{-\lambda x}$ and $x^2 e^{-\lambda x}$ tend to zero as $x \to +\infty$.

$$\mu_x = \int_0^\infty x\lambda e^{-\lambda x} dx = \left[x(-e^{-\lambda x})\right]_0^\infty - \int_0^\infty (1)(-e^{-\lambda x})dx = \frac{1}{\lambda}$$

$$\Psi_x{}^2 = \int_0^\infty x^2 \lambda e^{-\lambda x} dx = \left[x^2(-e^{-\lambda x})\right]_0^\infty - \int_0^\infty (2x)(-e^{-\lambda x})dx$$

$$= \left[-x^2 e^{-\lambda x} - 2x\left(\frac{1}{\lambda}e^{-\lambda x}\right)\right]_0^\infty + \int_0^\infty (2)\left(\frac{1}{\lambda}e^{-\lambda x}\right) dx$$

$$= \left[-x^2 e^{-\lambda x} - \frac{2x}{\lambda}e^{-\lambda x} - \frac{2}{\lambda^2}e^{-\lambda x}\right]_0^\infty = \frac{2}{\lambda^2}$$

$$\sigma_x{}^2 = \frac{2}{\lambda^2} - \left(\frac{1}{\lambda}\right)^2 = \frac{1}{\lambda^2}$$

Answer to example 9.1 (continued)

c. If m is the median, then $P(X > m) = \frac{1}{2}$.

$$\frac{1}{2} = \int_m^\infty \lambda e^{-\lambda x} dx = \left[-e^{-\lambda x} \right]_m^\infty = (0) - (-e^{-\lambda m}) = e^{-\lambda m},$$

i.e., $e^{\lambda m} = 2$, the median, $m = \frac{\ln 2}{\lambda}$.

d. If $x \geq 0$, the cdf is

$$F(x) = \int \lambda e^{\lambda x} dx = -e^{-\lambda x} + C.$$

Now $F(0) = P(X \leq 0) = 0$, so $0 = -1 + C$, i.e., $C = 1$.
Hence the cdf is

$$F(x) = \begin{cases} 0 & \text{if } x < 0 \\ 1 - e^{-\lambda x} & \text{if } x \geq 0 \end{cases}$$

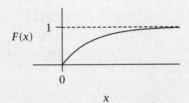

Note that as $x \to \infty$, $F(x) \to 1 - 0 = 1$.

9.2.3 Covariance function $C(\tau)$ and correlation function $R(\tau)$

The covariance and correlation functions furnish information about data in the *time domain*. They give the relationships between the probabilities of events occurring at a certain time and those occurring before and after that time.

1. *Covariance function* of a random process $x(t)$ is defined

$$C_x(\tau) = E\left[(x(t) - \mu_x(t))(x(t + \tau) - \mu_x(t + \tau)) \right],$$

where $\mu_x(t) = E[x(t)] = \int_{-\infty}^\infty x p(x) dx$ (mean value).

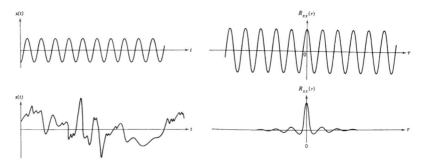

Figure 9.4   Time histories ($x(t)$) and correlation functions ($R(\tau)$) of a sine wave (upper traces) and random noise (lower traces). (From Bendat and Piersol 1986. Copyright © 1986 John Wiley & Sons, Inc. Reprinted by permission of John Wiley & Sons, Inc.)

2. *Correlation function* of a random process $x(t)$ is defined

$$R_x(\tau) = E[x(t)x(t+\tau)].$$

Thus,

$$
\begin{aligned}
C_x(\tau) &= E\left[(x(t) - \mu_x(t))(x(t+\tau) - \mu_x(t+\tau))\right] \\
&= \int_{-\infty}^{\infty} x(t)x(t+\tau)p(x)dx - \int_{-\infty}^{\infty} x(t)p(x)\mu_x s(t+\tau)dx \\
&\quad - \int_{-\infty}^{\infty} x(t+\tau)p(x)\mu_x(t)dx + \int_{-\infty}^{\infty} p(x)\mu_x(t)\mu_x(t+\tau) \\
&= R_x(\tau) - \mu_x{}^2 - \mu_x{}^2 + \mu_x{}^2.
\end{aligned}
$$

Therefore, $C_x(\tau) = R_x(\tau) - \mu_x{}^2$.

It is evident that the $R_x(\tau)$ (also called autocorrelation function) is equivalent to $C_x(\tau)$ when the mean value $\mu_x = 0$.

$R_x(\tau)$ [or $C_x(\tau)$] establishes the influences of values at any time over values at a future time. Sine functions (figure 9.4, first row) or any deterministic data have $R_x(\tau)$ that persists over all time displacements. Random data, on the other hand, have $R_x(\tau) \to 0$ for large displacements (figure 9.4, second row). Thus $R_x(\tau)$ or $C_x(\tau)$ measurements can be used as a tool for detecting deterministic data that might be masked in a random background.

9.2.4   Power spectral density function $S(f)$

$S(f)$ furnishes information of random data in the *frequency domain*. This function gives the relationship of the weights of noise (variance) at each frequency, and it can be obtained by two independent methods: the experimental and the theoretical approaches.

9.2.4.1 Experimental approach Since $S(f)$ of a random process describes the general frequency composition of the data in terms of the spectral density of its mean square value, the mean square value of a sample history in a frequency range between f and $f + \Delta f$ may be obtained by filtering the sample record with a bandpass filter having sharp cutoff characteristics, and computing the average of the squared output of the filter. This averaged square value approaches the exact mean square value Ψ_x^2 as $T \to \infty$. In equation form

$$\Psi_x^2(f, \Delta f) = \lim_{T \to \infty} \frac{1}{T} \int_0^T x^2(t, f, \Delta f) dt,$$

where $x(t, f, \Delta f)$ is the portion of $x(t)$ in the frequency range between f and $f + \Delta f$.

The power spectral density function (pdf) is defined:

$$S_x(f) = \lim_{\Delta f \to 0} \frac{\Psi^2(f, \Delta f)}{\Delta f} = \lim_{\Delta f \to 0} \frac{1}{\Delta f} \left[\lim_{T \to \infty} \frac{1}{T} \int_0^T x^2(t, f, \Delta f) dt \right].$$

Experimentally, $S(f)$ of random data $x(t)$ is estimated by the Fast Fourier Transform (FFT) algorithm. Detailed descriptions of FFT are given in Press et al. (1992).

9.2.4.2 Theoretical approach Another approach to obtain psd $S(f)$ is by the Wiener-Khinchin theorem, which states that the power spectral density function is the Fourier transform of the covariance function.

$$
\begin{aligned}
S(f) &= \Re \left[\mathcal{F}(C(\tau)) \right] \qquad \text{where } \mathcal{F} = \text{Fourier Transform} \\
&= 2 \int_{-\infty}^{\infty} C(\tau) e^{i2\pi f \tau} d\tau.
\end{aligned}
$$

Therefore, one can calculate $S(f)$ from the $C(t)$ or $R(t)$ (figure 9.5), and this is the theoretical approach to obtain the psd. Application of the power spectral density analysis on physiological channels will be described in chapter 10.

9.3 Statistical analysis of channel gating

Single-channel recordings have revealed that individual ion channels in excitable cells open and close randomly, and thus the behavior of populations of single channels in whole cells has to be dealt with statistically. In this section, we will discuss the statistical tools that can be used to describe the behavior of populations of N identical channels, each of which

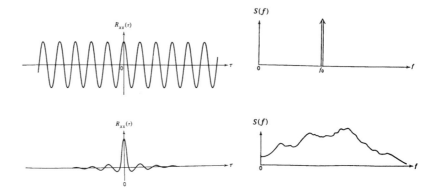

Figure 9.5   Correlation function ($R(\tau)$) and power spectral density function ($S(f)$) of a sine wave and random data. (From Bendat and Piersol 1986. Copyright © 1986 John Wiley & Sons, Inc. Reprinted by permission of John Wiley & Sons, Inc.)

opens and closes randomly. From this analysis, one can obtain information on individual channel properties, such as single-channel conductance (γ) and the number of channels (N), from data obtained from whole cells (population of single channels). This approach is important because it provides a tool to correlate whole-cell data with the single-channel data. Additionally, since it is sometimes difficult to record single-channel current in certain preparations, either because the channels are inaccessible to the patch electrodes or because the single-channel conductances are too small, this population approach may be essential for studying channel behavior.

Consider N independent two-state channels.

$$
\begin{array}{ccc}
 & \text{closed} \; \overset{\beta}{\underset{\alpha}{\rightleftharpoons}} \; \text{open} & \\
\text{state \#} & 2 \qquad\qquad 1 & \\
\text{probability} & P_2 \qquad\qquad P_1 &
\end{array} \tag{9.3.1}
$$

P_1 and P_2 are probabilities that the channel is at open (1) or closed (2) states, respectively. From equations 6.1.2 through 6.1.4, recall $y(\infty)$ in the gate model

$$
P_1(\infty) = \frac{\beta}{\alpha + \beta} \qquad P_2(\infty) = \frac{\alpha}{\alpha + \beta}. \tag{9.3.2}
$$

The probability that k channels out of N are open at time t is:

$$
P_1^k(t) = C_k^N \, [P_1(t)]^k \, [P_2(t)]^{N-k} \qquad k = 0, 1, \ldots, N. \tag{9.3.3}
$$

This is a *binomial distribution*, which is simply derived from the combinatorics for independent channels and the fact that these channels have two conductance states (open and closed). For those having three conductance states, a trinomial distribution would be appropriate.

When $N >> 1$, the binomial distribution can be approximated by the *Gaussian (normal) distribution.*

$$P_1^k(t) = C_k^N P_1^k P_2^{N-k} \cong \frac{1}{\sqrt{2\pi N P_1 P_2}} e^{-(k-NP_1)^2/2NP_1P_2}. \tag{9.3.4}$$

Let $\overline{k} = NP_1$ (mean number of open channels), $\qquad$ (9.3.5)

$\sigma_N^2 = \overline{(k - \overline{k})^2} = NP_1P_2$ (variance), and $\qquad$ (9.3.6)

$x = k - \overline{k},$ $\qquad$ (9.3.7)

Therefore,

$$P_1^x(t) = \frac{1}{\sigma\sqrt{2\pi}} e^{-x^2/2\sigma^2}. \tag{9.3.8}$$

When P_1 is small ($P_1 << 1$) and $\overline{k} = NP_1 << N$, then the binomial distribution can be approximated by the *Poisson distribution*:

$$P_1^k(t) = C_k^N P^k q^{N-k} \cong \frac{(\overline{k})^k}{k!} e^{-\overline{k}}. \tag{9.3.9}$$

Since mean ionic current μ_I through N channels is $I_1\overline{k}$, where I_1 is the current flow across one open channel, then

$$\mu_I = I_1 N P_1. \tag{9.3.10}$$

At steady state, i.e., $t \to \infty, P_1(t) = P_1(\infty) = \frac{\beta}{\alpha+\beta}.$

$$\mu_I = \frac{I_1 N \beta}{\alpha + \beta} = \frac{\gamma N(V - E_i)\beta}{\alpha + \beta}, \tag{9.3.11}$$

where $I_1(V) = \gamma(V - E_i)$, and γ = single-channel conductance.

The variance of current through N channels is

$$\begin{aligned} \sigma_{I_N}^2 &= \mathrm{Var}[I_N(t)] = I_1^2\sigma_N^2 = I_1^2 NP_1 P_2 = I_1^2 NP_1(1 - P_1) \\ &= I_1\mu_I - \mu_I^2/N. \end{aligned} \tag{9.3.12}$$

At steady state,

$$t \to \infty, \quad P_1(\infty) = \frac{\beta}{\alpha + \beta}, \quad P_2(\infty) = \frac{\alpha}{\alpha + \beta},$$

$$\sigma_{I_N}{}^2 = I_1{}^2 N P_1(\infty) P_2(\infty) = \frac{I_1{}^2 N \alpha \beta}{(\alpha + \beta)^2} = \frac{N \gamma^2 (V - E_i)^2 \alpha \beta}{(\alpha + \beta)^2}$$

$$= \frac{\mu_I \gamma (V - E_i) \alpha}{\alpha + \beta}. \tag{9.3.13}$$

$$\frac{d(\sigma_{I_N}{}^2)}{d\mu_I} = I_1 - 2\mu_I / N, \tag{9.3.14}$$

$$I_1 = \text{slope of } \sigma_{I_N}{}^2(\mu_I) \text{ at } \mu_I = 0, \tag{9.3.15}$$

and

$$\gamma = \frac{I_1}{V - E_i}.$$

At $\sigma_{I_N}{}^2$ maximum, $\dfrac{d}{d\mu_I}(\sigma_{I_N}{}^2) = 0$.

This yields

$$N = \frac{2\mu_I^m}{I_1}. \tag{9.3.16}$$

Sigworth (1980a, b) applied this approach to analyze the noise of whole-cell Na$^+$ current at the frog node of Ranvier. The whole-cell currents (I_{Na}) are given in the top portion of figure 9.6, the noise ($I_{Na} - \mu_I$) in the middle portion, and the variance (σ^2) in the lower portion. He then plotted σ^2 vs. μ_I, which is given in figure 9.7. By equation 9.3.14, the current carried by a single Na$^+$ channel I_1 can be estimated by the slope of the σ^2 / μ_I plot at $\mu_I = 0$, and the number of Na$^+$ channels contributing the whole-cell current (N) can be estimated by measuring μ_I^m and by equation 9.3.16, $N = 2\mu_I^m / I_1$.

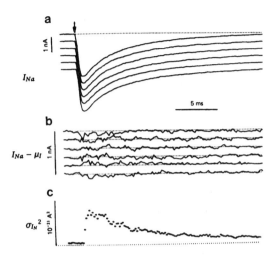

Figure 9.6 Na$^+$ current fluctuations at the frog node of Ranvier, using the nonstationary (or ensemble) analysis method. Trace (*a*) shows six successive current records produced by clamped depolarizations to −5 mV. (*b*) shows the deviations of the individual currents in (*a*) from their mean. (*c*) shows the variance of 65 such groups of records. (From Sigworth 1980a.)

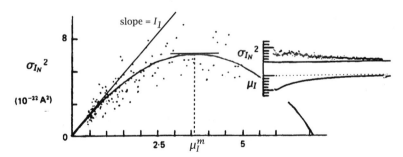

Figure 9.7 Variance-mean plot from nonstationary fluctuation analysis of the Na$^+$ currents at a frog node of Ranvier. The inserts show the mean Na$^+$ current I (lower trace, 1 nA per small division) and its variance var(I) (upper trace, $2 \times 10^{-22} A^2$ per small division). (From Sigworth 1980b.)

9.4 Probability density function of channel gating

$P_1^k(t)$ in the last section describes the probability that k channels out of N are open at time t. It is the probability function of the number of channels, but not of the time course of channel gating.

To express the probability functions as a function of time, we need to view them differently: Suppose time is divided into Δt-sized parcels and that the probability (or frequency) that an open channel will close during Δt is $\alpha \Delta t$ (note that α is the probability of an open channel being closed at unit time). If k successful transitions occur out of N trials (each molecular "stretch" of a channel driven by random thermal energy can be considered as one binomial trial) during Δt, then $\overline{k} = \alpha \Delta t$, and the probability that k successful transitions occur out of N trials during Δt is

$$P_1^k(\Delta t) = C_k^N [P_1(\Delta t)]^k [P_2(\Delta t)]^{N-k}$$

because

$$P_1(\Delta t) << 1, \text{ and } \overline{k} << N (N \approx 10^{12}/\text{sec}, \overline{k} \approx 10^3/\text{sec}).$$

$P_1^k(\Delta t)$ represents a Poisson process. Thus,

$$P_1^k(\Delta t) \approx \frac{(\alpha \Delta t)^k}{k!} e^{-\alpha \Delta t}.$$

Let $F(t)$ be the probability distribution function (cumulative) (cdf) of a channel of lifetime (dwell time) $\leq t$. Then,

$$
\begin{aligned}
F(t) \ &= 1 - \text{prob (channel open time} > t) \\
&= 1 - \text{prob (no closing transition occurs between 0 and } t) \\
&= 1 - \frac{(\alpha t)^0}{0!} e^{-\alpha t} = 1 - e^{-\alpha t}.
\end{aligned}
$$

The probability density function

$$f_1(t) = P_1(t) = \frac{dF(t)}{dt} = \frac{d}{dt}\left(1 - e^{-\alpha t}\right) = \alpha e^{-\alpha t}.$$

Therefore, $f_1(t) = P_1(t) = \alpha e^{-\alpha t} \quad (t \geq 0)$.

Similar procedures for pdf for channel closing yield

$$f_2(t) = P_2(t) = \beta e^{-\beta t} \quad (t \geq 0).$$

The pdf of channel opening $f_1(t) = \alpha e^{-\alpha t}$ and the pdf of channel closing $f_2(t) = \beta e^{-\beta t}$ give the probability density of the *time* that the channel is in the open and closed states, respectively. In other words, these pdfs give the dwell time distributions of the channel once it enters the open or closed states. This can be illustrated by the single-channel current record shown in figure 9.8.

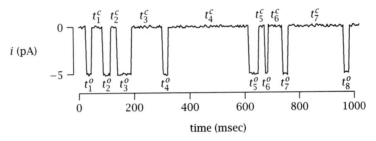

Figure 9.8 Single-channel current record illustrating the random open and closed times of the channel. t_n^o indicates open time, and t_n^c indicates closed time. $n = 1, 2, 3, \ldots$.

If one measures time lengths of each opening segment (t_n^o) and each closing segment (t_n^c) in the trace and plots a histogram for opening and one for closing, one usually observes the results shown in figure 9.9.

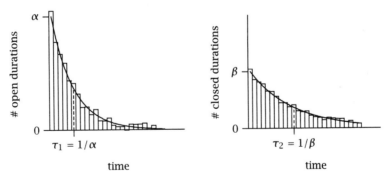

Figure 9.9 Probability density functions of the channel open lifetime $(f_1(t))$ and closed lifetime $(f_2(t))$. The continuous solid curves are $f_1(t) = \alpha e^{-\alpha t}$ and $f_2(t) = \beta e^{-\beta t}$.

As demonstrated in the example in section 9.2,

$$\int_0^\infty f_1(t)\,dt = \int_0^\infty \alpha e^{-\alpha t}\,dt = 1,$$

$$\int_0^\infty f_2(t)\,dt = \int_0^\infty \beta e^{-\beta t}\,dt = 1.$$

So the areas underneath the two pdfs are identical; that is, they both are equal to 1. Additionally, the same examples show that

$$\text{mean open time } = E[t] = \int_0^\infty t\alpha e^{-\alpha t}\,dt = \frac{1}{\alpha},$$

$$\text{mean closed time } E[t] = \int_0^\infty t\beta e^{-\beta t}\,dt = \frac{1}{\beta}.$$

Conceptually, it is quite obvious that if $\alpha > \beta$, that is, the transition rate from open to closed is higher than that of the reverse process, then the probability that the channel dwells in the open state is lower than that in the closed state. Thus, the mean open time $= \frac{1}{\alpha}$ is shorter than the mean closed time $= \frac{1}{\beta}$.

The fact that channel open and closed lifetimes follow exponential distributions indicates that channel gating is a *memoryless random process*. Mathematically, "memoryless" is expressed as follows:

Prob [channel open lifetime $> t + t_1$ | channel open lifetime $> t_1$]
= Prob [channel open lifetime $> t$], where

$$\text{Prob}[B|A] = \frac{\text{Prob}[A \text{ and } B]}{\text{Prob}[A]} \quad \text{(Conditional probability)}.$$

This equation states that the probability that a channel keeps open for an additional t seconds, given that the channel has already been open for t_1 seconds, equals the probability that the channel keeps open for t seconds starting from $t = 0$. Thus, *the probability that the channel keeps open for an additional t seconds is the same regardless of how long the channel has already been open.*

To prove that the exponential distribution of channel opening or closing is "memoryless," one can write

Prob [channel open lifetime $> t + t_1$ | channel open lifetime $> t_1$]

$$= \frac{\text{Prob } [(\text{channel open lifetime} > t + t_1) \cap (\text{channel open lifetime} > t_1)]}{\text{Prob [channel open lifetime} > t_1]}$$

$$= \frac{\text{Prob [channel open lifetime} > t + t_1]}{\text{Prob [channel open lifetime} > t_1]} \quad \begin{array}{l}\text{(Since second term in}\\ \text{numerator is redundant)}\end{array}$$

$$= \frac{1 - \text{Prob[channel open lifetime} \leq t + t_1]}{1 - \text{Prob [channel open lifetime} \leq t_1]}$$

$$= \frac{1 - \left[1 - e^{\alpha(t+t_1)}\right]}{1 - [1 - e^{-\alpha t_1}]} = e^{-\alpha t}. \quad \text{[See exponential cdf } F(t) \text{ above.]}$$

Also, Prob [channel open lifetime > t]

$= 1 - $ Prob [channel open lifetime $\leq t$]

$= 1 - \left[1 - e^{-\alpha t}\right]$ [See exponential cdf $F(t)$ above.]

$= e^{-\alpha t}$.

Therefore,

Prob[channel open lifetime > $t + t_1$|channel open lifetime > t_1]

$$= \text{Prob[channel open lifetime} > t]$$

provided that the channel open lifetime is exponentially distributed—that is, that $F(t) = 1 - e^{-\alpha t}$. A similar analysis can be done for the channel closed lifetime.

This section shows that the open and closed lifetime of the channel follows exponential random distribution and thus it is a memoryless random process. In other words, the probability of transition from one state to another (e.g., open → closed, which causes termination of the open lifetime, or closed → open, which causes termination of the closed lifetime) does not depend on the past history, but only depends on the state immediately before the transition (e.g., for open → closed, the transition probability requires *only* that the channel be at open state immediately before the moment of transition). This type of memoryless random process is also called the *Markov process*. The theory of the Markov process is the foundation of stochastic analysis for many computer and communication systems, and—more importantly for neuroscientists—it has been the main mathematical tool for analyzing behaviors of single-channel currents.

9.5 Review of important concepts

1. Single-channel data are nondeterministic, and stochastic analysis must be adopted to describe such data. The basic descriptive tools of stochastic processes are the mean square values, probability density function, cumulative probability function, covariance function, and power spectral density function.

2. Based on stochastic principles, the single-channel current and the number of channels in a cell can be estimated by the noise (variance vs. mean) of the whole-cell (contains N identical channels) current.

3. The transition of a single channel from one state to another is a *memoryless random process* or a *continuous-time Markov process:* The probability that a channel stays in a given state for an additional t seconds is independent of how long the channel has been in that state.

4. Because the transition is a Markov process, the probability density function of the dwell time of a channel in a given state is exponential.

9.6 Homework problems

1. A random variable x has the rectangular distribution between 2 and 6. Find, and sketch, the cdf $F(x)$.

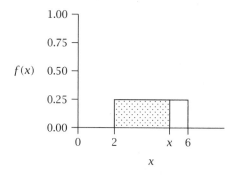

2. A random variable x has pdf

$$f(x) = \begin{cases} \frac{2}{9}x(3-x) & \text{if } 0 \le x \le 3 \\ 0 & \text{otherwise} \end{cases} .$$

Find the cdf $F(x)$, and use it to find $P(x > 2)$.

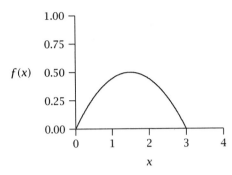

3. A random variable x has pdf

$$f(x) = \begin{cases} 0 & \text{if } x < 0 \\ \frac{6}{5}x & \text{if } 0 \le x \le 1 \\ \frac{6}{5x^4} & \text{if } x > 1 \end{cases} .$$

Find   a.   the cdf $F(x)$
         b. $P\left(\frac{1}{2} < x < 2\right)$
         c.   the median and the semi-interquartile range.

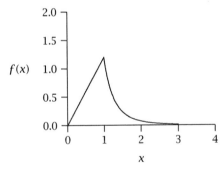

4. A neuron contains N identical channels that are gated by the neuro-transmitter glutamate. Glutamate opens these channels and results in an inward Na$^+$ current ($E_{Na} = +50$ mV). The glutamate-induced currents in this neuron under voltage-clamp conditions ($V_P = 10$ mV) are given in the following diagram:

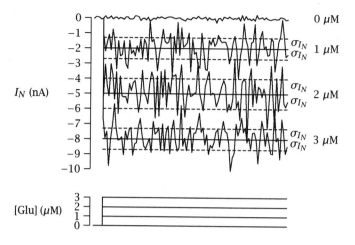

(a) Plot the variance $(\sigma_{I_N}{}^2)$ as a function of mean current μ_I on graph paper.

(b) Estimate the single-channel conductance and the total number of glutamate-gated channels in the neuron.

5. A neuron contains N voltage-gated Na^+ channels ($E_{Na} = +50$ mV) that follow the two-state transition scheme with voltage-dependent transition rate constants.

$$\text{closed} \underset{\alpha(V)}{\overset{\beta(V)}{\rightleftharpoons}} \text{open}$$

state number   2   1

$$\alpha(V) = 500e^{-V/50 \text{ mV}} \quad (\text{sec}^{-1}),$$
$$\beta(V) = 50e^{+V/25 \text{ mV}} \quad (\text{sec}^{-1}).$$

(a) What is the steady-state probability of each of these channels being open with the cell voltage-clamped at -100 mV, -50 mV, and 0 mV?

(b) If the main steady-state current of the whole cell is -100 nA when the cell is clamped at -50 mV, and if the conductance of a single open Na^+ channel is 25 pS, what is the minimum number of Na^+ channels in this cell?

(c) What is the steady-state Na^+ current of the whole cell when it is clamped at -100 mV and 0 mV?

(d) What are the amplitudes of noise for the whole-cell Na$^+$ current when it is clamped at -100 mV, -50 mV, and 0 mV? Noise amplitude $= |\sigma_N|$.

6. Explain, by using probability density functions for channel gating, why single-channel gating transitions can be treated as a memoryless Markov process.

10   Formulation of Stochastic Channel Mechanisms

10.1   Introduction

From the examples given in chapter 8, it is evident that all ion channels, either voltage-gated or ligand-gated, open and close randomly. Additionally, some channels exhibit bursting behavior and others do not. The random behaviors of ion channels require probabilistic descriptions. In this chapter, we derive a stochastic formulation for random channel behavior. The purpose of this effort is to provide a quantitative tool for physiologists to correlate measurable random data parameters such as mean channel open time, with channel mechanisms such as conformational states and transition schemes. A number of rules can be derived from the stochastic equations. These rules give explicit mathematical relationships between quantities obtained from single-channel or whole-cell data and transition coefficients of conformation transitions in individual channels. Consequently, one can gain insights on the molecular mechanisms of individual ion channels by analyzing random channel data in accordance with the rules derived from the stochastic formulation.

The probability distributions of channel opening and closing (i.e., the lifetime distributions) described in the last chapter show that channel gating is a memoryless Markov process. The time remaining at a given state once it is entered (lifetime) is exponentially distributed. This property of channel gating allows us to formulate a mathematical description of state transitions and channel kinetics in terms of the theory of continuous-time Markov chains: Each time the channel enters a state i (e.g., open state), an exponentially distributed state occupancy time T_i is selected. When the time is up, the next state j is selected (i.e., closed state) according to a discrete-time Markov chain (memoryless transition) with transition probabilities P_{ij}. Then the new exponentially distributed state occupancy time T_j is selected, and so on. This description scheme is suitable not only for a simple two-state process (for example, closed $\rightleftharpoons$ open), but also is applicable to complex n-state processes (e.g., $C_1 \rightleftharpoons C_2 \rightleftharpoons C_3 \dots C_k \rightleftharpoons 0_1 \rightleftharpoons 0_2 \dots$), which reflect the gating mechanisms of many channels in a biological system.

In this chapter we shall derive the major matrix differential equation (Chapman-Kolmogorov equation) for transition probabilities by using the simplest transition scheme, the two-state model. This avoids the tedious mathematical efforts in dealing with matrix differential equations of the nth order. (Readers who are interested in a detailed description of nth order derivations should consult Colquhoun and Hawkes, 1977 and 1981). Since the Chapman-Kolmogorov equation can be used to describe Markov processes of the nth order, we will list the general properties of the equation that are applicable to transition probabilities of the n-state model. These general properties (or rules) are extremely useful for analyzing the single-channel records in biological systems because they can help to extract kinetic parameters of single-channel transition states and to correlate single-channel parameters with whole-cell current records. The derivation of the stochastic equations in this chapter follow the conventions of Colquhoun and Hawkes (1983) and Tuckwell (1989).

10.2   Derivation of the Chapman-Kolmogorov equation

For channels that exhibit only two transition states, the open and closed states, the transition scheme can be written as follows.

Scheme I, two-state scheme:

$$
\begin{array}{ccc}
\text{closed} & \beta & \text{open} \\
C & \rightleftharpoons & O \\
 & \alpha &
\end{array}
\tag{10.2.1}
$$

$$
\begin{array}{ccc}
\text{state number:} & 2 & 1 \\
\text{probability} & P_2(t) & P_1(t).
\end{array}
$$

$P_1(t)$ and $P_2(t)$ are probabilities of the channel at state 1 (open) and state 2 (closed), respectively. α and β are rate constants in deterministic theories (e.g., first-order chemical reactions). However, in stochastic analysis, they are defined by the following equations:

$$\text{Prob[channel open at } t \rightarrow \text{closed at } t + \Delta t] = \alpha \Delta t + o(\Delta t), \tag{10.2.2}$$

Prob[channel closed at $t \rightarrow$ open at $t + \Delta t$] $= \beta\Delta t + o(\Delta t)$,   (10.2.3)

where $o(\Delta t)$ (remainder term) is the probability that more than one transition occurs during Δt, thus $\lim_{\Delta t \rightarrow 0} o(\Delta t) = 0$. Hence, equations 10.2.2 and 10.2.3 can be rewritten as

$\alpha\Delta t =$ [Prob(channel closed between t and $t + \Delta t$ | open at t)]   (10.2.4)

$\beta\Delta t =$ [Prob(channel open between t and $t + \Delta t$ | closed at t)]   (10.2.5)

where Prob[B|A] $= \dfrac{\text{Prob[A and B]}}{\text{Prob[A]}}$ (*conditional probability*).

Using the same definition and logic,

$1 - \alpha\Delta t$ = [Prob(channel open between t and $t + \Delta t$ | open at t)],

(10.2.6)

$1 - \beta\Delta t$ = [Prob(channel closed between t and $t + \Delta t$ | closed at t)].

(10.2.7)

If we define the *transition probability* $P_{ij}(t)$ as

$P_{ij}(t) =$ Prob [channel at state j at t | at state i at 0],   (10.2.8)

then equations 10.2.4-10.2.7 will give the transition probabilities during Δt for the two-state model:

$$
\begin{aligned}
P_{11}(\Delta t) &= 1 - \alpha\Delta t, \\
P_{12}(\Delta t) &= \alpha\Delta t, \\
P_{21}(\Delta t) &= \beta\Delta t, \\
P_{22}(\Delta t) &= 1 - \beta\Delta t.
\end{aligned}
\qquad (10.2.9)
$$

For transition between 0 and t,

$P_{11}(t) =$ Prob[open at t | open at 0]

For transition between 0 and $t + \Delta t$, the probability of a channel being open at 0 and $t + \Delta t$, $P_{11}(t + \Delta t)$, is equal to the probability of a channel open at 0, open at t, and open at $t + \Delta t$, plus the probability of the channel open at 0, closed at t, and open at $t + \Delta t$, i.e.,

$$
\begin{aligned}
P_{11}(t + \Delta t) \quad &= \text{Prob[open at } t + \Delta t \,|\, \text{open at 0]} \\
&= \text{Prob[open at } t + \Delta t \,|\, \text{open at } t] \cdot \\
&\quad \text{Prob[open at } t \,|\, \text{open at 0]} + \\
&\qquad \text{Prob[open at } t + \Delta t \,|\, \text{closed at } t] \cdot \\
&\qquad \text{Prob[closed at } t \,|\, \text{open at 0]} \\
&= [P_{11}(\Delta t)] \cdot [P_{11}(t)] + [P_{21}(\Delta t)] \cdot [P_{12}(t)].
\end{aligned}
$$

Equation 10.2.9 makes

$$
\begin{aligned}
P_{11}(t + \Delta t) \quad &= (1 - \alpha \Delta t) P_{11}(t) + \beta \Delta t P_{12}(t) \\
&= P_{11}(t) - [\alpha P_{11}(t) - \beta P_{12}(t)] \Delta t.
\end{aligned}
$$

Thus, as $\lim_{\Delta t \to 0}$,

$$
\lim_{\Delta t \to 0} \frac{P_{11}(t + \Delta t) - P_{11}(t)}{\Delta t} = -\alpha P_{11}(t) + \beta P_{12}(t).
$$

By the definition of derivative,

$$
\frac{dP_{11}(t)}{dt} = -\alpha P_{11}(t) + \beta P_{12}(t). \tag{10.2.10}
$$

Similar procedures can be applied to $P_{12}(t), P_{21}(t)$, and $P_{22}(t)$, and the results are:

$$
\frac{dP_{12}(t)}{dt} = \alpha P_{11}(t) - \beta P_{12}(t), \tag{10.2.11}
$$

$$
\frac{dP_{21}(t)}{dt} = -\alpha P_{21}(t) + \beta P_{22}(t), \tag{10.2.12}
$$

$$
\frac{dP_{22}(t)}{dt} = \alpha P_{21}(t) - \beta P_{22}(t). \tag{10.2.13}
$$

Equations 10.2.10–10.2.13 can be put in matrix form:

$$
\begin{aligned}
\frac{d}{dt}
\begin{bmatrix}
P_{11}(t) & P_{12}(t) \\
P_{21}(t) & P_{22}(t)
\end{bmatrix}
&=
\begin{bmatrix}
-\alpha P_{11}(t) + \beta P_{12}(t) & \alpha P_{11}(t) - \beta P_{12}(t) \\
-\alpha P_{21}(t) + \beta P_{22}(t) & \alpha P_{21}(t) - \beta P_{22}(t)
\end{bmatrix} \\[2mm]
&=
\begin{bmatrix}
P_{11}(t) & P_{12}(t) \\
P_{21}(t) & P_{22}(t)
\end{bmatrix}
\begin{bmatrix}
-\alpha & \alpha \\
\beta & -\beta
\end{bmatrix}. \tag{10.2.14}
\end{aligned}
$$

In matrix notation, equation 10.2.14 can be written as

$$
\frac{d\mathbf{P}(t)}{dt} = \mathbf{P}(t)\mathbf{Q} \quad \text{(Chapman-Kolmogorov equation)}, \tag{10.2.15}
$$

where $\mathbf{P}(t)$ is the transition matrix $[P_{ij}(t)]$. $\mathbf{Q} = [q_{ij}]$ is called the infinitesimal matrix, which is formally defined as

$$\mathbf{Q} = \lim_{\Delta t \to 0} \frac{\mathbf{P}(\Delta t) - \mathbf{I}}{\Delta t}, \qquad (10.2.16)$$

where $\mathbf{I}$ is an identity matrix,

$$\mathbf{I} = [\delta_{ij}] = \begin{matrix} 1 & i = j \\ 0 & i \neq j \end{matrix} \; .$$

Here we have derived the Chapman-Kolmogorov equation from the simple two-state model. The power of the Chapman-Kolmogorov equation is that it is applicable to the Markov process of n states (i.e., channel transition probabilities of multiple states). In the following sections, we will list the generalized rules derived from the Chapman-Kolmogorov equation for n-state channels and use the two-state model to verify these rules.

10.3 Chapman-Kolmogorov equation for n-state channels

Evidence in recent years has pointed out that the majority of ion channels in biological membranes, either voltage or ligand gated, do not follow the simple two-state transition scheme. Instead, most channels have more than two states. For example, some channels may go to blocked, inactivated, or desensitized states in addition to the open and closed states. Others may have multiple closed or open states. A good example for multiple-state channels is the K^+ channel in the squid axon. Although Hodgkin and Huxley use the two-state scheme to model each gating particle, their proposal of four gating particles for the channel is kinetically indistinguishable from the five-state scheme shown in figure 10.1.

$$C_0 \underset{\beta_n}{\overset{4\alpha_n}{\rightleftharpoons}} C_1 \underset{2\beta_n}{\overset{3\alpha_n}{\rightleftharpoons}} C_2 \underset{3\beta_n}{\overset{2\alpha_n}{\rightleftharpoons}} C_3 \underset{4\beta_n}{\overset{\alpha_n}{\rightleftharpoons}} O$$

Figure 10.1   Kinetic scheme for the Hodgkin-Huxley K^+ channel. C_{0-3} represent 4 closed states, and O is the open state of the channel.

Since all gating particles are independent and kinetically indistinguishable, all states with the same number of particles at closed states may be lumped together as one state with rate coefficients equal to the sum of individual rate coefficients. Therefore, the Hodgkin and Huxley n^4 model can be simplified to the state diagram in figure 10.1 with rate coefficients 4α, 3α, 2α, α (from left to right) and 4β, 3β, 2β, and β (from right to left).

The same argument leads to an eight-state scheme for the Hodgkin and Huxley Na$^+$ channel gating, which is kinetically indistinguishable from the Hodgkin and Huxley m^3h model (see figure 10.2).

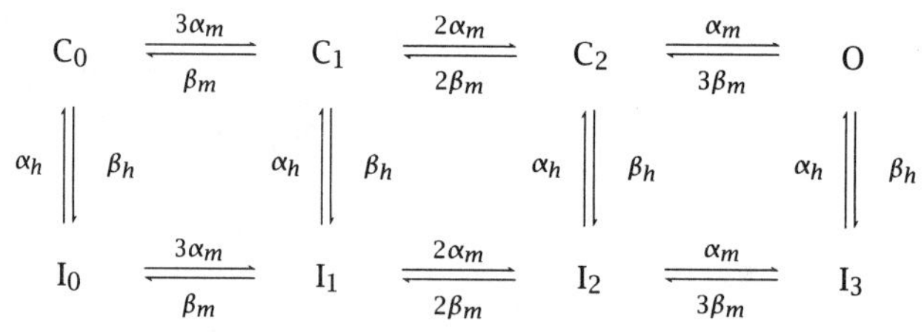

$$
\begin{array}{ccccccc}
C_0 & \underset{\beta_m}{\overset{3\alpha_m}{\rightleftharpoons}} & C_1 & \underset{2\beta_m}{\overset{2\alpha_m}{\rightleftharpoons}} & C_2 & \underset{3\beta_m}{\overset{\alpha_m}{\rightleftharpoons}} & O \\
\alpha_h \updownarrow \beta_h & & \alpha_h \updownarrow \beta_h & & \alpha_h \updownarrow \beta_h & & \alpha_h \updownarrow \beta_h \\
I_0 & \underset{\beta_m}{\overset{3\alpha_m}{\rightleftharpoons}} & I_1 & \underset{2\beta_m}{\overset{2\alpha_m}{\rightleftharpoons}} & I_2 & \underset{3\beta_m}{\overset{\alpha_m}{\rightleftharpoons}} & I_3
\end{array}
$$

Figure 10.2   Kinetic scheme for the Hodgkin-Huxley Na$^+$ channel. C_{0-2} represent 3 closed states, and O is the open state of the channel. I_{0-3} represent the 4 inactivated, nonconducting states.

It is therefore very important to develop mathematical tools that can help in analyzing multiple-state single-ion-channel data. The Chapman-Kolmogorov equation of the nth order can be used to describe transition probabilities and kinetic parameters of n-state single channels. The rules listed below are general properties of n-state Markov processes that satisfy the Chapman-Kolmogorov equations. We will list them and verify each of them with the two-state model. In the next section, we will apply these rules to more complex transition schemes involving more than two states. Because of the relative simplicity of the two-state model, it is important to understand the essence of the transition parameters and the relationship between whole-cell current and single-channel current associated with this model. Such understanding can serve as a conceptual guide for comprehending the parameters of the multiple-state channels, which are sometimes much more difficult to picture.

Rule 1: If $[x(t) \geq 0]$ represents the states of a Markov process of an n-state channel, $\{1, 2, 3, \ldots, n\} = S$, with the matrix $\mathbf{P}(t)$ of channel transition probabilities

$$P_{ij}(t) = \text{Prob}[(x(t_0 + t) = j) | (x(t_0) = i)],$$

where $i, j \epsilon S$, and $t_0 \geq 0$. $\mathbf{P}(t)$ will satisfy the Chapman-Kolmogorov equation

$$\frac{d\mathbf{P}(t)}{dt} = \mathbf{P}(t)\mathbf{Q}.$$

Verification with the two-state model: done in the previous section, equations 10.2.1–10.2.15.

Rule 2: The infinitesimal matrix of nth order will satisfy the following expressions:

$$\mathbf{Q} = \lim_{\Delta t \to 0} \frac{\mathbf{P}(\Delta t) - \mathbf{I}}{\Delta t},$$

$$\mathbf{Q} = [q_{ij}]$$

$$q_{ij} = - \text{ (sum of transition rates leading away from the state)} \quad i = j$$

$$= + \text{ (transition rate from state } i \text{ to state } j) \quad i \ne j.$$

Verification with the two-state model: From equations 10.2.9 and 10.2.16,

$$\mathbf{Q} = \lim_{\Delta t \to 0} \frac{\mathbf{P}(\Delta t) - \mathbf{I}}{\Delta t} = \lim_{\Delta t \to 0} \frac{\begin{bmatrix} 1 - \alpha\Delta t & \alpha\Delta t \\ \beta\Delta t & 1 - \beta\Delta t \end{bmatrix} - \begin{bmatrix} 1 & 0 \\ 0 & 1 \end{bmatrix}}{\Delta t}$$

$$= \lim_{\Delta t \to 0} \frac{\begin{bmatrix} -\alpha\Delta t & \alpha\Delta t \\ \beta\Delta t & -\beta\Delta t \end{bmatrix}}{\Delta t} = \begin{bmatrix} -\alpha & \alpha \\ \beta & -\beta \end{bmatrix}.$$

From rule 2,

$$C \underset{\substack{\beta \\ \alpha \\ 1}}{\overset{}{\rightleftharpoons}} O \qquad \mathbf{Q} = \begin{bmatrix} q_{11} & q_{12} \\ q_{21} & q_{22} \end{bmatrix} = \begin{bmatrix} -\alpha & \alpha \\ \beta & -\beta \end{bmatrix}.$$

Rule 3: The time interval spent by the channel in any state once entered is exponentially distributed with mean (mean lifetime in that state) $= -q_{ii}^{-1}$.
Verification with the two-state model: From chapter 9, the pdf for open lifetime $= \alpha e^{-\alpha t}$. Thus, mean open lifetime

$$\overline{\tau}_o = \int_0^\infty t\alpha e^{-\alpha t} = \frac{1}{\alpha} = -q_{11}^{-1}.$$

The pdf for closed lifetime $= \beta e^{-\beta t}$, thus mean closed lifetime

$$\overline{\tau}_c = \int_0^\infty t\beta e^{-\beta t} = \frac{1}{\beta} = -q_{22}^{-1}.$$

Rule 4: $\mathbf{Q}$ is a singular matrix. That is, determinant $\mathbf{Q}$ (det $\mathbf{Q}$) $= 0$.
Verification with the two-state model:

$$\det \mathbf{Q} = \begin{vmatrix} -\alpha & \alpha \\ \beta & -\beta \end{vmatrix} = (\alpha\beta - \alpha\beta) = 0.$$

Rule 5: Since $\mathbf{Q}$ is a singular matrix, it has only $n - 1$ nonzero eigenvalues. Eigenvalue λ is defined by $\det(\mathbf{Q} - \lambda\mathbf{I}) = 0$. (For readers unfamiliar with linear algebra, consult Noble 1969.)

Verification with the two-state model:

$$\det(\mathbf{Q} - \lambda\mathbf{I}) = \begin{vmatrix} -\alpha - \lambda & \alpha \\ \beta & -\beta - \lambda \end{vmatrix} = 0.$$

$$(\alpha + \lambda)(\beta + \lambda) - \alpha\beta = 0.$$

$$\alpha\beta + \alpha\lambda + \beta\lambda + \lambda^2 - \alpha\beta = 0.$$

$$\lambda(\alpha + \beta + \lambda) = 0.$$

$$\lambda = 0, \quad \lambda = -(\alpha + \beta).$$

Rule 6: The general solution of the Chapman-Kolmogorov equation can be written in two forms:

$$\mathbf{P}(t) = e^{\mathbf{Q}t} = \mathbf{I} + \mathbf{Q}t + (\mathbf{Q}t)^2/2! + \cdots,$$

and

$$P_{ij}(t) = p_j(\infty) + w_1 e^{\lambda_1 t} + w_2 e^{\lambda_2 t} + \cdots, \tag{10.3.17}$$

where $p_j(\infty)$ is the equilibrium probability that the channel is in state j. $\lambda_1, \lambda_2, \ldots,$ are nonzero eigenvalues of the matrix $\mathbf{Q}$.

Verification with the two-state model:

$$P_{11}(t) = p_1(\infty) + w_1 e^{\lambda t}.$$

Since only one nonzero eigenvalue exists, which is $-(\lambda + \beta)$, and since

$$P_{11}(0) = p_1(0) = p_1(\infty) + w_1 e^0,$$

then

$$w = p_1(0) - p_1(\infty).$$

Therefore,

$$P_{11}(t) = p_1(\infty) + (p_1(0) - p_1(\infty)) \, e^{-(\alpha+\beta)t}.$$

Since

$$\frac{dP_{11}(t)}{dt} = -\alpha P_{11}(t) + \beta P_{12}(t) \text{ (equation 10.2.10)},$$

at $t \to \infty$, steady state,

$$\frac{dP_{11}(\infty)}{dt} = -\alpha P_{11}(\infty) + \beta[1 - P_{11}(\infty)] = 0.$$

Therefore,

$$P_{11}(\infty) = P_1(\infty) = \frac{\beta}{\alpha + \beta},$$

and

$$P_{11}(t) = \frac{\beta}{\alpha + \beta} + \left(P_1(0) - \frac{\beta}{\alpha + \beta}\right)e^{-(\alpha+\beta)t}. \tag{10.3.18}$$

Similar procedures yield

$$P_{12}(t) = \frac{\alpha}{\alpha + \beta} + \left(P_{12}(0) - \frac{\alpha}{\alpha + \beta}\right)e^{-(\alpha+\beta)t}, \tag{10.3.19}$$

$$P_{21}(t) = \frac{\beta}{\alpha + \beta} + \left(P_{21}(0) - \frac{\beta}{\alpha + \beta}\right)e^{-(\alpha+\beta)t}, \tag{10.3.20}$$

$$P_{22}(t) = \frac{\alpha}{\alpha + \beta} + \left(P_2(0) - \frac{\alpha}{\alpha + \beta}\right)e^{-(\alpha+\beta)t}. \tag{10.3.21}$$

Rewrite equations 10.3.18–10.3.21 in matrix form:

$$
\mathbf{P}(t) = \begin{bmatrix} P_{11}(t) & P_{12}(t) \\ P_{21}(t) & P_{22}(t) \end{bmatrix}
$$

$$
= \begin{bmatrix} \frac{\beta}{\alpha+\beta} + \left(P_1(0) - \frac{\beta}{\alpha+\beta}\right)e^{-(\alpha+\beta)t} & \frac{\alpha}{\alpha+\beta} + \left(P_1(0) - \frac{\alpha}{\alpha+\beta}\right)e^{-(\alpha+\beta)t} \\ \frac{\beta}{\alpha+\beta} + \left(P_2(0) - \frac{\beta}{\alpha+\beta}\right)e^{-(\alpha+\beta)t} & \frac{\alpha}{\alpha+\beta} + \left(P_2(0) - \frac{\alpha}{\alpha+\beta}\right)e^{-(\alpha+\beta)t} \end{bmatrix}.
$$

Since $P_{12}(0) = P_1(0)$, $P_{21}(0) = P_2(0)$.

Note that all $P_{ij}(t)$ in the above matrix follow the form of

$$y(t) = y_\infty - \left[(y_\infty - y_0)e^{-(\alpha+\beta)t}\right].$$

This form is identical to the time dependence of the gating variable in the gate model of previous chapters, because they are all solutions of the first-order differential equations.

Graphically,

$$\mathbf{P}(t) = \begin{bmatrix} P_{11}(t) & P_{12}(t) \\ P_{21}(t) & P_{22}(t) \end{bmatrix} = \qquad\qquad (10.3.22)$$

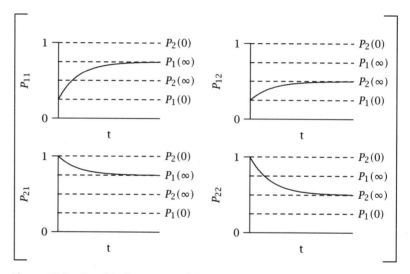

Figure 10.3   Graphic description of the time courses of P_{11}, P_{12}, P_{21}, and P_{22}.

Note that each $P_{ij}(t)$ transition follows the same exponential time course $\propto e^{-(\alpha+\beta)t}$ where $-\lambda = (\alpha + \beta) = \frac{1}{\tau}$. $\tau = \frac{1}{\alpha+\beta}$ is called the relaxation *time* constant, and $\alpha + \beta$ is called the relaxation *rate* constant. Therefore, the eigenvalues λ_i are negative reciprocals of the relaxation time constants. The relaxation time is also the *decay time* of the macroscopic current (whole-cell current, N channels of the same type), as illustrated by figure 10.4.

Rule 7: The power spectral density function $S(f)$ of the current noise mediated by N k-state channels ($\mathbf{Q}$ matrix has $k - 1$ eigenvalues) can be fitted by $k - 1$ Lorentzians.

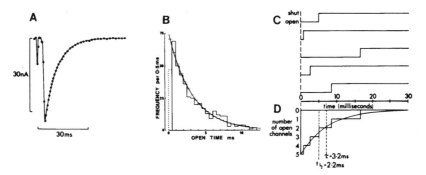

Figure 10.4 (A) Postsynaptic whole-cell current evoked by presynaptic nerve stimulation. The decay of the current can be fitted with a single exponential, which indicates that the postsynaptic channel kinetics can be approximated by a two-state scheme. (B) Probability density function of the channel open lifetime of a two-state channel. (C) Simulated behavior of five individual two-state channels that are open at $t = 0$. The channels stay open for a random (exponentially distributed) length of time with a mean of 3.2 msec. (D) Sum of the five records in (C). The total number of open channels decays exponentially with a time constant of 3.2 msec. (From Colquhoun 1981.)

Verification with the two-state model ($k = 2$):

Since the power spectral density function, according to the Wiener-Khinchin theorem, is the Fourier transform of the covariance function of the whole-cell current, we first have to derive the covariance function for one channel, then for N channels, and then take the Fourier transform.

For a stationary two-state channel, the mean current over a long period of time is equal to

$$\mu_{I_1} = E[I(t)] = I_1 P_1(\infty) = \frac{I_1 \beta}{\alpha + \beta},$$

and the variance is equal to

$$\sigma_I^2 = I_1 P_1(\infty) I_1 P_2(\infty) = \frac{I_1^2 \alpha \beta}{(\alpha + \beta)^2}.$$

The covariance function

$$C(\tau) = \frac{I_1^2 \alpha \beta}{(\alpha + \beta)^2} e^{-(\alpha+\beta)\tau}.$$

For N two-state channels, covariance function $C_N(\tau)$ is N times $C_1(\tau)$.

$$C_N(\tau) = \frac{N I_1^2 \alpha \beta}{(\alpha + \beta)^2} e^{-(\alpha+\beta)\tau} = \frac{\mu_I \gamma (V - E_i) \alpha}{\alpha + \beta} e^{-(\alpha+\beta)\tau}.$$

The power spectral density function for N two-state channels $S(f)$ can be obtained by the Fourier transform of $C_N(\tau)$, according to the Wiener-Khinchin theorem (chapter 9).

$$S(f) = \mathfrak{R}(\mathcal{F}[C_N(\tau)]) = \mathfrak{R}\left(2\int_{-\infty}^{\infty} \frac{\mu_I \gamma(V-E_i)\alpha}{\alpha+\beta} e^{-(\alpha+\beta)\tau} e^{-i2\pi f\tau}\,d\tau\right)$$

$$= \frac{2\mu_I\gamma(V-E_i)\alpha}{\alpha+\beta} \cdot \frac{\alpha+\beta}{(\alpha+\beta)^2 + (2\pi f)^2}.$$

$$S(f) = \frac{2\mu_I\gamma(V-E_i)\alpha}{\lambda^2 + (2\pi f)^2} = \frac{S(0)}{1 + (2\pi f/\lambda)^2}, \tag{10.3.23}$$

where

$$S(0) = \frac{2\mu_I\gamma(V-E_i)\alpha}{(\alpha+\beta)^2} \text{ and } \lambda = -(\alpha+\beta). \tag{10.3.24}$$

This form of $S(f)$ is called single *Lorentzian*.

One can define corner frequency (f_c) as the frequency at which $S(f)$ is equal to half of its maximum value, that is,

$$\frac{S(f_c)}{S(0)} = \frac{1}{1 + (2\pi f_c/\lambda)^2} = \frac{1}{2}.$$

Thus,

$$f_c = \frac{|\lambda|}{2\pi} = \frac{\alpha+\beta}{2\pi}. \tag{10.3.25}$$

The power spectral density analysis of physiological channels was first applied by Katz and Miledi (1970, 1972) while studying the ACh-induced noise in the neuromuscular junction. Anderson and Stevens (1973), using the voltage clamp technique, showed that the ACh-induced current noise in the frog neuromuscular junction could be fitted by a single exponential (figure 10.5). Based on equation 10.3.23, the single-channel conductance γ can be obtained by

$$\gamma = \frac{S(0)(\alpha+\beta)^2}{2\mu_I(V-E_i)\alpha} \approx \frac{S(0)\alpha}{2\mu_I(V-E_i)} \quad (\text{if } \alpha \gg \beta).$$

Since $S(0)$ and μ_I are determined by experimental data (figure 10.5), the value γ can be obtained. Additionally, from the corner frequency in figure 10.5 and equation 10.3.25, one can estimate the value of $\alpha + \beta$.

The ACh-gated current noise gives an example of channels whose transition can be approximated by the two-state model. For n-state channels, $S(f)$ can be fitted by $n-1$ order Lorentzians in the form of

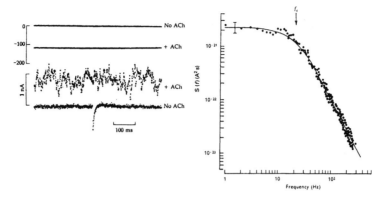

Figure 10.5 Currents measured from a frog muscle under voltage clamp. The currents are displayed at low gain (A) and at much higher gain (B). In the resting end plate, the low-gain record shows a zero net current. The high-gain record shows low noise and a single inward current transient, which is a miniature end-plate current from the spontaneous discharge of a single presynaptic transmitter vesicle. When a steady low concentration of ACh is applied iontophoretically to the end plate, the low-gain record shows a large, steady, inward end-plate current. The high-gain record reveals fluctuations due to the superimposed stochastic opening of many channels. (C) Spectral density curve (or power density spectrum) of current fluctuations produced at the end plate by acetylcholine. The membrane potential was clamped at -60 mV. (Adapted from Anderson and Stevens 1973.)

$$S(f) = \frac{S(0)}{\prod_{i=1}^{n-1}\left[1 + \left(\frac{2\pi f}{\lambda_i}\right)^2\right]},$$

where $\lambda_i (= -2\pi f_{c_i})$ are the nonzero eigenvalues of the $\mathbf{Q}$ matrix. The corner frequency (f_{c_i}) equals the nonzero eigenvalue of the $\mathbf{Q}$ matrix divided by -2π.

Example 10.1

A cell contains 10^5 voltage-gated Na^+ channels, and each of them follows the two-state transition scheme with voltage-dependent transition rate constants.

$$\text{closed} \underset{\alpha(V)}{\overset{\beta(V)}{\rightleftharpoons}} \text{open}$$

state number   2   1

The voltage-dependence of α is known to be

$$\alpha(V) = 380e^{-V/40 \text{ mV}} \quad (\text{sec}^{-1}),$$

but that of $\beta(V)$ is unknown. The power spectral density functions of the whole-cell Na^+ current at $V_c = -80$ mV and $+20$ mV are given below.

Example 10.1 (continued)

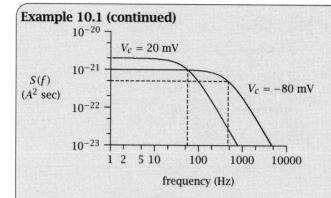

a. Write the **Q** matrices for this transition scheme (in sec^{-1}) at $V_c = -80$ mV and at $V_c = +20$ mV.

b. What are the steady-state whole-cell currents (assume the cell has Na^+ channels only) when the cell is voltage-clamped at -80 mV and at $+20$ mV? ($\gamma_{Na} = 20$ pS, $E_{Na} = +50$ mV.)

c. Draw the whole-cell current trace in appropriate units (of current and time) when the cell is stepped from -80 mV to $+20$ mV for 10 msec and then stepped back to -80 mV. Label the time constants (τ) (with appropriate values and units) of the current at the onset and cessation of the 10 msec voltage step.

Answer to example 10.1

a. $V_c = -80$ mV,

$\alpha(V) = 380e^{+80/40} = 2808 \ \ (sec^{-1})$.

From the figure, $f_c \approx 470$ Hz, taken from $S(f_c) = \frac{S(0)}{2}$.
Since $\alpha + \beta = 2\pi f_c$, therefore

$\beta = 2\pi f_c - \alpha = 2952 - 2808 = 144 \ sec^{-1}$.

Therefore,

$$\mathbf{Q} = \begin{pmatrix} -\alpha & \alpha \\ \beta & -\beta \end{pmatrix} = \begin{pmatrix} -2808 & 2808 \\ 144 & -144 \end{pmatrix} (sec^{-1}).$$

$V_c = +20$ mV,

$\alpha(V) = 380e^{-20/40} = 230 \ sec^{-1}$.

Answer to example 10.1 (continued)

From the figure, $f_c \approx 56$ Hz.

$\beta = 2\pi f_c - \alpha = 352 - 230 = 122$ sec^{-1}.

$$\mathbf{Q} = \begin{pmatrix} -\alpha & \alpha \\ \beta & -\beta \end{pmatrix} = \begin{pmatrix} -230 & 230 \\ 122 & -122 \end{pmatrix} \ (\text{sec}^{-1})$$

b. Steady-state whole-cell current

$$\mu_I = N p_1(\infty) I_1 = 10^5 \left(\frac{\beta}{\alpha + \beta} \right) \gamma (V - E_{Na}).$$

At $V_c = -80$ mV,

$$\mu_I = 10^5 \left(\frac{144}{2808 + 144} \right) 20 \times 10^{-12} (-80 - 50) \times 10^{-3} = -12.7 \text{ nA}.$$

At $V_c = +20$ mV,

$$\mu_I = 10^5 \left(\frac{122}{230 + 122} \right) 20 \times 10^{-12} (20 - 50) \times 10^{-3} = -20.8 \text{ nA}.$$

c. At $V_c = -80$ mV,

$$\tau = \frac{1}{\alpha + \beta} = \frac{1}{2808 + 144} = 0.00034 \text{ sec} = 0.34 \text{ msec}.$$

At $V_c = +20$ mV,

$$\tau = \frac{1}{\alpha + \beta} = \frac{1}{230 + 122} = 0.0028 \text{ sec} = 2.8 \text{ msec}.$$

Thus, the current trace

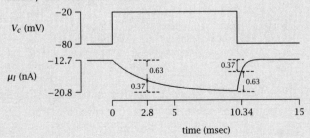

10.4   Stochastic analysis of n-state channels

10.4.1   Channels involving three-state transition schemes

In the last section, we described the rules derived from the Chapman-Kolmogorov equation that are applicable to n-state channels. In this section, we will apply these rules to two three-state transition schemes (schemes II and III). These two three-state schemes are extremely important in single-channel data analysis because numerous channels under physiological conditions can be described by them. These are the *blocked* and *agonist binding* schemes.

Scheme II, blocked scheme: (*Example:* ACh-gated channels in the presence of channel blocker gallamine. Colquhoun and Sheridan 1981.)

$$\begin{array}{ccccc} & & & x_B & \\ & & & + & \\ \text{closed} & \overset{\beta}{\underset{\alpha}{\rightleftharpoons}} & \text{open} & \overset{k_B}{\underset{k_{-B}}{\rightleftharpoons}} & \text{blocked} \\ \text{state number:} \quad 3 & & 1 & & 2 \end{array}$$

where x_B is the concentration of the blocker molecule B.

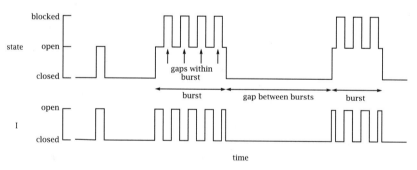

Figure 10.6   Schematic diagram illustrating transitions between various states (top) and observed single-channel currents (bottom) for the blocked scheme.

Scheme III, agonist binding scheme: (*Example:* ACh-gated channels in neuromuscular junction. Castillo and Katz 1957.)

$$A \quad + \quad R \quad \underset{k_{-1}}{\overset{k_{+1}}{\rightleftharpoons}} \quad AR \quad \underset{\alpha}{\overset{\beta}{\rightleftharpoons}} \quad AR^*$$

$$x_A$$

state number : $\qquad$ 3 $\qquad$ 2 $\qquad$ 1

x_A is the concentration of the agonist molecule A.

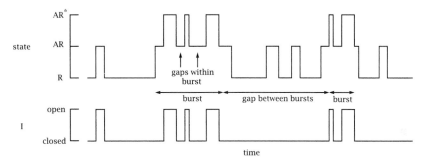

Figure 10.7 Schematic diagram illustrating transitions between various states (top) and observed single-channel currents (bottom) for the agonist binding scheme.

Next, we will use the generalized rules described in the last section to analyze schemes II and III. More rules will be derived during the analysis, and they can be applied to more complex schemes.

As stated in rule 1 (see section 10.3), the transition probabilities for both schemes should satisfy the Chapman-Kolmogorov equation. We do not derive this here because it is mathematically tedious. Readers interested in the complete derivation should consult Colquhoun and Hawkes (1981, 1982).

From rule 2, the infinitesimal matrices for the two schemes can be written as

$$\mathbf{Q}_{II} = \begin{bmatrix} -(\alpha + k_B x_B) & k_B x_B & \alpha \\ k_{-B} & -k_{-B} & 0 \\ \beta & 0 & -\beta \end{bmatrix}$$

(x_B : concentration of blocker B. Based on the law of mass action, the transition coefficient for open to blocked can be written as $k_B x_B$.)

$$\mathbf{Q}_{III} = \begin{bmatrix} -\alpha & \alpha & 0 \\ \beta & -(\beta + k_{-1}) & k_{-1} \\ 0 & k_1 x_A & -k_1 x_A \end{bmatrix}$$

(x_A : concentration of agonist A. Law of mass action gives $k_1 x_A$ for transition coefficient 3 → 1.)

From rule 3, the mean lifetimes in each state can be expressed as follows:

1. **Scheme II, blocked scheme:**

$$\text{Mean open lifetime } = m_o = -q_{11}^{-1} = \frac{1}{\alpha + k_B x_B}. \qquad (10.4.26)$$

$$
\begin{aligned}
\text{Mean blocked lifetime (gap within a burst)} \quad &= \quad m_w \\
&= \quad -q_{22}^{-1} \\
&= \quad \frac{1}{k_{-B}}. \qquad (10.4.27)
\end{aligned}
$$

$$\text{Mean closed lifetime (gap between bursts) } = m_b = -q_{33} = \frac{1}{\beta}.$$

The means of durations of various quantities characteristic of the burst can also be derived.

Number of openings per burst:

Define π_{ij} =Prob[channel at state i will, as its next transition, be at state j].

Then,

$$\pi_{12} = \frac{q_{12}}{q_{12} + q_{13}} = \frac{k_B x_B}{\alpha + k_B x_B},$$

$$\pi_{13} = \frac{q_{13}}{q_{12} + q_{13}} = \frac{\alpha}{\alpha + k_B x_B},$$

$$\pi_{21} = \frac{q_{21}}{q_{21} + q_{23}} = \frac{k_{-B}}{k_{-B} + 0} = 1.$$

Therefore,

$$
\begin{aligned}
p(1) \quad &= \text{Prob[one opening per burst]} \\
&= \text{Prob[channel once open, then closed]} \\
&= \pi_{13} \\
p(2) \quad &= \text{Prob[two openings per burst]} \\
&= \text{Prob[channel open once, then blocked]} \cdot \\
&\quad \text{Prob[blocked, then open]} \cdot \text{Prob[open, then closed]} \\
&= \pi_{12} \cdot \pi_{21} \cdot \pi_{13}.
\end{aligned}
$$

The extension of this argument gives

$$
\begin{aligned}
p(r) \quad &= \mathrm{Prob}[r \text{ openings per burst}] \\
&= (\pi_{12} \cdot \pi_{21})^{r-1} \pi_{13} \\
p(r) \quad &= (\pi_{12} \cdot 1)^{r-1}(1 - \pi_{12}).
\end{aligned}
$$

Mean number of openings per burst (m_r):

$$
m_r = \sum_{r=1}^{\infty} r p(r) = \frac{1}{1 - \pi_{12}} = \frac{\alpha + k_B x_B}{\alpha}. \tag{10.4.28}
$$

Combining equation 10.4.28 with equations 10.4.26 and 10.4.27, we can obtain

$$
\begin{aligned}
\text{\textit{Mean open time per burst}} \quad &= \quad m_r \cdot m_o \\
&= \quad \frac{\alpha + k_B x_B}{\alpha} \cdot \frac{1}{\alpha + k_B x_B} \\
&= \quad \frac{1}{\alpha}. \tag{10.4.29}
\end{aligned}
$$

$$
\begin{aligned}
\text{\textit{Mean closed time per burst}} \quad &= \quad (m_r - 1) \cdot m_w \\
&= \quad \left(\frac{\alpha + k_B x_B}{\alpha} - 1 \right) \left(\frac{1}{k_{-B}} \right) \\
&= \quad \frac{k_B x_B}{k_{-B} \alpha} = \frac{c_B}{\alpha}, \tag{10.4.30}
\end{aligned}
$$

where

$$
c_B = \frac{x_B}{k_B} = \frac{x_B k_B}{k_{-B}}, \quad K_B = \frac{k_{-B}}{k_B}, \text{ and} (m_r - 1) = \# \text{ of gaps/burst.}
$$

Adding equations 10.4.29 and 10.4.30 gives

$$
\begin{aligned}
\text{\textit{mean burst length}} \quad &= \quad \text{mean open time per burst} + \\
&\qquad\qquad \text{mean closed time per burst} \\
&= \quad \frac{1}{\alpha} + \frac{c_B}{\alpha} = \frac{1 + c_B}{\alpha}.
\end{aligned}
$$

2. **Scheme III, the agonist binding scheme:**

Mean open lifetime $= m_o = -q_{11}^{-1} = \frac{1}{\alpha}$.

Mean AR lifetime (gap within a burst) $= m_w = -q_{22}^{-1} = \frac{1}{\beta + k_{-1}}$.

Mean closed lifetime (gaps between bursts) $= m_b \neq -q_{33}^{-1} = \frac{1}{k_1 x_A}$
($\neq$ because closed time is made of time spent in *two* closed states)

Number of openings per burst

$$\pi_{21} = \frac{q_{21}}{q_{21} + q_{23}} = \frac{\beta}{\beta + k_{-1}} = 1 - \pi_{23}.$$

$$p(r) \quad = \text{Prob}[r \text{ openings per burst}]$$

$$= (\pi_{21})^{r-1} \pi_{23}.$$

Mean number of openings per burst m_r:

$$m_r = \sum_{r=0}^{\infty} r p(r) = \sum_{r=0}^{\infty} r (\pi_{21})^{r-1} (1 - \pi_{21}) = \frac{1}{1 - \pi_{21}} = \frac{\beta}{k_{-1}} + 1.$$

$$
\begin{aligned}
\text{\textit{Mean open time per burst}} \quad &= \quad m_r \cdot m_o \\
&= \quad \frac{k_{-1} + \beta}{k_{-1}} \cdot \frac{1}{\alpha} \\
&= \quad \frac{\beta + k_{-1}}{\alpha k_{-1}}. && (10.4.31)
\end{aligned}
$$

$$
\begin{aligned}
\text{\textit{Mean closed time per burst}} \quad &= \quad (m_r - 1) \cdot m_w \\
&= \quad \left(\frac{\beta}{k_{-1}}\right)\left(\frac{1}{\beta}\right) \\
&= \quad \frac{1}{k_{-1}}. && (10.4.32)
\end{aligned}
$$

Adding equations 10.4.31 and 10.4.32 gives

$$\text{\textit{mean burst length}} = \frac{\beta + k_{-1}}{\alpha k_{-1}} + \frac{1}{k_{-1}} = \frac{\alpha + \beta + k_{-1}}{\alpha k_{-1}}.$$

Rule 4 predicts $\det \mathbf{Q}_{II} = 0$, $\det \mathbf{Q}_{III} = 0$.

$$
\det \mathbf{Q}_{II} = \begin{vmatrix} -(\alpha + k_B x_B) & k_B x_B & \alpha \\ k_{-B} & -k_{-B} & 0 \\ \beta & 0 & -\beta \end{vmatrix} \tag{10.4.33}
$$

$$
= -(\alpha + k_B x_B) k_{-B} \beta + \alpha \beta k_{-B} + \beta k_B k_{-B} x_B = 0. \tag{10.4.34}
$$

$$\det \mathbf{Q}_{III} = \begin{vmatrix} -\alpha & \alpha & 0 \\ \beta & -(\beta + k_{-1}) & k_{-1} \\ 0 & k_1 x_A & -k_1 x_A \end{vmatrix} \tag{10.4.35}$$

$$= -\alpha k_1 x_A (\beta + k_{-1}) + \alpha k_{-1} k_1 x_A + \alpha \beta k_1 x_A = 0. \tag{10.4.36}$$

Therefore, rule 4 holds.

Eigenvalues for $\mathbf{Q}_{II}$ and $\mathbf{Q}_{III}$:

$$\det(\mathbf{Q}_{II} - \lambda_{II}) = \begin{vmatrix} -(\alpha + k_B x_B) - \lambda & k_B x_B & \alpha \\ k_{-B} & -k_{-B} - \lambda & 0 \\ \beta & 0 & -\beta - \lambda \end{vmatrix} = 0$$

$$-(\alpha + k_B x_B + \lambda)(k_{-B} + \lambda)(\beta + \lambda) \quad + \quad \alpha\beta(k_{-B} + \lambda) +$$
$$(\beta + \lambda)k_{-B}k_B x_B = 0$$

$$\lambda \left[(\beta k_{-B} + \alpha k_{-B} + k_B x_B \beta) + (k_{-B} + \alpha + \beta + k_B x_B)\lambda + \lambda^2 \right] = 0.$$

$$\lambda = 0.$$

$$\lambda^2 + (\alpha + \beta + k_B x_B + k_{-B})\lambda + \alpha k_{-B} \left(1 + \frac{\beta}{\alpha} + \frac{k_B \beta}{k_{-B}\alpha} x_B \right) = 0.$$

$$\lambda^2 + (\alpha + \beta + k_B x_B + k_{-B})\lambda + \alpha k_{-B} \left[1 + \frac{\beta}{\alpha} \left(1 + \frac{x_B}{k_B} \right) \right] = 0.$$

$$\lambda_1 + \lambda_2 = -b = \alpha + \beta + k_B x_B + k_{-B}. \tag{10.4.37}$$

$$\lambda_1 \lambda_2 = c = \alpha k_{-B} \left[1 + \frac{\beta}{\alpha} \left(1 + \frac{x_B}{K_B} \right) \right]. \tag{10.4.38}$$

Similarly,

$$\det(\mathbf{Q}_{III} - \lambda_{III}) = \begin{vmatrix} -\alpha - \lambda & \alpha & 0 \\ \beta & -(\beta + k_{-1}) - \lambda & k_{-1} \\ 0 & k_1 x_A & -k_1 x_A - \lambda \end{vmatrix} = 0.$$

$$\lambda_1 + \lambda_2 = \alpha + \beta + k_{-1}. \tag{10.4.39}$$

$$\lambda_1 \lambda_2 = \alpha k_{-1}. \tag{10.4.40}$$

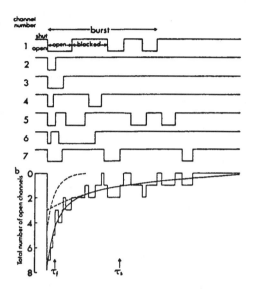

Figure 10.8 Schematic illustration to show how bursting behavior of single channels may result in biphasic relaxation. (A) Simulated behavior of seven individual ion channels in the presence of an ion-channel-blocking drug. Channels are supposed to be opened nearly synchronously at time zero. Each channel produces only one burst of openings before it finally shuts (as marked on channel 1, which has two blockages and therefore three openings before it shuts). (B) Sum of all seven records shown in (A). The initial decline is rapid (time constant τ_f) as open channels become blocked, but the current thereafter declines more slowly (time constant τ_s). The continuous line is the sum of two exponential curves (shown separately as dashed lines) with time constants τ_f and τ_s. The slow time constant, under these conditions, reflects primarily the burst length rather than the length of an individual opening. (From Sakmann and Neher 1983.)

The openings and closings of seven single channels following the blocked scheme and their ensemble averages are given in figure 10.8.

The general solutions for transition probabilities also follow the result in rule 6 (equation 10.3.17), $P_{ij}(t) = p_j(\infty) + w_1 e^{\lambda_1 t} + w_2 e^{\lambda_2 t}$, with eigenvalues λ_1 and λ_2 obtained in the previous section.

The relaxation time course of each transition $P_{ij}(t)$ is biphasic (two exponentials). Also, the decay time of the whole-cell current induced by ACh in the presence of blockers can be fitted by two exponentials (figure 10.9).

$$I(t) = I(\infty) + w_1' e^{\lambda_1 t} + w_2' e^{\lambda_2 t}.$$

The power spectral density function for the blocked scheme, for example, can be fitted by two Lorentzians, according to Rule 7. Experimental results show that $S(f)$ of the current noise induced by ACh in the presence of gallamine (blocker) can be fitted by two Lorentzians (figure 10.10).

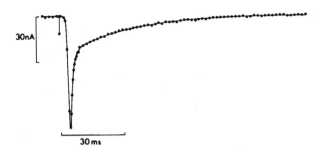

Figure 10.9 Postsynaptic current mediated by channels following a three-state transition scheme (ACh-gated channels in the presence of channel blocker gallamine). The decay of the current can be fitted with two exponentials. (From Colquhoun and Sheridan 1981.)

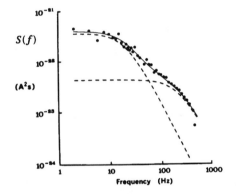

Figure 10.10 Power spectral density function of the fluctuation by the postsynaptic current mediated by channels following a three-state transition scheme (ACh-gated channels in the presence of channel blocker gallamine). Data points can be fitted with two Lorentzians (dashed curves). (From Colquhoun and Sheridan 1981.)

$$S(f) = \frac{S(0)}{\left[1 + \left(\frac{2\pi f}{\lambda_1}\right)^2\right]\left[1 + \left(\frac{2\pi f}{\lambda_2}\right)^2\right]} \quad \text{(double Lorentzian),}$$

where $-\lambda_1 = 2\pi f c_1$, and $-\lambda_2 = 2\pi f c_2$. The relationships between λ_1, λ_2, and α, β, k_B, and k_{-B} are given explicitly by equations 10.4.37 and 10.4.38.

Example 10.2

The figure below shows the transitions between three states of the agonist-binding scheme of a channel.

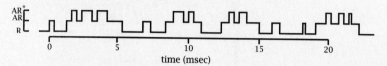

1. Draw the transition trace of a single-channel current trace.

2. Assume this record is representative of channel activity for long periods of time. What are the approximate values of β and k_{-1}?

Answer to example 10.2

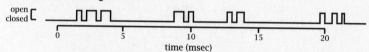

Mean AR lifetime (gaps within a burst) $= \frac{1}{\beta + k_{-1}}$.

$(0.3 + 0.4 + 0.3 + 0.5 + 0.5 + 0.2)\frac{1}{6}$

$= 0.37$ msec

$= \frac{1}{\beta + k_{-1}}$

$\beta + k_{-1} = 2703$ sec^{-1}.

Mean closed time per burst $= \frac{1}{k_{-1}}$ (equation 10.4.32).

$(0.7 + 0.3 + 0.5 + 0.7)\frac{1}{4} = 0.55$ msec

$= \frac{1}{k_{-1}}$.

$k_{-1} = 1818$ sec^{-1}.

$\beta = 2703 - 1818 = 885$ sec^{-1}.

10.4.2   Channels involving many-state transition schemes

The three-state schemes are useful for analyzing channels with simple binding or blocked states. Nevertheless, these schemes are often not adequate for describing channel mechanisms. For example, mechanisms involving more than one open state or several closed or inactivated states, or involving cyclic reactions, are quite common. These schemes usually involve as many as tens of discrete states, and it is tedious to extract kinetic parameters without the assistance of a computer. Since the same principles and rules apply to these higher-order transition schemes, we can write the **Q** matrix and determine the mean lifetimes of the channel at each state. It is much more difficult, however, to give explicit expressions for the eigenvalues, time constants, and the relationships between measured parameters (e.g., burst lifetime) and the rate coefficients. Numerical analysis with computers is needed to accomplish these, and the results are less intuitive than for the two- or three-state schemes.

10.4.2.1 Two-agonist binding scheme: A five-state scheme

Consider a channel that is bound by two agonist molecules sequentially, and the channel may open with either one or two molecules bound. The transition scheme for such channels can be written as follows:

$$
\begin{array}{ccc}
\text{state: } 5 & R & \\
& k_{+1} \updownarrow k_{-1} & \\
4 \quad AR & \underset{\alpha_1}{\overset{\beta_1}{\rightleftharpoons}} & AR^* \quad 1 \\
k_{+2} \updownarrow k_{-2} & & k^*_{+2} \updownarrow k^*_{-2} \\
3 \quad A_2R & \underset{\alpha_2}{\overset{\beta_2}{\rightleftharpoons}} & A_2R^* \quad 2
\end{array}
$$

Note that this scheme has two open states (states 1 and 2). If the two open states have identical conductances, it is difficult to distinguish between them. If the two conductances are different, then one may determine the mean lifetime of the two states individually. The **Q** matrix of this transition scheme can be written as follows:

$$
\mathbf{Q} = \begin{array}{c} 1 \\ 2 \\ 3 \\ 4 \\ 5 \end{array} \begin{bmatrix}
-(\alpha_1 + k^*_{+2} x_A) & k^*_{+2} x_A & 0 & \alpha_1 & 0 \\
2k^*_{-2} & -(\alpha_2 + 2k^*_{-2}) & \alpha_2 & 0 & 0 \\
0 & \beta_2 & -(\beta_2 + 2k_{-2}) & 2k_{-2} & 0 \\
\beta_1 & 0 & k_{+2} x_A & -(\beta_1 + k_{+2} x_A + k_{-1}) & k_{-1} \\
0 & 0 & 0 & 2k_{+1} x_A & -2k_{+1} x_A
\end{bmatrix}
$$

The mean lifetime of the channel at each of the five states can be written as q_{ii}^{-1}, as stated in rule 3. It is very tedious, but straightforward because the same rules apply, to calculate the eigenvalues, relaxation time constants, and other kinetic parameters. Special computer programs are developed to deal with this type of calculation.

10.4.2.2 The Hodgkin and Huxley K^+ and Na^+ channels

Hodgkin and Huxley modeled the K^+ and Na^+ channels in the squid axon with n^4 and $m^3 h$, respectively (see Chapter 6). Each of the gating particles can be either at open (1) or closed (0) states.

The K^+ channels   The Hodgkin and Huxley n^4 model for K^+ channels is described by the state diagram shown in figure 10.1.

The $\mathbf{Q}$ matrix of this transition scheme, according to rule 2, can be written as follows:

$$
\mathbf{Q}_{HHK} = \begin{array}{c c}
 & \begin{array}{c c c c c} 1 & \quad 2 & \quad 3 & \quad 4 & \quad 5 \end{array} \\
\begin{array}{c} 1 \\ 2 \\ 3 \\ 4 \\ 5 \end{array} &
\left[\begin{array}{c c c c c}
-4\beta & 4\beta & 0 & 0 & 0 \\
\alpha & -(\alpha+3\beta) & 3\beta & 0 & 0 \\
0 & 2\alpha & -(2\alpha+2\beta) & 2\beta & 0 \\
0 & 0 & 3\alpha & -(3\alpha+\beta) & \beta \\
0 & 0 & 0 & 4\alpha & -4\alpha
\end{array} \right]
\end{array}
$$

According to rule 3, the mean open lifetime of a single Hodgkin and Huxley K^+ channel equals $-q_{11}^{-1} = \frac{1}{4\beta}$. According to rules 5 and 6, there are four nonzero eigenvalues and thus 4 time constants (τ_n, $\tau_n/2$, $\tau_n/3$, and $\tau_n/4$). The Hodgkin and Huxley model has only one time constant $\tau_n = \frac{1}{\alpha+\beta}$, but it is raised to the 4th power. Kinetically, the two models are indistinguishable.

The Na^+ channels   Similarly, the Hodgkin and Huxley Na^+ channel is described by the state diagram shown in figure 10.2.

The $\mathbf{Q}$ matrix of this scheme can be written as:

$$
\mathbf{Q}_{HHNa} = \begin{array}{c c}
 & \begin{array}{c c c c} 1 & \qquad 2 & \qquad 3 & \qquad 4 \end{array} \\
\begin{array}{c} 1 \\ 2 \\ 3 \\ 4 \\ 5 \\ 6 \\ 7 \\ 8 \end{array} &
\left[\begin{array}{c c c c}
-(3\beta_m+\beta_h) & 3\beta_m & 0 & 0 \\
\alpha_m & -(\alpha_m+2\beta_m+\beta_h) & 2\beta_m & 0 \\
0 & 2\alpha_m & -(2\alpha_m+\beta_m+\beta_h) & \beta_m \\
0 & 0 & 3\alpha_m & -(3\alpha_m+\beta_h) \\
0 & 0 & 0 & 0 \\
0 & 0 & \alpha_h & 0 \\
0 & \alpha_h & 0 & 0 \\
\alpha_h & 0 & 0 & 0
\end{array} \right.
\end{array}
$$

$$
\begin{array}{c c c c} 5 & \qquad 6 & \qquad 7 & \qquad 8 \end{array} \\
\left. \begin{array}{c c c c}
0 & 0 & 0 & \beta_h \\
0 & 0 & \beta_h & 0 \\
0 & \beta_h & 0 & 0 \\
\beta_h & 0 & 0 & 0 \\
-(3\alpha_m+\alpha_h) & 3\alpha_m & 0 & 0 \\
\beta_m & -(2\alpha_m+\beta_m+\alpha_h) & 2\alpha_m & 0 \\
0 & 2\beta_m & -(\alpha_m+2\beta_m+\alpha_h) & \alpha_m \\
0 & 0 & 3\beta_m & -(3\beta_m+\alpha_h)
\end{array} \right]
$$

Again, from rule 2, the mean open lifetime of the Hodgkin and Huxley Na^+ channel $= -q_{11}^{-1} = \frac{1}{3\beta_m+\beta_h}$. There should be seven nonzero eigenvalues for the $\mathbf{Q}$ matrix and thus seven time constants. The Hodgkin and Huxley m^3h model only provides two time constants, τ_h and τ_m. For macroscopic currents, the eight-state model is kinetically indistinguish-able from the m^3h model. However, tests of the K$^+$ and Na$^+$ channels based on the state diagrams shown above show that the Hodgkin and Huxley model for Na$^+$ and K$^+$ channel gating is not correct in all details.

Limitations of the Hodgkin and Huxley model for the Na$^+$ channels

1. The deactivation time constant is slower than predicted from the Hodgkin and Huxley model.

   According to the Hodgkin and Huxley m^3h model, during deacti-vation (offset of a depolarizing voltage-clamp pulse), only one m particle is needed to be moved from permissive to nonpermissive state; thus the I_{Na} during deactivation should be three times faster than the off gating current (all three m particles are moved back to nonpermissive state, of course, under the condition of no inacti-vation). This is not supported by experimental data, which indicate that the off-I_{Na} is as slow as off gating current (figures 6.21 and 6.23 in chapter 6). This indicates that the movement of m particles may not be completely independent.

2. The gating current time course does not fit the Hodgkin and Huxley model.

   According to the Hodgkin and Huxley model, gating current is gener-ated by the movement of identical and independent gating particles. At the onset of a depolarizing pulse, I_g should rise instantaneously and then fall with a single exponential. However, experimental data on gating currents show a rising phase at the onset and multiple ex-ponential components at the falling phase. This indicates that Na$^+$ channel gating may be more complex than the Hodgkin and Huxley assumptions of identical independent gating particles.

3. Single-channel measurements give a longer channel open lifetime (in the absence of inactivation) than predicted by Hodgkin and Huxley's model. According to the Hodgkin and Huxley model (the eight-state scheme), the mean open lifetime of the Na$^+$ channel should be

$$\overline{\tau}_o = -q_{11}^{-1} = \frac{1}{3\beta_m + \beta_h} \quad \text{(rule 3)}.$$

If one removes inactivation (e.g., by pronase),

$$\overline{\tau}_o = \frac{1}{3\beta_m}.$$

Use the Hodgkin and Huxley original equation

$$\beta_m = \frac{1 - m_\infty}{\tau_m}$$

for $V_c = -40$ mV, $\quad \beta_m = 69$ sec^{-1}, $\quad \dfrac{1}{3\beta_m} = 5$ msec.

Experimentally, however, Patlak and Horn found that the mean open lifetime after removal of inactivation is about 30 msec. This indicates that the rate of deactivation is slower than what Hodgkin and Huxley imply, and it is consistent with the observation in (a) that the tail I_{Na} current is slower than the prediction of the Hodgkin and Huxley model (that deactivation is three times faster than the off gating current).

In conclusion, the three discrepancies of the Hodgkin and Huxley model mentioned above indicate that although the Hodgkin and Huxley formulation can explain most properties of the macroscopic Na$^+$ current, it falls short in explaining detailed gating mechanisms.

A transition scheme modified from the Hodgkin and Huxley model has been proposed by Patlak (1991). This scheme seems to be able to explain more single-channel and gating-current data than the original Hodgkin and Huxley scheme. The new scheme involves an additional closed state (C_1) prior to the m gates, and two inactivated states (I_1 and I_2).

One can use the rules we learned to write the **Q** matrix and calculate the kinetic parameters, and then compare the results with experimental data. So far, the question of how Na$^+$ channels are gated is not fully resolved. The model shown above is widely used, and it is one of the most useful developed to date. By obtaining more single-channel data, knowing more about the structure-function relationships, and fitting data with more realistic models, one can expect to determine the true transition mechanisms of the Hodgkin and Huxley Na$^+$ channels.

10.5 Conclusions: Analysis of ionic currents

This chapter provides mathematical and conceptual tools for analyzing the stochastic behaviors of single ion channels and of whole-cell current mediated by a large number of ion channels. The basic scheme of analysis for the single-channel and whole-cell currents can be summarized as follows:

For I_1 (single channel currents): By obtaining long periods of records ($T \rightarrow \infty$), one can observe burst behavior and guess the transition scheme. One can measure mean open, closed, burst lifetimes, gaps between and within bursts, etc., to determine the transition scheme and transition rate constants (α, β, k_1, k_{-1}, k_B, k_{-B}, etc.).

For I_N (N channels, whole-cell current): Measurements of $S(f)$ or relaxation time constants give estimates of the number of eigenvalues λ_i, which gives the number of transition states (number of nonzero $\lambda_i + 1$). One can also estimate γ from measuring $S(0)$, $\sigma_{I_N}^2$, and μ_I, the transition rate constants from the corner frequencies of the power spectral density function.

The stochastic analysis given above can be used to describe the behavior of single ionic channels or ensembles of channels. The time-dependent mechanisms of these channels cannot be described by explicit mathematical equations; they are random processes that can be determined only by statistical functions. This stochastic description of molecular mechanisms of membrane channels constitutes a new formulation for modern neurophysiological research. More and more new insights have been brought into this line of neurobiology as more and more mathematics is learned and applied by neuroscientists.

Dr. Carl Pantin, Director of Studies in Trinity, said, "You must continue to learn mathematics," and this I have endeavored to do during the rest of my life.—Alan Hodgkin

10.6 Review of important concepts

1. The transition probabilities of a channel following an n-state transition scheme (nth order Markov process) can be described by a matrix differential equation (Chapman-Kolmogorov equation).

2. The derivation of the Chapman-Kolmogorov equation is derived with the two-state transition scheme, which elucidates all basic concepts and general rules implicated in the equation.

3. Mathematical rules derived from the Chapman-Kolmogorov equation for n-state channels are given, and the applications of these rules on the two-state, three-state, and multiple-state channels are described.

4. The number of transition states of a channel can be estimated by the relaxation time course and the shape of the power spectral density function ($S(f)$) of the whole-cell current.

5. The transition rate coefficients in any given transition scheme can be estimated by measuring various parameters in single-channel current records, based on rules derived from the Chapman-Kolmogorov equation.

10.7 Homework problems

1. A cell contains 10^5 voltage-gated Na^+ channels, and each of them follows the two-state transition scheme with voltage-dependent transition rate constants.

$$\text{closed} \underset{\alpha(V)}{\overset{\beta(V)}{\rightleftharpoons}} \text{open}$$

$$\text{state number} \quad 2 \qquad\qquad 1$$

$$\alpha(V) = 400e^{-V/40 \text{ mV}} \quad (\text{sec}^{-1})$$
$$\beta(V) = 40e^{+V/20 \text{ mV}} \quad (\text{sec}^{-1})$$

(a) Write the $\mathbf{Q}$ matrices for this transition scheme (in sec^{-1}) at $V = -80$ mV and at $V = +20$ mV.

(b) What are the steady-state whole-cell currents when the cell is voltage-clamped at -80 mV and at $+20$ mV (assume the cell has Na^+ channels only)? ($\gamma_{Na} = 20$ pS, $E_{Na} = +50$ mV.)

(c) Draw the whole-cell current trace in appropriate units (of current and time) when the cell is stepped from -80 mV to $+20$ mV for 10 msec and then stepped back to -80 mV. Label the time constants (τ) (with appropriate values and units) of the current at the onset and cessation of the 10 msec voltage step.

2. Channels gated by extracellular ligand A follow the two-state transition scheme

$$A \; + \; R \; \underset{k_2}{\overset{k_1}{\rightleftharpoons}} \; AR$$

$\quad x_A \qquad$ closed $\qquad$ open.

If $k_1 = 2 \times 10^5 \; \text{sec}^{-1} \cdot M^{-1}, k_2 = 100 \; \text{sec}^{-1}, x_A = 10^{-5} \; M,$

(a) Write the infinitesimal matrix $\mathbf{Q}$ with numbers. What is the unit of each matrix element? What is det $\mathbf{Q}$?

(b) If one records one of these channels for a long time, what are the average open and closed lifetimes? About what percentage of the time is the channel open? What percentage of the time is the channel closed?

(c) A cell contains 10^4 of these channels in its plasma membrane, and the conductance of a single open channel is 10 pS. The open channel allows only Na^+ to go through ($E_{Na} = +50$ mV). What is the steady-state current of the cell when it is voltage-clamped at -50 mV and 10^{-5} M of ligand A is applied to the extracellular space?

(d) What is the approximate amplitude of the noise in the A-induced current in (c)?

(e) If x_A suddenly drops from 10^{-5} M to 0 M, draw the time course of the decay of the A-induced current in this cell. What is the value of τ?

(f) Sketch the power spectral density function $S(f)$ of this A-induced current. What is the value of $S(0)$? What is the value of f_c?

3. The membrane of a neuron at rest is permeable only to K^+, and the resting potential is -80 mV. In the presence of 10^{-4} M of glutamate, which is assumed to open Na^+ channels, the membrane potential of the neuron is -40 mV. The glutamate-gated Na^+ channels follow the two-state transition scheme

$$A \; + \; R \; \underset{k_2}{\overset{k_1}{\rightleftharpoons}} \; AR$$

$\quad x_A \qquad$ closed $\qquad$ open,

where x_A is the concentration of glutamate, and k_1 and k_2 are transition coefficients that are voltage-dependent, that is,

$$k_1(V) = 2000e^{V/40 \text{ mV}} \ (\text{sec}^{-1} \cdot \text{M}^{-1}),$$

$$k_2(V) = e^{-V/40 \text{ mV}} \ (\text{sec}^{-1}).$$

(a) Write the infinitesimal matrix $\mathbf{Q}$ in numbers (with appropriate units) for the glutamate-gated channels in the presence of 10^{-4} M glutamate.

(b) If one records one of these glutamate-gated Na$^+$ channels for a long period of time ($V_c = -40$ mV; assume the channel is stationary), about what percentage of the time is the channel open? What percentage of the time is the channel closed?

(c) Given that the input resistance of the neuron *at rest* is 10^7 Ω, $E_{Na} = +40$ mV, and the single-channel conductance of the glutamate-gated Na$^+$ channel is 20 pS, estimate the number of glutamate-gated Na$^+$ channels in the neuron activated by 10^{-4} M glutamate.

4. A voltage-dependent K$^+$ channel is gated by two identical gating particles y (valence + 1) that follow the following kinetic scheme

$$p \underset{\alpha(V)}{\overset{\beta(V)}{\rightleftharpoons}} 1-p$$

(Permissive)   (Nonpermissive),

where p and $1-p$ are the probabilities of particle y in the permissive and nonpermissive states, respectively. The two states are separated by a single energy barrier within the membrane.

(a) Draw a figure of the membrane with the energy barrier and the locations of the gating particle at its permissive and nonpermissive states so that the channel can be activated by membrane hyperpolarization.

(b) Draw the whole-cell current of a neuron that contains 10^4 of these voltage-dependent K$^+$ channels (and no other channels) when the neuron is stepped from 0 mV to -50 mV. ($E_K = -90$ mV; the activation voltage range for the K$^+$ channel is between 0 and -60 mV.) Does the time course of the onset of the K$^+$ current follow a single or multiple exponential?

5. The plasma membrane of a neuron contains 10^5 identical channels (C) that are gated by extracellular ligand A:

$$A + C \overset{k_{+1}}{\underset{k_{-1}}{\rightleftharpoons}} \underset{\text{closed}}{AC} \overset{\beta}{\underset{\alpha}{\rightleftharpoons}} \underset{\text{open}}{AC^*}$$

Given that k_{+1} and k_{-1} are *much* faster than α and β, the resting potential of the neuron is -80 mV, and channel C is permeable only to Na^+ ($[Na^+]_{out} / [Na^+]_{in} = +2$),

   (a) Sketch the whole-cell current evoked by ligand A when the neuron is voltage-clamped

      i. at the resting potential; and

      ii. at $+50$ mV.

   (b) What is (are) the approximate relaxation time constant(s) of the whole-cell current?

   (c) Sketch the power spectral density function ($S(f)$ vs. f in log-log scales) of the current noise during application of ligand A.

6. A channel follows the following transition scheme:

$$\underset{3}{\text{closed}} \overset{10}{\underset{100}{\rightleftharpoons}} \underset{1}{\text{open}} \overset{50}{\underset{5}{\rightleftharpoons}} \underset{2}{\text{blocked}} \; .$$

state number   3   1   2

All transition constants are in $\sec^{-1}$.

   (a) What are the eigenvalues of the $\mathbf{Q}$ matrix for this transition scheme?

   (b) What are the mean open lifetime, mean blocked lifetime, and mean closed lifetime of the channel?

   (c) Sketch the power spectral density function $S(f)$ of the current mediated by N of these channels. Label the cutoff frequencies (f_c) with appropriate values.

7. If a channel follows the following transition scheme:

$$\underset{4}{\text{closed}_1} \overset{k_1}{\underset{k_{-1}}{\rightleftharpoons}} \underset{3}{\text{closed}_2} \overset{k_2}{\underset{k_{-2}}{\rightleftharpoons}} \underset{2}{\text{closed}_3} \overset{k_3}{\underset{k_{-3}}{\rightleftharpoons}} \underset{1}{\text{open}} \; ,$$

state number   4   3   2   1

(a) Write the infinitesimal matrix $\mathbf{Q}$ for this channel.

(b) What are the mean lifetimes of the channel in each of the four states?

8. A channel follows the following transition scheme:

$$\text{closed} \underset{300}{\overset{100}{\rightleftharpoons}} \text{open} \underset{50}{\overset{50}{\rightleftharpoons}} \text{inactivated} .$$

state number   3   1   2

The numbers are transition rate constants (sec^{-1}).

(a) What are the eigenvalues of the $\mathbf{Q}$ matrix for this channel?

(b) Sketch the power spectral density function of the current mediated by N of these channels. What are the values of f_c?

9. The figure below is the single-channel current trace recorded from an ACh-gated channel in the presence of gallamine (ACh channel blocker). The reversal potential of this channel is about $+10$ mV, and the channel is voltage-clamped at -40 mV.

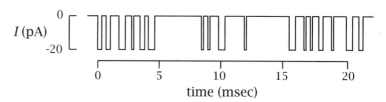

(a) What is the conductance of a single open channel if the *I-V* relation of the open channel is linear?

(b) Draw the single-channel current trace when the channel is clamped at $+60$ mV.

(c) If the trace given above is representative of current traces for a long period of time, estimate the values of α, β, $k_B x_B$, and k_{-B}.

10. A neuron contains N identical channels that can be activated by transmitter X. Each of these channels follows the two-state transition scheme

$$C \underset{\alpha}{\overset{\beta}{\rightleftharpoons}} O.$$

The macroscopic current (N channels) activated by a brief pulse of transmitter X is shown in figure 10.4A. Five single-channel current records and their sum are shown in figure 10.4C and 10.4D.

(a) Write the relaxation time constants of the macroscopic current (figure 10.4A) and of the sum of the five single-channel currents (figure 10.4B) in terms of the rate coefficients (α and β).

(b) Explain, by using the information in this chapter, why the two relaxation time constants are different.

11. The figure below shows the single channel current of a ligand-gated channel induced by 10^{-4} M ligand at various voltages (V_c). The transition scheme of this channel is two state, that is,

$$C \underset{\alpha}{\overset{\beta}{\rightleftharpoons}} O.$$

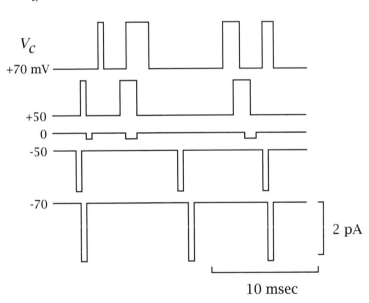

(a) If the records given are representative for current traces over a long period of time, estimate the values of α and β at each voltage.

(b) Plot, with appropriate values, the steady-state probability of channel opening with respect to membrane voltage. What is the steady-state probability of channel opening at $V_m = -30$ mV?

(c) What is the unitary conductance of this channel?

(d) If a neuron contains 10^4 of the channels shown in the figure, draw the *whole-cell current* induced by a brief "puff" of ligand when the cell is clamped at -70, -50, 0, $+50$, and $+70$ mV. Give the approximate values of the maximum current induced by 10^{-4} M ligand at each V_c and mark the relaxation time with appropriate time constants (values and units).

12. A channel follows the following transition scheme:

$$\text{closed}_1 \underset{100}{\overset{10}{\rightleftharpoons}} \text{open} \underset{5}{\overset{50}{\rightleftharpoons}} \text{closed}_2$$
$$\phantom{\text{closed}_1}\quad 2 \qquad\qquad 1 \qquad\qquad 3.$$

All transition coefficients are in $\sec^{-1}$.

(a) Write the **Q** matrix for this channel. What are the eigenvalues of the **Q** matrix?

(b) Sketch a histogram of the open lifetimes of this channel. What is the mean open lifetime?

(c) Sketch the power spectral density function $S(f)$ of the current mediated by N of these channels. Label the cutoff frequencies (f_c) with appropriate values, and explain how f_c's are determined.

13. We mentioned that Patlak's model gave the best description of the Hodgkin and Huxley Na^+ channel. Given the rate coefficients (in $\sec^{-1}$) of Patlak's model

(a) What are the values of k_1 and k_2 (in $\sec^{-1}$)?

(b) Write the infinitesimal matrix **Q** for this transition scheme.

(c) What is the mean open lifetime of this sodium channel?

(d) In the presence of pronase (which removes inactivation), what is the mean open lifetime of the sodium channel?

14. A cell contains 10^5 identical channels that are gated by transmitter A. While voltage-clamped at -40 mV, bath application of 10^{-4} M transmitter A increases the steady-state opening probability from 0 to 0.2 and results in a steady-state current. The power spectral density of this current is given below.

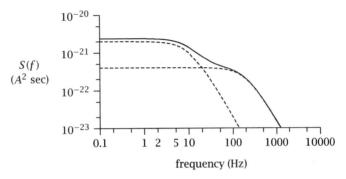

The reversal potential of these channels is $+20$ mV, and the single channel conductance is 20 pS.

   (a) What is the mean amplitude of the whole-cell current induced by transmitter A?

   (b) Draw the relaxation time course, with appropriate values of time constant(s), after transmitter A is suddenly dropped from 10^{-4} M to 0 mM.

   (c) Assume these channels follow the three-state agonist binding scheme:

   $$A + R \underset{k_{-1}}{\overset{k_1}{\rightleftharpoons}} AR \underset{\alpha}{\overset{\beta}{\rightleftharpoons}} AR^*,$$

   and $k_{-1} = 1000$ sec^{-1}, $k_1 = 100$ sec^{-1}, $\alpha = 10\,\beta$, $\lambda_1 + \lambda_2 = \alpha + \beta + k_{-1}$.

   Write the $\mathbf{Q}$ matrix for these channels (each matrix element should be in number of transitions per second).

15. Horn and Vanderberg (1984) showed that the Na$^+$ channels mediating action potentials (Hodgkin and Huxley's Na$^+$ channels) can be fitted with the following kinetic model:

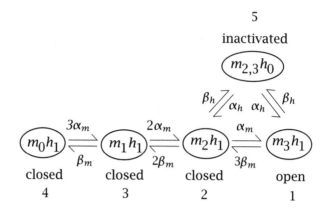

(a) Write the infinitesimal matrix (**Q** matrix) for the above transition scheme.

(b) Given that $\alpha_m = 100 \text{ sec}^{-1}$, $\beta_m = 1000 \text{ sec}^{-1}$, $\alpha_h = 50 \text{ sec}^{-1}$, and $\beta_h = 100 \text{ sec}^{-1}$, what is the mean open lifetime of the Na^+ channel?

(c) The peak current of a neuron containing 10^4 identical Na^+ channels with the above kinetic scheme when voltage is stepped from -100 mV to 0 mV is 2 nA. The conductance of a single open Na^+ channel (γ) is 100 pS. What is the opening probability of a single Na^+ channel? ($E_{Na} = +50$ mV.)

11   Synaptic Transmission I: Presynaptic Mechanisms

11.1   Introduction

Synapses[1] are the principal sites for communication among neurons. Although there can be electrical field interactions as well as communication via changes in ions in the extracellular space, synapses, both electrical and chemical, are specialized structures that have evolved for passing information from one neuron to another. Electrical synapses, which are direct electrical connections between neurons formed by way of *gap junctions*, play important roles in cell-to-cell communication, especially among glial cells and among neurons during development. In the adult mammalian central nervous system, however, electrical synapses among principal cell types (e.g., pyramidal neurons) are rare. The information passed from one neuron to another is therefore mainly in the form of chemicals called neurotransmitters released from the presynaptic terminals of chemical synapses.

In this and the next four chapters we will cover in some detail the physiology and biophyics of synaptic transmission. We will start with the presynaptic side of the synapse and derive some of the classical methods for analyzing the stochastic nature of transmitter release. Chapter 12 will discuss the role of presynaptic Ca^{2+} in the release process, then in chapter 13 we will move to the postsynaptic side of the synapse and discuss the mechanisms for secretion-excitation coupling. We will also discuss some of the biophysical principles associated with electrical synaptic transmission. In chapter 14 we will discuss the use of electrical field recordings for studying synaptic transmission, and in chapter 15 we will review some general principles and nomenclature for the study of synaptic plasticity.

---

[1] The word *synapse* comes from Greek, *syn*, together, and *haptein*, to fasten, and was first used by Sherrington.

As with any treatment of an important subject, something gets omitted. In the case of synaptic transmission, we will not be able to present much of the cell or developmental biology of the synapse, the biochemistry of the release process, or the neurochemistry of transmitter systems. For these important topics the reader is referred to several excellent books included in the reading list.

11.1.1 Why chemical synaptic transmission?

Although the prevailing view in the 1940s and early 1950s was in favor of electrical transmission and against the idea of released chemicals at synapses, it is fairly easy to show that the transmission of electrical signals between neurons in the absence of a synapse (either with a direct connection via gap junctions or with released chemicals) is highly unfavorable. Let's look at the situation of two neurons that have processes that are close to one another (refer to figure 11.1). Why can't the impulse from one neuron simply jump the gap and excite the next neuron, as was thought to be the case in the late 1940s? Assume that the resistance of the terminal membranes of neurons 1 and 2 is 1,000 $M\Omega$ (not unreasonable for a small area of membrane with a resistivity on the order of 10^4 to 10^5 Ω-cm^2), the input resistance of neuron 2 is 100 $M\Omega$, and the resistance of the extracellular space in the narrow gap between the processes of the two neurons is 1 $M\Omega$. Analysis of this relatively simple circuit yields an attenuation of potential from neuron 1 to neuron 2 by about 4 orders of magnitude, or about $1/10^4$ (the analysis of this circuit is left as a homework problem). Even if one reduced the gap between the two membranes such that the extracellular resistance increased to 10 $M\Omega$, the attenuation would still be about $1/10^3$, that is, for each 100 mV in neuron 1, about 100 μV would make it to neuron 2. Admittedly, this analysis assumes the steady state and ignores transient signals where membrane capacitance plays a role, but the point is that without some means of coupling one neuron to another, very little transfer of electrical events is likely to take place.

In the case of an electrical synapse, there is a direct connection from one neuron to another by way of a channel or gap junction. This channel reduces the resistance across the two terminal membranes that we had in the previous example, but even more importantly, it eliminates the electrical shunting by the extracellular space. The circuit representation of an electrical synapse is schematized in figure 11.2. The attenuation of potential from neuron 1 to neuron 2 is now only about 1/10.

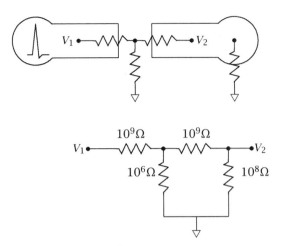

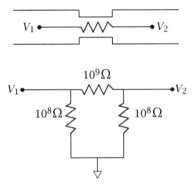

Figure 11.1 Schematic for the transfer of electrical signals between neurons in the absence of any direct electrical connections.

Figure 11.2 Transfer of electrical signals between neurons through a gap junction.

Obviously, with a direct connection there is a much better transfer of electrical events between neurons. The coupling, however, is relatively fixed, and this limits the capabilities (or modifiability) of the system. Also, transmission is sign conserving and the durations of pre- and postsynaptic events are roughly the same (the duration of the postsynaptic event depends on the membrane time constant). In general, electrical synapses are suitable for high-speed transfer of information and for synchronization of cells. They are also very important for signal processing in the retina. Some additional properties of electrical synapses will be covered in chapter 13.

11.2   Chemical transmission

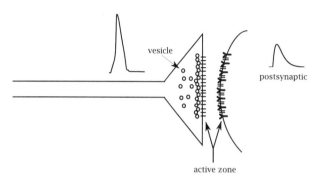

Figure 11.3   Schematic diagram of a synapse with an action potential in the presynaptic terminal and an EPSP in the postsynaptic cell.

For the remainder of this chapter and the next chapter, we will concentrate exclusively on synaptic transmission at chemical-releasing synapses. Most of what we know about synaptic transmission was gleaned from the study of two preparations: the neuromuscular junction (either the neuromuscular junction of a vertebrate such as the frog, or the invertebrate crayfish neuromuscular junction) and the squid giant synapse. The parts of a typical synapse are schematized in figure 11.3. The presynaptic axon can be myelinated or unmyelinated. The active zone contains the release sites for the neurotransmitter on the presynaptic side and the neurotransmitter receptors on the postsynaptic side. There is a narrow gap or cleft between the pre- and postsynaptic membranes. In general, the sequence of events at a synapse during neurotransmission is listed below:

1. **Action potential in nerve:** The action potential propagates down the axon and then invades the presynaptic terminal. The action potential in the axon is primarily dependent on Na^+ and K^+ channels.

2. **Action potential in terminal:** The action potential invades the terminal, presumably in a nondecremental fashion. There are many voltage-gated channels in the presynaptic terminal activated by the invasion of the action potential (e.g., Na^+, K^+, and Ca^{2+}) although the exact types of channels are largely unknown. The depolarization of the terminal by the action potential activates (at least) voltage-gated Ca^{2+} channels, allowing the entry of Ca^{2+} ions.

3. **Fusion of vesicle to membrane; release of chemical:** The entry of Ca^{2+} ions into the terminal near the release sites triggers some unknown sequence of events leading to the fusion to the plasma membrane of vesicles containing neurotransmitters. The fusion of vesicles causes release of one or more chemicals into the synaptic cleft. This process is also called *exocytosis*.

4. **Diffusion:** The neurotransmitters diffuse across the cleft and make contact with the postsynaptic membrane.

5. **Binding to receptor:** The neurotransmitter molecules bind to specialized receptors in the postsynaptic membrane.

6. **Gating of ion channels:** The binding of transmitter molecules to the receptors causes the rapid opening of ion channels. The opening of the channels causes a change in the membrane potential (either depolarization or hyperpolarization) of the postsynaptic neuron. At some synapses the binding of neurotransmitters to receptors does not directly gate ion channels. Instead, the activation of receptors triggers either the release of second-messenger molecules into the cytoplasm of the postsynaptic neuron, which then modulate ion channels, or the activation of GTP-binding proteins that couple to ion channels in the membrane and alter their function. The indirect coupling of neurotransmitters to ion channels through either second messengers or G proteins generally produces slower synaptic responses than those produced by the direct gating of channels by neurotransmitters.

7. **Recycling of vesicles:** After exocytosis the vesicle membrane gets pinched off from the plasma membrane to reform vesicles in a process called *endocytosis*.

Steps 1–6 above can take place in 0.5–1 msec at many synapses. In fact, at some insect neuromuscular junctions transmission can take place in less than 100 μsec. As mentioned above, there are two classical preparations for studying synaptic transmission: the neuromuscular junction and the squid giant synapse. Each has its own advantages and disadvantages. The vertebrate neuromuscular junction has the advantage of allowing one to record the spontaneous release of transmitters. Experiments at the neuromuscular junction led to the idea that transmitter release is quantal in nature. We will develop the theory for the quantal release of neurotransmitters in the following sections.

11.3   Experiments at the neuromuscular junction

If one places a microelectrode into a muscle fiber near the end plate, one can record spontaneous depolarizations, as depicted in figure 11.4. These spontaneous depolarizations are called miniature end-plate potentials or mEPPs, and they appear to occur at random intervals that average around 1 per sec. A single mEPP results from the release of a neurotransmitter-containing vesicle from the presynaptic terminal. If the amplitudes of all of the mEPPs recorded over a given time period (say 10 min) are measured and used to construct an amplitude histogram, one obtains a simple, unimodal distribution such as that illustrated in figure 11.5. The mean amplitude of all of the mEPPs recorded over that time period in this fictitious experiment was around 0.4 mV. The variance in the histogram may be due to many factors, such as the variance in the amount of transmitter in each vesicle, the amount of transmitter molecules binding to receptors, the number of postsynaptic receptors, and the probability of channel opening after binding a molecule of transmitter.

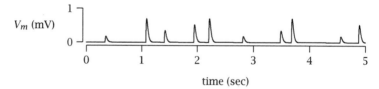

Figure 11.4   Examples of spontaneous, miniature end-plate potentials (mEPPs).

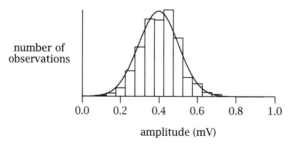

Figure 11.5   Amplitude histogram of spontaneous mEPPs.

Let us now do a slightly different experiment at this fictitious neuromuscular junction. First, we will add a high concentration of Mg^{2+} ions to the bath and at the same time lower the Ca^{2+} concentration. This change in the divalent cation concentrations reduces neuromuscular transmission (the reasons for this will become clear in the next chapter). Stimulation of the nerve innervating the muscle fiber elicits depolarizations (called end-plate potentials or EPPs) that fluctuate in amplitude from trial to trial. Such an experiment is illustrated in figure 11.6. In the early 1950s Castillo and Katz did exactly the kinds of experiments described here. They observed that the fluctuation in amplitude of the EPPs appeared to average around 0.4 mV, and suggested that the EPP was composed of unit multiples of mEPPs. They hypothesized that the fluctuation of EPP amplitude was due to a variation from trial to trial in the number of mEPPs (or *quanta*) released. This so-called *quantum hypothesis* is one of the cornerstones of our understanding of synaptic transmission. We will present in the following sections the theory and mathematics underlying this extremely important hypothesis.

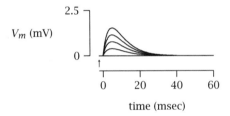

Figure 11.6 Evoked responses (EPPs) are superimposed to illustrate the variability in amplitude from trial to trial. The arrow indicates the time of nerve stimulation.

11.4 Statistical treatment of quantum hypothesis

In presenting the theory underlying the quantum hypothesis, we will first derive the equations describing the spontaneous release of transmitter in the form of mEPPs. We will then show how these equations can be extended to the case where there is a transient increase in probability of release following an action potential in the presynaptic fiber. Many of the equations and much of the theory (i.e., probability theory) are similar to those presented for single channels in chapter 9 (see also Stein 1980). As always, there are a number of assumptions, which are listed below.

1. n units (or sites) are available for release of transmitter with only a single release (i.e., single vesicle) occurring at each site. Without an action potential there is an extremely small but finite probability of release at each release site. This very small probabililty of release

is what leads to the infrequent appearance of a mEPP. Following an action potential, however, the probability of release is greatly increased for a brief period of time. This probability is assumed to be uniform over all release sites and has an average value of $\overline{p}$ (per action potential). The mean number of units released per action potential is $m = n\overline{p}$. If the probability of release at any release site is p and the probability is uniform over all release sites, then $\overline{p} = p$.

2. The amount of transmitter per unit of release is rate limiting. In other words, there are many more receptors on the postsynaptic membrane than there are transmitter molecules per unit of release. This also leads to the assumption that the variability in the amplitude of the mEPP is due to the variability in the number of transmitter molecules that make up each unit of release. (As we will see, this is an important assumption that may not be true in all cases.)

3. With low Ca^{2+} and high Mg^{2+} in the bath, p is small.

4. For now we will assume that n is large.

5. The release of one unit of transmitter is independent of the release of any other units of transmitter. This assumption leads to the idea that the release process is the summation of a number of independent events. *This is an important assumption.*

6. One unit of transmitter (vesicle) release will be called a quantum.

As was illustrated above, transmitter release occurs spontaneously as well as being evoked by nerve stimulation. From the assumption of independent events for release, we will first derive an equation that describes spontaneous release of transmitter. We will then extend that equation to one that describes evoked release.

11.4.1 Spontaneous release

Let us assume an average release rate of mEPPs of r/sec. Referring to the time line in figure 11.7, a release occurs at $t = 0$. We want to derive an equation that describes the probability of another release occurring, as a function of time, following this release. As we will see, this equation will describe the distribution of intervals between successive mEPPs.

The probability of one release in Δt is $r\Delta t$ (for small r and small Δt) while the probability of no release is $1 - r\Delta t$. Using the terminology presented in chapter 9, the probability of 0 quanta released in $t + \Delta t$ is

$$P(0, t + \Delta t) = P(0, t) \cdot (1 - r\Delta t). \tag{11.4.1}$$

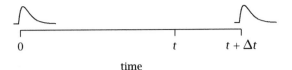

time

Figure 11.7 Time between successive mEPPs. One mEPP occurs at time 0 and another at time $t + \Delta t$.

Rearranging,

$$\frac{P(0, t + \Delta t)}{\Delta t} = \frac{P(0, t)}{\Delta t} - \frac{rP(0, t)\Delta t}{\Delta t} \qquad (11.4.2)$$

and

$$\frac{P(0, t + \Delta t) - P(0, t)}{\Delta t} = -rP(0, t). \qquad (11.4.3)$$

As $\Delta t \to 0$,

$$\lim_{\Delta t \to 0} \frac{P(0, t + \Delta t) - P(0, t)}{\Delta t} = \frac{d[P(0, t)]}{dt} = -rP(0, t). \qquad (11.4.4)$$

This is a simple first-order differential equation where we can solve for $P(0, t)$ and obtain

$$P(0, t) = e^{-rt}, \qquad (11.4.5)$$

where $P(0, 0) = e^0 = 1$.

We have derived this equation so that we can use it to derive what we are really interested in, that is, the probability of 1 quantum released in $t + \Delta t$, as Δt approaches zero. This can be determined by

$$P(1, t + \Delta t) = P(1, t) + P(0, t) \cdot P(1, \Delta t). \qquad (11.4.6)$$

Remember from chapter 9 that the probability density function is simply the derivative of the probability. The probability density can be obtained from equation 11.4.6 by

$$\begin{aligned} \frac{d[P(1, t)]}{dt} &= \lim_{\Delta t \to 0} \frac{P(1, t + \Delta t) - P(1, t)}{\Delta t} = P(0, t)\frac{P(1, \Delta t)}{\Delta t} \\ &= e^{-rt} \cdot \frac{r\Delta t}{\Delta t} \\ &= re^{-rt} = f_1(t). \end{aligned} \qquad (11.4.7)$$

The probability density function for the intervals between releases is therefore $f_1(t)$ (see [2]).

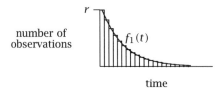

Figure 11.8   Probability density for intervals between successive mEPPs.

Figure 11.9   Measurement of intervals between successive mEPPs.

Equation 11.4.7 is plotted in figure 11.8 and is an important test of our assumption of independent release sites (assumption 5). If we measure the intervals between successive releases for a large number of mEPPs (depicted in figure 11.9), then these intervals should be distributed exponentially with the rate constant equal to r, the mean rate of release. If not, then our assumption of independence may be inappropriate.[3]

In addition to the intervals between successive releases, we may want to know something about the intervals between multiple releases—for example, the time from one release to the second, third, or some other, later release. This can easily be derived from equation 11.4.7, and the time line for this situation is illustrated in figure 11.10.

Figure 11.10   Time between multiple mEPPs. One mEPP occurs at time 0, a second at time u, and a third at time t.

To determine the probability density function for the time to the second release, we must consider all possible values of u, the time between successive releases. In other words, one release must occur in time u (with probability $f_1(u)$) and another in time $t - u$ (with probability $f_1(t - u)$), and the probability of these two occurring is the product of the individual probabilities. But since the first release can occur at any u, we must take

---

[2]This was obtained by letting the beginning of each mEPP be $t = 0$ and measuring the time to the next release.

[3]The concept of independent events was defined as a random, memoryless process in chapter 9.

into account all possible values of u. To do this we integrate (or summate over all u)

$$f_2(t) = \int_0^t f_1(t-u)f_1(u)du. \tag{11.4.8}$$

To extend this to the time to the third release, we have the probability of two events occurring in time u times the probability of one event occurring in time $t - u$. Again, however, we must take into account all possible values of u by integrating

$$f_3(t) = \int_0^t f_1(t-u)f_2(u)du. \tag{11.4.9}$$

This can be stated more generally as

$$f_k(t) = \int_0^t f_1(t-u)f_{k-1}(u)du. \tag{11.4.10}$$

These integrals are called *convolution integrals*. One method of solution, which is convenient in this case, uses Laplace transforms.

In general, the Laplace transform of $f_k(t) = \mathcal{L}[f_k(t)] = \mathbf{F}_k(s)$ and the inverse Laplace transform of $\mathbf{F}_k(s) = \mathcal{L}^{-1}[\mathbf{F}_k(s)] = f_k(t)$. The Laplace transform of equation 11.4.8 is

$$\mathbf{F}_2(s) = \mathbf{F}_1(s) \cdot \mathbf{F}_1(s) = [\mathbf{F}_1(s)]^2,$$

where this is the well-known property of the Laplace transform of a convolution integral (see book on Laplace transforms, e.g., Bracewell 1978), and, similarly, for equation 11.4.9

$$\mathbf{F}_3(s) = \mathbf{F}_1(s) \cdot [\mathbf{F}_1(s)]^2 = [\mathbf{F}_1(s)]^3,$$

which, for equation 11.4.10, leads to

$$\mathbf{F}_k(s) = [\mathbf{F}_1(s)]^k. \tag{11.4.11}$$

From a table of Laplace transforms,

$$\mathcal{L}\left[re^{-rt}\right] = r/(s+r), \text{ so}$$

$$\mathbf{F}_k(s) = r^k/(s+r)^k.$$

From a table of inverse Laplace transforms, we get $f_k(t)$ by looking up $\mathcal{L}^{-1}\left[r^k/(s+r)^k\right]$. This leads to the solution

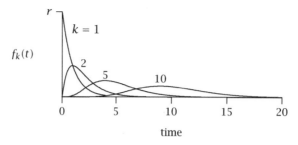

Figure 11.11 Gamma distribution for different k with an average rate of release (r) of 1/sec. Gamma distribution represents the distribution of intervals between some number (k) of mEPPs.

$$f_k(t) = \frac{r^k t^{k-1} e^{-rt}}{(k-1)!},$$
(11.4.12)

where $f_k(t)$ is the probability density function. ($f_k(t)$ is also called a *gamma distribution*, because the denominator $(k-1)!$ is the gamma function of k or $\Gamma(k)$.) $f_k(t)$ is the probability of the kth release happening in time t. (Remember that $\int_{-\infty}^{\infty} f_k(t)dt = 1$.) $f_k(t)$ is plotted in figure 11.11 for several different k's. The curve for $k = 1$ is the same as that plotted in figure 11.8.

As k gets large, $f_k(t)$ becomes a Gaussian or normal distribution. This would represent the probability density for the time for a large number of releases. For example, suppose you wanted to know how long you would have to wait for 10 releases to occur. Looking at the curve for $k = 10$ in figure 11.11, it would take, on average, slightly less than 10 sec for 10 releases. Sometimes it would take as little as 5 sec or as long as 15 sec, and so forth. Over any 20 sec period, however, the probability is very high that you would always observe at least 10 releases.

11.4.2 Evoked release

11.4.2.1 Poisson model We will now use some of the equations derived above for spontaneous release to explore quantitative aspects of evoked release. Evoked release occurs when multiple quanta of transmitter are released in a brief interval of time following an action potential. During this period the probability of release and the average rate of release are very high (i.e., r is large). In the case of evoked release, we are no longer interested in the distributions of time intervals between some number of releases, because essentially all of the releases we are interested in take place in a very brief period of time. Instead, given a high rate of release following an action potential, we want to know the probability of a certain number of quanta (k) being released. We can now define a new variable, m, which is the mean number of quanta released following an action potential, or, over many trials, $m = rt$, where t is the time following an action potential. The probability that k quanta are released, or $P(k)$, can be obtained by first substituting for rt into equation 11.4.12 and then integrating from 0 to t (total probability), or

$$
\begin{aligned}
P(k) &= \int_0^t \frac{r^k t^{k-1} e^{-m}}{(k-1)!} dt \\
&= \frac{r^k t^k e^{-m}}{k!}.
\end{aligned}
\tag{11.4.13}
$$

Again letting $m = rt$ = mean number of quanta released in time t (remember that r is much higher immediately following an action potential than under spontaneous release conditions), we obtain the well-known Poisson distribution

$$
P(k) = \frac{m^k e^{-m}}{k!},
\tag{11.4.14}
$$

where the variance = mean, or $\sigma^2 = m$. Also,

$$
P(k) = \frac{N_k}{N},
$$

where N_k is the number of times one observes k quanta released and N is the total number of trials.

The above is only one of many ways for deriving the Poisson distribution. What is particularly illustrative with this derivation is that we started with a description of the stochastic nature of spontaneous release and merely extended that description to release immediately following an action potential. The only difference between the two cases is that for evoked release the rate of release is very high following an action potential and the time window over which we are observing release is small. $P(k)$ is the probability that k quanta are released over a period of time (following an action potential) when the mean number released per action potential is m. m is called the *quantal content*.

The Poisson distribution (for different m) is illustrated in figure 11.12. It should be noted that it is very similar to that in figure 11.11. Equation 11.4.14 and figure 11.12 tell us the total probability that some number of quanta will be released following an action potential.

11.4.2.2 Binomial model In order for the Poisson distribution to describe transmitter release, there must be a large number of quanta that

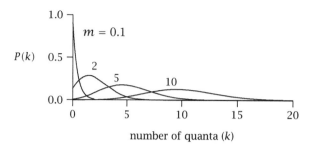

Figure 11.12   Poisson distribution for different m. The ordinate represents the total probability that some number of quanta (k) will be released following an action potential for a given m.

can be released (large n), each with low probability for release (small p). This is seldom the case except under special low-release conditions (e.g., low Ca^{2+}, high Mg^{2+}). If this is not the case, then we must use a binomial model. This can be derived quite easily.

Let the number of quanta (or number of release sites) = n, and let p = average probability of a quantum being released at any one site. Assume that p is uniform over all release sites and that n and p do not vary with time (that is, they are *stationary*). The probability of k quanta being released requires that there be k successes with probability p^k and $n - k$ failures with probability $(1 - p)^{n-k}$. From elementary probability theory, the number of ways to combine k and $n - k$ is given by

$$C_k^n = \frac{n!}{k!(n - k)!},$$ (11.4.15)

so

$$P_n(k) = \frac{n!}{k!(n - k)!} \cdot p^k (1 - p)^{n-k}.$$ (11.4.16)

This is the binomial distribution (where $m = np$ and $\sigma^2 = m(1 - p)$). $P_n(k)$ is the probability that k quanta will be released, where the probability is p for each of n release sites. In figure 11.13, the binomial distribution is plotted for $p = 0.1$ and $n = 1$ to 100.

The binomial model is generally more representative of the release process, especially for higher values of quantal content (m). The binomial model also reduces to the Poisson model as $n \to \infty$ (and $p \to 0$). The proof of this is left as a homework exercise.

The assumptions of uniformity and stationarity, however, provide somewhat severe restrictions. In simulations of quantal release, Brown et al. (1976) have shown that estimates of m using simple binomial models can

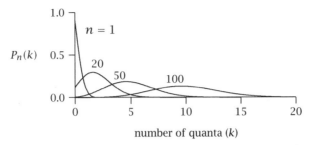

Figure 11.13 Binomial distribution. The ordinate represents the total probability that some number of quanta will be released following an action potential. $p = 0.1$ and n varies from 1 to 100.

be significantly in error if there is nonuniformity or nonstationarity at the release sites. In case of nonuniformity or nonstationarity, a compound binomial model must be used in which the equation includes the individual values of p at each of n release sites.

11.4.2.3 Examples Now that we have derived equations for spontaneous and evoked release, we will work through some examples. These examples follow closely from the actual experiments of Fatt and Katz (1952), Castillo and Katz (1954a), and Boyd and Martin (1956) at the neuromuscular junction.

Assume that you are recording at the muscle end plate and stimulating the motor nerve under low-release conditions. If the EPPs from all of the trials are superimposed, results similar to those illustrated in figure 11.14 might be obtained. From inspection of the figure, it appears that sometimes there is no release (a failure), while at other times the amplitudes of the EPPs appear to fluctuate among values separated by a nearly constant amount.

Let us also assume that you have previously measured spontaneous mEPPs and found that the interval histogram is exponential (i.e., tested for independent release) and that the mean amplitude is 0.4 mV. You want to determine the mean quantal content, m. It would be desirable to use the Poisson model, as it has only one unknown (m, as opposed to the binomial model which has at least two unknowns, n and p), but it will also be necessary to test the applicability of this model.

A direct calculation for m can be obtained from the assumption that the EPP is composed of unit multiples of the mEPP, which are assumed to be released at the same time and to add linearly with each other. This can be stated mathematically as follows:

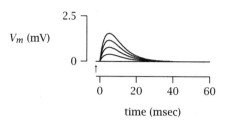

Figure 11.14 Superimposed EPPs from successive trials. The arrow indicates the time of nerve stimulation.

$$\overline{V} = m \cdot \overline{q}, \tag{11.4.17}$$

where $\overline{V}$ is the mean amplitude of the EPP and $\overline{q}$ is the mean amplitude of the mEPP (i.e., *quantal size*). Such a direct calculation for m does not depend on any assumption of the statistic used to describe the release process (i.e., Poisson or binomial). m, by this direct method, is simply

$$m_d = \frac{\text{mean EPP}}{\text{mean mEPP}} = \frac{\overline{V}}{\overline{q}}. \tag{11.4.18}$$

First test for Poisson model   As before, N_k = number of observations at a given amplitude and N = total number of observations. One way to test for the Poisson model is to make a calculation for m using the Poisson model and compare this to a calculation using the direct method. This can be done by using the Poisson equation to predict the number of failures. The number of failures (N_0) by the Poisson equation is

$$P_0 = \frac{m^0 e^{-m}}{0!} \doteq \frac{N_0}{N}, \text{ where}$$

$$N_0 = Ne^{-m}, \text{ or}$$

$$m_f = \ln(N/N_0). \tag{11.4.19}$$

If the Poisson model is appropriate for this experiment, then m_d should equal m_f, or

$$\frac{\overline{V}}{\overline{q}} = \ln(N/N_0).$$

Taking data from a number of different experiments, a plot of m_f from the method of failures vs. m_d from the direct method should be a straight line if release follows a Poisson model. Such a plot is illustrated in figure 11.15. The linear relationship will hold only when m is small and there is linear summation of individual quanta.

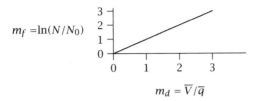

$m_f = \ln(N/N_0)$

Figure 11.15 Ideal relationship between m from method of failures and m from direct method for different experiments.

Second test for Poisson model Figure 11.16 illustrates an amplitude histogram of spontaneous mEPPs. This histogram is reasonably well fit by a Gaussian of the form

$$f(x) = \frac{1}{\sigma\sqrt{2\pi}}e^{-(x-\mu)^2/2\sigma^2}, \qquad (11.4.20)$$

where μ is the mean and σ^2 is the variance.

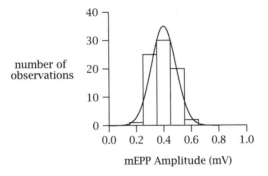

number of observations

mEPP Amplitude (mV)

Figure 11.16 Amplitude histogram of spontaneous mEPPs.

Given a number of trials of evoked EPPs such as those illustrated in figure 11.14, an amplitude histogram of evoked release can be constructed. If the release process follows Poisson statistics, then we should be able to predict the number of times we would measure 1 quantum (N_1), 2 quanta (N_2), 3 quanta (N_3), and so forth. We can use these calculations as well as the information we already have about mEPPs to try and fit the histogram with a theoretical curve such as illustrated in figure 11.17. The methods for this are outlined below.

The number of times one quantum is released is (calculated using equation 11.4.2.3)

$$N_1 = (me^{-m})N = mN_0. \qquad (11.4.21)$$

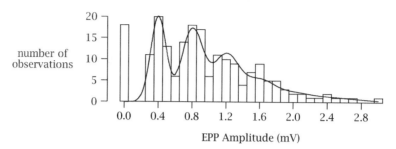

Figure 11.17   Histogram of evoked release. (Adapted from Boyd and Martin 1956.)

If the Poisson model is valid, then the shape of the Gaussian curve that should fit the first peak in the histogram will have its mean centered at 0.4 mV, its variance equal to the variance of the Gaussian used to fit the mEPP histogram (σ^2), and its area should overlap a portion of the histogram whose total number of observations are equal to the calculation of N_1 above.

Similarly for N_2 and N_3,

$$N_2 = \frac{m^2 e^{-m}}{2} \cdot N = \frac{m}{2} \cdot N_1; \quad \sigma_2{}^2 = 2\sigma^2,$$

and

$$N_3 = \frac{m^3 e^{-m}}{6} \cdot N = \frac{m}{3} \cdot N_2; \quad \sigma_3{}^2 = 3\sigma^2.$$

The second and third peaks in the histogram should be fit by Gaussians with their means centered at 0.8 mV ($2 \times \overline{q}$) and 1.2 mV ($3 \times \overline{q}$), their variances equal to 2 and $3 \times \sigma^2$, respectively, and their areas should encompass numbers of observations given by the calculations for N_2 and N_3 shown above. This process is continued for as many peaks as possible in the evoked histogram and is illustrated in figure 11.18. The total curve used to fit the histogram in figure 11.17 is the sum of the individual Gaussians in figure 11.18. The ability of the curve to fit the histograms is a second test of the Poisson model.

One assumption of this analysis, however, is that the variance at each of the release sites is equal to the variance of the population of mEPPs. For the neuromuscular junction this assumption appears to hold, and therefore the variance of each peak in the histogram is equal to the number of quanta that make up the peak times the variance of the mEPPs and thus gets bigger for a larger number of releases. In fact, when m gets large the variance can actually be greater than the mean amplitude of the EPP,

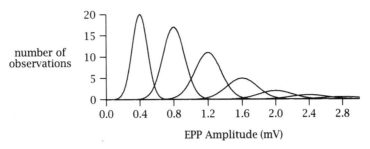

Figure 11.18 Gaussian curves used to fit the histogram of evoked release shown in figure 11.17. (Adapted from Boyd and Martin 1956.)

which would make the observation of peaks in the histogram difficult (see figure 11.17). This may or may not be true for other synapses. One could easily imagine the variance being different at individual release sites, in which case the variance of the peaks in the histogram would be less than the multiple of the mEPP variance.

Third test for Poisson model Remember that for a Poisson distribution, $\sigma^2 =$ mean. When we applied the Poisson model to transmitter release, the mean was equal to m, the mean quantal content. Therefore,

$$\sigma^2 = m, \text{ or } \sigma = \sqrt{m}. \tag{11.4.22}$$

Also, from elementary statistics, the *coefficient of variation*, CV, is defined as:

$$CV = \text{standard deviation/mean} = \sigma/\text{mean}. \tag{11.4.23}$$

For the Poisson model we can therefore define the CV as follows:

$$CV = \sigma/m, \tag{11.4.24}$$

or

$$CV = \frac{1}{\sqrt{m}} = \frac{1}{\sigma}. \tag{11.4.25}$$

Remember that $\overline{V} = m \cdot \overline{q}$, and therefore CV is also given by

$$CV = \frac{\sigma}{\overline{V}}, \tag{11.4.26}$$

where σ is the standard deviation around the mean of the EPP amplitude distribution, and thus

$$m_{cv} = \frac{1}{CV^2} = \frac{\overline{V}^2}{\sigma^2}. \tag{11.4.27}$$

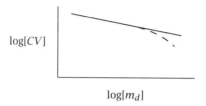

Figure 11.19   Third test of Poisson model and effect of nonlinear summation at high release rates. The solid line is the theoretical curve while the dotted line represents experimental data. The deviation of the two at large m is due to nonlinear summation (see text for further details). (Adapted from Castillo and Katz 1954a.)

The third test of the Poisson model is to calculate the σ and mean amplitude from each peak (or class) of the amplitude distribution of EPPs and compare to the m calculated for each peak by the direct method

$$m_d = \frac{\overline{V}}{\overline{q}}.$$

From equation 11.4.25 we can derive the following:

$$\log(CV) = -(1/2) \cdot \log(m).$$

A plot of $\log(CV)$ vs. $\log(m_d)$ for different peaks should be a straight line with slope of $-1/2$. An example is illustrated in figure 11.19.

The dotted line (representing the data points) in figure 11.19 deviates from the theoretical curve at high rates of release. This deviation is caused by nonlinear summation of the individual quanta that make up the EPP. Nonlinear summation of synaptic responses is due to the fact that the driving force for the EPPs decreases with the synaptic depolarization. (This will be discussed more fully in chapter 13.) The error associated with nonlinear summation at high release rates can be significant. A method of correcting for nonlinear summation was derived by Martin (1955) and is given by

$$V_{EPP}/(E_s - V_{EPP}) = mG_{\text{peak}}/G_N, \tag{11.4.28}$$

where V_{EPP} is the EPP amplitude, E_s is the synaptic equilibrium potential, G_{peak} is the peak synaptic conductance change, G_N is the input conductance of the cell, and m is the quantal content. The above equation will make the curve in figure 11.19 approximately linear at high rates of release. A more accurate method of correction is quite complicated (Stevens 1976), and so the best way to avoid the problem of nonlinear summation is to use voltage-clamp techniques so that the driving force for the synaptic current stays constant, regardless of quantal content (see also chapter 13).

Another source of nonlinear summation and error in estimates of m can occur if individual quanta are not released at the same time. This leads to what are called *latency fluctuations.* Latency fluctuations can also affect the rise time of the synaptic response, particularly the rise time of the synaptic current when using a voltage clamp.

In summary, if one wishes to determine m at a particular synapse using the Poisson model, one can use three methods: the *direct* method, the method of *failures*, and the *coefficient of variation (CV)* method. If the Poisson model is appropriate, then the m calculated with each method should be equivalent. Remember, however, that the direct method is not dependent on the Poisson model. For review, the three methods and their equations for the Poisson model are

1. **Direct method:** $m_d = \frac{\overline{V}}{\overline{q}}$.
2. **Failure method:** $m_f = \ln(N/N_0)$.
3. *CV* **method:** $m_{cv} = \frac{1}{CV^2} = \frac{\overline{V}^2}{\sigma^2}$.

Binomial model The assumptions necessary for the use of Poisson statistics (large n small p resulting in small m) hold only under very special conditions. It was originally believed that n might be equal to the number of vesicles at a release site, which is often quite large. p, however, is usually small only under abnormal conditions (e.g., with high Mg^{2+} and low Ca^{2+}). It is now generally believed that n = the number of active zones at a synapse (see figure 11.20). This idea was derived from experiments where the morphology of the synapse was correlated with the physiology. The number of active zones at a synapse can be a relatively small number, thus necessitating the use of binomial statistics.

There are several ways of doing a quantal analysis when the Poisson model cannot be assumed. Four methods are outlined below.

1. **Direct method:**

$$m_d = \frac{\overline{V}}{\overline{q}}.$$

Remember that this method does not depend on the release statistic. For large m, however, nonlinear summation can be significant and thus voltage clamping should be used if possible.

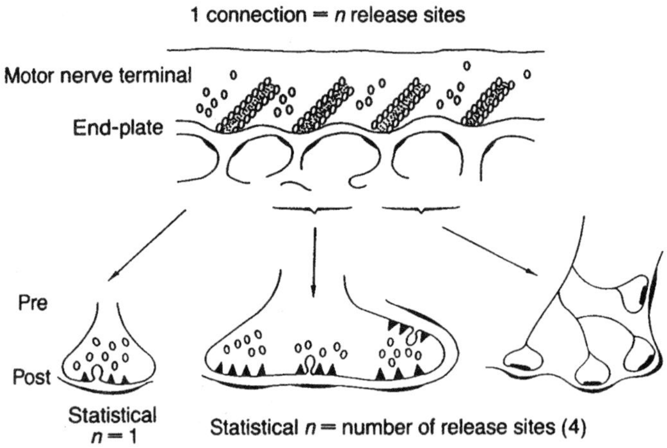

1 connection = *n* release sites

Figure 11.20   Number of release sites equals number of active zones. Upper diagram is that of a neuromuscular junction. The bottom diagrams represent three central synapses. The statistical parameter *n* would be 1 for the left bouton and 4 for the middle and right boutons. (From Korn and Faber 1991.)

2. **Method of failures:** From the binomial model,

$$m_f = \frac{p}{\ln(1-p)} \ln(N_0/N). \qquad (11.4.29)$$

Since there are now two unknowns in this equation, the use of this method requires an independent measure of p. Unfortunately, there are few satisfactory ways of obtaining p.

3. *CV* **method:** From the binomial model,

$$m_{cv} = (1-p)/CV^2 = (1-p)\frac{\overline{V}^2}{\sigma^2}. \qquad (11.4.30)$$

4. **Histogram fitting:** With this method the evoked histogram is best fitted using a simple or compound binomial model with Gaussian curves representing multiple units of transmitter release. The statistical methods for obtaining a best fit include maximum likelihood and deconvolution methods (see Redman 1990).

11.4.2.4   Requirements for quantal analysis   In general, there are a number of requirements for doing a proper quantal analysis (after Korn and Faber 1991). These include

1. Stimulation of a single axon and the measurement of the spike in the presynaptic terminal. This will avoid uncertainties associated with variability of spike generation or failure of spikes to invade a particular terminal.

2. A sufficiently low noise level so that miniature events and the first peak in evoked histograms are clearly separate from the noise (no overlap in histograms). Otherwise, a decrease in failures could be due to the emergence from the noise of previously undetectable evoked events. Also, direct methods become possible.

3. Resolution of single quantal events. Recording must be done sufficiently close to release sites to avoid filtering due to cable properties.

4. Quantal events all at the same electrotontic location. If events are filtered, they must at least be at the same electrotonic distance.

5. Direct information on variation in quantal amplitude at single release sites. If postsynaptic receptors are saturated by the released transmitter (see section 11.6), data may be difficult to interpret.

6. Morphological identification of synaptic active zones, the potential sites of release.

7. Information on whether p is uniform at all release sites. This will determine whether a simple or compound binomial model should be used.

There is a large literature on quantal analysis. The subject is interesting and important, but at the same time quite complicated. The interested reader is referred to many excellent reviews in the reading list.

11.5 Use-dependent synaptic plasticities

11.5.1 Facilitation, post-tetanic potentiation, and depression

Use-dependent synaptic plasticity refers to the change in the strength of a synaptic connection depending on the prior use or activity at that synapse. Three common forms of short-term plasticity are called *facilitation, post-tetanic potentiation* or *PTP*, and *depression*.

Facilitation Facilitation can be studied by giving a pair of stimuli to the motor nerve terminal under low-release conditions (low Ca^{2+}, high Mg^{2+}) and plotting the fractional increase in amplitude of the second response

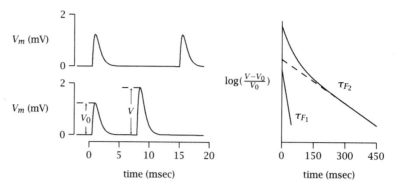

Figure 11.21 Paired-pulse facilitation (left) and a semi-log plot of the change in amplitude of the EPPs vs. time (right). The time constants for facilitation can be obtained by peeling (see chapter 4).

as a function of the interval between the pair of stimuli. For obvious reasons, this is also called *paired-pulse facilitation*.

At the motor end plate, where facilitation has been studied most extensively, a semilog plot of the fractional increase in amplitude of the EPP yields a double exponential curve indicating that there are two phases to facilitation characterized by different time constants. These were called F_1 and F_2 (see figure 11.21). Magleby and colleagues (reviewed in Magleby 1987) have done careful studies of these different phases of facilitation using stimulus trains of different frequencies and measuring the amplitude of the EPP (compared to the initial EPP) as a function of time during the train. They found that the time constants were:

$$\tau_{F_1} \simeq 50 \text{ msec and } \tau_{F_2} \simeq 300 \text{ msec.}$$

Facilitation can be expressed mathematically as

$$F(t) = \frac{\text{EPP}(t) - \text{EPP}_0}{\text{EPP}_0}, \tag{11.5.31}$$

where $F(t)$ is the amount of facilitation and $\text{EPP}(t)$ is the facilitated EPP at time t following the control EPP (EPP_0).[4]

For relatively small amounts of facilitation ($F \leq 2$), $F(t)$ is approximately the linear sum of the facilitation from each action potential, or

$$F(t) = \sum_{i=1}^{n} f(t_i), \tag{11.5.32}$$

---

[4]The same equation might hold for any synapse by merely replacing EPP with EPSP, IPSP, EPSC, or IPSC.

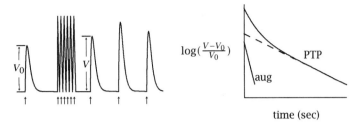

Figure 11.22   Measurement and time course of augmentation and PTP. (Note that the first EPP after the train is reduced in amplitude because of depression and is thus not included in measurements of PTP.)

where $f(t_i)$ is the incremental facilitation contributed by each action potential. At the neuromuscular junction $f(t)$ is given by

$$f(t) = F_1 e^{-(t/50)} + F_2 e^{-(t/300)}, \tag{11.5.33}$$

where F_1 and F_2 are the maximum amounts of each of the phases of facilitation, and t is time (in msec, in this example).

Castillo and Katz (1954b) were the first to perform a quantal analysis during facilitation and found that there was a significant *decrease* in the number of failures associated with the second response—consistent with a hypothesis of an increase in the number of quanta (m) released during the second stimulus. They suggested that facilitation is due to an increased probability of release (p) at each release site and thus represents a presynaptic change at the synapse.

PTP   One can study PTP by giving a train of stimuli and comparing the amplitude of the EPP before to the EPP at various times after the train. As with facilitation, a semilog plot of the fractional increase in amplitude of the EPP with time after the train often yields a double exponential decay. The earlier phase has been called *augmentation*, while the later phase is what most people call PTP. The decay time constant of augmentation is $\simeq 7$ sec; that for PTP is $\simeq 1$ min. The two phases to PTP are described by

$$\frac{\text{EPP}(t) - \text{EPP}_0}{\text{EPP}_0} = Ae^{-t/7} + Pe^{-t/60}, \tag{11.5.34}$$

where $\text{EPP}(t)$ and EPP_0 are the same as in equation 11.5.31, A and P are the maximum values of augmentation and PTP, respectively, and t is in seconds.

Facilitation, augmentation, and PTP are all believed to be due to an increase in quantal content (m), possibly through an increase in p, and thus represent presynaptic changes in release mechanisms. All three also involve presynaptic Ca^{2+} in some way. Katz and Miledi (1968) proposed what has been called the *residual Ca^{2+} hypothesis* to explain these forms of synaptic plasticity. Simply stated, this hypothesis suggests that residual Ca^{2+} in the terminal from the conditioning train or impulse adds non-linearly to the Ca^{2+} influx during the test impulse to increase release. This hypothesis will be discussed more fully in the next chapter when the role of Ca^{2+} in the release process is presented. For reasons that are not understood, Ba^{2+} and Sr^{2+}, when replacing Ca^{2+} in the bath, have differential effects on these various components of synaptic plasticity. Sr^{2+} increases the magnitude and decay time of F_2, while Ba^2 increases the magnitude of augmentation (Zengel and Magleby, 1980, 1981).

Depression   Depression in the amplitude of the EPP occurs during repetitive stimulation under conditions of normal or increased rates of release. Castillo and Katz also studied quantal properties of synaptic depression and found that mEPP amplitude was unchanged during the period of depression, suggesting that depression is due to a decrease in the number of released quanta (i.e., quantal content, m). For a variety of reasons, they and others suggest that there is a decrease in the number of release sites (n). Synaptic depression has both fast and slow components. Following brief trains, or even a single impulse under greatly elevated Ca^{2+}, a test EPP 100 msec after the conditioning impulse can be depressed to 15–25% below control values; recovery follows an exponential time course with a time constant of about 5 sec. This recovery has the form

$$\text{EPP}(t)/\text{EPP}_0 = 1 - D_0 e^{-t/\tau}, \tag{11.5.35}$$

where $\text{EPP}(t)$ and EPP_0 are the depressed and control amplitude EPPs, respectively, D_0 is the depression immediately following the conditioning train or impulse (e.g., 0.15–0.25), and τ is the recovery time constant. After a longer train of impulses to the motor nerve (e.g., 3 min), a test EPP can be depressed to about 10% of control values and recover exponentially with a time constant of about 4 min. Equation 11.5.35 can similarly describe this slow phase of depression.

   Although all of these forms of synaptic plasticity have been studied most extensively at the neuromuscular junction, they appear to also be present at other types of synapses as well, including synapses in the central nervous system. The quantitative properties of synaptic plasticities and the similarities and differences between plasticities at the neuromus-

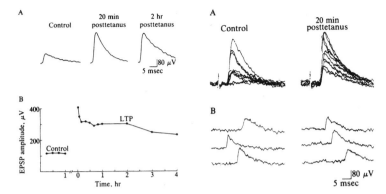

Figure 11.23 Quantal analysis of LTP at crayfish neuromuscular junction (from Baxter et al. 1985). The traces in the upper left are the EPSPs before and after high-frequency stimulation. The graph in the lower left is the change in amplitude of the EPSP vs. time. The superimposed traces in the upper right illustrate the fluctuation in amplitudes (variance) of the EPSPs and the decrease in failures during LTP. The traces in the lower right show spontaneous mEPSPs before and during LTP.

cular junction and those at central synapses, however, remain to be determined.

11.5.2 Long-term potentiation

Another form of synaptic plasticity is called *long-term potentiation*, or LTP. This is a long-term change in synaptic strength observed under certain conditions at "integrating"-type synapses. (Note: An integrating synapse means a synapse for which there is not one-for-one firing of pre- and postsynaptic neurons. To reach threshold at integrating synapses, there must be some form of summation of synaptic potentials. The frog neuromuscular junction is an example of a *nonintegrating* synapse because an action potential and muscle contraction are achieved each time the motor nerve fires. Examples of *integrating* synapses include most excitatory synapses in the central nervous system.) LTP is similar to PTP in that there is an increase in the amplitude of the EPSP after a brief train of stimuli. Whereas PTP decays within a few minutes, LTP decays over the course of several hours or, under certain conditions, up to a month or more. LTP is the best candidate mechanism available for aspects of memory, and will be discussed more fully in chapter 15.

The first quantal analysis of LTP was done at a crayfish synapse. All three methods were employed (direct, failures, and *CV*) to test for changes in quantal content during LTP. This work nicely illustrates the use of the Poisson model for doing a quantal analysis. In figure 11.23 (bottom right)

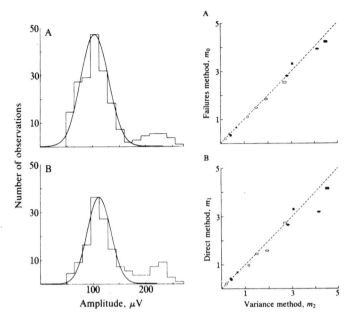

Figure 11.24   Quantal analysis of LTP (from Baxter et al. 1985). The graphs on the left are frequency histograms of mEPSP amplitudes before (top) and during (bottom) LTP. The graphs on the right are plots of m_f (m_0, top) and m_d (m_1, bottom) against m_{cv} (m_2) before and during LTP (different symbols). Data are consistent with a change in m and no change in q.

the amplitudes of miniature excitatory postsynaptic potentials (mEPSPs) are illustrated before and during LTP. There is no apparent change in the mean amplitude (figure 11.24, left). In figure 11.24 (right) the use of the Poisson model, in which the quantal content, m, is calculated using the three methods, is illustrated. The m calculated from the three methods agrees favorably, justifying the use of Poisson statistics for the release process.

The results are quite clear. There is an increase in m during LTP with no change in $\overline{q}$. LTP, at least at this synapse, is due to a presynaptic change that results in an increase in transmitter release.

11.6   Synaptic transmission between central neurons

Quantal analysis of LTP in hippocampus has also been performed by a number of groups. As with the crayfish, most reports suggest an increase

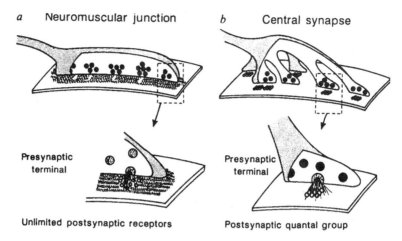

Figure 11.25 The saturation of receptor hypothesis (from Edwards 1991, used by permission of *Nature*, copyright © 1991 Macmillan Magazines Limited). The diagram on the left (a) is that of a neuromuscular junction in which there are many postsynaptic receptors at each release site. The diagram on the right (b) illustrates the hypothesis for central synapses in which there might be a limited number of receptors at each release site, leading to the saturation of the receptors during each quantal event. See text for further explanation.

in quantal content with LTP. The principal methods utilized for these studies were the method of failures and the *CV* method. These reports also discuss some of the additional problems in doing a quantal analysis when single quantal events are either difficult to resolve or difficult to identify with the synapses being stimulated. As mentioned previously, the method of failures requires knowledge of the amplitude of a single quantal event. Otherwise, increases in the amplitudes of quantal events could appear as a decrease in failures simply because they were previously buried in the noise and appeared as failures in evoked release. Also, a recently proposed hypothesis for saturation of receptors raises other concerns about interpreting results from traditional methods of quantal analysis.

Specifically, it was first suggested by Jack, Redman, and Wong and later by others that postsynaptic receptors are saturated by released neurotransmitter at central synapses (see figure 11.25). Under this hypothesis the number of receptors at the subsynaptic membrane, rather than for the amount of transmitter in each vesicle, is rate limiting for the amplitude of the quantal event. This would lead to little or no quantal variance at each release site. It has also been suggested that there may be a relatively fixed number of receptors at all active release sites on any particular neuron

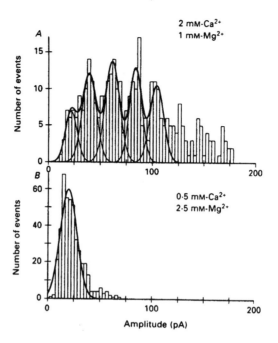

Figure 11.26   Small quantal variance across release sites (from Edwards et al. 1990). The upper graph is a frequency histogram of evoked IPSPs from a dentate granule cell in normal saline; the bottom graph is from the same cell with low extracellular Ca^{2+} and high Mg^{2+}. The Gaussian curve that fits the bottom histogram also fits each of the five peaks in the top histogram. The variance of each peak is small and does not appear to sum over multiple releases.

(see figure 11.26). This would lead to little or no quantal variance *across* release sites. These possibilities have important functional implications for both the method of failures and the *CV* method and will require further study. For example, there could be "silent" synapses where release occurs but with no postsynaptic receptors. The insertion of receptors would reduce the "failures" at this synapse and yet represent a postsynaptic change. Also, it is possible to show that if the quantal variance has different values at different release sites, a purely postsynaptic change can lead to changes in *CV*.

Another hypothesis with functional implications for central synapses is that the conductance associated with an individual synapse may vary depending on its electrotonic distance from the soma (see Jack et al. 1990). This hypothesis was derived from data in which it appeared that the *charge* measured in the soma for proximal and distal synapses was approximately the same. One explanation for this would be that distal synapses produce larger conductance changes than proximal synapses, so that the same amount of charge ultimately reaches the soma from the two inputs. Another possibility is that distal synapses get amplified by their activation of voltage-gated channels in the dendrites.

Many of the problems associated with studying the properties of central synapses arise from the fact that it is difficult to isolate and study individual synapses in central neurons. It can be seen from just this rather brief discussion that the quantal properties of synaptic transmission in central neurons with complex dendritic trees will be the subject of intense investigation for many years to come. Reviews of some of the issues can be found in Stevens (1993), Jack et al. (1990), and Redman (1990).

11.7 Summary of important concepts

1. Poisson model.
2. Tests for Poisson model.
3. Binomial model.
4. Tests for binomial model.
5. Problems associated with quantal analyses.
6. Facilitation and post-tetanic potentiation.

11.8 Homework problems

1. Show that for the binomial model

$$m_f = \frac{p}{\ln(1-p)} \cdot \ln(N_0/N),$$

and

$$m_{cv} = (1-p)/CV^2.$$

2. (a) In an experiment in which the nerve was stimulated 500 times, how many observations of failure in neuromuscular transmission would be expected? Assume Poisson statistics and a mean quantal content of 5.

(b) How many observations of two quanta released would be expected during the same experiment? Use the same assumptions.

3. Prove that as $n \to \infty$, but $m = np$ = constant, the binomial distribution approaches the Poisson distribution.

4. While recording at a frog neuromuscular junction, you measure the amplitudes of spontaneous synaptic events (mEPPs) at the times indicated below.

mEPP #	amplitude (mV)	total time (sec)	interval (sec)
1	1.0	0	–
2	1.0	0.5	0.5
3	0.6	1.0	0.5
4	1.1	2.0	1.0
5	0.9	2.5	0.5
6	0.8	4.0	1.5
7	1.3	4.5	0.5
8	1.0	5.5	1.0
9	1.2	8.0	2.5
10	1.0	9.0	1.0
11	0.7	12.0	3.0
12	1.0	13.0	1.0
13	1.0	13.5	0.5
14	1.1	15.5	2.0
15	0.9	17.5	2.0
16	0.9	20.0	2.5
17	1.1	23.5	3.5
18	1.0	27.0	3.5
19	1.0	27.5	0.5
20	1.1	28.0	0.5
21	0.9	30.0	2.0
22	1.2	31.0	1.0
23	0.8	33.0	2.0
24	0.7	34.5	1.5

mEPP #	amplitude (mV)	total time (sec)	interval (sec)
25	1.4	36.0	1.5
26	1.3	38.0	2.0
27	1.0	40.5	2.5
28	0.9	43.0	2.5
29	1.1	46.0	3.0
30	1.2	49.0	3.0
31	0.8	50.0	1.0
32	1.0	53.5	3.5
33	1.0	54.0	0.5
34	1.1	55.0	1.0
35	0.9	58.5	3.5
36	1.0	60.5	2.0
37	1.1	63.0	2.5
38	1.0	66.0	3.0
39	1.0	67.5	1.5
40	1.2	69.0	1.5
41	0.8	72.0	3.0
42	1.0	73.0	1.0
43	1.0	74.5	1.5
44	1.1	75.0	0.5
45	0.9	76.5	1.5
46	0.9	82.5	6.0
47	1.0	88.5	6.0

(a) What is the mean amplitude of the mEPPs?

(b) Is the occurrence of each event independent of the occurrence of the others? Verify your answer quantitatively with the appropriate histograms.

(c) If you could record for a long time (much longer than for the sample given in the table), what would be the mean rate of the mEPPs?

5. A neuromuscular junction is stimulated 25 times. The amplitudes of the EPPs for each of the 25 trials are (in mV): 0.3, 0, 0.5, 0.7, 0, 1.1, 1.5, 0, 1.1, 0.9, 0, 0, 1.3, 1.1, 0.5, 0.5, 0.7, 0, 0.7, 1.1, 0.5, 0.5, 0, 0.9, 1.3.

(a) Plot an amplitude histogram of the responses using a bin width of 0.2 mV.

(b) In other experiments you determined that the mean amplitude of the mEPP was 0.5 mV. Calculate m by at least two methods, assuming Poisson statistics for the release process.

(c) Given your value for m, how many times would you predict that 3 quanta would be released in an experiment in which the nerve was stimulated 200 times?

6. You are voltage-clamping a neuromuscular junction at -80 mV (perfect space clamp), and you measure the following end-plate currents in response to low-frequency nerve stimulation: (EPCs, in nA) 0.3, 0.5, 0.7, 1.1, 1.5, 1.1, 0.9, 1.3, 1.1, 0.5, 0.5, 0.7, 0.7, 1.1, 0.5, 0.5, 0.9, 1.3.

    (a) If you assume that $E_S = 0$ mV, what is the approximate synaptic conductance?

    (b) Without any knowledge about miniature EPCs, what is the mean number of quanta released per stimulus?

    (c) From your answer in (b), what is the predicted number of failures during a 1000-stimulus experiment?

    (d) What are two reasons why you measure fewer failures than predicted from your calculation in (c)?

7. You wish to use methods of quantal analysis to determine if the increases in synaptic transmission associated with LTP result from pre- or postsynaptic changes. The frequency histograms shown here illustrate measurements of spontaneous mEPSCs during a fixed 10 min interval before and during LTP.

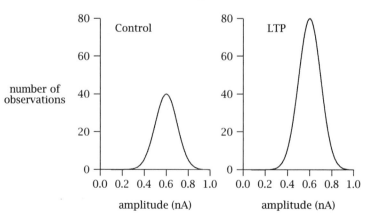

(a) Using only the histograms, can you make any tentative conclusions regarding pre- vs. postsynaptic changes during LTP? Be sure to explain your answer and to give all of the reasons you can think of.

(b) If the mean evoked EPSC was 2.0 nA before and 3.0 nA during LTP, calculate the quantal content before and during LTP.

(c) If you assumed that the number of release sites was fixed at 500, what would be the average probability of release at each site before and during LTP?

(d) On average, how many trials would be necessary to observe a single failure before and during LTP? Does this depend on the type of release statistic used to determine your answer?

8. Given that $P(0, t) = e^{-rt}$, where r = average rate of spontaneous mEPSPs, derive the probability density function for intervals between successive (single) releases.

9. (a) Given the following amplitudes for spontaneous mEPSPs, calculate their mean amplitude: (in mV) 0.4, 0.5, 0.6, 0.7, 0.3, 0.7, 0.7, 0.9.

(b) Given the following amplitudes for evoked synaptic potentials, calculate their mean evoked amplitude and σ: (in mV) 2.0, 0.5, 2.5, 0.5, 2.5, 2.0, 0, 1.0, 0.5, 2.5, 2.5, 1.0, 0.5, 2.5.

(c) Determine m by three methods (you can assume Poisson statistics *for this question*). Why might the results differ for the three methods?

(d) Calculate the expected number of failures during 25 trials using a binomial model. Assume $n = 10$.

10. Using the data illustrated below, plot synaptic plasticity as a function of time and derive an equation that suitably describes the relationship between plasticity and time.

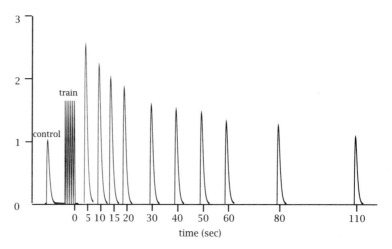

12   Synaptic Transmission II: Ca^{2+} and Transmitter Release

12.1   Introduction

It has been known for some time that Ca^{2+} is somehow necessary for synaptic transmission. It is known, for example, that elevated $[Ca^{2+}]_{out}$ increases transmission while elevated $[Mg^{2+}]_{out}$ decreases transmission, as was mentioned in the previous chapter. Furthermore, Ca^{2+} appears to be unique among divalent cations in promoting synaptic transmission. Only Sr^{2+} and Ba^{2+}, as substitutes for Ca^{2+} in the medium, will allow release at all, and they do so with much less efficiency (i.e., $Ca^{2+} > Sr^{2+} > Ba^{2+}$) and somewhat differently than Ca^{2+}. Transmitter release is typically asynchronous with these other divalent cations, as illustrated in figure 12.1.

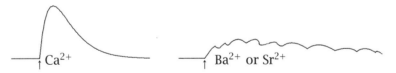

Figure 12.1   Typical synaptic response with normal Ca^{2+} in the bath compared to that with either Ba^{2+} or Sr^{2+}. The arrows indicate time of nerve stimulation.

Again, most of what we know about the mechanisms of transmitter release has come from the study of two preparations, the neuromuscular junction and the squid giant synapse. Study of the vertebrate neuromuscular junction led to the *quantum hypothesis*, which was presented in the previous chapter. Experiments on the squid giant synapse, in particular, but also on the neuromuscular junction, led to the *Ca^{2+} hypothesis* for transmitter release. This hypothesis will be discussed at some length in the rest of this chapter.

12.2   Formulation of the Ca^{2+} hypothesis

The squid giant synapse is a rather unique preparation. Because of its large size it allows direct access to the presynaptic terminal for microelectrode recording and injection of drugs. The species of squid most frequently used for studies of synaptic transmission is *Loligo pealii*, available at the Marine Biological Laboratory, Woods Hole, Massachusetts. Several giant synapses are located in the stellate ganglion of the squid. These were first discovered by the anatomist J. Z. Young in 1939. The most medial and largest of the synapses is the one used for electrophysiological studies. This synapse also gives rise to the famous giant axon, which has been exploited for studies of nerve conduction and was discussed in chapter 6.

The squid stellate ganglion and giant synapse are shown in figures 12.2 and 12.3. The synapse is typically over 1 mm in length and about 50 μm in diameter. It is possible to place 3 or more microelectrodes into the presynaptic terminal along with 2 or more into the postsynaptic axon. This permits simultaneous voltage clamping of both sides of the synapse, something that is generally not possible with any other synaptic preparation.

In spite of its large size, however, the giant synapse is a very difficult experimental preparation and has several disadvantages for the investigation of synaptic transmission. First, squid do not live very long in captivity, so the experiments must be done near where they are captured. Second, very few squid have synapses with the postsynaptic axon in the proper orientation for impaling the presynaptic axon. Third, the synapse is very fragile and is often damaged by the insertion of multiple microelectrodes. And fourth, because of the large size and low input resistance of the postsynaptic axon, miniature postsynaptic potentials are very small and difficult to record. Nevertheless, the squid synapse has proven to be a valuable preparation for the study of presynaptic mechanisms of neurotransmitter release and for testing the Ca^{2+} hypothesis.

Some of the more important experiments that led to the Ca^{2+} hypothesis were performed by Bernard Katz and Ricardo Miledi in the late 1960s on the squid synapse and the vertebrate neuromuscular junction. The results of these experiments, which formed the foundation for the Ca^{2+} hypothesis for synaptic transmission, will be described briefly, in rough chronological order, in this and the following few sections.

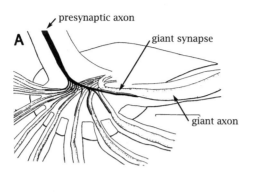

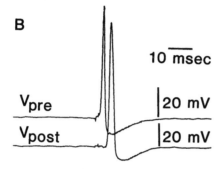

Figure 12.2 Diagram of squid stellate ganglion and giant synapse. (A) The presynaptic axon is shown in black making synaptic contacts with each of the stellate nerves. The giant axon is in the upper right. (B) Intracellular recordings from the pre- and postsynaptic axons. (Adapted from Zucker 1991.)

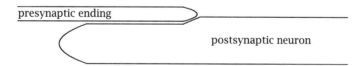

Figure 12.3 Enlarged diagram of squid giant synapse.

a. When the action potential in the presynaptic terminal was blocked with TTX, no EPSP in the postsynaptic axon was recorded. This experiment suggested that a presynaptic action potential was necessary for release.

b. Using the squid giant synapse, depolarization of the presynaptic terminal resulted in an EPSP, even in the presence of TTX. This further elaboration of the previous result showed that an applied depolarization to the terminal could substitute for an action potential.

c. Again with the squid, when Na$^+$ channels were blocked with TTX and K$^+$ channels were blocked with TEA, depolarization of the presynaptic terminal resulted in spike-like or regenerative events in the terminal. These "spikes," which presumably were mediated by Ca^{2+}, elicited EPSPs in the postsynaptic axon. If the terminal was depolarized to extreme positive potentials, near the expected Nernst potential for Ca^{2+}, there was no EPSP. This experiment suggested that Ca^{2+} influx occurring during these spikes was necessary for release.

d. In other experiments at the frog neuromuscular junction, Ca^{2+} was removed from the bath so that stimulation of the motor nerve elicited no response at the muscle end plate. Ca^{2+} was then iontophoresed onto the junction at the time of stimulation, and this resulted in an EPP. If Ca^{2+} was iontophoresed a few msec *after* the stimulation, no response was observed. This very important experiment indicated that not only was Ca^{2+} required for release, but it had to be present almost exactly when the action potential arrived at the presynaptic terminal.

12.3 Cooperative action of Ca^{2+} ions on transmitter release

The experiments outlined above, and many others, gave firm support for the hypothesis that Ca^{2+} is necessary for transmitter release. In 1967 Dodge and Rahamimoff performed an extremely important set of experiments at the frog neuromuscular junction. The results of these experiments, which were not fully accepted for almost 20 years, suggested a particular quantitative relationship between Ca^{2+} and transmitter release. Specifically, they proposed that transmitter release is dependent upon the 4th power of the Ca^{2+} concentration in the extracellular bath, or, in more molecular terms, four Ca^{2+} ions act in a cooperative manner to cause release.

Without knowing any details about how transmitter release is dependent upon Ca^{2+}, or how Mg^{2+} inhibits release, Castillo and Katz (1954), Jenkinson (1957), and then Dodge and Rahamimoff assumed that Ca^{2+} must bind to some critical site X at the presynaptic terminal in order for transmitter release to occur. This assumption leads to the following kinetic equation:

$$Ca^{2+} + X \rightleftharpoons CaX \text{ (with dissociation constant } K_1 \text{)}. \qquad (12.3.1)$$

Release would occur only upon the formation of CaX. Using the law of mass action, Dodge and Rahamimoff derived the following:

$$[CaX] = \frac{W\left[Ca^{2+}\right]_{out}}{1 + \frac{[Ca^{2+}]_{out}}{K_1}},$$ (12.3.2)

where W is a constant. This equation makes a number of interesting predictions. If release is directly proportional to [CaX], then the amplitude of the EPP should be directly proportional to [CaX]. If release is instead dependent on the formation of 2 CaXs, then the EPP would be proportional to $[CaX]^2$, and so forth. More generally,

$$EPP = k\,[CaX]^n = k\left(\frac{W\left[Ca^{2+}\right]_{out}}{1 + \frac{[Ca^{2+}]_{out}}{K_1}}\right)^n,$$ (12.3.3)

where k is a proportionality constant and n is a positive integer. If we let $k = K_1 = 1$, then the predicted relationship between EPP and $[Ca^{2+}]_{out}$ for different values of n can be observed in figure 12.4. Note that at high external Ca^{2+} concentrations the curves are highly nonlinear for all values of n. If we expand the scale between 0 and 1.5 mM $[Ca^{2+}]_{out}$, however, the initial portion of the curve for $n = 1$ is actually fairly linear, while those for $n = 2, 3, 4$, and 5 are still quite nonlinear. With the fairly simple assumptions made by Dodge and Rahamimoff, equation 12.3.3 predicts that at low concentrations of external Ca^{2+} the relationship between EPP amplitude and external Ca^{2+} will be linear if release is dependent on the formation of a single [CaX], or nonlinear if release is dependent on the formation of more than one [CaX]. In other words, if there is *cooperativity* in the action of Ca^{2+} ions on the release process, the relationship should be nonlinear.

Also in figure 12.4, equation 12.3.3 is plotted on log-log axes. The initial portion of each of the curves is linear, and the slope varies for different n's. The experiments performed by Dodge and Rahamimoff were simply to vary the external concentration of Ca^{2+}, measure the amplitude of the EPP, and compare the relationship obtained in this way to the theoretical curves of figure 12.4. The results of their experiments, which are replotted in figure 12.5, are most consistent with $n = 4$. In other words, transmitter release appeared to be dependent on the 4th power of external Ca^{2+} or on this complex, CaX.

There are several significant aspects to these findings, to which we will return later. First, the 4th power relationship sets constraints on the biochemical mechanisms involved in release. The release of a single quantum

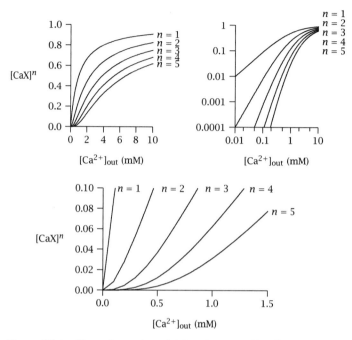

Figure 12.4   Plots of equation 12.3.3 on linear and log-log coordinates. (After Dodge and Rahamimoff 1967.)

of transmitter requires the cooperative action of Ca^{2+}; that is, the action (or binding) of four Ca^{2+} ions is required for release.[1] Second, the relationship between the EPP and external Ca^{2+} is highly nonlinear. Referring to figure 12.5, a small increase in Ca^{2+} of around 20–25% from an initial level of 0.2 mM leads to a doubling in the amplitude of the EPP. A more detailed description of the cooperative action of external Ca^{2+} on transmitter release will be discussed in the next section.

12.4   Biophysical analysis of Ca^{2+} and transmitter release

Further support for the Ca^{2+} hypothesis came from Miledi in 1973. He found that direct injection of Ca^{2+} into the presynaptic terminal of the squid synapse led to transmitter release. Also, Llinás and Nicholson (1975), using the Ca^{2+}-sensitive photoprotein aequorin, which emits light when

---

[1]Mathematically, it makes no difference whether 4 Ca^{2+} ions bind to a single molecule or whether 4 molecules of CaX form before release occurs.

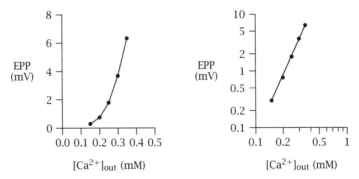

Figure 12.5   Relationship between external Ca^{2+} and the amplitude of the EPP. Data are plotted on linear and log-log coordinates. (Adapted from Dodge and Rahamimoff 1967.)

it binds Ca^{2+}, demonstrated that a rise in preterminal Ca^{2+} occurs during neurotransmission. They injected aequorin into the presynaptic terminal and measured light flashes upon stimulation of an action potential in the presynaptic axon. The results of these experiments indicated that a rise in Ca^{2+} in the terminal could cause release and that a rise in Ca^{2+} occurred under physiological conditions when an action potential was elicited in the preterminal. By this time, voltage-gated Ca^{2+} currents had been measured from a variety of preparations, and thus the rise in Ca^{2+} following an action potential in the presynaptic terminal was assumed to occur from an influx through voltage-gated Ca^{2+} channels.

By the late 1970s, the Ca^{2+} hypothesis was obviously well formulated. Little or nothing was known, however, about the details of Ca^{2+} entry into the presynaptic terminal or about the relationship between Ca^{2+} entry and transmitter release. We will now discuss the work of two groups, which was published over the period from 1981 to 1987. This work provided the first description of Ca^{2+} current (I_{Ca}) in the squid presynaptic terminal and the first definition of a quantitative relationship between I_{Ca} and transmitter release. The two groups were Llinás, Walton, and Steinberg and Augustine, Charlton, and Smith, and both groups worked during summers at the Marine Biological Laboratory.

12.4.1   Voltage clamping the squid giant synapse

Although the size of the squid synapse appears favorable for inserting multiple microelectrodes and voltage clamping, one serious problem has to do with something called *space clamp*.

As mentioned in chapter 6, voltage clamping requires that the voltage across the membrane from which current is measured be held constant throughout the time interval of interest. In other words, the voltage in *time* and across *space* must be *clamped* to a given value, or

$$\frac{\partial V_m}{\partial t} = \frac{\partial V_m}{\partial x} = 0.$$

Only under these conditions will the clamp current (I_{cl}) equal the membrane ionic current, or

$$I_{cl} = I_{ionic} = \sum g_{ion}(V_m - E_{ion}),$$

where the *ion* subscript represents the different ionic conductances in the membrane.

Hodgkin and Huxley and other physiologists achieved these conditions in the squid by inserting a silver wire inside of the axon and then measuring current from a region along the length of this wire. The internal wire had the effect of *short-circuiting* the axon so that the length of axon along the wire was isopotential. The axon was said to be *space clamped* throughout this region because the potential was clamped not only in time but also across space or distance. Unless a space clamp is achieved along with a voltage clamp, the current measured (the clamp current) will not be an accurate reflection of the underlying membrane conductance change. The necessity of space clamping, and the errors associated with voltage clamping synaptic inputs that occur on dendrites, will be discussed again in chapter 13.

In the case of the squid giant synapse, it is not possible to insert a silver wire to short-circuit the presynaptic terminal. Voltage clamping with two microelectrodes will maintain the voltage constant in a small region of the terminal, but, because of its cable-like structure, regions of the terminal a short distance away from the microelectrodes will not be at the same potential as those near the microelectrodes. The current injected by the current microelectrode will be the sum of the current flowing across the clamped portion of the terminal and the current flowing across the rest of the terminal for which the potential is not being held constant in either time or distance. In other words, the clamp current will be

$$
\begin{aligned}
I_{cl} \;=\; & \sum g_{ion}(V_m - E_{ion}) \text{ (for the clamped region)} \\
& +\frac{\partial I_i}{\partial x} \text{ (for the rest of the axon),}
\end{aligned}
\tag{12.4.4}
$$

where I_i is the axial current flow (refer to chapter 4). Under normal conditions, there is no way of separating the current across the clamped region

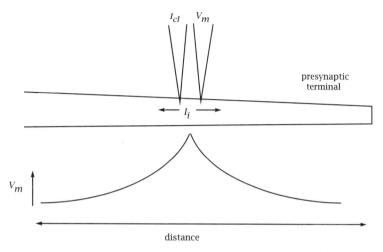

Figure 12.6   Diagrammatic representation of the lack of space clamp in a presynaptic terminal using a two-microelectrode voltage clamp. The membrane potential is clamped to a depolarized value only in the vicinity of the electrodes.

of the terminal from that across the rest of the axon. This is illustrated in figure 12.6.

A method for partially separating these currents was devised by Adrian, Chandler, and Hodgkin (1970) for use on axons or other cable-like structures where multiple microelectrodes can be inserted. It is called the *three-microelectrode voltage clamp* and was used by Augustine, Charlton, and Smith to minimize the errors associated with the lack of space clamp at the squid giant synapse. As an additional procedure for minimizing space-clamp errors, they perfused a Ca^{2+}-containing solution across the tip of the terminal while blocking Ca^{2+} channels (and all other voltage-gated channels) elsewhere.

Referring to figure 12.7, V_1 was the voltage-sensing electrode and I was the current-passing electrode for a two-microelectrode voltage clamp. A voltage clamp was really achieved only at point V_1. Current (I_{cl}) was passed through electrode I to maintain the potential constant at V_1 (via the electronic feedback circuitry of the voltage clamp; refer to appendix A). Only a portion of the current injected through I, however, actually reaches point V_1. The rest flows either across the membrane between electrodes I and V_1 or in the completely opposite direction of the axon toward the cell body. The current flow up the axon will produce a potential drop that can be measured by the second microelectrode V_2. The difference in potential between V_2 and V_1 will represent the current that flows to V_1 and ultimately across the distal portion of the terminal. This is diagrammatically represented in figure 12.7C. The total injected current (I_{cl}) flows either up the axon toward the cell body or toward the end of the terminal (some of it leaks out across the membrane along the way). The current flowing up the axon is represented by I_2 in the diagram, and that flowing toward the terminal by I_1. The idea behind this method of voltage clamping is that the difference in potential between V_2 and V_1 will be a more accurate representation of the current at the end of the terminal, I_m, than would be the total current, I_{cl}. Since Ca^{2+} is present only in the external medium near the end of the terminal, $V_2 - V_1$ will be a close approximation of I_m and thus I_{Ca} across the end of the terminal. Although this method does not produce a perfect space clamp, it does greatly reduce the errors associated with imperfect space clamp. One also begins to appreciate how significant these errors must be to motivate Augustine, Charlton, and Smith to go to such extreme lengths to try to avoid them.

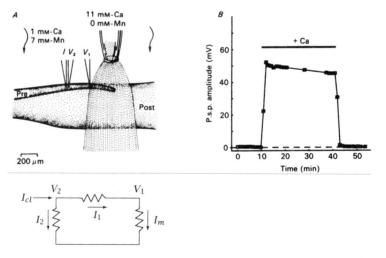

Figure 12.7   (A) A voltage clamp was applied between I and V_1 while Ca^{2+} was present only near the end of the terminal. (B) The change in postsynaptic response (P.s.p.) as Ca^{2+} is added to the perfusing saline is shown on the right. (A and B from Augustine et al. 1985a.) (C) Lumped equivalent circuit representation of three-microelectrode voltage clamp. The terminal is not space clamped, (i.e., $I_{cl} \neq I_m$ at V_1) but a better approximation can be obtained for I_m near V_1 by using $V_2 - V_1$ as a measure of I_m. $V_2 - V_1 = I_1 \propto I_m = I_{Ca}$.

12.4.2 Presynaptic Ca^{2+} currents

Before describing results obtained with this three-microelectrode voltage clamp, we use figure 12.8 to illustrate some of the first voltage-clamp results obtained from the squid synapse preparation. This figure nicely illustrates the separation of the Na^+ and K^+ currents from I_{Ca} and their relative magnitudes. In figure 12.8A the total ionic current in response to a step command is depicted in trace (a), while trace (b) was obtained after applying TTX and a general K^+-channel blocker 3-aminopyridine. The current in trace (b) presumably represents I_{Ca}. In figure 12.8B, current traces from another synapse (also in TTX and 3-aminopyridine) before and after applying Cd^{2+} to block I_{Ca} are illustrated. The relatively small amplitude of I_{Ca} in comparison to the Na^+ and K^+ currents is even more apparent in B if one notes the difference in calibration bars between A and B. (Also note the inflection of the trace in B(a). This is sometimes characteristic of a less then optimum space clamp.)

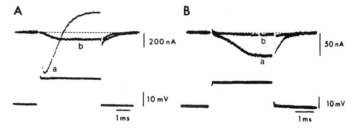

Figure 12.8 Voltage-clamp records from squid presynaptic terminal. (From Llinás et al. 1981a by permission of the Biophysical Society, copyright © 1981 Biophysical Society.) (A) A depolarizing command in normal saline (a) elicits an inward Na^+ and an outward K^+ current. After TTX and 3-aminopyridine are applied, these currents are blocked, leaving a small inward Ca^{2+} current (b). (B) Inward Ca^{2+} currents from another terminal before (a) and after (b) applying Cd^{2+}.

Figure 12.9 illustrates I_{Ca} from a squid synapse using the three-microelectrode voltage clamp described earlier. A number of important features of these current records should be emphasized and understood. First, the amplitude of the current increases with the depolarizing command, reaching a maximum around 0 mV, and then decreases with further depolarization. This is typical for a channel that is activated by depolarization and that has an apparent Nernst potential at very positive values. Second, there is very little inactivation of the current during the time course of the step commands shown in figure 12.9 (6 msec). Third, the onset of the current is slow and delayed from the onset of the step command. And fourth,

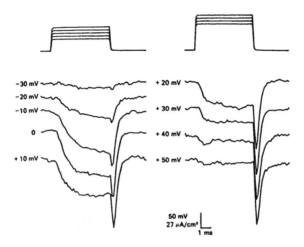

Figure 12.9   Presynaptic Ca^{2+} currents. (From Augustine et al. 1985a.) The command potentials are shown above while their values are given at the left of each current trace.

there is a large tail current at the end of each command. The tail current would be expected for a channel that is activated during the command and for which there is an increased driving force after the step command. (Remember from chapter 6 that the amplitude of the tail current will depend on the magnitude of the conductance activated during the command and the driving force ($V_m - E_{Ca}$) for current through the open channels, where E_{Ca} is the apparent Nernst potential for Ca^{2+}.) The *I-V* relationship for this experiment is illustrated in figure 12.10. Note that the current is not activated until about –40 mV, and that, as mentioned above, the peak inward current occurs at around 0 mV.

12.4.3   Relationship between I_{Ca} and transmitter release

The simultaneous measurement of I_{Ca} and the postsynaptic response is illustrated in figure 12.11. In this experiment, as well as for that illustrated in figure 12.12, the postsynaptic axon was also voltage clamped so that the postsynaptic current could be measured. Again, several very important features of the release process are illustrated in figure12.11. First, in part A and as seen in the figure 12.9, there is a delay in the onset of I_{Ca} from the onset of the step depolarization. Second, the rise in I_{Ca} is relatively slow. This delay and slow rise in I_{Ca} is reflected in the postsynaptic current, which does not begin to appear until nearly the middle of the step command. And third, the peak of the postsynaptic current doesn't occur until *after the end* of the step command. In other words,

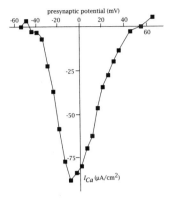

Figure 12.10 *I-V* curve of presynaptic Ca^{2+} currents. (Adapted from Augustine et al. 1985a.)

the turning on of I_{Ca} by the onset of the voltage step produces little or no postsynaptic current while the *turning off* of the voltage step produces most of the current. Because release occurs mostly with the turning off of the voltage, it is called an *off* response. This characteristic of the release process is also observed in figure 12.12. It can be explained by remembering the characteristics of I_{Ca} noted in figure 12.9. The onset of the current is slow and delayed so that release, which is nonlinearly related to the rise in intracellular Ca^{2+}, is also slow to begin. The large off response can be understood by noting the large tail current at the end of the command. The turning off of the command produces a large and fast I_{Ca}, which elicits a large release of transmitter.

Another extremely important feature of the release process is illustrated in part B of figure 12.11. In this instance a very large depolarizing command was given such that little or no I_{Ca} flowed during the command, even though the channels were opened by the depolarization. After the end of the command, however, a large rapid tail current was observed, leading to release of transmitter and a postsynaptic current. Given that there was no I_{Ca} during the command, and thus no postsynaptic response, the activation kinetics of the I_{Ca} was not a factor in the release of neurotransmitter. The large rapid tail current and the resulting postsynaptic response allow one to determine the minimum time between the entry of Ca^{2+} into the presynaptic terminal and a postsynaptic response. From experiments such as these, that interval has been estimated to be about 200 μsec. In other words, from the entry of Ca^{2+} into the terminal it takes no more than 200 μsec for Ca^{2+} to trigger release of transmitter and for

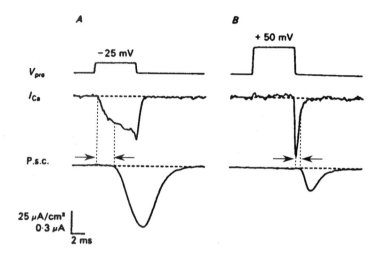

Figure 12.11   Simultaneous presynaptic Ca²⁺ currents (I_{Ca}) and postsynaptic currents (P.s.c.). (Adapted from Augustine et al. 1985b.) In (A) an inward Ca²⁺ current elicits a postsynaptic current during the command (V_{pre}). In (B) there is no Ca²⁺ current or postsynaptic current until after the end of the command. The dotted lines indicate the delay from the Ca²⁺ current to the postsynaptic current, which is about 200 μsec in (B).

the transmitter to diffuse across the cleft and activate postsynaptic receptors. Transmitter release can obviously be an extremely rapid process.

The full relationship between presynaptic Ca²⁺ and the postsynaptic response can be obtained from the experiment illustrated in figure 12.12. Step commands from a holding potential of −70 mV to between −33 mV and +57 mV were given to the voltage clamped presynaptic terminal of the squid giant synapse. The I_{Ca} is present at −33 mV, it increases with depolarization up to −3 mV, and then it decreases with further depolarization, in accordance with the *I-V* curve for Ca²⁺ shown in figure 12.10. In general, the postsynaptic response also increases with depolarization of the presynaptic terminal until the maximum I_{Ca} is obtained, and then the postsynaptic response also declines as the presynaptic terminal is further depolarized.

Looking carefully at these traces, there are again a number of important features to the data that should be emphasized. First, at −33 mV and −28 mV there is a significant amount of I_{Ca} flowing during the command, and yet there is essentially no postsynaptic response. As we will see in figure 12.13, this can be most easily understood on the basis of the nonlinearity in the relationship between I_{Ca} and transmitter release.

Second, for all of the commands, the largest postsynaptic response oc-

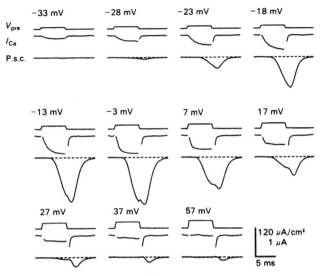

Figure 12.12 Presynaptic Ca²⁺ currents (I_{Ca}) and postsynatic currents (P.s.c.) elicited by different amplitude depolarizing commands (V_{pre}) given to the presynaptic terminal. (From Augustine et al. 1985b.)

curs after the end of the command. As discussed above, transmitter release is essentially an off response. Although the tail currents have been mostly removed from the traces in figure 12.12, we saw in figure 12.11 that at the end of the command the increase in driving force for Ca²⁺ produces a large rapid I_{Ca} through Ca²⁺ channels activated by the depolarization and thus a large rapid release of transmitter.

Third, with step commands of +27, +37, and +57 mV, the off response decreases with increasing depolarization even though the Ca²⁺ conductance should have been maximally activated, and the tail current should thus have been constant, for each of these commands. These results can be understood by recognizing that facilitation of transmitter release will occur from the small I_{Ca} that flows during the command (this will be discussed further in section 12.5.2). For example, for the command to +27 mV, the I_{Ca} during the command causes little or no release itself, but it does facilitate the release that occurs in response to the tail current flowing at the end of the command. At +37 mV there is less I_{Ca} during the command and thus less facilitation of the off response, and, finally, at +57 mV no I_{Ca} flows during the command and no facilitation of release occurs during the tail current. The release that occurs during the tail current at the end of the command to +57 mV is presumably the amount of release that would have occurred at +27 and +37 mV if there had been no current during the command. How the results of these experiments relate to transmitter release following an action potential will be discussed in section 12.4.5.

12.4.4 Cooperative action of Ca^{2+} at the squid synapse

Using data such as those illustrated in figure 12.12, the relationship between presynaptic I_{Ca} and postsynaptic response is shown in figure 12.13. The data are plotted on both linear and log-log scales, and the curves should be compared to those shown in figure 12.5. The relationship is highly nonlinear, and the slope of the log-log plot is between 2 and 4. The data from the squid synapse thus supports the cooperativity among Ca^{2+} ions in promoting release, as predicted by the more indirect experiments of Dodge and Rahamimoff some 20 years earlier. Careful examination of the nonlinear relationship should also make the explanation of the results in the previous two figures more understandable. For example, at the low end of the curve, a small I_{Ca} (but actually up to almost 20% of the maximum current) elicits essentially no postsynaptic response whereas a 20% increase in the current starting in the middle of the curve (at about 0.5 on the abscissa) can lead to a doubling of the postsynaptic response.

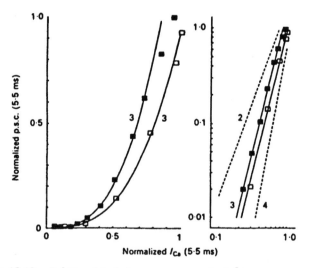

Figure 12.13 Relationship between presynaptic Ca^{2+} current (I_{Ca}) and postsynaptic response (P.s.c.) plotted on linear (left) and log-log (right) coordinates. (From Augustine et al. 1985b.) The numbers next to the lines refer to the power function that provides the best fit.

The data illustrated in the previous two figures support the idea of a cooperative relationship among Ca^{2+} ions and transmitter release with a power function of between 2 and 4. The above experiments, however, did not entirely mimic those of Dodge and Rahamimoff. The relationship in which Dodge and Rahamimoff found a power function of 4 was between extracellular Ca^{2+} and postsynaptic response as opposed to the relationship illustrated above between presynaptic I_{Ca} and postsynaptic response. What is missing for a direct correspondence with Dodge and Rahamimoff is a relationship between extracellular Ca^{2+} and presynaptic I_{Ca}. Augustine and Charlton did investigate just such a relationship, and this is illustrated in figure 12.14.

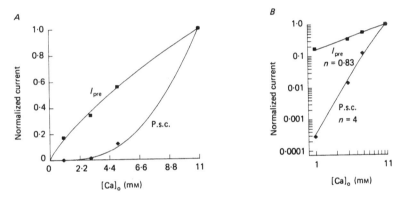

Figure 12.14   Relationship between either presynaptic Ca^{2+} current (I_{pre}) or postsynaptic response (P.s.c.) and $[Ca^{2+}]_{out}$ (from Augustine et al. 1986). Data are plotted on linear (left) and log-log (right) coordinates. The best-fitting power function (n) is indicated for each curve on the right.

In part A of figure 12.14, the relationships between $[Ca^{2+}]_{out}$ and either postsynaptic current (PSC) or I_{Ca} are plotted on linear scales while in B they are plotted on log-log scales. As expected, there is a very nonlinear relationship between $[Ca^{2+}]_{out}$ and PSC. This relationship yields a power function of about 4 when plotted in B. The relationship between $[Ca^{2+}]_{out}$ and I_{Ca} was found to be less than linear, or sublinear, yielding a power function of around 0.8. The overall relationship between $[Ca^{2+}]_{out}$ and PSC can be viewed as the combination of these two relationships, as follows (Augustine and Charlton 1986):

$$I_{Ca} \propto ([Ca^{2+}]_{out})^{n_1},$$
$$PSC \propto ([Ca^{2+}]_{out})^{n_2},$$

and therefore

$$\text{PSC} \propto (I_{Ca})^{n_2/n_1}. \tag{12.4.5}$$

With n_1 being about 0.9 and n_2 about 3.5, the overall relationship yields an exponent of about 4, which is internally consistent with the different experimental results. One of the main conclusions from the experiments illustrated in figure 12.14 is that the source of the cooperativity of Dodge and Rahamimoff and that of Augustine and colleagues illustrated in figure 12.13 cannot be due to cooperativity between $[Ca^{2+}]_{out}$ and presynaptic I_{Ca}, but instead between I_{Ca} and the mechanisms of release.

12.4.5   A model for transmitter release at the squid synapse

Using the voltage-clamp data for presynaptic I_{Ca} as a function of voltage, and the relationship between I_{Ca} and release, Llinás and coworkers developed a computer model for synaptic transmission at the squid synapse. This model is extremely useful for predicting the general features of I_{Ca} during the presynaptic action potential and is illustrated in figure 12.15.

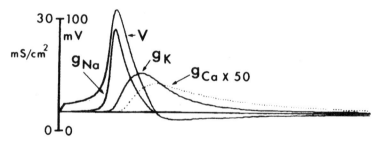

Figure 12.15   Relationships among presynaptic action potential (*V*) and the Na$^+$, K$^+$, and Ca^{2+} conductances that are responsible for the action potential. (From Llinás et al. 1981a by permission of the Biophysical Society, copyright © 1981 Biophysical Society.) The Ca^{2+} conductance is enlarged 50 times to fit on the same scale.

The model used the Hodgkin-Huxley formulation for Na$^+$ and K$^+$ (see chapter 6). I_{Ca} was derived from a conductance equation of the form

$$g_{Ca} = m^5 \overline{g}_{Ca},$$

with rate constants determined from experimental data and with no inactivation. Note that, as determined from the voltage-clamp data, the Ca^{2+} conductance is slow to turn on and does not begin until near the peak of the action potential. The I_{Ca} follows the activation of the conductance, although it is even slower to rise because the driving force is low at the

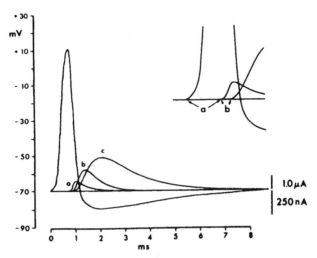

Figure 12.16 Relationships among presynaptic action potential, presynaptic Ca²⁺ current (a), postsynaptic current (b), and postsynaptic potential (c). All currents and potentials are plotted upwards to facilitate comparison of their time courses. The inset is a blowup of the curves to indicate the delay from the action potential to the onset of the Ca²⁺ current (a) and the delay from the Ca²⁺ current to the onset of the postsynaptic potential (b). (From Llinás et al. 1981b by permission of the Biophysical Society, copyright © 1981 Biophysical Society.)

peak of the action potential—the current doesn't actually peak until the end of the action potential. The Ca²⁺ current during the action potential is truly an off response, in that it flows primarily as the action potential is repolarizing and while the driving force for Ca²⁺ is increasing. These simulations are also useful for illustrating the time courses of g_{Na} and g_K during the action potential.

The relationships among the action potential, I_{Ca}, excitatory postsynaptic potential (EPSP), and excitatory postsynaptic current (EPSC) in the model and those found experimentally are illustrated in figure 12.16. In part A of the figure the action potential is followed by curves a, b, and c. These are the presynaptic I_{Ca}, the EPSC, and the EPSP, respectively. Some of the features noted in the voltage-clamp records from previous figures and their significance for neurotransmission can be appreciated more fully from these curves. First, the slow onset of I_{Ca} is responsible for the delay between the beginning of the action potential and I_{Ca} (*a* in the inset). Second, there is very little delay (~200 μsec) between the onset of I_{Ca} and the EPSC. (Note: The rapid off response that we have been emphasizing in this section has particular functional significance with respect to synaptic transmission. The large, fast off response is what is responsible for the relatively brief period of synchronous transmitter release and the brief postsynaptic response. If the Ca^{2+} current were instead spread out over a longer period of time, then transmitter release would also be more prolonged and would produce a longer and smaller peak response in the postsynaptic axon.) Third, the majority of I_{Ca} flows during the falling phase of the action potential. And fourth, the EPSP rises slower and peaks later than the EPSC (this will be discussed further in the next chapter).

The Ca^{2+} hypothesis continues to be widely supported. Additional experiments have used flash photolysis of compounds that normally bind Ca^{2+} and then release Ca^{2+} with light (see appendix B). When these molecules are injected into the squid terminal, a rapid increase in $[Ca^{2+}]_{in}$ can be obtained after a light flash. Such experiments have demonstrated rapid release of transmitter and suggest that Ca^{2+} is necessary and sufficient for release. A minority view, however, holds that release is dependent on membrane potential in addition to a rise in Ca^{2+}. This is the so-called Ca^{2+}-voltage hypothesis. Despite intense investigation there is still much to be learned about neurotransmitter release mechanisms and the role of Ca^{2+}.

12.4.6   Synaptic delay

From what has been presented thus far, the synaptic delay of 0.5–1.0 msec from an action potential in the presynaptic axon to an EPSP in the postsynaptic cell can be separated into two components. The first delay is that between the action potential and the onset of I_{Ca} in the presynaptic terminal. From figure 12.16, this delay is around 800 μsec and accounts for by far the larger amount of delay in synaptic transmission. The second delay is that between the onset of I_{Ca} and the postsynaptic current. From figures 12.12 and 12.16, this delay is about 200 μsec. The overall synaptic delay will vary from synapse to synapse, but if the results from the squid can be generalized to other preparations, then we can assume that the most significant delay is that associated with the kinetics of I_{Ca}. The synaptic delay will also be temperature dependent, because Ca^{2+} channel kinetics are temperature dependent, while the other steps in the release process (e.g., diffusion) do not vary much with temperature.

(The diffusion time of transmitter across the synaptic cleft is quite short. It can be estimated by the equation

$$t = \frac{\Delta x^2}{2D},$$

where t is the diffusion time, Δx is the distance across the cleft, and D is the diffusion coefficient for the transmitter. Using the diffusion coefficient for ACh of 7.6×10^{-6} cm$^2 \cdot$sec^{-1} and a cleft thickness of 5×10^{-6} cm, the diffusion time is 1.6 μsec. This delay due to diffusion of transmitter across the cleft is therefore negligible compared with delays of the other processes associated with release.)

From a molecular standpoint the 200 μsec delay associated with the release process itself may be the more important and the more interesting. This delay sets severe constraints on the molecular mechanisms of release. Even this small delay of ~ 200 μsec must be broken up into several steps. From the time of Ca^{2+} entry to some 200 μsec later, Ca^{2+} ions must diffuse to their internal binding sites and bind to the Ca^{2+} *receptor*, vesicles must fuse to the membrane and release their contents, transmitter molecules must diffuse across the synaptic cleft, and transmitter molecules must bind to their postsynaptic receptors and open ion channels.

12.5 Ca^{2+} and synaptic plasticity

As mentioned in the previous chapter, the forms of synaptic plasticity discussed thus far, namely depression, facilitation, post-tetanic potentiation (PTP), and long-term potentiation, appear to involve Ca^{2+} in some way. (Long-term potentiation will be discussed more fully in chapter 15.) Moreover, depression, facilitation, and PTP are also thought to be presynaptic phenomena; that is, each is thought to involve changes in transmitter release. Given the results presented in this chapter concerning the relationship between Ca^{2+} and transmitter release, it is appropriate to review some of the leading hypotheses for the role that Ca^{2+} might play in these short-term changes in transmitter release.

12.5.1 Synaptic depression

Synaptic depression is a use- or activity-dependent decrease in synaptic strength. The two main hypotheses proposed for synaptic depression are the decrease of Ca^{2+} influx hypothesis and the transmitter depletion hypothesis. The first proposed that Ca^{2+} influx is depressed following intense stimulation, leading to a reduction in transmitter release. There are a number of ways in which intense stimulation could lead to a reduction in Ca^{2+} influx. For example, changes in amplitude of the presynaptic action potential or inactivation of Ca^{2+} channels would lead to a decrease in Ca^{2+} influx. This hypothesis was tested carefully by a number of workers. One of the more convincing experiments was done by voltage clamping the presynaptic terminal of the squid synapse and showing that depression could be obtained with brief, repetitive depolarizations during which a constant amplitude I_{Ca} was recorded (Charlton et al. 1982). The results of these and other experiments demonstrate that this hypothesis is very unlikely.

The second hypothesis proposed that the amount of transmitter available for release is depleted following high rates of release. This hypothesis is essentially one of exclusion. In other words, most of the other hypotheses that have been proposed for depression, such as that discussed above and also hypotheses regarding changes in extracellular potassium and desensitization of transmitter receptors, have been disproven, leaving as the most reasonable remaining hypothesis, the transmitter depletion hypothesis.

12.5.2   Facilitation and PTP: The residual Ca^{2+} hypothesis

The most influential hypothesis for a mechanism to explain facilitation and PTP was originally proposed by Katz and Miledi (1968) for facilitation. This hypothesis has been called the *residual Ca^{2+} hypothesis* and is based on the nonlinear relationship between Ca^{2+} and transmitter release. As alluded to in previous sections, this hypothesis proposes that a portion of the Ca^{2+} that enters during the first stimulus (or during a train of stimuli) is present in the terminal during the second (or subsequent) stimulus. This *residual* Ca^{2+} is too little to evoke release itself (i.e., at the low end of the curve in figure 12.14) but will add nonlinearly to the Ca^{2+} influx occurring during the next stimulus. The result will be a greater release of transmitter during the second stimulus than that during the first. The residual Ca^{2+} hypothesis has been an extremely important one and has been tested in many ways during the past 15 years. Although a detailed explanation of the mechanisms for facilitation and PTP is not yet available, at least the role of residual Ca^{2+} in these forms of plasticity is a little better understood.

12.5.2.1   Facilitation   One of the experimental tests for the residual Ca^{2+} hypothesis was done by Milton Charlton and his colleagues. They proposed that if Ca^{2+} buffers were added to the presynaptic terminal such that the residual Ca^{2+} following the control stimulus would be buffered and thus reduced, then the duration of facilitation would decrease (faster decay). The experiments were done by introducing a Ca^{2+} chelator, BAPTA, into the presynaptic terminal of either the frog or crayfish neuromuscular junction. Their results were exactly opposite to the prediction of the residual Ca^{2+} hypothesis. At the frog neuromuscular junction the time course of facilitation was increased (slower decay) while at the crayfish junction there was no change in time course. These results are incompatible with the residual Ca^{2+} hypothesis.

Zucker and his colleagues have used fluorescent imaging of changes in $[Ca^{2+}]_{in}$ and computer modeling of Ca^{2+} influx in the vicinity of release sites (see the next section and appendix B) and also reach the conclusion that the residual Ca^{2+} hypothesis, at least in its simplest form, is not adequate to explain facilitation. It has been proposed instead that Ca^{2+} entering during the control stimulus acts at some site other than that which directly triggers release. Furthermore, the action of Ca^{2+} at this other site is noncooperative in that facilitation may be linearly dependent on $[Ca^{2+}]_{in}$ rather than on a 4th power relationship such as would be predicted from the residual Ca^{2+} hypothesis.

Another argument against the residual Ca^{2+} hypothesis for facilitation is that the time course for decay of the fast Ca^{2+}-dependent K^+ current, which is activated in the presynaptic terminal (crayfish) by the first stimulus, is much faster than the time course for facilitation. The decay of the Ca^{2+}-dependent K^+ current would presumably reflect the decay of residual Ca^{2+} in the terminal and, by this argument, is too fast to account for facilitation.

12.5.2.2 PTP As mentioned in the previous chapter, the time course of PTP is on the order of 1 minute. The residual Ca^{2+} hypothesis, at least as proposed for facilitation by Katz and Miledi and involving the nonlinear summation of Ca^{2+}, is unlikely to explain PTP because of the relatively slow decay of PTP. That is, the rise in Ca^{2+} near the release sites at the inner surface of the membrane would be expected to fall very rapidly ($\sim$10–100 msec) following a stimulus train and be too small to account for PTP that lasts up to a minute. The influx of Ca^{2+} during the train, however, will diffuse throughout the volume of the terminal. The residual Ca^{2+} averaged over the entire volume of the terminal could indeed play a role in PTP.

This idea was tested by George Augustine and his colleagues using a similar logic as that of Charlton for facilitation. They found that the time course of decay of the average Ca^{2+} in the terminal measured by fluorescent imaging techniques matched closely the time course of decay of

PTP. Furthermore, reducing the residual Ca^{2+} by injection of EGTA reduced PTP. (EGTA is different from BAPTA in that it is a relatively slow buffer. Injection of EGTA has no effect on release during single stimuli, whereas BAPTA will reduce such release.) Similar conclusions have been reached by others. It appears therefore that residual Ca^{2+} does play a role in PTP (and augmentation), but not by way of nonlinear summation of Ca^{2+}. Instead the residual Ca^{2+} must somehow play a role in mobilizing transmitter vesicles or sensitizing the release machinery to Ca^{2+} so that more transmitter gets released for a period of time following a stimulus train.

It is not entirely clear what controls the rate of decay of residual Ca^{2+} in a presynaptic terminal. The possible routes for removal of Ca^{2+} include uptake and storage by intracellular organelles, active transport across the plasma membrane, and removal by Na^+-Ca^{2+} exchange. All of these removal processes are likely to have different rate constants, and therefore any transmitter release mechanisms dependent on $[Ca^{2+}]_{in}$ will be dependent on these rate constants. It is tempting to speculate that facilitation (F_1, F_2, see chapter 11), augmentation, and PTP are each dependent on a different Ca^{2+} removal process, but this speculation is probably too simplistic (see Magleby 1987).

12.5.3 Presynaptic modulation of transmitter release

Many presynaptic terminals also contain receptors for neurotransmitters, sometimes including receptors for the transmitter released at that same synapse—so-called autoreceptors. The list of well-known presynaptic receptors includes γ-aminobutyric acid ($GABA_B$), muscarinic acetylcholine, serotonin, α-adrenergic, opioid peptide, and adenosine receptors. There are undoubtedly many more. From the mechanisms of release discussed in this chapter, it is obvious that there are many ways in which transmitter release could be modulated by presynaptic receptors. One of the more important and prominent sites of modulation is on the presynaptic Ca^{2+} current itself. As mentioned in chapter 7, there are multiple types of Ca^{2+} channels. At vertebrate synapses two types of channels have been suggested to take part in transmitter release: the N- and P-type channels. (One characteristic of the presynaptic Ca^{2+} channel, whatever it might be, is that it shows little or no inactivation during short depolarizations; see figure 12.9.) For the N-type channel in particular, the activation of a number of the above transmitter receptors has been demonstrated to inhibit N-type channel activity. These include $GABA_B$, muscarinic cholinergic,

α-adrenergic, serotonin, opioid, and adenosine receptors.

In addition to direct modulation of I_{Ca}, anything that alters the potassium current in the presynaptic terminal would alter the duration of the action potential and indirectly modulate I_{Ca}. For example, drugs that block the A-, D-, and C-type K^+ channels cause a significant increase in synaptic transmission in most preparations. Many of the above list of transmitter receptors have been suggested to modulate at least one of these types of K^+ currents. Moreover, most of the above list of receptors also couple to one or more second messenger systems, raising the possibility of altering release through direct effects on the biochemical mechanisms of release in addition to, or instead of, altering of release through changes in I_{Ca}. Just such a modulation of release has been demonstrated for serotonin at several invertebrate synapses (see Delaney et al. 1991).

12.6 Molecular mechanisms of release

12.6.1 Early hypotheses for Ca^{2+}-dependent exocytosis

The short delay (< 200 μsec) from the entry of Ca^{2+} to the release of neurotransmitter places some theoretical constraints on what the molecular nature of the release mechanism can be. The bulk of available evidence still favors the idea that the transmitter-containing vesicles observed at the presynaptic terminal are the units of transmitter release, with one vesicle equal to one quantum.

Although similar data are not available at other synapses, from the synaptic delays one can estimate that release must occur within a similar time frame, including at the vertebrate neuromuscular junction and at glutamatergic, cholinergic, and GABAergic synapses in the central nervous systems. Not all exocytotic processes are this fast, however. For example, the Ca^{2+}-dependent secretion of histamine from mast cells and of catecholamines from adrenal chromaffin cells takes place on the order of seconds instead of microseconds. Release of hormones, catecholamines, and neuropeptides from various nerve terminals in the peripheral and central nervous system are also thought to take place much more slowly than the fast neurotransmission discussed here for the squid synapse and neuromuscular junction. Many of these slower-releasing systems have nevertheless proven to be useful preparations for the study of the general mechanisms of Ca^{2+}-dependent exocytosis. It is important to keep in mind, however, that the molecular mechanisms of release at these slower-secreting systems may have some similarities but also some significant differences from the mechanisms underlying fast exocytosis.

Many hypotheses have been proposed for the rapid Ca^{2+}-dependent events leading to neurotransmission. Some of these are listed below along with some of the evidence for and against—mostly against.

1. **Charge neutralization:** One early idea was that the external surface of the vesicle membrane and the inner surface of the synaptic membrane were both negatively charged and that the entry of Ca^{2+} neutralized their electrostatic repulsion and allowed the two membranes to fuse. Although this could be a rapid process, there also would be no special requirement for Ca^{2+}. Any divalent cation (e.g., Mg^{2+}) would serve equally well to neutralize the negative charge. Because there *is* a special requirement for Ca^{2+} in the release process, this simple hypothesis is very unlikely.

2. **Ca^{2+}-dependent K^+ channel:** It was proposed that if there were Ca^{2+}-dependent K^+ channels in the vesicles they would bind incoming Ca^{2+} and open channels permeable to K^+ and H_2O. If the vesicles were normally hypertonic, the resulting influx of H_2O would cause the vesicles to swell and fuse with the plasma membrane. This hypothesis went out of favor when the injection of blockers for Ca^{2+}-dependent K^+ channels into the squid synapse had no effect on release. There is also no evidence that vesicles swell prior to fusion. They do, however, appear to swell *after* fusion.

3. **Actin/tubulin:** Another idea was that Ca^{2+}-dependent contractile proteins such as actin and myosin might be attached to the vesicle and synaptic membranes and contract upon the entry of Ca^{2+}, bringing the two in contact for fusion. Also, tubulin, a constitutive protein of microtubules, was suggested to be phosphorylated upon entry of Ca^{2+} and somehow to trigger release. These hypotheses have been rejected because depolymerizing drugs such as cytochalasin, colchicine, and vinblastine fail to inhibit release (reviewed in Augustine et al. 1987).

4. **Miscellaneous:** There have been many other proposals. These include other mechanisms for vesicle swelling, metalloendoproteases that would cleave certain proteins, and numerous other molecules that seemed attractive for an involvement in the release process. The apparent requirement for 3 or 4 Ca^{2+} ions makes any molecule that binds Ca^{2+} with a stoichiometry of near 4 a likely candidate.

One of the more prominent of these is calmodulin. Unfortunately, there is little evidence for a direct role of calmodulin, or of any of these other suggested substances, in the release process. One must also separate the biochemical steps that are involved in the preparation of the vesicles for release from the final, rapid Ca^{2+}-dependent step of vesicle fusion. There is much interest in proteins that are associated with vesicles. Some of these proteins may be involved in vesicle trafficking and docking as well as in the final fusion step. These vesicle-associated proteins will be discussed in section 12.6.3.

12.6.2 Fusion pores

The actual membrane fusion event, at least for the slower-secreting systems, can be measured electrically by monitoring the total input capacitance of the cell. As a vesicle fuses to the plasma membrane, there is a small increase in membrane area and thus in total capacitance of the cell (remember that total capacitance= $C_m\times$ area). This is illustrated in figure 12.17.

In addition to the change in capacitance due to the added membrane from the vesicle, if one assumes that the potential across the vesicle is initially different from that of the cell (for example, chromaffin granules have an inside positive potential of +50 mV; see figure 12.17), then when fusion occurs a small current will flow from the vesicle to the exterior of the cell. From the magnitude of this current, one can determine the conductance and size of the initial opening.

It has been suggested that the rapid fusion of the bilayers from the vesicle and plasma membrane may be mediated by a macromolecule that spans the two membranes and forms a pore or channel upon the influx of Ca^{2+}. The opening of this *fusion pore*, which would allow the diffusion of transmitter from the vesicle to the extracellular space, could be an extremely rapid event. When the current associated with the initial fusion event was measured, it turned out that it was produced by a fairly constant conductance of around 230 pS (at least for mast cells), which then increased gradually as the pore dilated. This finding lent support to the idea that there is a pore or channel in the vesicle and plasma membrane that opens with the entry of Ca^{2+}. Candidate molecules for the fusion pore include the annexins, synaptophysin, synaptotagmin, and syntaxin (see section 12.6.3). Other more recent data, however, suggest that these proteins instead form a "scaffolding" between vesicle and plasma membrane to facilitate the fusion event and that the pore itself is a single lipid bilayer (Monck and Fernandez, 1994).

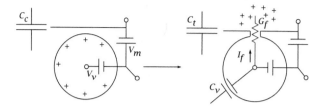

Figure 12.17   Diagram of vesicle fusion. When fusion occurs (right) there is an increase in capacitance ($C_t = C_c + C_v$). If the initial potential across the vesicle, V_v, is different from that of the cell, V_m, then a current will flow, $I_f = G_f(V_m - V_v)$, where I_f and G_f are the current and conductance of the fusion pore, respectively. (Adapted from Breckenridge and Almers 1987. Used by permission of *Nature*, copyright © 1987 Macmillan Magazines Limited.)

12.6.3   Vesicle-associated proteins: Possibilities for the Ca^{2+} receptor

There is much interest in the proteins associated with vesicles because of their possible role in preparing vesicles for release and in mediating the final, rapid fusion event. There must be some site near the Ca^{2+} channel and vesicle that binds Ca^{2+} and triggers release. This Ca^{2+} *receptor* or Ca^{2+} *trigger molecule* may be one of the vesicle-associated proteins. In this section we will summarize some of the better studied examples of vesicle-associated proteins, with emphasis on their putative physiological functions.

1. **Synapsins:** The synapsins family contains the major phosphoproteins present in nerve terminals. The synapsins are associated with vesicles, and they bind to cytoskeletal elements. In addition, they are phosphorylated by a number of protein kinases including cAMP-dependent kinase, Ca^{2+}/calmodulin-dependent kinases I and II (CaM kinase I and II), and proline-directed kinase. The possibility that the synapsins play a role in transmitter release was tested using injection of phosphorylated and dephosphorylated synapsin into the squid synapse. Dephosphorylated but not phosphorylated synapsin inhibited release. Also, the injection of CaM kinase II increased release, while none of these injections affected the influx of Ca^{2+}. These and other results are consistent with the idea that dephosphorylated synapsin binds to cytoskeletal proteins and immobilizes vesicles. The vesicles are then released from the cytoskeleton when synapsin is phosphorylated by CaM kinase II. The synapsins are therefore believed to play a role in the mobilization of vesicles and regulating the number of vesicles available for release.

2. **Synaptotagmin:** Synaptotagmin is an abundant membrane protein of synaptic vesicles that interacts with phospholipids and proteins in the surface membrane. It also has one or two Ca^{2+} binding sites and is present in the membrane as a dimer (hence a possible stoichiometry of four Ca^{2+}-binding sites). It is an attractive candidate molecule for the Ca^{2+} receptor and may also play a role in vesicle docking and fusion, although it does appear that fusion can occur in the absence of synaptotagmin.

3. **Synaptophysin:** Synaptophysin is the most abundant vesicle membrane protein. It is a Ca^{2+}-binding protein that, for a number of reasons, is an attractive candidate molecule for the "scaffolding" near the fusion pore.

4. **Others:** Other molecules that have recently been suggested to be either associated with vesicles or with the plasma membrane and to be involved in the release process include synaptobrevin; rab3A (reviewed in Südhof and Jahn 1991), a protein suggested to target vesicles to the active zone; syntaxin, a presynaptic membrane protein that is associated with Ca^{2+} channels and interacts with synaptotagmin; and the neurexins, a family of cell surface proteins, at least one of which may interact with synaptotagmin. Undoubtedly, others will be identified in the near future that also have some function in transmitter release.

12.6.4 Calcium domains, active zones, and calcium buffering

The active zone of a synapse is a specialized structure spanning both the pre- and postsynaptic sides of the synapse (see figure 12.18). The postsynaptic side of the synapse contains a high density of postsynaptic receptors. The synaptic cleft also shows a distinctive staining pattern, suggesting molecules that physically hold the pre- and postsynaptic membranes together. On the presynaptic side, there is a high density of transmitter-containing vesicles in close proximity to the membrane and an array of cytoskeletal elements surrounding these vesicles. In freeze-fracture micrographs of the active zone, distinct particles can be observed in the membrane that are suggested to be the Ca^{2+} channels responsible for release.

Active zones are characteristic only of terminals of fast-releasing synapses. At slower-releasing synapses, such as neuropeptide-secreting terminals, no apparent active zone is observed. This may also indicate that the Ca^{2+} channels for slow secretion are more distant from the vesi-

A

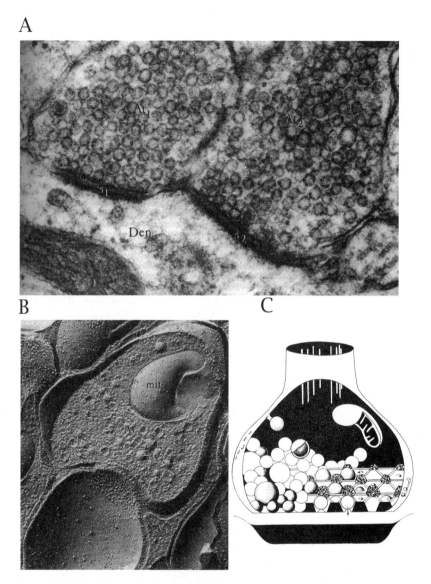

Figure 12.18   (A) Transmission EM of synaptic terminal. (Den = dendrite, and s$_1$ and s$_2$ are active zones for synapses At$_1$ and At$_2$.) (B) Scanning EM of freeze-fractured terminal. (mit = mitochondrion in synaptic terminal At, and sv = synaptic vesicle.) (C) Diagram of synapse. Note the specialized structures on both the pre- and postsynaptic sides of the synapse (A and B) and the suggested grid for vesicle release in the presynaptic terminal (in C). (A) and (B) are from Peters, A., Palay, S. L., and Webster, H., *The Fine Structure of the Nervous System*, used by permission of Oxford University Press, copyright © 1991 Oxford University Press; (C) is from Pappas and Purpura 1972.

cles and explain the finding that neuropeptide secretion usually requires higher frequencies of stimulation (which brings in more Ca^{2+}) than does transmitter release at fast-secreting synapses.

The possible close proximity of the Ca^{2+} channels to the vesicles, at least for the fast-releasing synapses, raises a number of issues. These include the concentration of Ca^{2+} in the terminal necessary for release, the buffering and diffusion of Ca^{2+} in the terminal, the concept of Ca^{2+} *domains* near the inner surface of the Ca^{2+} channels, and the proximity of the vesicles to the Ca^{2+} channels.

The concentration of Ca^{2+} in the presynaptic terminal necessary for release has been estimated to be on the order of 10-100 μM. Given that the normal resting level of Ca^{2+} is about 50-100 nM, a rise of $[Ca^{2+}]_{in}$ of roughly 3 orders of magnitude must take place to trigger release. How is this possible, and why do fluorescent measurements of Ca^{2+} in the terminal during trains of action potentials indicate rises of only about a few hundred nM? The answer appears to lie in the concept of a Ca^{2+} domain near the mouth of an open Ca^{2+} channel, in which the Ca^{2+} concentration can be very high and yet fall off very steeply with distance away from the channel. This would also necessitate that the Ca^{2+} *receptor* that facilitates release must be very close to a Ca^{2+} channel (i.e., ~10 nm). The fluorescent changes associated with an increase in $[Ca^{2+}]_{in}$ averaged over a large area would merely reflect the diffusion of Ca^{2+} into the entire volume of the terminal resulting from the influx through a few open Ca^{2+} channels. This idea is illustrated in figure 12.19. It is also estimated that there is one Ca^{2+} channel associated with each release site and that the domains from adjacent Ca^{2+} channels do not overlap.

The buffering of Ca^{2+} in the presynaptic terminal is also of great interest. Because so many cellular processes depend on Ca^{2+}, the concentration of Ca^{2+} in cells is tightly regulated. Free Ca^{2+} can bind to endogenous intracellular buffers, be transported across the membrane by pumps or Na^{+}-Ca^{2+} exchange, or be sequestered by intracellular organelles. The buffering of Ca^{2+} is believed to occur via both mobile and immobile molecules that bind Ca^{2+}. The buffering of Ca^{2+} in a hypothetical cell is illustrated in figure 12.20. The immobile buffer is believed to have a high capacity for binding Ca^{2+} and thus to be essentially nonsaturable with a low affinity for Ca^{2+}. Approximately 98% to 99% of the entering Ca^{2+} rapidly binds to this immobile endogenous buffer.

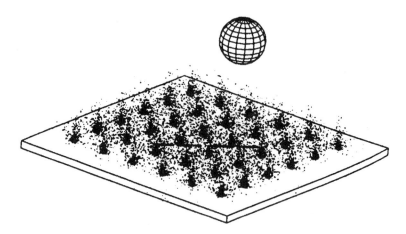

Figure 12.19 Ca^{2+} domains in presynaptic terminal. (From Smith and Augustine 1988.) Each cluster of dots represents Ca^{2+} ions entering the presynaptic terminal through a single channel. The sphere is roughly the size of a transmitter-containing vesicle in comparison to the Ca^{2+} channels.

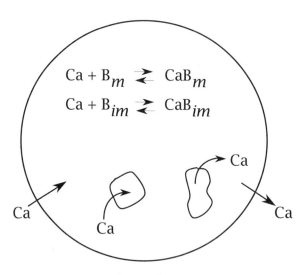

Figure 12.20 Diagram of Ca^{2+} homeostasis in a cell. B$_m$ and B$_{im}$ are the mobile and immobile buffers, respectively. Ca^{2+} can enter the cell through channels in the plasma membrane, be transported out of the cell by active transport or Na$^+$-Ca^{2+} exchange, be taken up by intracellular organelles, or be released into the cytoplasm by intracellular stores.

12.7 Summary of important concepts

1. Cooperativity of Ca^{2+} and release.
2. Space clamp.
3. Relationships among presynaptic action potential, I_{Ca}, and EPSC.
4. Origin of synaptic delay.
5. *Off* response.
6. Residual Ca^{2+} hypothesis.

12.8 Homework problems

1. Reproduced below is part of figure 12.16. It depicts different aspects of synaptic transmission as studied at the squid giant synapse. Label and describe all parts of the figure.

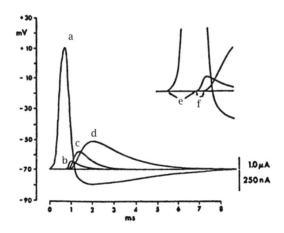

2. Recalling the Ca^{2+} hypothesis for transmitter release, briefly answer (one page or less for each) the following questions:

   (a) What is an *on* response?
   (b) What is an *off* response?
   (c) What is the significance of each?

(d) Describe the origin of the synaptic delay.

(e) What is the nature and significance of the nonlinear relationship between $[Ca^{2+}]_{in}$ and transmitter release?

(f) Describe the residual Ca^{2+} hypothesis for faciliation, augmentation, and PTP.

3. (a) Given the three-microelectrode voltage clamp shown below, label the three electrodes and give the equation for I_m in terms of whatever you label the three electrodes. What is the relationship between I_{clamp} and I_m?

(b) List all of the information about Ca^{2+} and transmitter release that can be obtained from the data illustrated in the following figure.

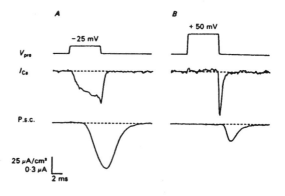

13 Synaptic Transmission III: Postsynaptic Mechanisms

13.1 Introduction

The last two chapters discussed various presynaptic mechanisms involved in synaptic transmission, including the quantal nature of neurotransmitter release and the role of Ca^{2+} in release. In this chapter we move to the postsynaptic side of the synapse. As in previous chapters, the focus will be on physiological mechanisms. We will assume knowledge of previous chapters, in particular those concerning cable properties and the analysis of single ion channels. This chapter will cover some of the basic principles associated with the events at the postsynaptic membrane following the release of neurotransmitter, and the functional properties of excitatory and inhibitory synapses. We will also briefly discuss electrical synapses, synaptic events involving a conductance decrease, and certain unique aspects of different neurotransmitter responses. For a discussion of the neurochemistry of neurotransmitter systems and the molecular biology of transmitter receptors, the reader is referred to several excellent texts (Cooper et al. 1986; Hall 1992; Siegel et al. 1989).

13.2 General scheme for ligand-gated channels

The general kinetic scheme for ligand-gated channels is illustrated below (see also chapters 8–10).

$$
\begin{array}{ccccccc}
\text{(hydrolysis or uptake)} & & & (C) & & (O) & \\
\uparrow & & & k_1 & & \beta & \\
T & + & R & \rightleftharpoons & T \cdot R & \rightleftharpoons & T \cdot R^* \\
\downarrow & & & k_2 & & \alpha & \\
\text{diffusion} & & & & & &
\end{array}
$$

Neurotransmitter molecules are released from the presynaptic terminal. The concentration of transmitter molecules (T) in the synaptic cleft is dependent upon the concentration of transmitter in vesicles, the number of vesicles released, and the geometry of the cleft. The decline in the concentration of transmitter in the cleft is dependent on diffusion and either hydrolysis or reuptake by the pre- or postsynaptic elements and glial cells. One or more of these transmitter molecules will bind to a receptor (R) on the postsynaptic membrane, forming a bound but closed state of the channel (C). The closed state makes a transition to the open state (O) in a probabilistic manner (see chapters 8–10).

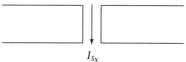

I_{s_x}

Figure 13.1   Diagram of current flow through a single, open channel.

During the open state of the channel the amount of current flow through the channel is dependent on the single-channel conductance and the *driving force* for current through the channel. The current per channel is given by

$$I_{s_x} = G_{s_x}(V_m - E_s),\tag{13.2.1}$$

where I_{s_x} is the single-channel current, G_{s_x} is the conductance of a single channel, and $(V_m - E_s)$ is the driving force. Equation 13.2.1 is simply Ohm's law, where the driving force is the difference in potential across the membrane for that particular ion. Using the familiar parallel conductance model to represent the postsynaptic membrane and, for the moment, considering each channel as an element of the model, the situation just described is depicted in figure 13.2.

If each channel opened at the same time and stayed open, then the total synaptic conductance would be the sum of all the channel conductances, or

$$G_s = \sum_{x=1}^{N} G_{s_x}.$$

As we saw in chapters 8–10, however, channel opening is a stochastic process, and one must take into account the probability of channel opening as well as the number of channels in the postsynaptic membrane available for opening. A better equation for synaptic conductance is therefore

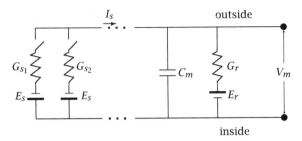

Figure 13.2 Parallel conductance model of a synapse with multiple ligand-gated channels. The switch associated with each conductance element depicts opening or closing of the channel (i.e., closed switch = open channel and current flow). G_r and E_r are the resting conductance and resting potential of the postsynaptic cell, respectively. All other elements are as described elsewhere.

$$G_s = y_s(N \cdot P), \tag{13.2.2}$$

where y_s is the more conventional symbol for single-channel conductance, N is the number of channels available for opening, and P is the probability of a channel being open. P will depend on the concentration of transmitter molecules in the cleft, the kinetic properties of the channel, and, for some receptors, the postsynaptic membrane potential. Combining equations 13.2.1 and 13.2.2, we have

$$I_s = y_s(N \cdot P)(V_m - E_s), \tag{13.2.3}$$

which is the equation for synaptic current in terms of single-channel properties. For most of the discussion in the rest of this chapter, however, we will ignore the stochastic properties of the ligand-gated channels and use the total (peak) synaptic conductance G_s from equation 13.2.2, or

$$I_s = G_s(V_m - E_s). \tag{13.2.4}$$

Referring to figure 13.2, when the channels are open (switches closed) current flows in accordance with equation 13.2.4. I_s charges the membrane capacitance, C_m, and flows across the membrane conductance, G_r. When the transmitter concentration is low and the channels close (open switches), the potential developed across C_m decays with a rate determined by τ_m. For synaptic events faster than τ_m, the decay of the synaptic potential is governed by τ_m, whereas for events slower than τ_m, the time course of the synaptic potential is governed by the time course of the conductance change itself. The direction of I_s, and thus the polarity of the potential developed across C_m during the flow of I_s, is determined by E_s relative to E_r.

13.3 Synaptic conductances and reversal potentials

The release of a neurotransmitter from a presynaptic ending produces a conductance change in the postsynaptic neuron. The strength of the synaptic connection will depend, in part, on the magnitude of this conductance change and the driving force for the synaptic current. This section will discuss these and other issues related to the properties of synaptic events.

13.3.1 Definitions of excitatory and inhibitory responses

Orthodromic activation of a neuron occurs when an action potential travels in the direction from cell body to synaptic ending. This is in contrast to *antidromic* activation, in which an action potential travels in the direction toward the cell body, away from the synapse. If the stimulation is done

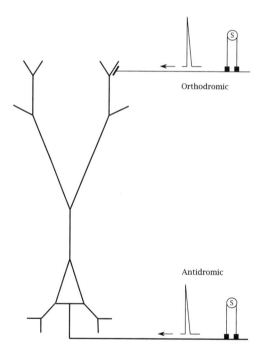

Figure 13.3   Diagram of orthodromic and antidromic stimulation. Orthodromic and antidromic refer to the direction of travel of the action potential, that is, orthodromic = AP from soma to synapse; antidromic = AP from synapse or axon to soma.

somewhere along the axon, then action potentials travel in both directions, one orthodromically toward the synapse and one antidromically toward the cell body. Orthodromic and antidromic stimulation are illustrated in figure 13.3. The terms *afferent* and *efferent* fibers are also useful to define in this context. Afferents are axons in which action potentials travel *toward* a neuron by way of a synapse (e.g., the axon at the top of figure 13.3), and efferents are axons in which action potentials travel *away* from the soma of the neuron (e.g., the axon at the bottom of figure 13.3).

Orthodromic stimulation of an excitatory synaptic input produces excitatory postsynaptic potentials (EPSPs) in the postsynaptic neuron. Examples of EPSPs are shown in figure 13.4. If one changes the membrane potential of the postsynaptic neuron by passing steady current (either inward or outward), the amplitude of the EPSPs will change. If the postsynaptic membrane is completely passive, with no voltage-gated conductances, then the amplitude of the EPSPs will increase with hyperpolarization and decrease with depolarization in accordance with the change in driving force (i.e., change in V_m). At some membrane potential depolarized from rest (~ 0 mV in figure 13.4), the polarity of the EPSP will reverse sign and become hyperpolarizing. This is called the *reversal potential*, V_{rev}.

Inhibitory inputs are typically (but not always) hyperpolarizing from rest. Examples of inhibitory postsynaptic potentials (IPSPs) are also illustrated in figure 13.4. Note the change in amplitude of IPSPs at different membrane potentials and also the reversal potential for the IPSPs (~ -75 mV in figure 13.4).

A synapse is considered *excitatory* if it increases the probability of a neuron firing an action potential. Likewise, a synapse is considered *inhibitory* if it decreases the probability of firing an action potential. These definitions may seem straightforward, except that they lead to a situation that may seem counterintuitive at first: *A synaptic input could produce a depolarization of the membrane potential and still be inhibitory.* What determines whether a synaptic input is excitatory or inhibitory is the relationship between its equilibrium potential (E_s) and the threshold for firing an action potential. If we assume a normal resting potential of around -60 mV and a threshold for the action potential that is positive to this value (i.e., $V_{\text{rest}} < V_{th}$), then we can put the definition for excitatory and inhibitory synapses in more quantitative terms with the following:

If $E_s > V_{th}$, then the synapse is excitatory.
If $E_s < V_{th}$, then the synapse is inhibitory.

E_s is the synaptic equilibrium potential and V_{th} is the voltage threshold for an action potential (an action potential is elicited 50% of the time at this potential). As we will see later, the above definitions hold only for synapses that produce an increase in conductance in the postsynaptic membrane—so-called conductance *increase* PSPs. For conductance *decrease* PSPs, the inequalities are reversed.

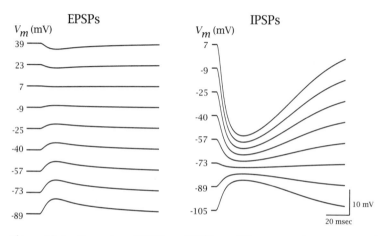

Figure 13.4   Examples of EPSPs and IPSPs at different membrane potentials in a passive neuron.

One might naturally (but incorrectly) think that excitatory synapses produce transient depolarizations from rest and inhibitory synapses produce transient hyperpolarizations from rest. As stated above, what determines whether a synapse is excitatory or inhibitory is the relationship between threshold and synaptic equilibrium potential. Excitatory synaptic inputs are indeed depolarizing from rest, but inhibitory inputs could be hyperpolarizing, depolarizing, or produce no change at all in the membrane potential. For example, one can imagine a situation in which $V_{rest} = -60$ mV, $V_{th} = -50$ mV, and $E_s = -55$ mV. Such a synaptic input would be depolarizing from rest yet would still be classified as inhibitory by these definitions. The answer to this apparent paradox may become clearer when we discuss current-voltage relationships for synaptic inputs. In the meantime, consider an extreme case of an inhibitory synapse with the above parameters that produces a *huge* increase in conductance, some 10 times the normal resting conductance of the cell. During this synaptic input, the membrane potential would essentially be "clamped" to -55 mV, a value below threshold. To reach threshold the neuron would have to receive a very large excitatory input—much larger than would be necessary in the absence of the inhibitory input—to depolarize the membrane potential to threshold. The inhibitory input, even though depolarizing, keeps the membrane potential below threshold and lowers the probability of firing an action potential. It can therefore be properly classified as inhibitory (see also example 13.1).

Another example would be an inhibitory input that produces no change in membrane potential. Such an input would have an equilibrium potential near rest, or −60 mV in the above example. During such a synaptic event, the membrane would now be "clamped" to the resting potential and again resist depolarization to threshold. These types of synaptic inputs have also been called *shunting* inputs because they shunt current flow from a simultaneous excitatory input and attenuate its amplitude. Although this is certainly true, the terminology is a bit misleading when applied only to a synaptic input with no potential change, because essentially all inhibitory inputs are "shunting," regardless of whether they are depolarizing, hyperpolarizing, or produce no change in membrane potential. In each case they will shunt current flow from an excitatory input and reduce its amplitude. A few additional points about shunting inhibition include: (a) The shunting lasts only for the duration of the conductance change, which is not necessarily the same as the duration of the inhibitory potential; (b) The shunting is fairly localized in space and occurs only near the site of the inhibitory input; and (c) The shunting is not subtractive. In other words, the shunting does not merely subtract off a fixed amount the way a hyperpolarization might subtract from a depolarization. Instead, shunting can reduce the depolarizing event by an amount that depends on the relative magnitudes of the inhibitory and excitatory conductance changes (i.e., a large shunting conductance change will reduce a small EPSP by a large amount, but a large EPSP by a lesser amount—the equation for this is given in example 13.1). Inhibition that is not shunting, for example, inhibition taking place at a site far from the synaptic input or at a time following the underlying conductance change, can be considered *subtractive*.

13.3.2 Voltage-clamp analysis of synaptic parameters (*I-V* curves)

Although much can be learned from studying EPSPs and IPSPs such as those shown in figure 13.4, it is difficult to determine the conductance change associated with the synaptic input by using current-clamp techniques. An important method for analyzing synaptic inputs is once again the voltage clamp. Using voltage-clamp methods, one can determine the

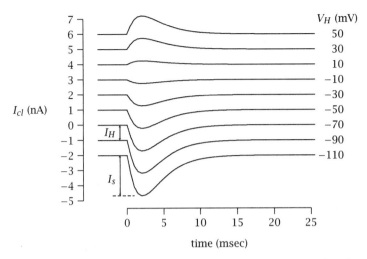

Figure 13.5 Voltage-clamp analysis of a synaptic input. Examples of excitatory synaptic currents (EPSCs) measured at different holding potentials to illustrate the measurement of the holding current, I_H, and the peak synaptic current, I_s. I_H is measured from the 0 current level, while I_s is measured from the holding current. See text for further explanation.

peak conductance change, G_s, the time course of the conductance change, g_s, and the equilibrium potential, E_s. One can also determine some of the resting properties of the neuron. Using the equivalent circuit of figure 13.2, the following equation can be derived:

$$I_{cl} = I_H + I_s = G_r(V_m - E_r) + G_s(V_m - E_s),   (13.3.5)$$

where I_{cl} is the total clamp current and I_H is the *holding* current. The use of a voltage clamp to determine properties of a synaptic input is illustrated in figure 13.5. In this figure the membrane potential of the postsynaptic neuron is clamped at various values while the synaptic input is activated at each potential. The holding current, I_H, is the steady current necessary to keep the neuron at the intended holding potential. By definition, the resting potential is the holding potential at which $I_H = 0$. When the synapse is activated, the current associated with the synapse, I_s (also called an EPSC or IPSC, depending on whether it is excitatory or inhibitory), is added to I_H. From data such as these one can measure I_H and I_s as a function of V_m and plot the associated current-voltage (*I-V*) curves (figure 13.6).

The slope of the line for I_H vs. V_m (particularly in the hyperpolarizing direction where the effect of voltage-gated conductances on the holding current is minimal) gives the resting conductance of the neuron, or G_r.

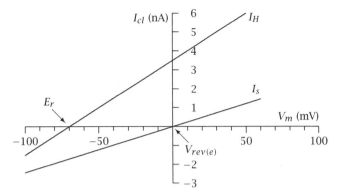

Figure 13.6 *I-V* curves from a voltage-clamp experiment of an excitatory synaptic input. I_s vs. V_m gives the synaptic conductance and I_H vs. V_m gives the resting conductance of the neuron. E_r is the resting potential, and $V_{rev(e)}$ is the reversal potential of the excitatory response.

The slope of the line for I_s vs. V_m gives the peak synaptic conductance, G_s. I_s could be measured at any point in time after the onset of the synaptic current. If the measurement is made at the peak, as illustrated in figure 13.5, then the conductance is the *peak* conductance. A measurement of I_s at some other time point would yield an *I-V* curve with a different slope and thus a different value for G_s.

As mentioned above, the voltage intercept of the I_H vs. V_m line gives the value for the resting potential. The intercept of the I_s vs. V_m line gives the value for the reversal potential, V_{rev}, because it is the potential at which the polarity of the synaptic current *reverses* its sign. The reversal potential is also equal to the synaptic equilibrium potential, E_s, when the synaptic input is isopotential with the location of the voltage clamp (see also section 13.3.3).

Example 13.1
Suppose we are recording from a neuron receiving an excitatory synaptic input with properties the same as those illustrated in figure 13.6. At any potential between rest and threshold, the excitatory input will produce an inward current that depolarizes the neuron toward (or beyond) threshold. In the figure below, an *I-V* curve for a hypothetical inhibitory input is illustrated. The conductance of this input is 5 times the excitatory input (5 times the slope) with an equilibrium potential (-60 mV) depolarized from rest (-70 mV) but hyperpolarized from threshold ($V_{th} = -45$ mV). Is this input really inhibitory? At any

Example 13.1 (continued)

potential between rest and its equilibrium potential, the synaptic input will produce an inward current that depolarizes the neuron toward threshold—as would an excitatory input. At potentials between E_s and threshold, however, the synaptic input will produce an outward current that hyperpolarizes the neuron away from threshold. This would indeed be inhibitory.

But what if the excitatory and inhibitory inputs are activated together? Such a composite synaptic response will have an *I-V* curve that is the sum of the individual *I-V* curves. This is illustrated by the dotted line in the figure. In this case, the combined inputs will still be inhibitory because the composite equilibrium potential is negative to threshold. The inhibitory input has thus overpowered the excitatory input and shunted the amplitude of the excitatory current.

The conductance (slope of *I-V* curve, G_{S_T}) and equilibrium potential (voltage intercept, E_{S_T}) of the combined synaptic response is given by $G_{S_T} = G_{S_e} + G_{S_i}$, and $E_{S_T} = \frac{G_{S_e} E_{S_e} + G_{S_i} E_{S_i}}{G_{S_T}}$.

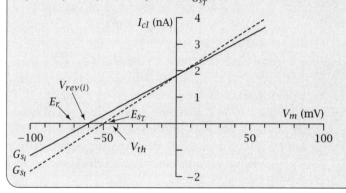

13.3.3 Conductance and reversal potentials for nonisopotential synaptic inputs

Most excitatory synapses in the CNS terminate on dendrites that are electrically remote from the cell body. Even inhibitory synapses, which were once believed to be restricted to the soma, are now known to have contacts throughout the dendritic tree. In such cases, the preceding discussion of voltage clamping synaptic inputs becomes complicated by a lack of space clamp, because the subsynaptic membrane at the synapse will not be isopotential with the site of recording.

The lack of space clamp has a number of effects (see also chapter 12). First, the measured reversal potential (measured, for example, from the soma) will no longer be equal to the synaptic equilibrium potential (i.e., $V_{rev} \neq E_s$). Second, the synaptic conductance change measured from the soma will not be equal to the conductance change occurring at the synapse. And third, the time course of the synaptic current measured from the soma will be slower than that occurring at the synapse because of the low-pass filtering properties of the dendrites (see also chapter 4). We will discuss each of these errors in the following three sections and in section 13.4.2.

13.3.3.1 Reversal potentials and conductance ratios: General Most fast inhibitory responses in the CNS are mediated by Cl^-. In other words, the inhibitory neurotransmitter (usually either GABA or glycine) opens channels that are permeable to Cl^-. The equilibrium potential for these inhibitory synapses is equal to that for Cl^- or $E_s = E_{Cl}$. For fast excitatory synapses, the ligand-gated channels (usually either glutamate or nicotinic ACh) are nonselective for monovalent cations (Na^+ and K^+, although some are also permeable to Ca^{2+}). The equilibrium potential (E_s) for excitatory synapses is around 0 mV.

For the case of a channel permeable to two or more ions, measuring the reversal potential as a function of different ionic concentrations in the bath will provide information about the relative conductance of the channel to different ions as well as, perhaps, the error associated with the measurement of the reversal potential for synapses not isopotential with the recording site. The situation for two ions (Na^+ and K^+) is depicted in figure 13.7.

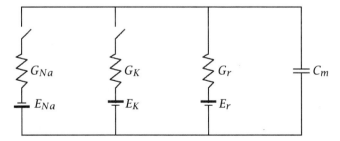

Figure 13.7 Parallel conductance model for an excitatory synapse in which the ligand-gated channels are permeable to both Na^+ and K^+. When the synapse is active, the switches are closed, and the G's represent constant parameters. Note, however, that although Na^+ and K^+ are depicted here as separate current pathways, they actully flow through the same channel.

The synaptic equilibrium potential for this synapse is

$$E_s = \frac{G_K E_K + G_{Na} E_{Na}}{G_K + G_{Na}}. \tag{13.3.6}$$

If we change the concentration of K^+ in the bath and assume that the intracellular concentration of K^+ does not change, then

$$\begin{aligned}
\Delta E_K &= E_K^1 - E_K^2 = \frac{RT}{F} \ln \frac{[K^+]_{out}^1}{[K^+]_{in}} - \frac{RT}{F} \ln \frac{[K^+]_{out}^2}{[K^+]_{in}} \\
&= \frac{RT}{F} \ln \frac{[K^+]_{out}^1}{[K^+]_{out}^2},
\end{aligned} \tag{13.3.7}$$

where $[K^+]_{out}^1$ and $[K^+]_{out}^2$ are two different bath concentrations. Using equations 13.3.6 and 13.3.7, the change in synaptic equilibrium potential is given by

$$\Delta E_s = \frac{G_K}{G_K + G_{Na}} (\Delta E_K), \tag{13.3.8}$$

which is plotted in figure 13.8. Similar equations are obtained if the concentration of extracellular Na^+ in the bath is varied, and this is illustrated in figure 13.9.

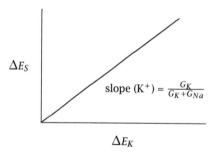

Figure 13.8 Hypothetical experiment in which the synaptic equilibrium potential varies as a function of extracellular K^+.

From these two experiments one can estimate the relative conductance of the channels to the two ions by looking at the ratio of the slopes in the figures, or

$$\frac{\text{slope } (K^+)}{\text{slope } (Na^+)} = \frac{G_K}{G_{Na}}, \tag{13.3.9}$$

which is approximately 1 for most EPSPs. This analysis can easily be extended if there are more (or less) than two permeable ions present in the

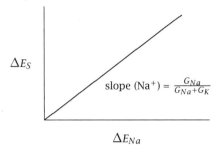

Figure 13.9 Hypothetical experiment in which the synaptic equilibrium potential varies as a function of extracellular Na$^+$.

bath solution or if one wishes to determine the relative conductance of the channel to an ion not normally present in the extracellular medium. From the above equations one can also show that if there are only two ions in the bath that can permeate the channels, then

slope (K$^+$) + slope (Na$^+$) = 1,

and

$$\frac{G_K}{G_{Na}} = \frac{\text{slope (K}^+)}{1 - \text{slope (K}^+)}.$$

13.3.3.2 Reversal potentials and conductance ratios: Nonisopotential synapses

In the above example we were assuming that the measurements were made right at the synapse so that the synaptic equilibrium potential was equal to the reversal potential. This type of analysis is very powerful for identifying the selectivity of channels to different ions. But what if the synapse is not located at the site of recording? What is the relationship between V_{rev} and E_s? The problem is illustrated once again in figure 13.10 (see also chapter 4).

The polarity of the synaptic current reverses when the membrane potential at the synapse is equal to the synaptic equilibrium potential. The membrane potential at the site of recording, however, will not be the same as that at the synapse. For example, one may have to depolarize the soma to +20 mV in order for the synaptic site to be depolarized to 0 mV. This difference represents the amount of attenuation of potential from the soma to the synapse. In this case the reversal potential measured from the soma will not be the same as the reversal potential (equilibrium potential) at the synapse. Using the ball-and-stick model of chapter 4, this difference is illustrated in figure 13.11.

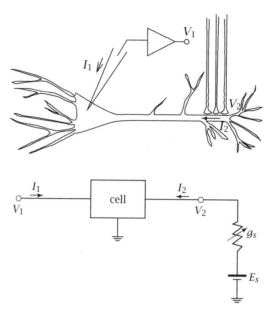

Figure 13.10   Synapses remotely located from the site of recording. (Adapted from Carnevale and Johnston 1982.)

The amount of error associated with the measurement of the reversal potential in the soma can be calculated for an equivalent cylinder representation of the neuron by (see chapter 4)

$$(E_s - E_r) = (V_{rev} - E_r)\frac{\cosh(L - X)}{\cosh(L)}, \tag{13.3.10}$$

where V_{rev} is the reversal potential measured at the soma, L is the electrotonic length of the (finite-length) equivalent cylinder, and X is the electrotonic distance of the synapse from the soma. The attenuation of potential from the soma to the synaptic site would obviously be useful to know. It is not always possible, however, to assume an equivalent cylinder representation of the neuron. The two-port analysis introduced in chapter 4 can sometimes be used instead. The experimental situation depicted in figure 13.10 is shown in electrical terms in figure 13.12 where the "cable" merely represents that portion of the neuron or dendrites that renders the synaptic site nonisopotential with the soma.

In general, $(V_{rev} - E_r) \neq (E_s - E_r)$ and, actually, $(V_{rev} - E_r) \geq (E_s - E_r)$. As in previous chapters, if we define *gain* as the output/input ratio, where V_{rev} is the input and E_s is the output, then

$$(V_{rev} - E_r) \cdot \text{gain} = (E_s - E_r).$$

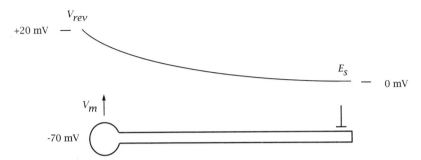

Figure 13.11 Equivalent cylinder representation of the soma and dendrites—the ball-and-stick model. The difference between V_{rev} measured in the soma and E_s at the synapse is illustrated. See text for further explanation.

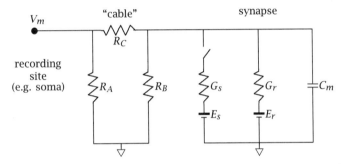

Figure 13.12 Equivalent circuit representation for recording from a remotely located synaptic input. The "cable" merely represents the part of the neuron between the recording site and the synapse.

Let this *gain* be equal to k. Then

$$(V_{rev} - E_r) = (E_s - E_r) \cdot 1/k, \tag{13.3.11}$$

where $k \leq 1$ and is the *gain* of the cable. Note that for a finite-length equivalent cylinder,

$$k = \frac{\cosh(L - X)}{\cosh(L)}.$$

A similar analysis to that shown in the previous section, where the synaptic equilibrium potential varied with the concentration of extracellular ions, now leads to some interesting and potentially useful relationships for remotely located synapses. Varying the concentrations of extracellular K^+ and Na^+ and using equation 13.3.11 leads to the relationships shown in figure 13.13.

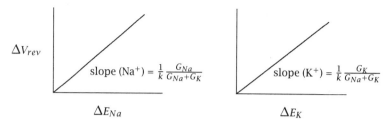

ΔV_{rev}

slope $(Na^+) = \frac{1}{k}\frac{G_{Na}}{G_{Na}+G_K}$

slope $(K^+) = \frac{1}{k}\frac{G_K}{G_{Na}+G_K}$

ΔE_{Na} ΔE_K

Figure 13.13 Hypothetical experiment in which the synaptic equilibrium potential varies as a function of extracellular K^+ and Na^+.

From figure 13.13, the sum of the slopes will now give an estimate of k, or

$$\sum slopes = \frac{1}{k}.$$

Further discussion of this type of two-port analysis is given in Carnevale and Johnston (1982).

13.3.3.3 Synaptic conductance change for nonisopotential synapses
As discussed in section 13.3.2, the synaptic conductance of an isopotential synapse can be obtained from the slope of the I-V curve when measured under voltage-clamp conditions. In most cases, however, the synaptic input is remotely located from the site of the voltage clamp. Under these conditions, what is the relationship between the slope of the I-V curve measured, for example, in the soma, and the conductance change at the synapse? The answer to this question is somewhat complicated, and the interested reader should refer also to more complete treatments given elsewhere (Carnevale and Johnston 1982; Johnston and Brown 1983; Rall and Segev 1985; and Spruston et al. 1993).

The fact that the synaptic conductance change occurs remotely from the voltage clamp means that the potential at the synaptic site is not constant during the synaptic input. In other words, although the potential at the soma may be constant, the potential at the synapse is not *clamped*, but changes with time. This also means that the driving force for the synaptic current $(V_m - E_s)$ changes during the synaptic event, and I_s will actually be less than if V_m had been held constant. In addition, the synaptic current will attenuate in amplitude from the site of input to the soma due to the cable properties of the dendrites. The amount of attenuation will depend on the electrotonic distance of the synapse from the soma as well as on the kinetics of the synaptic current (the faster the kinetics of the current, the greater the attenuation). These two factors (decreased I_s due to decreased driving force and decreased I_s due to attenuation along the cable) mean that the *I-V* curve of the synapse as measured from the soma is quite different from that which one would have obtained under space-clamped conditions. But, how bad is it?

A quantitative answer to this question requires knowing the specific parameters associated with the synaptic input and the postsynaptic neuron. An example for a hippocampal pyramidal neuron is illustrated in figure 13.14. In general, the errors associated with measuring the synaptic equilibrium potential for a remote synapse, such as those discussed in the previous section, are small—perhaps 10–20 mV for most excitatory synapses. In contast, the error associated with estimating the synaptic conductance change can be quite large. For the example shown in figure 13.14, the conductance change measured from the soma for a synapse located about 20% of the total distance out the dendritic tree is only about 1/5 of that at the synapse!

13.4 Synaptic kinetics

The simple two-state kinetic scheme for a ligand-gated channel presented in section 13.2 and in chapters 8–10 is given again below.

$$
\begin{array}{ccccccc}
\text{hydrolysis} & & & & & & \\
\text{(or uptake)} & & & (C) & & (O) & \\
\uparrow & & & k_1 & & \beta & \\
T & + & R & \rightleftharpoons & T \cdot R & \rightleftharpoons & T \cdot R^* \\
\downarrow & & & k_2 & & \alpha & \\
\text{diffusion} & & & & & &
\end{array}
\qquad (13.4.12)
$$

The model assumes that one molecule of transmitter binds to one receptor and that there are no interactions among receptors (independent events). As discussed in chapter 10, this kinetic model can be easily modified for the case of two molecules binding to the receptor by adding an additional closed state to the channel representing a singly bound, but closed channel. Nicotinic ACh and NMDA glutamate receptors appear to bind two molecules of transmitter before opening. The stoichiometry of many other receptors, however, is not known.

The model can also be extended to include *desensitization*. Desensitization is the inactivation of the receptor/channel due to prolonged action of the transmitter. Although the mechanism of desensitization is not well

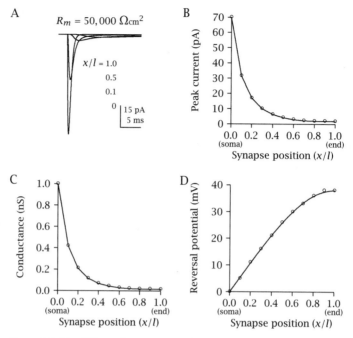

Figure 13.14 Effect of poor space clamp on measured synaptic parameters in a finite-length, equivalent cable (from Spruston et al. 1993). (A) Synaptic currents measured from the soma for synapses at different electrotonic distances out the dendrites. (B) Plot of peak amplitude of synaptic current measured in the soma as a function of the electrotonic distance of synapse from the soma. (C) Plot of synaptic conductance measured from the soma as a function of the electrotonic distance of synapse from the soma. (D) Plot of reversal potential measured from the soma as a function of the electrotonic distance of synapse from the soma.

understood, it may represent another closed state to the channel that is reached from the open state. The receptor, once desensitized, usually has a higher affinity for the transmitter than do nondesensitized receptors.

For the neuromuscular junction where ACh is released and binds to nicotinic receptors, the decay of the end-plate current (EPC) is voltage dependent. At hyperpolarized potentials the decay is slower than at depolarized potentials. This is illustrated in figure 13.15. The rate of decay of the EPC is exponentially related to membrane potential and is described by the following equation:

$$\alpha = be^{aV_m},\tag{13.4.13}$$

where α is the rate of decay of the EPC, V_m is membrane potential, and a and b are constants.

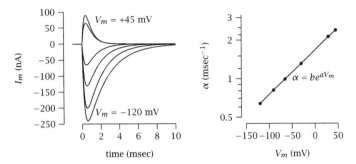

Figure 13.15 Voltage-dependent decay of end-plate currents at the neuromuscular junction. Superimposed currents at different membrane potentials are shown on the left, and a plot of rate of decay for the currents as a function of V_m is on the right. (After Magleby and Stevens 1972a.)

13.4.1 Theory for channel kinetics and the time course of synaptic currents

The observed voltage dependence of the decay of the EPC inspired Magleby and Stevens to develop a theory for the time course of the synaptic conductance change with respect to channel kinetics. This theory holds quite well for the end plate and makes some important predictions for the operation of this synapse. The theory may also apply to some of the glutamate receptors (e.g., NMDA; see section 13.5) at glutamate-releasing synapses, although this is currently an active area of research.

It had originally been assumed that the time course of the EPC reflected the concentration of ACh in the cleft so that the decay time of the EPC was due to the decay of the ACh concentration by diffusion or hydrolysis. The finding that the decay rate of the EPC was voltage dependent at the end plate made this idea less likely because neither diffusion nor hydrolysis should be dependent on the membrane potential. Magleby and Stevens proposed instead that the decay of the EPC was due to the kinetics of the ACh-gated channels. The theory can be derived as follows.

Assume a total of N receptors at the end plate with x of them bound with transmitter and in the open configuration and y of them bound with transmitter but closed. The synaptic conductance is simply

$$g_s = \gamma x, \tag{13.4.14}$$

where γ is the single-channel conductance. The change in x with time can be derived from equation 13.4.12 as

$$\frac{dx}{dt} = -\alpha x + \beta y, \tag{13.4.15}$$

while that for y is given by

$$\frac{dy}{dt} = \alpha x + k_1 c(t)(N - x - y) - (\beta + k_2)y, \tag{13.4.16}$$

where $c(t)$ is the concentration of transmitter in the cleft. If the binding of transmitter to receptor is rapid compared to diffusion, then the binding is essentially at equilibrium and $dy/dt \approx 0$ in equation 13.4.16. We can then solve for y as follows:

$$y = \frac{k_1 c(t)N + x(\alpha - k_1 c(t))}{\beta + k_2 + k_1 c(t)}. \tag{13.4.17}$$

Again, assuming that binding is rapid compared to the opening and closing of the channel (i.e., k_1 and k_2 are large compared to α and β) and that N is large compared to x, then

$$y = \frac{c(t)N}{K + c(t)}, \tag{13.4.18}$$

with $K = k_2/k_1$. Combining equations 13.4.14, 13.4.15, and 13.4.18,

$$\frac{dx}{dt} = -\alpha x + \beta \frac{c(t)N}{K + c(t)} \tag{13.4.19}$$

$$\frac{dg_s}{dt} = -\alpha g_s + \gamma \beta \frac{c(t)N}{K + c(t)} \tag{13.4.20}$$

$$\frac{dg_s}{dt} = -\alpha g_s + \beta W(t), \tag{13.4.21}$$

where $W(t) = \gamma \frac{c(t)N}{K+c(t)}$. Equations 13.4.19, 13.4.20, and 13.4.21 were derived by Magleby and Stevens and provide a description of the number of receptors in the open configuration, and the synaptic conductance produced by those open receptor-channels, as a function of time and transmitter concentration.

Magleby and Stevens proposed two alternative schemes to explain the single exponential decay of the EPC. First, the conformational change of the channel from the closed to open configuration could be rapid compared with the change in ACh concentration in the cleft. Under such a situation equation 13.4.12 would be essentially in equilibrium, and $dg_s/dt = 0$ in equation 13.4.21, yielding

$$g_s(t) = \frac{\beta}{\alpha} W(t).$$

The exponential decay of the EPC would then reflect the exponential decay of the ACh concentration as described by $W(t)$. The second possibility is that the concentration of ACh in the cleft rises and falls rapidly compared with the time for conformational changes of the receptor-channels (i.e., rapidly with respect to $1/\alpha$). In this situation $c(t)$ would go to zero near the beginning of the decay phase of the EPC, and equation 13.4.21 would reduce to $dg_s/dt = -\alpha g_s$. This can be solved to yield

$$g_s(t) = g_s(0)e^{-\alpha t}.$$

Magleby and Stevens favored the second scheme for several reasons. First, the rate of decay is voltage dependent, and neither diffusion nor hydrolysis is likely to be dependent on the membrane potential of the postsynaptic cell. Second, the rate of the decay of the EPC is very temperature dependent, with a Q_{10}[1] of about 3. Diffusion has a Q_{10} of about 1, and although it is possible that hydrolysis could have a Q_{10} of 3, Magleby and Stevens argued that hydrolysis does not participate in the decay of the EPC. Their reasoning for this argument was that treatment of the end plate with anticholinesterases that block hydrolysis of ACh do not affect the voltage sensitivity of the decay rate. It was also shown subsequently by Gage and McBurney that the decay rate still has a Q_{10} of 3 after removing ACh hydrolysis with anticholinesterases. None of these findings are consistent with the rate of decay or the voltage sensitivity of the rate of decay being dependent on the ACh concentration in the cleft.

The conclusion that the duration of the synaptic conductance is dependent on channel kinetics and not on the concentration of ACh in the cleft is an extremely important concept. Although this conclusion and the resulting theory were derived from studies of the muscle end plate, there are many aspects to the theory that are generally applicable to other fast releasing synapses as well. The situation for glutamate and GABA will be discussed a little later. There are also aspects of the theory that don't quite fit the experimental findings even at the end plate. For example, although anticholinesterases do not affect the voltage sensitivity to the decay rate, they do prolong the duration of the EPC. This is not predicted by the theory, in which diffusion and hydrolysis are assumed to be rapid processes, and Magleby and Stevens suggested that there may be some direct action of the anticholinesterase enzyme on the channel. Another possible explanation is that in the absence of hydrolysis, ACh from neighboring quanta interact nonlinearly at adjacent receptor domains. This and other anomalies, however, have not yet been fully resolved.

---

[1]Q_{10} is the ratio of reaction rates for a $10°C$ increase in temperature, or $Q_{10} = r_2/r_1$, where r_1 is the rate of reaction at a given temperature and r_2 is the rate at $10°C$ higher in temperature.

With the conclusion that EPC decay is dependent on α, Magleby and Stevens derived equations that provide some physical interpretation for the processes underlying its voltage dependence. From Eyring rate theory, discussed in chapter 5,

$$\alpha(V_m) = \nu \exp\left(\frac{-U(V_m)}{kT}\right),\qquad(13.4.22)$$

where ν is a vibration frequency, k is Boltzmann's constant, T is absolute temperature, and U is the Helmholtz free energy difference between the open and closed states of the channel. Assuming that the voltage dependence results from an electric dipole associated with the channel that moves in the direction of the applied electrical field, the energy difference is given by

$$U(V_m) = U_c - E\mu_c,$$

where U_c is the height of the energy barrier for the closing of the channel in the absence of a field, E is the electric field, and μ_c is the change in the dipole moment normal to the field associated with closing of the channel. Note also that $V_m = E \cdot M$, where M is the membrane thickness. The equation for α can be rewritten as

$$\alpha(V_m) = \nu \exp\left(\frac{-U_c}{kT}\right) \exp\left(V_m \frac{\mu_c}{MkT}\right).\qquad(13.4.23)$$

Letting $A_c = \mu_c/MkT$ and $B_c = \nu \exp(-U_c/kt)$, we obtain

$$\alpha(V_m) = B_c e^{A_c V_m},\qquad(13.4.24)$$

which is the identical equation to that determined from the experimental data.

A similar series of equations can be derived for the rising phase of the EPC so that

$$\beta = B_o e^{A_o V_m},\qquad(13.4.25)$$

where $A_o = \mu_o/MkT$ and $B_o = \nu \exp(-U_o/kt)$, with μ_o as the change in the normal component of the dipole moment associated with the opening of the channel and U_o as the height of the energy barrier for opening the channel.

Equations 13.4.21, 13.4.24, and 13.4.25 can be used to predict quite accurately the time course and voltage dependency of EPCs from neuromuscular junctions. Furthermore, the curvature of the *I-V* curve of the EPC in the depolarizing direction can be explained on the basis of the voltage dependency of channel closing.[2] Although the single-channel *I-V* curve is linear, according to the theory of Magleby and Stevens, at depolarized potentials channels will spend less time in the open state, resulting in less total current from a population of channels.

The data supporting the theory for the rising phase of the EPC is not as strong as that for the decay phase. The rising phase is not as voltage or temperature dependent as predicted by the theory. It may be that the rising phase of the EPC is indeed limited by diffusion time of the transmitter. Nevertheless, the general concept that the time course of transmitter concentration in the cleft is brief and that channel kinetics is the major determinant of the duration of synaptic responses is of fundamental importance to our understanding of synaptic transmission.

As mentioned in chapter 10, single-channel recordings have demonstrated that the mean channel open time is much shorter than the rate constant for decay of the EPC. The two-state model given by equation 13.4.12 is therefore not adequate, and at least a three-state (two closed states and one open state) model may be necessary. In this case the decay time of the EPC would be related to the channel burst duration rather than the mean channel open time.

There is much less information about glutamate- and GABA-releasing synapses than about ACh at the neuromuscular junction. One reason is that there are few experimental preparations as favorable for study as the neuromuscular junction. It does appear, however, that the general principle outlined above—that changes in transmitter concentration in the cleft are brief compared to the time course of the synaptic conductance change—holds for other synapses. In particular, the concentration of glutamate is brief compared to the time course of the NMDA component of the synaptic response (see section 13.5 for discussion of NMDA-receptor-mediated synaptic responses), and the rate of decay of the NMDA response probably reflects channel kinetics. For the non-NMDA component of the synaptic response, there is little effect of membrane potential on the rate of decay to the EPSC, leading some to speculate that, in contrast to the end plate, rapid desensitization or reuptake of transmitter may be rate-limiting processes for the decay of current. Recent work, however, suggests that at many synapses the concentration of glutamate is brief even compared to the time course of the non-NMDA component, and so the decay of the non-NMDA current may also be due to channel kinetics. In

---

[2]The *I-V* curve of the EPC shows outward rectification in that there is less outward current flow at depolarized potentials than inward current at corresponding hyperpolarized potentials.

the case of non-NMDA receptors, there just may be no significant electric dipole associated with the receptor molecule to affect decay at different membrane potentials. In any event, the general theory first developed by Magleby and Stevens guides much research into the mechanisms of synaptic transmission.

13.4.2   Kinetics of nonisopotential synapses

As discussed in previous sections, the analysis of synaptic events is complicated when the input occurs at a site remote from the soma or site of measurement. This is true when one wants to determine the synaptic equilibrium potential from a measurement of reversal potential (section 13.3.3.2), the synaptic conductance from a measurement of the *I-V* curve (section 13.3.3.3), or the kinetics of the synaptic current from waveform measurements in the soma. The cable properties of a neuron and its associated dendrites will attenuate and distort the waveform of a synaptic event (see also chapter 4).

The dendrites act as a low-pass filter and attenuate the higher-frequency components of a synaptic input. This means that, in general, the rising phase of a brief synaptic current will be distorted and slowed more than its falling phase. The falling phase will follow an exponential time course when the synapse is isopotential with the recording site and become more multiexponential at increasing electrotonic distances from the soma. The amount of slowing of the rising and falling phases depends on the cable properties of the neuron, the kinetics of the ligand-gated channels, and the electrotonic distance of the synapse from the recording site. Examples from a representative hippocampal pyramidal neuron are shown in figures 13.16 and 13.17.

Note that the rising phase of a synaptic response is typically characterized by its rise time, which is most conveniently measured as the time for the response to go from 10% to 90% (or sometimes 20% to 80%) of its peak value. (The 10% value is used in order to avoid uncertainties associated with measuring the onset of the response, which is usually buried in the baseline noise. The 90% value is used to avoid uncertainties associated with measuring the time of the peak of the response when the waveform is relatively flat or rounded. Other measures of the rising phase include the time-to-peak (see chapter 4). Time-to-peak is useful for simulations in which there is no noise or as a measure from a fixed time point such as from a stimulus artifact.)

The falling phase of a synaptic response can be characterized either by determining the exponential that best fits the decaying phase of the response or by using the measure of half width. The half width is simply the total duration of the response at its half amplitude (see chapter 4).

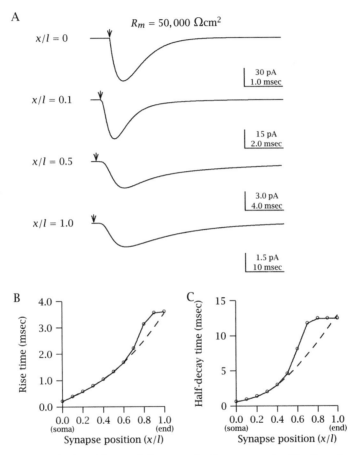

Figure 13.16 Effects of electrotonic distance on the kinetics of a synapse measured from the voltage-clamped soma in a finite-length ($L = 1$), equivalent cable (from Spruston et al. 1993). (A) Synaptic currents measured in the soma for synapses located at different electrotonic distances from the soma. The arrows indicate the onset of the conductance change. (Note the different amplitude and time scales for each trace.) (B) Plot of rise time (10% to 90%) of synaptic current measured from the soma as a function of its electrotonic distance from the soma. (C) Plot of half-decay time of synaptic current measured from the soma as a function of its electrotonic distance from the soma. The dashed lines in (B) and (C) represent simulations for a semi-infinite cylinder.

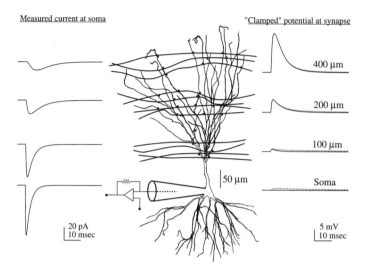

Measured current at soma   "Clamped" potential at synapse

400 μm

200 μm

100 μm

50 μm   Soma

20 pA / 10 msec   5 mV / 10 msec

Figure 13.17   Voltage- and space-clamp errors in currents measured with a somatic voltage clamp in a realistic compartmental model of a hippocampal pyramidal neuron. Fast excitatory synapses modeled using an identical conductance were simulated at four different locations in the model. The measured currents (left) are clearly filtered and attenuated in comparison to the perfectly-clamped synaptic current in the soma (bottom left). The ability of the somatic electrode to clamp the membrane potential at the synapse is shown by the records of synaptic escape voltage (solid lines in right column; mean distances of the synapses from the soma are indicated). The unclamped EPSPs (dashed lines in the right column) are also shown for comparison. Note that the EPSPs in the distal dendrites are essentially unaffected by the somatic clamp (from Spruston et al. 1994).

13.5   Excitatory amino acid receptors

Most fast excitatory synapses in the central nervous system use glutamate as the neurotransmitter. Glutamate receptors can be separated broadly into two main types—those sensitive to *N*-methyl-D-aspartic acid (NMDA) and those that are not (non-NMDA). The non-NMDA class can be subdivided further, based on agonist selectivity, into kainic acid (KA) and α-amino-3-hydroxy-5-methyl-4-isoxazole-proprionic acid (AMPA) receptors. Many glutamate receptors have now been cloned, and an intense effort is being made to investigate the structure-function relationships among the many different subunits that make up the individual receptors.

The KA and AMPA classes of receptors share many of the functional properties of nicotinic ACh receptors in that they mediate fast excitatory synaptic responses. The NMDA class of receptors, however, has several unique properties with important physiological consequences that will be discussed briefly in this section.

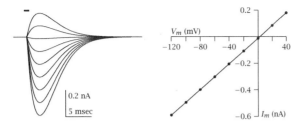

Figure 13.18 Responses of a neuron to rapidly applied glutamate measured under voltage clamp. The responses on the left are due to KA/AMPA receptors because no NMDA receptors are present. The bar indicates the time of glutamate application. The *I-V* curve for the responses is shown on the right.

If brief pulses of glutamate (or KA or AMPA) are applied to a responsive neuron under voltage-clamp conditions, the responses and resulting *I-V* curves are illustrated in figure 13.18.[3] The current responses reverse near 0 mV and exhibit a fairly linear relationship with membrane potential.

If a similar experiment is done with NMDA, however, the result is quite different (figure 13.19). The responses are small at negative potentials, and the *I-V* curve is very nonlinear with a prominent region of negative slope. This region of negative slope is sensitive to the extracellular concentration of Mg^{2+}: The same experiment done after removing Mg^{2+} from the bath yields a fairly linear *I-V* curve (see figure 13.19). The chord conductance ($I_s / (V_m - E_s)$) as a function of voltage is also plotted in figure 13.19 and nicely illustrates the fact that the *conductance* of the NMDA response is voltage dependent and that this voltage dependency is sensitive to external Mg^{2+}. The voltage dependency of the NMDA receptor is quite unique and interesting. As we will see in chapter 15, the NMDA receptor plays an important role in certain types of long-term potentiation. It also may be involved in *excitotoxicity*, where excessive release of glutamate (e.g., during ischemia) can lead to cell death.

EPSPs or EPSCs that occur from the evoked release of glutamate and that are mediated exclusively by NMDA receptors (e.g., with an antagonist to the non-NMDA receptors present in the bath) are relatively slow compared to those mediated by KA or AMPA receptors. The slow rise and decay times of the responses appear to be due to the slow kinetics of the NMDA receptor-channels. This is illustrated in figure 13.20. It is generally

---

[3]Brief pulses of KA or AMPA can be achieved by filling a blunt glass electrode with a concentrated solution and then either applying a brief pressure pulse to the electrode or a brief current pulse. The former method is just pressure application while the latter method is called *iontophoresis*. Another method for fast application of a drug to an excised patch is to have a continuous stream of solution flowing from a large pipette and rapidly move the patch pipette into and out of the stream.

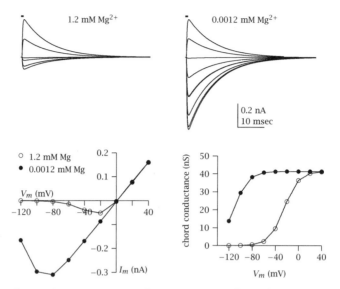

Figure 13.19   Responses of a neuron to rapidly applied NMDA measured under voltage clamp (the bar indicates the time of application). The currents above were obtained at two different Mg^{2+} concentrations, and the resulting *I*-*V* curves are shown in the lower left. The chord conductance (calculated using $G_{NMDA} = I_s / (V_m - E_s)$) is shown in the lower right for the same two Mg^{2+} concentrations.

thought that NMDA and non-NMDA receptors are located together in the subsynaptic membrane. Thus when glutamate is released from the presynaptic ending, it will normally bind to both receptors. Near the resting membrane potential, however, NMDA receptor-channels are blocked by Mg^{2+} and will not pass much current. A single EPSP evoked from such a glutamate synapse will therefore be due mainly to non-NMDA receptor-channels. At more depolarized potentials, such as occur when the membrane is depolarized by injected current or when a train of stimuli is given to summate synaptic responses, then the evoked response becomes a mixture of NMDA and non-NMDA responses (figure 13.21).

Another feature of the NMDA receptor, although not quite so unique as its voltage dependency, is that the receptor-channel is permeable to Ca^{2+}. Nicotinic ACh receptors and certain types of KA and AMPA receptors are also permeable to Ca^{2+}, but the NMDA receptor appears to be the most permeable of all. The permeability of the NMDA receptor can be demonstrated by measurement of the change in the reversal potential as a function of extracellular Ca^{2+} (see section 13.3.3.2) and also by the use of fluorescent imaging techniques to measure changes in intracellular Ca^{2+} during activation of the receptors.

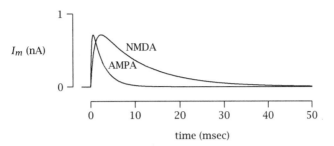

Figure 13.20 Comparison of the time course of reversed NMDA and non-NMDA EPSCs at a potential depolarized to 0 mV.

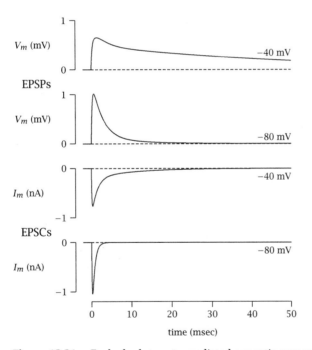

Figure 13.21 Evoked, glutamate-mediated synaptic responses at different membrane potentials. The upper two traces are EPSPs at −40 and −80 mV. The NMDA component of the response is revealed at −40 mV and prolongs the decay of the EPSP. The bottom two traces are EPSCs at the same two membrane potentials. Again, the slower decay of the current at −40 mV is due to the NMDA component of the response. At −80 mV the NMDA component is blocked by Mg^{2+} resulting in a pure KA- or AMPA-mediated EPSC (see text for further explanation). The current and voltage scales are arbitrary.

13.6 Functional properties of synapses

If the time course of the synaptic conductance change is brief compared to τ_m, EPSPs will decay with the membrane time constant. This is illustrated in figure 13.22. The rising phase of g_s can be seen to precede the rise of the EPSP while g_s has essentially decayed to zero during the decay of the EPSP.

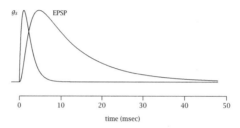

Figure 13.22   Relative time courses of a typical fast EPSP and its associated conductance change, g_s.

Referring to the parallel conductance model in figure 13.23, when synapse A is active, the switch is closed and current flows as a function of the time course of g_s and its driving force. Assuming that the rising phase of g_s is essentially instantaneous, then the initial portion of the rising phase of the EPSP will be described by a single exponential with time constant of $C_m/(G_{s_A} + G_r)$. When the synapse is inactive, the switch is open and the potential decays with a slower time constant described by C_m/G_r. If both A and B are active together, then the time constant of the rising phase would be $C_m/(G_{s_A} + G_{s_B} + G_r)$ while the time constant of the falling phase would still be C_m/G_r.

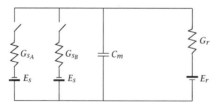

Figure 13.23   Parallel conductance model representing the separate synaptic inputs, A and B.

13.6.1 Spatial summation

What happens to the amplitude of the response when two or more synapses are active together? This is illustrated in figure 13.24. The activation of synapse *A* alone produces an EPSP of amplitude *a*, while activation of synapse *B* alone produces an EPSP of amplitude *b*. *Spatial summation* refers to the simultaneous activation of two or more spatially separated synaptic inputs and the resulting summation of their individual responses. In general, the summation of individual synaptic potentials is not linear. As shown in figure 13.24, the amplitude of the summated EPSP is smaller than the linear sum of $a + b$. The explanation for the nonlinear summation is that the driving force for synaptic current changes during the EPSP. The larger EPSP that occurs when both synapses are active together produces a smaller driving force than that produced when the synapses are active individually. This means that I_{s_A} and I_{s_B} are less during the combined EPSP than during the individual EPSPs. Nonlinear summation of synaptic events was discussed in chapter 11 in the context of quantal analysis. Remember that the amount of nonlinear summation depends, in part, on the change in driving force during the summated synaptic response. For IPSPs, where the driving force is small to begin with, the amount of nonlinear summation will usually be greater than for EPSPs.

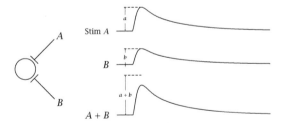

Figure 13.24 Spatial summation of synaptic inputs. Summation is typically nonlinear (see text for details).

The amount of nonlinear summation of synaptic inputs will also depend on the electrotonic distance between the synaptic inputs. For example, the amount of nonlinear summation seen at the soma for two synaptic inputs on different dendritic branches will be less than from two inputs on the same branch (see figure 13.25).

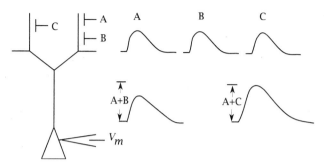

Figure 13.25   Nonlinear summation from synapses on the same and different dendritic branches. In general, $A + C$ will summate more linearly than $A + B$.

13.6.2   Temporal summation

In addition to the summation of different synaptic inputs, summation also occurs for the same input when it is repeatedly activated. This is called *temporal summation* and is illustrated in figure 13.26. The amount of summation will depend on the decay time course of the synaptic response (which is usually τ_m) and on the interval between the successive responses (frequency of stimulation). Like spatial summation, temporal summation is nonlinear. The maximum amount of temporal summation (assuming no facilitation or depression; see chapter 11) that would be achieved at a high frequency of stimulation or at synapses that release neurotransmitter in the steady state (e.g., certain synapses in the retina) is given by (see figure 13.27)

$$E_{sum} = \frac{G_s E_s + G_r E_r}{G_s + G_r}.$$

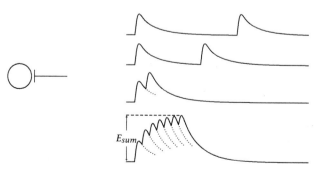

Figure 13.26   Temporal summation of synaptic responses. E_{sum} represents the maximum summated response, which is dependent on the frequency of stimulation.

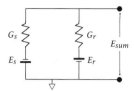

Figure 13.27 Schematic diagram for determining the maximum temporal summation (see text for derivation).

13.7 Slow synaptic responses: Conductance-decrease PSPs

In addition to the fast, conductance-increase EPSPs and IPSPs discussed in previous sections, there are numerous examples in the central and peripheral nervous system of relatively slow PSPs. Many of these have been shown to be due to conductance-decrease mechanisms whereby the binding of a neurotransmitter leads to a *decrease* in membrane conductance. For these conductance-decrease PSPs the change in conductance is typically mediated by an intermediate GTP-binding protein and reflects several indirect steps between the ligand binding to its receptor and the gating of the ion channel. An example of a conductance-decrease EPSP is shown in figure 13.28.

Two of the more prominent neurotransmitters that mediate conductance-decrease EPSPs in the CNS are ACh, through muscarinic receptors, and glutamate, through quisqualate receptors. These types of receptors

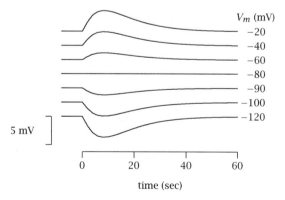

Figure 13.28 Examples of slow, conductance-decrease EPSPs. The reversal potential is −80 mV, and note that the voltage responses move away from the reversal potential. Contrast this with conductance-increase EPSPs, shown in figure 13.4, in which the voltage responses (for both EPSPs and IPSPs) move towards the reversal potential.

are called *metabotropic* because they represent an indirect, metabolic action on an ion channel as opposed to *ionotropic* receptors, such as the KA, AMPA, and NMDA receptors discussed in the previous section, which directly gate an ion channel. The muscarinic ACh and metabotropic glutamate receptors both lead to a decrease in a voltage-dependent potassium conductance (see chapter 7 and discussion of the M-current) in part by releasing intracellular Ca^{2+} from internal stores. As mentioned in section 13.3.1, even though their equilibrium potential is negative to threshold (i.e., $E_s = E_K$), they are still classified as excitatory, because at all potentials between rest and threshold the responses are depolarizing and tend to bring the membrane potential closer to threshold.

13.8   Diversity of neurotransmitters in the central nervous system

There are many different types of neurotransmitters in the CNS activating a variety of receptors. There are ionotropic receptors, which directly gate ion channels, and metabotropic receptors, which gate ion channels indirectly through coupling to a GTP-binding protein (G-proteins) or through second-messenger systems activated by G-proteins. The vast majority of neurotransmitters bind to metabotropic receptors, whereas relatively few neurotransmitters are involved in fast synaptic transmission through ionotropic receptors. There is great molecular diversity within each family of ionotropic receptors, but surprisingly few neurotransmitters actually mediate such responses. The list of neurotransmitters mediating fast transmission includes nicotinic ACh, glutamate, serotonin, GABA, and glycine. A list of putative neurotransmitters in the CNS is given in table 13.1. Many more are likely to be added to the list in the future.

Table 13.1   Neurotransmitter candidates in the central nervous system

Neurotransmitter	Ionotropic Receptor	Ion
Glutamate	AMPA	$Na^+/K^+/Ca^{2+}$ (some)
	Kainate	$Na^+/K^+/Ca^{2+}$ (some)
	NMDA	$Na^+/K^+/Ca^{2+}$
Acetylcholine (ACh)	nicotinic	$Na^+/K^+/Ca^{2+}$ (some)
Serotonin (5-HT)	5-HT_3	Na^+/K^+
ATP	Purine P1	Na^+/K^+
γ-aminobutyric acid (GABA)	A	Cl^-
Glycine		Cl^-

Table 13.1 (continued)

Neurotransmitter	Metabotropic	
	Receptor	Ion
Glutamate	Quisqualate	G-coupled $\downarrow K^+$
ACh	muscarinic (M1–5)	G-coupled $\downarrow K^+$ (M-current), $\downarrow K^+$ (AHP) $\uparrow K^+$ (Inward rectifier) $\downarrow Cl^-$ $\downarrow Ca^{2+}$ (N & L), $\uparrow Ca^{2+}$ (T)
GABA	B	G-coupled $\uparrow K^+$, $\downarrow Ca^{2+}$ (N)
Norepinephrine (NE) (α, β)	β α α_2	G-coupled $\downarrow K^+$ (AHP), $\uparrow Ca^{2+}$ (L & N) $\downarrow Ca^{2+}$ (N) $\uparrow K^+$
Dopamine (DA)	$(D_1, D_2, \ldots)$	G-coupled $\downarrow K^+$ (AHP)
5-HT	 5-HT$_2$ 5-HT$_{1A}$	G-coupled $\downarrow K^+$ (M-current) $\downarrow K^+$ $\uparrow K^+$
Histamine	$(H_1, \ldots)$	G-coupled $\downarrow K^+$ (AHP)
Adenosine	$(A_1, \ldots)$	G-coupled $\uparrow K^+$, $\downarrow Ca^{2+}$
Opioids (μ, δ, κ)	μ μ κ	G-coupled $\uparrow K^+$ (inward rectifier) $\uparrow K^+$ (voltage-dependent) $\downarrow Ca^{2+}$
Substance P		G-coupled $\downarrow K^+$ (M-current)
Somatostatin		G-coupled $\uparrow K^+$ (M-current)
Bradykinin		G-coupled $\downarrow K^+$ (M-current), $\downarrow K^+$ (AHP)
VIP		G-coupled
Cholecystokinin		G-coupled
NPY		G-coupled $\downarrow Ca^{2+}$ (N)
Neurotensin		G-coupled
TRH		G-coupled
Vasopressin		G-coupled
Oxytocin		G-coupled
CRF		G-coupled
LHRH		G-coupled $\downarrow K^+$ (M-current)

13.9   Electrical transmission

13.9.1   Electrical synapses

In addition to chemical transmission, there are many examples of direct electrical connections between cells. These connections occur via special channels that span the pre- and postsynaptic membranes and are called *gap junctions*. Electrical synapses play important roles in many invertebrate nervous systems. In mammalian nervous systems electrical synapses are a prominent form of synaptic transmission in the retina and other sensory end organs and among interneurons and glial cells in the CNS. It is not clear what role they play in communication among principal neurons in the brain (e.g., pyramidal neurons). In addition to allowing ions to pass from one cell to another, gap junctions are also permeable to small molecules so that they provide a limited means of chemical communication between cells.

Each gap junction is composed of many individual channels (see figure 13.29). The elementary conductance of each channel is about 100 pS, and the opening of the channels is regulated by pH, $[Ca^{2+}]_{in}$, second messengers, and, to a small extent, voltage. The channels are called *connexons*, and the six subunits that make up each channel are called *connexins*. Some of the electrophysiological properties of electrical synapses will be discussed briefly below.

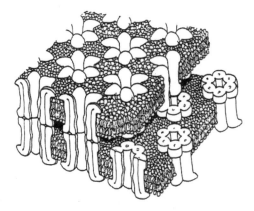

Figure 13.29   Diagram of a gap junction illustrating the individual channels or connexons (from Nicholls et al. 1992).

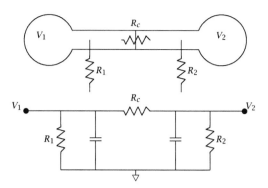

Figure 13.30   Diagram and electrical model of an electrical synapse (see text).

A diagram of two neurons connected via a gap junction is given in figure 13.30. The potential in neuron 1 will be coupled to neuron 2 through the junctional or coupling resistance, R_c. If a voltage is applied in neuron 1, the potential that appears in neuron 2 will depend on R_c and R_2. If a voltage is applied in neuron 2, the potential that appears in neuron 1 will depend on R_c and R_1. This can be stated mathematically as

For a voltage at V_1 : $\dfrac{V_2}{V_1} = \dfrac{R_2}{R_2 + R_c} = K_{12}.$

For a voltage at V_2 : $\dfrac{V_1}{V_2} = \dfrac{R_1}{R_1 + R_c} = K_{21}.$

K_{12} and K_{21} are called the *coupling coefficients* and, in general, $K_{12} \neq K_{21}$. An extreme example of unequal electrical coupling is illustrated in example 13.2.

Example 13.2
The following is an extreme example of asymmetrical electrical coupling. The figure below depicts two neurons of unequal size coupled via a gap junction.

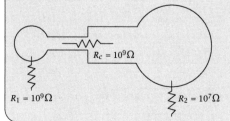

Example 13.2 (continued)

The electrical coupling from neuron 1 to neuron 2 is given by

$$K_{12} = \frac{10^7}{10^9 + 10^7} = \frac{1}{101} = 0.01.$$

In other words, if there were a potential of 100 mV in neuron 1, only 1 mV would couple to neuron 2, very weak coupling.

The electrical coupling from neuron 2 to neuron 1 is given by

$$K_{21} = \frac{10^9}{10^9 + 10^9} = 0.5.$$

A potential of 100 mV in neuron 2 would provide 50 mV in neuron 1, very strong electrical coupling.

This would obviously be an example of an almost unidirectional electrical synapse.

In addition to the asymmetrical coupling arising from different input resistances of the coupled neurons (depicted in example 13.2), asymmetrical coupling can also occur if the input resistance of one of the coupled neurons is more voltage dependent than the other. Moreover, highly rectifying electrical coupling is possible because hyperpolarizing potentials (in either direction) might couple better than depolarizing potentials because of the larger decrease in input resistance of the neurons during depolarizing potentials. A comparison of some of the properties of electrical and chemical synapses is given in table 13.2.

Table 13.2   Properties of chemical and electrical transmission

Chemical	Electrical
Unidirectional	Can be bidirectional or unidirectional
Excitatory or Inhibitory	Sign conserving. Depolarization (or hyperpolarization) in one neuron leads to depolarization (or hyperpolarization) in the other
Delay of ~0.5–1.0 msec	No delay other than from low pass filtering. Ideal for rapid communication
Amplification. Uses energy from ion gradients and can prolong response	Dissipative. Signals usually smaller in the coupled neuron. Time course limited by membrane time constant
Efficient for impedance mismatches	Inefficient when coupling between neurons of different input resistances
Plasticity (use dependent)	Coupling can be modulated by chemicals
Chemical communication primarily by exocytosis	Small molecules can pass between neurons

Experimentally, chemical transmission is distinguished from electrical transmission most easily if a reversal potential can be identified. No reversal potential will exist for an electrical synapse. The most direct test for electrical coupling between cells is to simultaneously record from the cells and demonstrate electrical responses in one cell following depolarization or hyperpolarization of the other. Dye coupling has also been used to identify gap junctions between cells, but in tissue where the neurons are densely packed the results can be misleading. A comparison of the physiological tests for chemical and electrical transmission is given in table 13.3.

Table 13.3 Physiological tests for type of transmission

Chemical	Electrical
Measurement of a reversal potential	Direct electrical coupling using simultaneous recordings between two or more cells
Small synaptic delay	No delay
Pharmacological experiments. Use antagonists of neurotransmitter receptors	Fluorescent dye injected into one neuron diffuses and fills other neurons
High Mg^{2+}/low Ca^{2+} will block transmission	High Mg^{2+}/low Ca^{2+} will not block transmission
Morphology. Separation of pre- and postsynaptic membranes	Presence of gap junctions

13.9.2 Ephaptic coupling

In the next chapter we will discuss the electrical fields that are generated by active neurons. These electrical fields can induce current flow in adjacent neurons without any direct electrical coupling via gap junctions. The coupling is indirect and occurs from the electrical fields in the extracellular space. An active neuron (either firing an action potential or a synaptic potential) generates a field that induces current flow in any conductor (e.g., another neuron) present within this field. This type of coupling is called *ephaptic* coupling and is another form of electrical transmission among neurons. Ephaptic interactions occur mostly in regions of the brain where the neurons are tightly packed together in high density. Ephaptic coupling is particularly effective in synchronizing the firing of action potentials among neighboring neurons. For an understanding of how a potential change in one neuron can induce current flow in another, refer to the discussion of figure 11.1. From the calculations for the case with no direct electrical connection between neurons (i.e., no gap junction), the electrical coupling was shown to be quite small. If there are large numbers of neurons tightly packed together and each is firing an action potential, however, one can imagine how the summation of these small responses from many cells could become significant.

13.10   Compartmental models for a neuron

Throughout this book we use mathematical and electrical circuit models to represent particular concepts in neurophysiology. This chapter on postsynaptic mechanisms would not be complete without a discussion of the neuron models used to simulate synaptic inputs and their effects on the postsynaptic neuron. To do this we will combine some of the models for electrical cables and synapses presented in this and previous chapters and introduce the idea of *compartmental modeling.* An understanding of compartmental models will also be useful for the analysis of dendritic spines presented in the next section (section 13.11).

As the name implies, compartmental modeling involves the representation of various parts of a neuron as individual compartments or groups of electrical circuit elements. For example, a complex dendritic tree can be represented as hundreds (or thousands) of small segments, each of which is considered an isopotential compartment made up of a parallel resistor and capacitor. Each of these segments or compartments is then interconnected with the others through a single resistor placed in series between any two compartments (figure 13.31). The values for the parallel resistors and capacitors in each compartment are derived from the membrane surface area and the specific membrane properties of that segment of dendrite. For example, for a small segment of a dendrite, the parallel resistor of the compartment representing that segment would have a value of $R_m \div S$ (where S is the surface area of the segment), and the parallel capacitor would have a value of $C_m \times S$. The values of the series resistors interconnecting the compartments are derived from the diameters of the segments and the internal resistivity of the cytoplasm. For example, the total internal resistance (R_{int}) of one compartment is calculated by

$$R_{int} = R_i \times (\text{length of segment}) \div (\text{cross sectional area of segment}),$$

while the series resistor interconnecting two compartments would equal 1/2 the internal resistance of one compartment plus 1/2 the internal resistance of the next compartment.

Within each compartment one can also add mathematical representations for voltage- and ligand-gated channels as needed (e.g., Hodgkin-Huxley equations). The spatial resolution and accuracy of the model will depend, in part, on the number of compartments and on the amount of information one has to constrain the parameters of the model. The con-

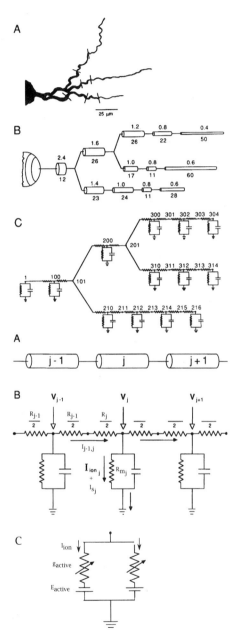

Figure 13.31 Simplifying a complex neuron (A, top) into a series of isopotential compartments (B and C, top). The transmembrane current at each node is the sum of active currents, synaptic currents, and leak currents. This is illustrated in the circuit diagrams below (A, B, and C, bottom). See text for further explanation. (Adapted from Koch and Segev 1989, chapter 3).

struction of a compartmental model from a neuron and the calculation of the values of the circuit elements is illustrated in figure 13.32.

For investigating cable properties, this type of model is particularly useful when the anatomy does not lend itself to the assumptions and simplifications inherent in the Rall model (i.e., 3/2 power rule, all dendrites at same L, uniform R_m and R_i; see chapter 4). Furthermore, compartmental modeling greatly simplifies the mathematics necessary to simulate realistic membrane properties on a computer by substituting ordinary differential equations for the partial differential equations of cable theory (chapter 4). The equations describing current and voltage in individual compartments lend themselves well to solutions using standard numerical methods. Many of the examples in this book were derived from compartmental models of neurons using a computer program called NEURON, which was written by Michael Hines.

Referring to figure 13.31, current flow in compartment j is described by

$$I_{m_j} = I_{j-1,j} - I_{j+1,j},$$ (13.10.26)

where $I_{j-1,j}$ is the current flow from compartment $j-1$ into compartment j, and $I_{j+1,j}$ is the current flow from compartment j into compartment $j + 1$. The membrane current is described by

$$I_{m_j} = C_{m_j} \frac{dV_j}{dt} + \frac{V_j}{R_{m_j}} + I_{\text{ion}_j} + I_{s_j},$$ (13.10.27)

where I_{ion_j} consists of the current from all of the voltage-gated channels present in compartment j, and I_{s_j} consists of the current from all of the synaptic inputs to compartment j. I_{ion} would be calculated from Hodgkin-Huxley-type equations for the different channels, and I_s would be calculated from equations such as

$$I_s = \alpha t e^{(1-\alpha t)} \cdot G_s(V_j - E_s),$$ (13.10.28)

which is a small modification of the alpha function first presented in chapter 4 where the need for a scaling constant has been eliminated by using $1 - \alpha t$ as the exponent. (Another equation that could be used to represent synaptic current is

$$I_s = K \left(e^{-t/\tau_2} - e^{-t/\tau_1} \right) G_s(V_j - E_s),$$ (13.10.29)

where τ_1 and τ_2 represent the rise and decay time constants, respectively, of the synaptic current, and K is a scaling constant. Equation 13.10.29 provides a better representation of a synaptic current than the alpha function, because the rise and decay times of equation 13.10.29 can be adjusted independently to fit a particular synaptic waveform.)

Equation 13.10.26 can be rewritten so that the current into and out of compartment j is in terms of the internal resistance, or

$$I_{m_j} = \frac{V_{j-1} - V_j}{R_{int_{j-1,j}}} - \frac{V_j - V_{j+1}}{R_{int_{j,j+1}}}, \qquad (13.10.30)$$

where R_{int} is the internal resistance between two compartments. Equations 13.10.27 and 13.10.30 are then combined and solved numerically for all compartments in the model. The more compartments, the more ionic and synaptic currents included in each compartment, and the smaller the time step needed for accuracy in the simulations the greater is the time necessary to solve the equations.

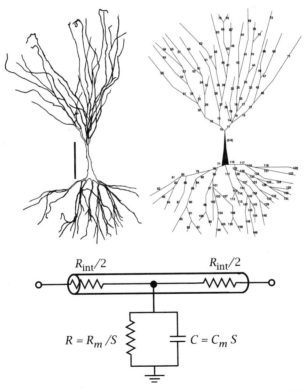

Figure 13.32 Compartmental model of a hippocampal CA3 pyramidal neuron and the calculation of circuit elements for one compartment (bottom). The diagram in the upper left is the reconstructed CA3 neuron, while the diagram in the upper right illustrates the numbering of the different compartments. Scale bar is 100 μm. (Adapted from Spruston et al. 1993.)

13.11   Dendritic spines and their effects on synaptic inputs

Most excitatory synapses in the adult CNS terminate on dendritic spines. Examples of dendritic spines from hippocampal neurons are illustrated in figure 13.33. As discussed in chapter 4, there have been numerous ideas expressed over the years for the function of dendritic spines. These notions have included that spines increase the surface area of dendrites, that spines somehow electrically modify the synaptic signal on the end of the spine, and that spines provide for chemical isolation or compartmentation.

Recent anatomical work has demonstrated that the dendritic surface area is more than adequate to support the number of synaptic endings on a typical pyramidal neuron in the cortex. Therefore, the idea that the surface area added by spines is necessary to accommodate the large number of synaptic inputs to a neuron is not likely to be true.

The idea that spines, by modifying the electrical signals from synaptic inputs, might be a substrate for learning has been an attractive one for many years. Wilfrid Rall was one of the first to recognize that changes in the shape of spines (for example, increases in spine neck diameter) might increase the effectiveness of a particular synaptic input and be a mechanism for learning. An increase in the effectiveness, efficiency, or, as it is more commonly termed, the *efficacy* of a synapse simply means that the activity of the synapse is more likely to cause the neuron to fire, in the case of an excitatory synapse, or is more likely to prevent a neuron from firing, in the case of an inhibitory synapse. After our presentation of cable theory in chapter 4 and of synaptic transmission in chapters 11–13, we are now in a position to address quantitatively the question of how spines might modify synaptic inputs. It seems reasonable to believe that spines are somehow important for synaptic transmission, and possibly even for learning (see also chapter 15). As we will see in the next section, spines can certainly affect synaptic transmission. They are also likely to provide some degree of chemical isolation from the rest of the dendrites.

13.11.1   Attenuation of potential between dendritic shaft and spine head

One important question is whether a spine provides any degree of electrical isolation from the rest of the neuron. For example, will depolarizations due to the activity of synaptic inputs produce a local depolarization at a

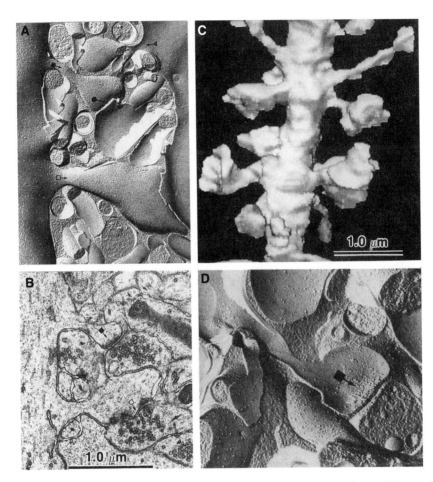

Figure 13.33 Dendritic spines and their synapses in hippocampal area CA1. (A) Cytoplasmic profile (P-face) of a small, thin, or 'pedunculated' dendritic spine (filled square) revealed by freeze-fracture electron microscopy to be near a large, mushroom-shaped dendritic spine (open square) of the same dendritic segment. (B) Thin-section view of two spines with similarly diverse shapes that also have different types of postsynaptic densities (PSDs). The smaller spine has a continuous, macular-shaped PSD (filled square), while the larger spine has an electron-lucent perforation in the PSD (open square). (C) Three-dimensional reconstruction of a segment of CA1 pyramidal cell dendrite revealing multiple spine shapes along its length. (D) Particle aggregate on the extracellular half of the membrane (E-face) at the site of a synapse on the head of a thin dendritic spine (filled square). (From Lisman and Harris 1993.)

different, inactive spine head? This issue is obviously important because such a depolarization could affect the driving force for a synaptic input on the spine, activate voltage-gated conductances if any are present in the spine, and affect the behavior of any ligand-gated channels that are voltage sensitive (e.g., NMDA receptors).

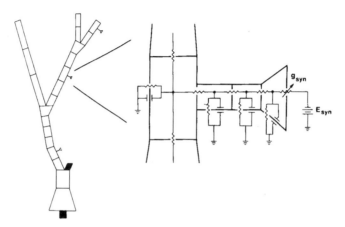

Figure 13.34   Compartmental model of a spine attached to a dendrite. The spine is represented as three compartments and the dendrite as one compartment. A synaptic input at the head of the spine is also modeled. (From Brown et al. 1988.)

As illustrated in figure 13.34, a spine can be represented as a finite-length cable with one end attached to the dendritic shaft and the other end sealed. With any reasonable estimates for R_m and R_i, the actual electrotonic length of this cable is quite short, because the spine is so short anatomically. In chapter 4 it was demonstrated that the attenuation of potential along the length of a sealed-end, finite cable depends on its electrotonic length. For a short L (in this case probably $L < 0.1$), the attenuation of potential from the dendritic shaft to the spine head would be negligible. This implies that *the spine is not isolated electrically from the potential of the local dendrite and that, whatever is the potential change in the dendrite, the resulting potential change in the spine head will be nearly identical.*

In addition to knowing about attenuation of potential from the dendrite to the spine head, it is also of interest to know whether there is attenuation of potential from the spine head to the dendritic shaft. If a synaptic potential is generated in the spine head, it is obviously important to know whether or not it will attenuate upon reaching the local dendritic shaft. As discussed in chapter 4, the attenuation of potential along a finite cable with different end terminations is, in general, asymmetrical. In this case, one end of the cable is sealed at the spine head while the other end is attached to the dendritic shaft. Because the diameter of the dendrite is typically much larger than that of the spine and because the dendrite represents a long cable compared to the spine, the dendrite provides a large load for that end of the spine. In other words, the input resistance of the dendrite at the attachment point of the spine is much lower than if the dendritic end of the spine had been sealed. It is this difference in end terminations that renders the voltage attenuation from spine head to dendrite *greater* than that from dendrite to spine head. Thus, *the spine can, in principle, act as an attenuator of synaptic potentials.* The implications of this are dealt with further in the next section.

13.11.2 Synapse on the head of a spine

As discussed above, the spine has the capability of attenuating the synaptic potential from the head to the dendritic shaft. What may seem a bit contradictory, however, is that the spine can also act to amplify or boost the synaptic potential. More precisely, *the amplitude of a synaptic potential on the spine head will be greater than if the same synaptic conductance change had occurred on the dendritic shaft.* Whatever the amplitude of the synaptic potential at the head of the spine, however, it may still be attenuated upon reaching the dendrite, as discussed in the previous section. This is illustrated in figure 13.35.

The implications of this boosting or amplifying of the local synaptic potential are at least twofold. First, the larger amplitude of the local potential might activate any voltage-gated channels present in the spine or affect any ligand-gated channels that are voltage sensitive (e.g., NMDA receptors). Second, the larger potential would result in a decrease in the driving force and less current flow from the conductance change.

13.11.3 Spines represented as series resistors

In figure 13.35 the spine was represented by three compartments, each of which consisted of a parallel resistor and capacitor along with the appropriate internal resistor. Because of the extremely small surface area of each compartment, the value of each of these parallel resistors is very large and the value of each of the capacitors is very small. The current flow (either resistive or capacitive) across the spine membrane and into the extracellular space is therefore negligible. This is illustrated in figure 13.36. The result of this analysis is that the rather complex model of

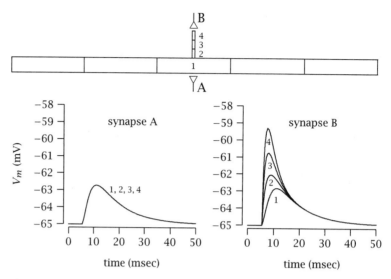

Figure 13.35   Differences in synaptic potentials in response to a synapse on the dendritic shaft compared to an identical synapse on the head of a spine. The spine is represented as 3 compartments (2, 3, and 4) attached to the dendrite (compartment 1) (see also figure 13.34). Note that when the synapse is on the dendrite (synapse A), there is no attenuation of potential to the spine head. The amplitudes of the potentials in all compartments (1, 2, 3, 4) are essentially the same (curves in lower left). When the synapse is on the spine head (synapse B), however, the amplitude of the synaptic potential in 4 is greater than when the same synapse was on the dendrite. Also, the amplitude of the potential attenuates upon reaching the dendrite. The degree of boosting and attenuation depends on the properties of the spine and on the magnitude of the synaptic conductance change ($G_s = 1$ nS, $R_i = 100$ Ω-cm, and $R_m = 50,000$ Ω-cm^2 for this simulation). See text for further explanation. (After Brown et al. 1988.)

the spine used in figure 13.35 can be greatly simplified to that of a single resistor, representing the internal resistance of the spine (R_{sp}), in series with the synapse. (While this simplification may be appropriate for most spines, it might not be justified for some extremely complex spines.)

13.11.4   The attenuation of synaptic inputs by spines

The consequences of the above simplification become apparent when one considers trying to measure the synaptic conductance with a voltage clamp somewhere in the cell. Suppose that the voltage clamp is as close as possible to the spine (ideally in the dendrite itself), then the *I-V* curve of the synaptic input, obtained as described in chapter 13, will have a slope determined by both the synaptic conductance *and* the conductance of the spine. Referring to figure 13.36, this can be stated formally as

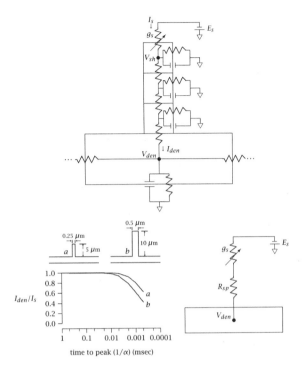

Figure 13.36 Attenuation of synaptic current from spine head to dendrite as a function of the kinetics of the synaptic conductance change. The spine model is indicated in the top half of the figure. The ratio of the synaptic current measured by a voltage clamp in the dendrite (I_{den}) to the peak current at the spine head (I_s), as a function of the time to peak ($1/\alpha$) of the synaptic conductance change, is plotted in the lower left. For a time to peak as short as 0.01 msec, the current reaching the dendrite is almost identical to that entering the spine head. There is thus no capacitance and resistance current flow across the spine membrane for reasonable values of time to peak, and the complex three compartment model of the spine can be reduced to a simple resistor, as shown in the lower right. ($G_s = 1$ nS, $R_i = 100$ Ω-cm, $R_m = 50,000$ Ω-cm^2, and I_s = alpha function for this simulation.)

$$G_T = \frac{G_{sp}G_s}{G_{sp} + G_s},$$

(13.11.31)

where $G_{sp} = 1/R_{sp}$ and G_T is the slope of the *I-V* curve measured by a voltage clamp in the dendrite. From this relationship it can be seen that *the total conductance measured by the voltage clamp, or perceived by the dendrite, is less than the actual synaptic conductance.* As R_{sp} gets small, G_{sp} gets large and G_T approaches G_s. Therefore, the effect of the spine depends on the relative magnitudes of G_s and $1/R_{sp}$.

One should note from figure 13.37 that although the slope of the *I-V* curve may be affected by the spine, the reversal potential is essentially un-

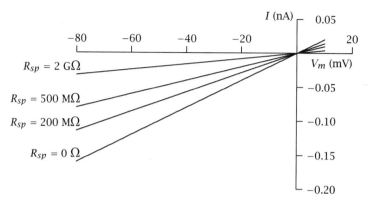

Figure 13.37   The effect of a spine on the synaptic *I-V* curve measured by a voltage clamp in the dendrite. G_s at the synapse is 2 nS for this simulation while the measured G_s in the dendrite (as determined from the slope of the *I-V* curve) decreases for increasing values of R_{sp}. (After Brown et al. 1988.)

changed. This fact is another demonstration that the voltage attenuation from dendrite to spine head is negligible.

13.11.4.1   Attenuation of synaptic current by a spine: Synaptic charge transfer

As shown above, the effect of the spine will be to reduce the synaptic current measured by a voltage clamp. It follows from equation 13.11.31 that the overall synaptic conductance, as determined from voltage-clamp measurements, will also be reduced by the spine. In other words, the current flowing into the dendrite from a synaptic input at the spine head will be less than for an identical synapse directly on the dendrite. The amount of this reduction will again depend upon the relative magnitudes of G_s and $1/R_{sp}$. The equation for this is

$$I_{den} = \frac{(V_{den} - E_s)}{1/G_s + R_{sp}},$$ (13.11.32)

where I_{den} and V_{den} are the current and potential in the dendrite, respectively. If R_{sp} is small compared to $1/G_s$, then the reduction in I_{den} by the spine will be negligible.

It is often convenient to consider the transfer of charge from a synaptic input into a neuron, or from one region of a neuron to another, rather than the flow of current. There are several reasons for this. First, the ability of a neuron to reach threshold from combined synaptic inputs is dependent more on the total charge injected by the synapses than on the peak amplitudes of their currents or potentials. Second, peak amplitudes of summated synaptic inputs are highly dependent on latency fluctuations of the individual synapses, whereas the charge injected is insensitive to such fluctuations. Third, by the same reasoning, peak amplitudes are more sensitive to noise in the measurement system than is charge. And finally, as discussed in more detail below, measurements of synaptic charge from electrotonically remote synapses are distorted much less by the cable properties of the neuron than are measurements of the time course and peak amplitudes of synaptic potentials (or currents).

Synaptic charge can be calculated from the current by

$$q_s = \int_0^\infty PSC \, dt, \tag{13.11.33}$$

or from the potential by

$$q_s = \frac{\int_0^\infty PSP \, dt}{R_N}, \tag{13.11.34}$$

where q_s is the synaptic charge. The amount of charge injected by the synapse may be different under current- and voltage-clamp conditions because of differences in driving force (i.e., there may be a greater depolarization at the subsynaptic membrane during the synaptic response under current clamp than under voltage clamp, leading to a smaller driving force and less injected charge).

It is important to recognize that the charge measured in the soma, by either of the above two equations, for a synaptic response in the dendrites may not be equal to the total charge injected by the synapse. The more distal the input, the more it is filtered by the membrane properties of the dendrites. It has been shown (Carnevale and Johnston 1982) that the attenuation of charge from point A to point B in a neuron is the same as the steady-state attenuation of potential from point B to point A. Thus the attenuation of charge is determined by the DC electrotonic distance of the synapse. (Remember that charge is stored by capacitors. Even though there can be current flow across a capacitor, there is no loss of charge. The loss of charge results only from current flow across a resistor—hence the dependence of charge transfer on DC properties.) As an example of charge attenuation from a distal synapse, consider a synapse located at the end of the dendritic tree. The amount of attenuation of charge from the end of the dendrites to the soma is the same as the amount of attenuation of a steady-state potential from the soma to the end of the dendrites—this can be quite small for dendrites with short L. In contrast, the amount of attenuation of an EPSP or EPSC from the end of the dendrites to the soma is determined by the AC properties of the neuron. This means that the faster the synaptic conductance change or the faster the membrane time constant, the more the time course of the synaptic potential or current will be slowed and the more its peak amplitude will be attenuated. Thus, the measurement of charge rather than peak amplitude of distal synaptic inputs has the advantage of being less affected by the cable properties of the neuron.

What is the effect of a spine on synaptic charge transfer? From the principle that charge transfer is dependent on steady-state voltage attenuation, it follows that the loss of charge from spine head to dendrite will be equal to the voltage attenuation from dendrite to spine head. Because of the short electrotonic length of a spine, this will be nearly zero. Therefore, a spine will not attenuate the transfer of charge from a synapse to the dendrite. It is possible, however, that the amount of charge injected by a synapse on the head of a spine will be less than if the synapse were directly on the dendrite. The reduction in charge will again be dependent on the relative magnitudes of G_s and $1/R_{sp}$, in an identical way as given by equation 13.11.32.

13.11.4.2 Examples of voltage boosting, voltage attenuation, and charge transfer for a typical spine In this section examples of voltage boosting, voltage attenuation, and charge transfer for a typical spine will be illustrated. Figure 13.38 depicts the conceptual model to be used for these calculations. All potentials will be referenced to the resting potential so that $E_s = 60$ mV. For the purpose of this example, assume that the synaptic conductance $G_s = 1$ nS (or $R_s = 1$ GΩ), the spine resistance $R_{sp} = 100$ MΩ (or $G_{sp} = 10$ nS), and the input resistance of the neuron at the attachment point of the spine to the dendrites $R_{den} = 100$ MΩ. We will also assume that the spine can be adequately represented by a single resistor. The figure below is a circuit model for a synapse at the head of a spine attached to a dendrite.

1. **Synapse directly on dendrite.** The amplitude of the EPSP for a synapse with the above values attached directly to the dendrite (i.e., $R_{sp} = 0$) can be calculated using the simple voltage divider equation derived in appendix A:

$$\text{EPSP}_{den} = 60 \text{ mV} \frac{R_{den}}{1/G_s + R_{den}} = 60 \text{ mV} \frac{100}{1100}$$
$$= 5.5 \text{ mV},$$

where EPSP_{den} is the amplitude of the EPSP in the dendrite.

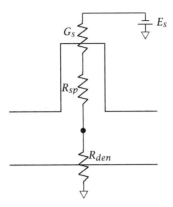

Figure 13.38 Diagram and circuit model for a synapse at the head of a spine attached to a dendrite. R_{den} represents the total input resistance of the neuron as measured at the base of the spine. Dendritic capacitance is ignored for this example.

2. **Synapse on spine.** If the synapse is at the head of the spine, the EPSP in the dendrite will now be

$$\begin{aligned} \text{EPSP}_{den} &= 60 \text{ mV} \frac{R_{den}}{1/G_s + R_{sp} + R_{den}} \\ &= 60 \text{ mV} \frac{100}{1200} \\ &= 5.0 \text{ mV}. \end{aligned}$$

Adding the spine therefore reduces the amplitude of the EPSP in the dendrite by about 9%.

The amplitude of the EPSP at the head of the spine (EPSP$_{sp}$) can be similarly calculated by

$$\begin{aligned} \text{EPSP}_{sp} &= 60 \text{ mV} \frac{(R_{den} + R_{sp})}{1/G_s + R_{sp} + R_{den}} \\ &= 60 \text{ mV} \frac{200}{1200} \\ &= 10.0 \text{ mV}. \end{aligned}$$

Adding the spine has thus increased the amplitude of the local EPSP by 83% while the amplitude of the EPSP in the dendrite has actually decreased by 9%.

3. **Charge transfer.** Since we have represented the spine as a single resistor for this example, there is clearly no loss of charge from the spine head to the dendrite. In other words, all of the charge entering the spine will transfer to the dendrite. What is illustrated instead

in this example is the effect of the spine on the amount of charge injected by the synapse.

Charge is calculated by integrating the synaptic current, or

$$q_s = \int_0^\infty I_s dt$$

If we let the current be represented by the alpha function, then

$$q_s = \int_0^\infty (\alpha t e^{1-\alpha t}) \cdot G_s(V_m - E_s)dt,$$

or

$$q_s = (e/\alpha) \cdot G_s(V_m - E_s).$$

Remember that current is in coulombs (C)/sec and charge is in C. Let $\alpha = 1000$ sec^{-1} and $1/\alpha = 0.001$ sec, which represents a synaptic waveform with a time-to-peak of 1 msec.

(a) *Synapse on dendrite.* The charge injected into the dendrite for a synapse on the dendrite ($R_{sp} = 0$) is

$$
\begin{aligned}
q_s &= \int_0^\infty I_s dt \\
&= (e/\alpha) \cdot \frac{60 \text{ mV}}{1/G_s + R_{den}} \\
&= 2.72 \times 10^{-3} \text{ sec} \cdot \frac{60 \times 10^{-3}}{1100 \times 10^{-6}} \\
&= 1.5 \times 10^{-13} \text{ C.}
\end{aligned}
$$

(b) *Synapse on spine.* The charge injected into the dendrite for a synapse at the head of a spine is

$$
\begin{aligned}
q_s &= \int_0^\infty I_s dt \\
&= (e/\alpha) \cdot \frac{60 \text{ mV}}{1/G_s + R_{sp} + R_{den}} \\
&= 2.72 \times 10^{-3} \text{ sec} \cdot \frac{60 \times 10^{-3}}{1200} \\
&= 1.4 \times 10^{-13} \text{ C.}
\end{aligned}
$$

The difference between the charge injected for a synapse on the dendrite and that for a synapse on the spine is therefore only about 8%.

The above examples illustrate that placing a synapse on a spine instead of directly on a dendrite can boost the amplitude of the local EPSP at the subsynaptic membrane while having much less effect on the amount of charge injected by the synapse. The attenuation of potential from the spine head to the dendrite is also illustrated. The reader is encouraged to try other values for G_s, R_{sp}, and R_{den} to see how these principles are dependent on the chosen parameters.

13.12 Summary of important concepts

1. Difference between excitation and inhibition.
2. Analysis of *I-V* curves.
3. Use of reversal potential for determining conductance ratios.
4. Effects of poor space clamp.
5. Theory of Magleby and Stevens.
6. Spatial summation.
7. Temporal summation.
8. Characteristics of electrical synapses.
9. Unique properties of NMDA receptors.
10. Compartmental models.
11. Charge transfer.
12. Effects of dendritic spines on synaptic signals.

13.13 Homework problems

1. The conductance of an inhibitory synapse is 100 nS with a reversal potential of -70 mV. The conductance of an excitatory synapse is 20 nS with a reversal potential of 0 mV.

   (a) Draw the *I-V* curves (on graph paper) for these two synaptic inputs.

(b) If the two inputs were stimulated together, draw the resultant *I-V* curve one would obtain under voltage clamp. Is this combined response excitatory or inhibitory? Assume that threshold is -50 mV and that the EPSP and IPSP have the same time course.

2. You are recording from a pyramidal neuron in the olfactory cortex and stimulating the main excitatory inputs to this neuron (nerves from the olfactory bulb). You give a high-frequency stimulus train to the nerves and discover that the amplitudes of the EPSPs are increased for a period of time after the train. The increase in the EPSPs lasts longer than facilitation and post-tetanic potentiation but less time than long-term potentiation, so you call this phenomenon medium-term potentiation or MTP. The "raw" data from your experiments are in the figure below. Use these to answer the following questions.

(a) Using the current-clamp data, describe quantitatively the time course of decay of MTP.

(b) Calculate G_s and E_s from the voltage-clamp results given for control and MTP.

(c) Assuming that the postsynaptic channels responsible for the EPSP are permeable only to Na^+ and K^+, calculate values for G_K and G_K/G_{Na} during control and MTP.

(d) What are the likely mechanisms responsible for MTP? Would you describe these mechanisms as being pre- or postsynaptic?

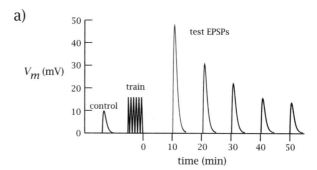

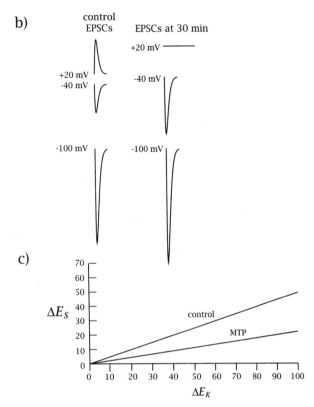

3. At a postsynaptic cell, application of transmitter A results in changes in Na$^+$ conductance (ΔG_{Na}), and application of transmitter B results in changes in K$^+$ conductance (ΔG_K). Both transmitters generate synaptic currents according to Ohm's law. The equilibrium potentials for Na$^+$ and K$^+$ of the cell are $E_{Na} = +50$ mV, $E_K = -100$ mV.

(a) At rest, the ratio of ionic conductances is $G_K : G_{Na} : G_{Cl} = 1 : 0.8 : 15$. $\frac{[Cl^-]_{out}}{[Cl^-]_{in}} = 15$. What is the resting potential of this postsynaptic cell?

(b) Plot the current-voltage relations (ΔI vs. V) of the synaptic current when (label the axes with the appropriate units):

   i. transmitter A is applied, making $\Delta G_{Na} = 0.02$ S/cm^2

   ii. transmitter B is applied, making $\Delta G_K = -0.02$ S/cm^2

   iii. transmitters A and B are applied together, making $\Delta G_{Na} = 0.02$ S/cm^2 and $\Delta G_K = -0.02$ S/cm^2

(c) What are the reversal potentials of the postsynaptic signals for stimuli (i), (ii), and (iii) described in part (b)?

4. Postsynaptic voltage responses to presynaptic stimuli (s) are recorded from synapses A and B under current-clamp conditions (see figure below). Postsynaptic current responses to presynaptic stimuli are recorded from synapses C and D under voltage-clamp conditions (parts C and D in figure). Numbers on the left of each trace indicate the membrane voltage of the postsynaptic cell.

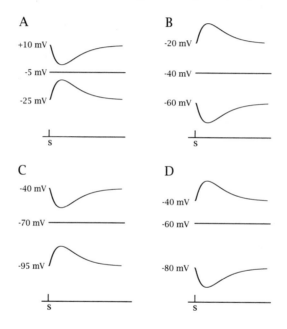

(a) What are the reversal potentials for synapses A–D?

(b) For each synapse (A–D), is the postsynaptic signal accompanied by a conductance increase or a conductance decrease? Explain your answers briefly.

5. The traces shown in the figure below represent synaptic currents measured under voltage clamp. The given voltages are the holding potentials with respect to the resting potential.

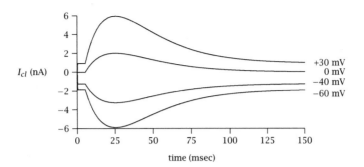

(a) What are the conductance and the reversal potential for this synaptic input?

(b) Is the synapse likely to be excitatory or inhibitory?

(c) What is the approximate input resistance of the cell?

(d) What is the approximate decay time constant of the synaptic current?

(e) Is the decay time constant voltage dependent? Show the calculations that led to your answer.

(f) If V_{rev} is not equal to E_s, what does this tell you about this synapse?

6. The *I-V* curves shown were determined for two different synapses on the same cell. Both have identical kinetics and are located on the soma, the site of the recording.

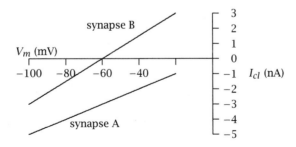

(a) What are the conductances and reversal potentials for synapses A and B?

(b) If threshold is at −50 mV, are the synapses excitatory or inhibitory?

(c) Can you determine the resting conductance of the cell from the above *I-V* curves, and if so, what is it?

(d) If A and B were stimulated *simultaneously*, draw the *I-V* curve that would have been obtained, and calculate the conductance and reversal potential. Is it excitatory or inhibitory?

(e) Is synapse B governed by a conductance-increase or conductance-decrease mechanism?

(f) If your answer to (e) is conductance increase, draw an *I-V* curve for a conductance decrease mechanism with the same reversal potential. If your answer to (e) is conductance decrease, draw an *I-V* curve for a conductance-increase mechanism. Whichever your answer, is the *I-V* curve you draw for an excitatory or an inhibitory synapse?

7. At a postsynaptic membrane, we may think of an EPSP as a small increase in Na$^+$ conductance (ΔG_{Na}) that generates a synaptic current according to Ohm's law. Therefore, if we measure the synaptic current ΔI_{Na} as a function of clamped membrane voltage V_m, $\Delta I_{Na} = (V_m - E_{Na})\Delta G_{Na}$, which is plotted here.

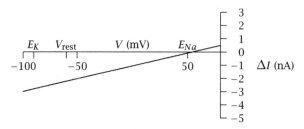

(a) What does the slope of the line represent? What is the reversal potential for this synaptic current?

(b) Now let us add a simultaneous K$^+$ conductance increase of the same size ($\Delta G_K = \Delta G_{Na}$). Trace the figure and add to it a curve for the K$^+$ current, ΔI_K, vs. membrane voltage. Place an x on the V_m axis at the reversal potential for the *combined* Na$^+$ and K$^+$ current and explain your choice of location. At the resting potential, would this current produce an IPSP or an EPSP?

(c) Now suppose that the K$^+$ conductance increase is ten times larger than the Na$^+$ increase ($\Delta G_K = 10\Delta G_{Na}$). Again trace the figure and add to it a curve for the K$^+$ current, ΔI_K vs. membrane voltage. Place an x on the V_m axis at the reversal potential for the *combined* Na^+ and K^+ current and explain your choice of location. At the resting potential, would this current produce an IPSP or an EPSP?

8. You have found that two putative neurotransmitters (X and Y) cause depolarization of isolated retinal horizontal cells and that both responses reverse at 0 mV. You want to determine whether X and Y use the same or different ligand-gated channels. You obtain the following results. In voltage clamp at V_{rest}, a saturating concentration of X alone causes a 2 nA inward current, and a saturating concentration of Y alone also causes a 2 nA inward current. Also, $G_{rest} = 10\,nS$ and $V_{rest} = -100\,mV$. Assume that there is no desensitization and that the neuron is passive (no voltage-dependent conductances).

   (a) If X and Y use the same channels, calculate the expected membrane potential (under current clamp) with X alone, with Y alone, and with X and Y together. Similarly, calculate the expected total current under voltage clamp at V_{rest} when X and Y are applied together.

   (b) Repeat part (a) for the expected results if X and Y use different channels.

9. An investigator from another lab reports that in addition to the effect you have shown, drug X causes an increase in measured conductance and a shift in the reversal potential of excitatory synaptic currents on mature cerebellar neurons in culture. He suggests that an increase in transmitter release presynaptically would increase the conductance with no effect on the reversal potential and proposes that drug X is likely to have a postsynaptic effect on the transmitter-gated ion channels. You suspect he has misinterpreted the data and do some experiments to test your idea. Your data from a complete anatomical and physiological analysis of a typical neuron are shown below.

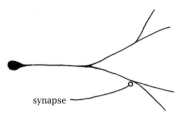

synapse

compartment	length (μm)	diameter (μm)
0: soma	—	10
1: 1° dendrite	530	2.50
2: 2° dendrite	210	1.57
3: 2° dendrite	210	1.57
4: 3° dendrite	120	0.99
5: 3° dendrite	120	0.99
6: 3° dendrite	120	0.99
7: 3° dendrite	120	0.99

(a) Use the voltage-clamp data shown to calculate the measured synaptic conductance (G_s), synaptic reversal potential (V_{rev}), and input resistance (R_N) before and after the addition of drug X.

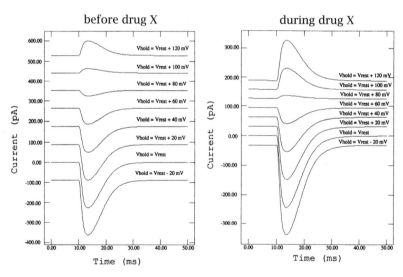

(b) You know that in this cell, drug X causes a change in the membrane time constant (τ_m) from 17 msec to 57 msec. Calculate the space constant (λ) for each segment before and after the addition of drug X. See the table above and use $R_i = 70$ Ω-cm.

(c) Assume that the neuron can be represented by an equivalent cable. List the requirements for this to be the case (consider all assumptions of the model). Calculate the electrotonic length (L) and electrotonic distance of the synapse (X) before and after addition of drug X (use the data in the table).

(d) Do your data support or refute the hypothesis that drug X causes a change in reversal potential of the synaptic response in these cells? Explain briefly. (*Hint:* Calculate the synaptic equilibrium potential (E_s) before and after drug X.)

(e) Explain briefly how drug X could alter the synaptic conductance measured from the soma if its only effect is to block voltage-insensitive (non-ligand-gated) potassium channels.

10. Draw the expected I-V curve for a conductance-*decrease* PSP that has a $\Delta G_s = 10$ nS and $E_s = -90$ mV in a neuron with a resting potential of -70 mV. Is this PSP likely to be excitatory or inhibitory?

11. You are recording intracellularly from a pyramidal neuron and stimulating an excitatory synaptic input to this neuron. You depolarize the cell and find that the EPSP reverses at around 0 mV. Also, with the neuron at the resting potential, you give brief current pulses and measure a decrease in the input resistance of the neuron during the EPSP. You think that you understand the properties of this EPSP until you do the next experiment, the results of which confuse you momentarily. You find that the amplitude of the EPSP *decreases* when you hyperpolarize the neuron from rest. You quickly recognize that there are at least two explanations for this result. Describe two hypotheses that you believe can account for these results and outline briefly the additional experiments you would do to test them. (*Hint:* You don't need to propose anything exotic to answer this question. The outlined results could easily be obtained (and probably have been) from many types of CNS neurons. You can assume a single presynaptic fiber releasing a single neurotransmitter and with the neuron having fairly common properties for a neuron in the CNS.)

12. You are voltage-clamping a neuron that has an input resistance (completely passive) of 100 MΩ and a resting potential of -60 mV. It has two separate synaptic inputs that occur on the cell body: Input A has a conductance of 10 nS and a reversal potential of 0 mV while input B has a conductance of 20 nS and a reversal potential of -100 mV.

(a) Draw on graph paper the *I-V* curves of the resting neuron, input A, and input B.

(b) If inputs A and B have identical kinetics and are activated together, what is the total *I-V* curve for inputs A and B? What is the conductance and reversal potential of this combined synaptic response? What would be the total input resistance of the *neuron* during the peak of this synaptic response?

(c) If input A has a time to peak of 1 msec and a decay time constant of 4 msec while input B has a time to peak of 3 msec and a decay time constant of 10 msec, and they are activated together as above, draw the *I-V* curves for the combined synaptic responses at 1 msec, 3 msec, and 6 msec from the onset of the synaptic responses. Assume a linear rise time and a single exponential decay for the synaptic currents.

13. (a) From the figure, calculate the slope conductance of the neuron before and during the application of transmitter X.

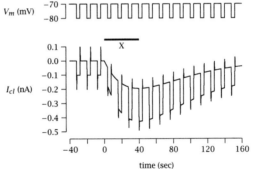

(b) Is the response to the transmitter a conductance increase or a conductance decrease?

(c) What (if anything) can you say about the reversal potential of this response and whether it is likely to be excitatory or inhibitory?

14. Refer to the following figure.

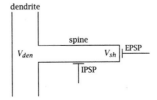

(a) If for the excitatory synapse $G_s = 1$ nS, $R_{den} = 100$ MΩ, and $R_{sp} = 500$ MΩ, calculate V_{sh} and V_{den}. Assume steady-state conditions, $E_s = 0$ mV, and $E_r = -70$ mV.

(b) If the inhibitory synapse is halfway between the spine head and the dendrite, $G_{s_i} = 5$ nS, and $E_{s_i} = -70$ mV, what are the new values for V_{sh} and V_{den}?

14 Extracellular Field Recordings

14.1 Introduction

Most of what we have presented up to now in this book has involved the principles of cellular neurophysiology from the perspective of transmembrane electrical events. The electrical signals we have discussed have been measured from the inside of neurons with respect to the outside (or from the outside with respect to inside in the case of cell-attached patch recordings of single channels). An important topic that once dominated the field of neurophysiology and is still quite important is that of electrical signals measured in the *extracellular space*. These signals are generated by the electrical fields produced by the activity of single neurons or groups of neurons and are typically measured between two points in the extracellular space rather than across the membrane. The signals are called *extracellular field potentials*. Examples of extracellular field potentials include the relatively simple case of extracellular recordings of action potentials along a nerve fiber, the more complex case of field potentials associated with activity of a group of neurons within a particular region of the brain, and, the most complex case of all, electroencephalographic or EEG recordings of gross brain activity from the scalp.

A full quantitative treatment of this subject would be enormously complex and is well beyond the scope of this book. Our goal in this chapter is to provide the conceptual framework along with a semiquantitative analysis of a few specific examples. Because the focus of this book is on cellular neurophysiology, the examples discussed will be those in which extracellular field recordings are used to infer the activity of neurons and synapses rather than the activity of large brain regions (i.e., the EEG).

14.2 Potentials in a volume conductor

The extracellular fluid constitutes a conductive medium surrounding neurons. When neurons are at rest, the membrane is uniformly polarized and there is no net current flow anywhere. An active neuron, however, is nonuniformly polarized. That is, the dendrites may be at a different potential from the soma, the soma may be at a different potential from the axon, and, as the action potential propagates along the axon, different regions of the axon may be at different potentials from one another. These spatial nonuniformities in potential produce current flow from one part of a neuron to another through the extracellular space. The extracellular fluid can therefore be considered a *volume conductor*. The flow of current in this volume conductor establishes electrical fields that can exert a force on electrical charge in the conductor. This force is what we measure in the form of a potential difference (see appendix A).

In previous chapters we assumed that the extracellular space was at ground, or zero, potential, implying that the extracellular fluid has zero resistance. This was an approximation that proved useful for the derivations and was justified because the resistivity of the extracellular fluid is typically much lower than that of the membrane. If, however, the fluid were actually at zero potential everywhere, then the entire extracellular environment of the brain would be isopotential, there would be no field potentials, and there would be no EEG. Taking the other extreme, if the extracellular space were a perfect insulator, then there would be no transmembrane current flow (nothing to carry current) and there would be no nervous activity whatsoever. This would also produce a flat EEG but for different reasons. Clearly then, there is current flow in the extracellular fluid during activity of neurons and this flow of current produces spatial gradients of potential.

14.2.1 Action potential along a nerve fiber

Referring to figure 14.1, assume initially that the axon is uniformly polarized. If an action potential is somehow elicited at A, then the membrane potential at A will momentarily reverse polarity. A will now be at a different potential from the rest of the axon, and local currents will flow as indicated in the diagram. Because the current flow is into the axon at A, it is called a current *sink*, while the site where the current exits is called a current *source*. If we were recording extracellularly at point A (with

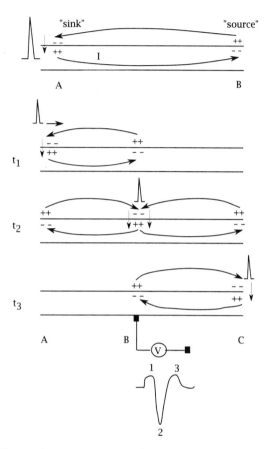

Figure 14.1 Action potential propagation along an axon. In the upper diagram an action potential is initiated at A and the current flow (*I*) from A to B is indicated. The bottom three diagrams illustrate a time sequence ($t_1 < t_2 < t_3$) for the propagation of an action potential from A to B to C. The recording at B indicates a positivity (1), a negativity (2), and a positivity (3) corresponding to this time sequence of AP propagation (see text for further explanation).

respect to a distant ground electrode), we would record a negative potential (negativity) during the action potential, while at point B we would record a positivity.

In this example the action potential is not stationary but will propagate along the axon from left to right. The location of current sinks (inward currents) and sources (outward currents) will therefore change with time. As discussed above, an action potential initiated at A will produce an initial source or positivity at B. As the action potential propagates, point B eventually becomes a sink and exhibits an extracellular negativity. With further propagation of the action potential, point B becomes a source again (positivity) for the action potential located at another site. A recording at B during the propagation of the action potential will therefore display a triphasic waveform (positivity→negativity→positivity), which is frequently seen with extracellularly recorded action potentials.

14.2.2 Active neuron is an electric dipole

In figure 14.2 we have illustrated the situation for a neuron with apical dendrites. Let us assume for the moment that the dendrites are passive and that an action potential is initiated in the soma. As before, the soma would be a sink while the dendrites would be a source. Because there is a conductance change occurring in the soma to produce the sink, it is termed an *active* sink (by active, we mean a change in conductance, and the reason for this added terminology will become clearer shortly), while the dendrites are a *passive* source. The neuron can be considered an *electric dipole*,[1] because during the action potential the dendrites are positive with respect to the soma.

Synaptic inputs to various locations on a neuron will also set up electric dipoles. For example, excitatory synaptic inputs to the apical dendrites will produce an active sink in the dendrites and a passive source in the soma (figure 14.2). In contrast, inhibitory input to the soma produces an active source in the soma and a passive sink in the dendrites. Because these extracellular signals are not stationary in either time or space and because many neurons may be active together or in various phase relationships, extracellularly recorded field potentials can be quite complicated.

14.2.3 Volume conductor theory

The mathematical derivation for volume conductor theory can be obtained from a number of excellent sources (Plonsey 1969; Stevens 1966) and will not be repeated here. Although the derivations are elegant in a mathematical sense, they have less practical value for neurophysiologists, because it is very difficult to apply the theory in a rigorous way to specific biological problems. For example, to use the mathematics to predict quantitatively the electrical fields in a volume conductor consisting of active neurons would require knowing the precise geometry and conductivity (as a function of time) of each neuron in the volume as well as the geometry and

---

[1]An electric dipole is simply a quantity of positive and negative charges separated by some distance.

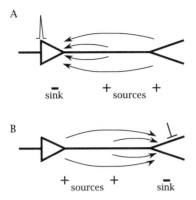

Figure 14.2   Current sources and sinks in a neuron with dendrites. An action potential is stationary in the soma in (A); an excitatory synaptic input to the dendrites is depicted in (B). The arrows indicate the direction of extracellular current flow (see text for further explanation).

conductivity of the extracellular space. Such data are seldom, if ever, available, so gross simplifications are usually made. The result is simply a qualitative application of the theory to a particular situation and thus a qualitative prediction of the expected electrical fields. The latter is all that we will attempt here.

From an observation point in a volume conductor (see figure 14.3), the measured potential (with respect to a distant ground) will depend on the *solid angle* made with the dipole. The solid angle is directly analogous to its two-dimensional counterpart. It depends on the physical size of the dipole and the distance between the observation point and the dipole. Remember also that the dipole need not be one-dimensional but could be a sheet of dipoles or a *dipole layer*. In the simple example shown in figure 14.3, the measured potential at the observation point could be positive, negative, or zero depending on its position with respect to the dipole. If there are many dipoles in the volume conductor, the superposition principle (see appendix A) can be invoked and will lead to the prediction that the measured potential at the observation point would represent the linear sum of that contributed by each dipole individually. This of course assumes that the individual dipoles are independent of each other, which is probably a reasonable assumption. In any event, as the geometry of the dipoles in the volume conductor gets more complicated, it is easy to see how the predictions of what the potential would be at any given observation point could become difficult to make. In fact, if the observation point is on the surface of the scalp and the dipoles are from neurons in

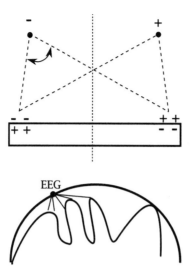

Figure 14.3 Electrical field in a volume conductor. Top of figure represents some type of nervous tissue in which there is a separation of charge (negative on the left, positive on the right) and two observation points (filled circles). The dashed lines represent the solid angle (assuming 3-dimensions) subtended by the dipole and the observation point. The observation point on the left would measure a negative potential with respect to a distant ground while the one on the right would measure a positive potential. The dotted line in the middle represents the zero potential axis. The bottommost part of figure schematizes a measurement point on the skull and the summation of numerous such dipoles from the underlying cortex.

the underlying cortex, which is folding and curving, the predictions are indeed *extremely* difficult to make.

14.3   Classification of fields

If one is recording extracellularly from a particular area of the brain, what would be the electric fields associated with the activation of a group of neurons by a simultaneous synaptic input? This is a common problem in neurophysiology. There are three main types of geometrical arrangements of neurons that produce characteristic field potentials: the *open field, the closed field,* and *the open-closed field.* They will be discussed separately in the following sections (see also Hubbard et al. 1969).

14.3.1 Open field

The open field is encountered when neurons are organized in a laminar array in which the dendrites are facing in one direction and the somata in the other. This type of field is typically encountered in the cerebellum, neocortex, and hippocampus. If this array of neurons is activated by a synchronous synaptic input (for example, by stimulation of an afferent pathway), a dipole will be established between the dendrites and the somata. This is illustrated in figure 14.4. As discussed in section 14.2.3, the potential measured at an observation point will depend on the solid angle made with the dipole layer. In the case of an action potential in the cell body layer, the field would be negative in the soma and positive in the dendrites. Recordings from an open field will be discussed further in section 14.4.1.

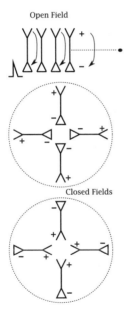

Figure 14.4 Diagram of open and closed fields. Dotted lines represent zero potential lines for each type of field (see text for explanation). (After Hubbard et al. 1969.)

14.3.2 Closed field

The closed field is also illustrated in figure 14.4. It consists of a spherical array of neurons in which the somata are either at the center of the sphere with the dendrites directed toward the periphery or at the periphery with

Open-Closed Field

Figure 14.5   Diagram of an open-closed field. (After Hubbard et al. 1969.)

the dendrites directed toward the center. The synchronous activation of such a field produces dipoles with spherical symmetry (for example, a positive core and negative periphery). The polarity of a measurement made within the sphere of symmetry would depend on its location within this spherical dipole. A measurement from outside the field, however, would record zero potential—the dipoles within the sphere would cancel.

14.3.3   Open-closed field

The open-closed field (see figure 14.5), as expected from its name, is a combination of the open and closed fields. It consists of a spherical array of neurons embedded in a laminar array. The resulting field potentials would also be a combination of the two types of fields. Recordings outside the sphere would be similar to those from the open field, whereas recordings within the sphere would combine the fields generated from the neurons in both the open and closed arrays.

14.4   Semiquantitative theory for extracellular fields

In this section we will attempt a semiquantitative analysis that should help explain the origin of the dipoles and the approximate relationship between the extracellularly recorded potentials and the transmembrane electrical events that give rise to the extracellular potentials.

In figure 14.6, a circuit diagram is illustrated that depicts the simplified or lumped features of an active neuron and the measurement of potential in the extracellular space. At point A a conductance change has occurred (an action potential) that produces an inward current (I_m). R_{int} is the lumped internal resistance of the neuron between A and B, and R_{ext} is

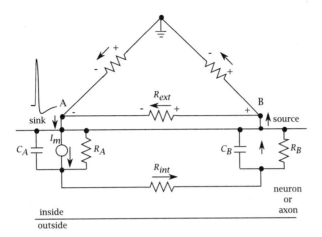

Figure 14.6 Lumped circuit diagram for the origin of electrical fields. "Inside" and "outside" refer to that of a hypothetical neuron or axon. An action potential is initiated at A and the resulting current flow is indicated (see text for further explanation). (After Hubbard et al. 1969.)

the lumped extracellular resistance between A and B ($R_{A,B}$ and $C_{A,B}$ are the usual membrane elements). Because current must always flow in a complete circuit (or loop), the inward current at A flows through the inside of the neuron and exits at B. To complete the circuit, the current flows in the extracellular space through R_{ext}. Assuming that we have a distant ground electrode in the extracellular fluid, current will also flow from B to ground and from ground to A.

From this diagram one can easily obtain the polarity of the extracellular field potentials by observing the direction of the current flow and remembering that the potential drop across a resistor (positive to negative) is in the direction of current flow. Point B will be positive with respect to ground, point A will be negative with respect to ground, and B will be positive with respect to A. This simple diagram, then, is a useful tool with which to determine the expected polarity of extracellular field potentials.

But what about the time course of the extracellular field potentials? Because the transmembrane current I_m draws current from the extracellular space, the time course of the inward current at A will dictate the time course of the extracellular potential across R_{ext}. The time course of the extracellular field potential will be roughly proportional to I_m, or

$$V_A \propto I_m \propto -V_B. \tag{14.4.1}$$

For electrical events that are relatively fast with respect to the membrane time constant (e.g., the action potential and fast EPSPs), most of brane time constant (e.g., the action potential and fast EPSPs), most of the current exiting at B will be capacity current across C_B. Thus for fast conductance changes (Δg fast with respect to τ_m), the time course of the extracellular field potential is roughly proportional to the first derivative of the transmembrane potential, or

$$I_m \propto \frac{dV_m}{dt} \propto -V_B.$$

This would obviously not be true for slower events, because a steady current flow would produce a steady extracellular potential.

The above analysis yields the important concept that *the extracellular field potential has a time course that is approximately equal to the transmembrane current.* Furthermore, for fast events the extracellular field potential is roughly equal to the first derivative of the transmembrane potential.

14.4.1 Typical extracellular recordings in an open field

Extracellular field potentials have probably been used in the hippocampus more than anywhere else to infer the activity of neurons and synapses. The orderly anatomical arrangement of the neurons in the hippocampus (figure 14.7) and the highly laminated pattern of inputs and outputs makes the hippocampus an almost perfect open field structure. Examples of extracellular field recordings from the CA1 region of the hippocampus are shown in figure 14.8. Stimulation of the Schaffer collaterals produces an EPSP in the dendrites and, if sufficiently strong, an action potential in the soma. The corresponding field potentials are illustrated in figure 14.8. The field potential in the dendrites corresponding to the EPSP is called the population EPSP (or pEPSP) while the potential corresponding to the action potential is called the population spike (or pSpike).

Remember from the previous section that the time course of the pEPSP is roughly the same as that of the synaptic *current*. (This should be obvious if one compares the time course of the pEPSP in figure 14.8 to that of intracellularly recorded EPSPs illustrated in chapter 13.) In figure 14.9, superimposed traces of pEPSPs resulting from increasing stimulus intensities are illustrated. At higher intensity the peak of the pEPSP is contaminated by a source due to the occurrence of a pSpike. If time-to-peak of the synaptic current is constant, then the peak of the synaptic current (and thus the pEPSP) will be roughly equal to the slope of the rising phase *times* the time-to-peak, or

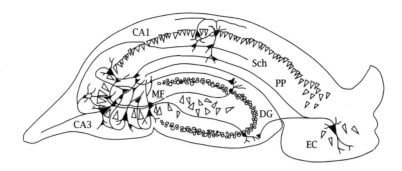

Figure 14.7 Diagram of hippocampal slice. Entorhinal cortex (EC), dentate gyrus (DG), and CA3 and CA1 subfields are labeled. The synaptic pathways—perforant path (PP), mossy fibers (MF), and Schaffer collaterals (Sch)—are also labeled.

peak of field EPSP = slope $\times$ time-to-peak

and thus for a constant time-to-peak

slope of field EPSP $\propto$ peak I_m.

This is a very useful relationship. Peak measures of field EPSPs are often contaminated with pSpikes, pIPSPs (or population IPSPs), and polysynaptic events. They would thus be poor reflections of synaptic current. A measure of the initial slope, however, can be made quite easily and, as the above analysis illustrates, will be roughly proportional to the peak current. Furthermore, because the slope is measured from the initial part of the field, the synaptic current that it represents will be dependent on the difference between E_s and the initial resting potential. This means that the slope is less sensitive to the effects of nonlinear summation (see also chapter 13) than that of the peak. In fact, one can show that the slope of the field pEPSP is linearly related to synaptic conductance whereas the relationship between peak current and synaptic conductance is quite nonlinear (see figure 14.10).

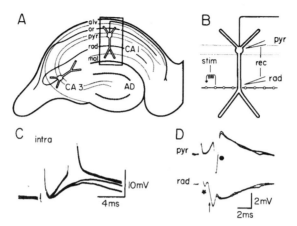

Figure 14.8 Field potentials in the hippocampus due to stimulation of Schaffer collaterals. (A) Diagram of hippocampal slice (alv = alveus; or = oriens; pyr = pyramidal cell layer; rad = radiatum; mol = molecular layer). (B) Enlarged view of a pyramidal neuron. Recordings are indicated from pyr and rad layers. Stimulation of Schaffer collaterals in rad is also indicated. (C) Intracellular recording at increasing stimulus intensity. At highest intensity an AP is elicited (truncated). (D) Field potentials in pyr and rad layers. • indicates pSpike; ↑ indicates pEPSP; and * indicates field potential associated with APs in presynaptic axons (called a fiber volley). Note the sink associated with the pEPSP recorded in the dendrites (rad) and the sink associated with the pSpike recorded near the soma (pyr). (From Langmoen and Andersen 1981.)

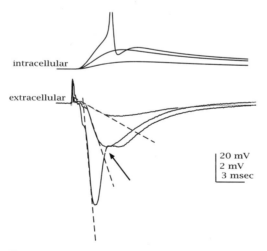

Figure 14.9 Measurement of the slope of pEPSP. Simultaneous intracellular and extracellular recordings. The extracellular recording is from rad (see figure 14.8), and the slopes of the pEPSPs are indicated for responses at the three different stimulus intensities. Note the positive going inflection in the largest pEPSP (arrow). This corresponds to the source in the dendrites associated with the AP in the soma.

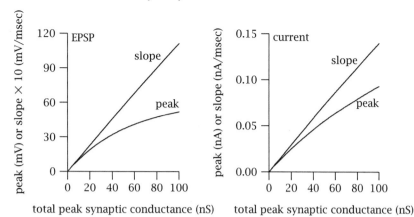

Figure 14.10 Plot of peak and slope of EPSP (left) and plot of peak and slope of EPSC (right) vs. synaptic conductance. Note that the slope is almost linearly related to the synaptic conductance whereas the peak voltage and current associated with the EPSP show pronounced nonlinear summation. (From simulations by Erik Cook.)

14.5 Current source-density analysis

The presentation in the previous sections illustrated the concepts underlying field potentials and how they can be used to get rough approximations of intracellular events. Additional methods of analysis are available, however, that allow more precise information regarding neuronal activity to be extracted from field potentials. Although field potentials, when measured at a single location, are useful for determining *changes* in intracellular events, they are poor at determining the location of actual current flow into and out of a neuron. The *current source-density analysis* of field potentials, however, allows one to determine the net extracellular current flow into and out of active neuronal tissue as a function of distance.

Recalling from chapter 4, the transmembrane current along a nerve fiber is equal to the second spatial derivative of membrane potential, or

$$i_m = K \frac{\partial^2 V_m}{\partial x^2}, \tag{14.5.2}$$

where K is a proportionality constant. The membrane current i_m is in units of current/area or current density.

Recall that for an action potential propagating along the nerve fiber (a uniform, nondecrementing wave, see chapter 6)

$$i_m = K \frac{\partial^2 V_m}{\partial x^2} = \frac{1}{\theta^2} \frac{\partial^2 V_m}{\partial t^2}. \tag{14.5.3}$$

For this propagating wave in one dimension, the extracellular field potential (which is proportional to i_m from equation 14.4.1) would be proportional to the second derivative of the action potential and would thus be the triphasic waveform, which we also determined on purely qualitative grounds in section 14.2.1.

If we are interested in the analyzing neural activity in a volume of tissue with current flowing in three dimensions, equation 14.5.3 must be expanded to

$$i_m = K \left[\frac{\partial^2 \phi}{\partial x^2} + \frac{\partial^2 \phi}{\partial y^2} + \frac{\partial^2 \phi}{\partial z^2} \right], \tag{14.5.4}$$

where ϕ is now the extracellular field potential. The beauty of the open field structure presented in the preceding sections is that it allows us to make an approximation to the above equation that greatly simplifies the analysis. If the volume of tissue is laminated, consisting of two-dimensional planes of uniform anatomy (for example, a layer of somata and several layers of uniform synaptic inputs), we can assume that the change in potential within each of these planes or laminae is zero and

$$\frac{\partial^2 \phi}{\partial y^2} = \frac{\partial^2 \phi}{\partial z^2} = 0.$$

Equation 14.5.4 can then be reduced to one dimension, where the x dimension is along the length of the neuron from soma to dendrites. This equation states that the membrane current (as a function of time) is equal to the second spatial derivative of the field potential (also as a function of time). Let's see how this can be determined.

Figure 14.11 illustrates a typical open field arrangement of neurons in which field potentials are measured at a number of locations (filled circles) along the x axis, each separated by a constant Δx. The second spatial derivative of the field potentials can be determined from these measurements using standard finite difference formulas. Take measurements at point V_0, V_a, and V_b; they are related to each other by

$$V_0 = V(x, t),$$
$$V_a = V(x - \Delta x, t),$$

and

$$V_b = V(x + \Delta x, t).$$

The first spatial derivative can be obtained by

$$\frac{\partial V(t)}{\partial x} = \frac{V_b - V_a}{2\Delta x},$$

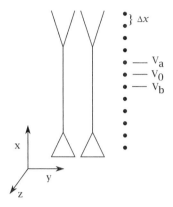

Figure 14.11 Diagram for current source-density analysis. The dots represent points for measuring the extracellular fields, and Δx represents the distance between them.

and the second spatial derivative is obtained by

$$
\begin{aligned}
\frac{\partial^2 V(t)}{\partial x^2} &\simeq \frac{\frac{V_b - V_0}{\Delta x} - \frac{V_0 - V_a}{\Delta x}}{\Delta x} \\
&= \frac{V_b - V_0 - V_0 + V_a}{\Delta x^2} \\
&= \frac{V_b + V_a - 2V_0}{\Delta x^2} = i_m(t),
\end{aligned}
\tag{14.5.5}
$$

where $i_m(t)$ is called the current source-density (CSD) function. The above analysis assumes constant conductivity of medium.

The relationship between the field potentials measured in the hippocampus and the corresponding CSD function is illustrated in figure 14.12. Note that the location of synaptic input would be difficult to localize from the field potentials in figure 14.12A. From the CSD function in figure 14.12B, however, the location of maximum synaptic current can be easily identified by the first maximum inward current density (denoted by the ∗ in radiatum). Similarly for the action potential, the maximum current sink can be easily localized from the CSD function (denoted by the ∗ in the cell body layer). The CSD function thus provides considerable information regarding the location of sources and sinks (current flow in and out of the neuron) and is quite useful for locating the sites of synaptic inputs in dendrites or the sites of action potential initiation.

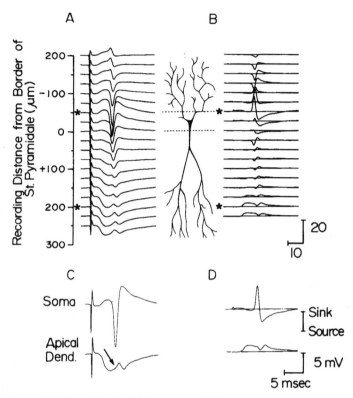

Figure 14.12 Relationship between field potentials and CSD in hippocampus (from Richardson et al. 1987). (A) Field potentials at different distances from the cell body layer (which is at 0 μm) in response to stimulation of the Schaffer collaterals. (B) CSD function calculated from the field potentials in (A) as described in the text. (C) Field potentials at two locations (the *'s in (A)). The arrow indicates the positive going pSpike in radiatum. (D) CSD at same two locations as in (C).

14.6   Summary of important concepts

1. Sources and sinks.

2. Analysis of extracellular current flow.

3. Relationships between extracellular fields and intracellular potentials.

4. Current source-density analysis.

14.7 Homework problems

1. Assume that you are recording extracellularly in the hippocampus:

   (a) If your recording electrode is in the cell body layer, would you measure a sink or a source during the arrival of an antidromically activated action potential? Would it be active or passive? What would be the potential with respect to a distant ground electrode when the action potential arrives?

   (b) If your electrode is in the dendrites near the site of the synaptic inputs and you stimulate these inputs, what is the shape of the field potential approximately equal to? Why would you want to measure the slope of the field potential? What is the slope proportional to?

   (c) Briefly (and qualitatively) describe how and why you would do a current source-density analysis?

2. (a) Draw the expected field potentials at sites A and B for a fast GABAergic IPSP to the soma. Label each site as a sink or source, active or passive. Show the expected relationship between the field potential at the soma and the intracellular potential measured with a microelectrode.

   (b) Draw the expected field potentials at A and B for a conductance decrease EPSP at the soma.

15   Cellular Neurophysiology of Learning and Memory

15.1   Introduction

In this chapter we will introduce some of the general concepts and basic principles in the field of synaptic plasticity and cellular mechanisms of learning and memory. This is an exciting and rapidly changing area of neuroscience, but this chapter is not intended to be a review of the field. Instead, the concepts and principles that we will cover will be those that are well established and which form the underpinnings of the field. We will not cover much in the area of invertebrate learning, but will use examples mostly from the mammalian literature, particularly the hippocampus. As there is much current interest in trying to use the results from cellular studies of synaptic plasticity to understand behavioral learning, we will also discuss briefly some of the data linking the two. The chapter will end with a description of how, using certain rules for changes in synaptic strengths, a simple neural network can store information.

15.1.1   Spine shape changes as a substrate for synaptic plasticity

We begin this chapter by repeating the idea proposed by Rall and others many years ago that changes in the shape of a spine (in particular, an increase in the diameter of the neck of a spine) might form the basis for synaptic plasticity and learning (see also chapters 4 and 13). In this section we will use the theories developed in chapters 4 and 13 to determine whether changes in the shape of a spine are likely to produce changes in a synaptic input. The principles derive directly from those discussed in chapter 13. Whether or not changes in spine shape can affect the synaptic response depends on the relative magnitudes of G_s and $1/R_{sp}$. We will first examine several different values for these parameters and then consider which combinations are the most realistic.

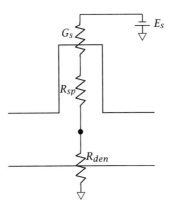

Figure 15.1   Diagram and circuit model for a synapse at the head of a spine attached to a dendrite. R_{den} represents the total input resistance of the neuron as measured at the base of the spine. This model is only valid for the steady-state analysis of spines, because the capacitance of the dendrites is not included.

From the simplified model of a spine shown in figure 15.1, the total resistance (R_T) of the spine with an active synapse on the head is given by

$$R_T = R_{sp} + 1/G_s,$$

while the total conductance is

$$\frac{1}{G_T} = \frac{1}{G_{sp}} + \frac{1}{G_s}, \text{ or,}$$

$$G_T = \frac{G_{sp}G_s}{G_{sp} + G_s}. \tag{15.1.1}$$

From equation 15.1.1 it can be seen that if G_{sp} is small relative to G_s (e.g., a large R_{sp} might be associated with a long, skinny spine), then the effectiveness of the synapse is negligible. That is, there is little or no synaptic current to measure. On the other hand, if G_{sp} is large relative to G_s, then changes in G_{sp} as a function of learning would be relatively ineffective for altering the efficacy of the synapse. These principles will be illustrated more clearly in the examples below.

15.1.1.1   Examples for spine shape changes and synaptic plasticity

1. First, assume that the synaptic conductance and the spine conductance are matched, or $G_s = 1$ nS and $G_{sp} = 1$ nS. During synaptic activity the total conductance will be

   $$G_T = 0.5 \text{ nS.}$$

If during learning R_{sp} is cut in half, or $G_{sp} = 2$ nS (as would be expected if the diameter of the spine neck increased), then

$$G_T = 0.67 \text{ nS},$$

and the measured synaptic conductance is increased by 34%, a significant increase in the synaptic response.

2. Second, assume that the synaptic conductance is much smaller than the spine conductance (i.e., large neck diameter), or $G_s = 1$ nS and $G_{sp} = 10$ nS. Thus during synaptic activity

$$G_T = \frac{10}{11} = 0.9 \text{ nS}.$$

If during learning R_{sp} is cut in half, or $G_{sp} = 20$ nS, then

$$G_T = \frac{20}{21} = 0.95 \text{ nS},$$

and the measured synaptic conductance is increased by only 5.6%. In this example changes in the shape of the spine have relatively little effect on the synaptic response. Achieving a larger synaptic response requires an increase in G_s.

3. Third, assume that the spine conductance is much smaller than the synaptic conductance (i.e., long, skinny spine), or $G_s = 1$ nS and $G_{sp} = 0.1$ nS. Thus during synaptic activity

$$G_T = 0.091 \text{ nS}.$$

If during learning the synaptic conductance increases so that now $G_s = 2$ nS while G_{sp} remains the same, then

$$G_T = 0.095 \text{ nS},$$

and the measured synaptic conductance is increased by only 4.8%. In this example changes in the synaptic conductance have relatively little effect on the synaptic response because the spine is restricting so much of the current flow to begin with. In order for a significantly larger synaptic response to occur, there must be an increase in G_{sp}.

So which of the above examples (two of which are extreme cases) is most likely to reflect physiological behavior? Based on anatomical reconstructions of variously shaped spines, example #2 is most likely to reflect the case for spines in the CNS. In other words, using any of a wide range of possible membrane properties for spines and their synaptic conductances, the conclusion is that $G_{sp} \gg G_s$. If this is true, then it is unlikely that changes in the shapes of spines are the substrate for learning—at least not in the purely electrical sense as envisioned by Rall. Changes in the shapes of spines may still be associated with synaptic plasticity or with development, but most likely for other reasons.

15.1.2 Summary of the possible effects of dendritic spines (refer also to chapter 13)

1. There is very little attenuation of potential from dendrite to spine head.

2. There may be attenuation of potential from spine head to dendrite.

3. A spine may increase the amplitude of the local synaptic potential at the subsynaptic membrane. This could be important if there are voltage-gated or voltage-sensitive channels in spines.

4. There is very little loss of synaptic charge from spine head to dendrite, but there may be a small difference in the amount of injected charge for a synapse on a spine vs. a synapse on a dendrite.

5. It is unlikely that changes in the diameter of the spine neck alter the amplitude of a synaptic response.

6. Spines may provide chemical isolation from the dendrite.

15.2   Long-term synaptic plasticity

The two most prominent forms of long-term synaptic plasticity are long-term potentiation (first introduced in chapter 11) and long-term depression. Each will be discussed in turn in the following sections.

15.2.1   Long-term potentiation

Long-term synaptic potentiation, or LTP, is a long-lasting increase in the amplitude of a synaptic response following brief, high-frequency activity of a synapse. LTP was first described in hippocampus by Bliss and Lømo (see Bliss and Lynch 1988) and has subsequently been observed at most excitatory synapses in the CNS. LTP represents a set of mechanisms (some of which may be present at all chemical synapses) responsible for increasing and maintaining the strength of a synaptic connection. LTP was discussed briefly in chapter 11 with respect to the use of methods of quantal analysis for exploring pre- vs. postsynaptic changes during LTP. In this chapter we will describe some of the properties of LTP and why it is an attractive cellular mechanism for certain aspects of memory.

15.2.1.1 Definitions LTP is loosely defined as an enduring, activity-dependent increase in synaptic efficacy. Short-term forms of plasticity such as facilitation and post-tetanic potentiation are considered a different class of phenomena because they last on the order of milliseconds to minutes, compared with LTP, which persists on the order of an hour or more. There may be multiple mechanisms associated with LTP—for example, mechanisms responsible for the early vs. the later parts of the potentiation—but for the most part all are currently considered under the general label of LTP. The very early part of potentiation, lasting from a few minutes up to ~30 min, has been given its own label of *short-term potentiation* or STP,[1] because it can be mechanistically separated from LTP. No doubt more such separations will be made in the future as the different mechanisms involved in the phenomena of what is now called LTP become known.[2]

For ease of discussion, it is useful to separate LTP into two phases: the *induction* and *maintenance* phases. The induction phase is that which occurs during and shortly after the high-frequency stimulation used to initiate LTP. It consists of all the steps and mechanisms that lead to the long-lasting changes that are associated with LTP. The maintenance phase is that period following the induction in which the change in synaptic efficacy is *expressed.* The onset kinetics of LTP are not known reliably because of the difficulty in separating LTP from STP and PTP, but all indications are that for certain forms of LTP the expression begins within minutes or less of the induction phase. Some forms of LTP, however, may take up to 30 min or more to be expressed. The duration of LTP is also not known, although current data suggest that LTP is not permanent but decays over a time course of days to weeks. It is also worth noting that the mecha-

---

[1]STP was also called at one point *decremental* LTP.

[2]A different terminology for LTP has been fostered by B. L. McNaughton and colleagues. They prefer to use the term *long-term enhancement* or LTE to distinguish it clearly from post-tetanic potentiation or PTP. Although this distinction makes sense, few people have adopted it and so the most common term in the literature is LTP. LTP and LTE are just different acronyms for the same set of phenomena.

nisms responsible for the early expression of LTP may not be the same as those at longer times following induction and that the characteristics, as well as the mechanisms, of LTP may be different at different synapses.

15.2.1.2 Induction paradigms   As presented in chapter 11, LTP is typically induced by a short train or burst of stimulation to a set of afferent fibers. In the hippocampus this may consist of a 1 sec train of 100 Hz stimulation repeated several times or some variation thereof (e.g., 50 Hz for about 20 sec or up to 200 Hz for about 0.5 sec). Other stimulation paradigms that have been used to try to better mimic physiological firing patterns consist of shorter trains of high-frequency stimulation repeated several times. A popular paradigm is a 100 Hz train for 50 msec (5 pulses) repeated 8–10 times in succession at intervals of 200 msec. Because hippocampal neurons in vivo frequently fire in short bursts of high-frequency activity and because 200 msec is the approximate period of the theta rhythm in the EEG, this stimulus paradigm is quite attractive as an input pattern that the neuron may actually receive during physiological activity. Interestingly enough, it is also quite effective in eliciting LTP in hippocampal neurons. As we will see in section 15.2.1.7, some forms of LTP can also be induced by pairing depolarization of the post-synaptic neuron with single stimuli to the presynaptic fibers. Whatever the stimulus paradigm used, however, one measures LTP by recording the amplitude of the synaptic response before and after the period of high-frequency stimulation. This is illustrated in figure 15.2.

15.2.1.3 EPSP vs. E-S potentiation   Most studies of LTP in hippocampus have been done using field potential recordings (see chapter 14). Measurements of the pEPSP (actually the slope of the pEPSP, for the reasons explained in chapter 14) and the pSpike are made before and after high-frequency stimulation. An observation made during some of the earliest studies of LTP was that the amplitude of the pSpike appeared to increase by a greater amount than did that of the pEPSP. In fact, in many experiments there was LTP of the pSpike without any change in the pEPSP. This was curious because the EPSP is what triggers the action potential in each neuron. An increase in the pSpike should reflect an increase in the number of neurons firing in response to increases in the EPSPs in the individual neurons. Because in some experiments the pSpike appeared to increase by an amount more than that expected from the increase in the pEPSP, it was felt that some process in addition to LTP of the EPSP must be taking place, and this process was called *EPSP-to-spike potentiation* or E-S potentiation.

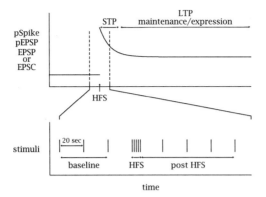

Figure 15.2 Typical induction paradigm for LTP and time course for different phases of potentiation. HFS (high-frequency stimulation) is used to represent any of a number of induction paradigms mentioned in the text and is not drawn to scale. After HFS the amplitude of the pSpike, pEPSP, EPSP, or EPSC typically increases. The initial increase can be due to a combination of PTP, STP, or LTP. The duration of PTP is about 1 min, STP about 30 min, and LTP hours or longer (see text for further explanation).

If one plots the amplitude (or slope) of the pEPSP vs. stimulus intensity, one obtains a monotonic, roughly sigmoidal-shaped, curve called an input-output or I/O curve. The shape of the I/O curve basically reflects the number of fibers activated at the different stimulus intensities and the pEPSPs that result from these activated fibers.[3] During LTP, the I/O curve shifts to the left along the stimulus intensity axis, suggesting that at a given intensity, and with the same number of fibers activated, the resulting pEPSP is larger in amplitude (see figure 15.3). One can obtain a similar relationship if one plots pSpike vs. stimulus intensity before and during LTP (dotted line in figure 15.3).

One can also plot the pSpike vs. pEPSP. This is a kind of I/O curve for the pSpike in that it reflects the relationship between the amplitude of the pEPSP, which drives the pSpike, and that of the pSpike itself. If during LTP the larger amplitude pEPSP was all that accounted for the larger pSpike, then one would predict that there would be no change in the pSpike vs. pEPSP curve. That is, the larger pEPSP produced by activating the same number of fibers would simply elicit a larger pSpike. This is not the case, however, as there is also an upward shift in this curve during LTP, suggesting that for any given amplitude pEPSP there are more neurons firing. This is what is called E-S potentiation. The time course of E-S potentiation

---

[3]It is important to remember that a larger pEPSP resulting from a higher stimulus intensity is due to a larger number of activated fibers (i.e., spatial summation of their individual EPSPs), and *not* to larger EPSPs at individual synapses.

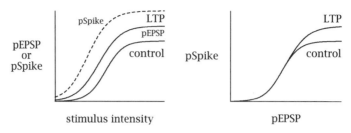

Figure 15.3   Hypothetical input-output (I/O) curves for pEPSPs and pSpikes before and during LTP (left). The control curve is the same for both the pEPSP and the pSpike. Examples of I/O cuves to illustrate the concept of E-S potentiation are shown at the right.

is similar to that for LTP of the pEPSP, but the mechanisms underlying E-S potentiation are not entirely clear. One possibility is that changes in the excitability or threshold of the postsynaptic neurons may be partly involved so that for a given EPSP a neuron is more likely to fire. One interesting finding is that E-S potentiation may sometimes be restricted to the inputs being stimulated (i.e., synapse specificity, discussed in section 15.2.1.5) suggesting that if there are changes in excitability, they may be localized to limited regions of the dendrites.

15.2.1.4   Neuromodulation   As would be expected, LTP can be modulated by a number of neurotransmitter systems. These include norepinephrine, acetylcholine, serotonin, dopamine, adenosine, and many of the neuroactive peptides. One of the interesting features of the modulation is that it appears to be different at different synapses. For example, in the hippocampus norepinephrine enhances LTP at the perforant path synapses in the absence of high-frequency stimulation, enhances LTP at mossy fiber synapses with high-frequency stimulation, and has no effect on LTP at Schaffer collateral synapses (see chapter 14 for description of synaptic pathways in hippocampus). Another interesting example is acetylcholine (ACh). ACh enhances LTP at Schaffer collateral synapses and inhibits LTP at mossy fiber synapses. Combinations of these neurotransmitters may also have important effects on LTP, as has been suggested for norepinephrine and ACh in visual cortex.

The detailed mechanisms by which LTP is modulated by these neurotransmitters are not known. Given the complex actions of these neurotransmitter systems, however, their effects on LTP induction and maintenance are likely to involve multiple second-messenger systems.

15.2.1.5   Synapse specificity and homosynaptic vs. heterosynaptic LTP
One of the important features of LTP is that the enhancement in the am-

plitude of the EPSP is, for the most part, confined to the stimulated pathway and is thus input specific or *synapse specific*. In other words, the fibers given high-frequency stimulation are the only inputs to a neuron that show potentiation (illustrated in figure 15.4). This result implies that a generalized change in the postsynaptic neuron is not responsible for LTP, because that would produce potentiation at most if not all synaptic inputs. Using another term to describe this feature, LTP is generally considered to be *homosynaptic*.

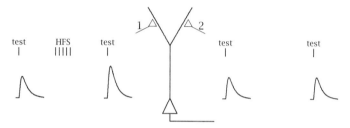

Figure 15.4 Input specificity for LTP. Both synapses 1 and 2 are tested before and after HFS, but only synapse 1 receives HFS and only synapse 1 displays LTP.

Heterosynaptic LTP describes a condition in which LTP induced by stimulation in one pathway results in LTP being induced in another, nonstimulated, pathway. For example, LTP induced at the mossy fibers in CA3 of the hippocampus could induce LTP at commissural fibers under certain conditions. In most cases, though, LTP is homosynaptic.

As mentioned previously, E-S potentiation can show input specificity, but it is likely to be less restricted spatially than LTP of the EPSP. For example, E-S potentiation may occur for all inputs to a limited region of the dendrites whether or not they are stimulated during the induction phase, while synapses separated by some distance on the dendrites (for example, apical vs. basilar) may show synapse specificity of E-S potentiation. This would suggest that if there are generalized changes in the postsynaptic neuron associated with E-S potentiation, then they are confined to certain regions of the neuron.

15.2.1.6 Cooperativity A stimulus intensity threshold usually exists for the induction of LTP. That is, in the absence of other manipulations a minimum number of input fibers must be activated to induce LTP. McNaughton and his colleagues concluded from this finding that there is a requirement for some type of *cooperative* interaction among afferent fibers. What that cooperative action is will become clearer in later sections when induction mechanisms are discussed. A plot of LTP magnitude vs. stimulus intensity (figure 15.5) reveals a curve that begins at some nonzero value of stimulus intensity. This requirement for the stimulus intensity to exceed some threshold level for LTP induction has been called *cooperativity*.

In addition to a threshold for LTP induction, the magnitude of LTP also shows saturation. At extremely high stimulus intensities, or after repeatedly inducing LTP at lower stimulus intensities, a maximum amount of potentiation is achieved such that further episodes of high-frequency stimulation fail to elicit more LTP. This is called LTP *saturation*.

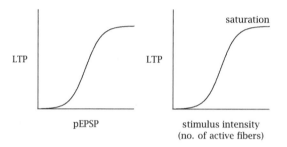

Figure 15.5 LTP vs. stimulus intensity. The amount of LTP as a function of the amplitude of the pEPSP used during HFS is shown on the left. The fact that there is a stimulus threshold for the induction of LTP has been called *cooperativity*. The amount of LTP will also *saturate* (right) at very high stimulus intensities or with repeated episodes of HFS at lower stimulus intensities.

15.2.1.7 Associativity If LTP represents a cellular mechanism for learning and memory, then one might expect that there would be a cellular analog to Pavlovian or classical conditioning involving LTP. Levy and Steward were among the first to use an induction paradigm for LTP analogous to that for classical conditioning (see Bliss and Lynch 1988). They stimulated two separate input pathways to the same set of postsynaptic neurons (i.e., the contralateral and ipsilateral perforant pathways to the dentate gyrus studied in vivo) and demonstrated that LTP could be induced in the weaker contralateral pathway only if its stimulation was paired with stimulation of the stronger ipsilateral pathway. The weak pathway was considered analogous to the conditioned stimulus (CS) and the strong pathway to the unconditioned stimulus (UCS) in a typical classical conditioning experiment (refer to section 15.3.2). A similar experiment involving two different sets of Schaffer collateral fibers was later performed in vitro by Barrionuevo and Brown. The LTP induced in this type of experiment was called associative LTP, and the general phenomenon is called *associativity* (see figure 15.6).

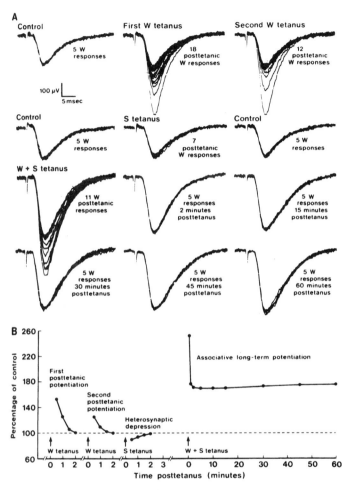

Figure 15.6 Associative LTP. S represents a strong (high-stimulus intensity) input and W a weak intensity input from the Schaffer collaterals. The corresponding pEPSPs measured in stratum radiatum of CA1 are given in (A). In the upper row of traces, 5 W responses are superimposed on the left followed by 18 W responses after a tetanus (HFS) was given to W (middle). A second, identical tetanus to W again elicits only PTP (right). In the second row of traces, 5 W responses (left) are followed by 7 W responses after a tetanus was given to S (middle). Note that there is no potentiation of W, only a small depression. 5 W responses are superimposed again after returning to control levels (right). In the third row of traces, 11 W responses are superimposed (left) after a tetanus was given simultaneously to S and W. 5 W responses 2 min (middle) and 15 min (right) after the tetanus indicate that LTP in W was induced. In the bottom row of traces, 5 W responses are shown at 30 (left), 45 (middle), and 60 min (right) after tetanus to indicate the duration of LTP. The amplitudes of the pEPSPs as a function of time for all of the above traces are plotted in (B). Note that LTP in W was only induced when W received a tetanus at the same time as S. (From Barrionuevo and Brown 1983.)

It was also found that LTP was not induced if stimulation of the strong pathway preceded that of the weak pathway. Stimulation of the weak pathway could precede stimulation of the strong, but only by a fairly short period of time. The *temporal contiguity* requirements are therefore fairly strict, and the maximum LTP is induced when stimulation of the weak and strong pathways overlaps at least partly in time (see figure 15.7).

A distinction should be drawn between cooperativity and associativity. Cooperativity implies that there is a minimum number of input fibers that must be activated for the induction of LTP *in those fibers.* Associativity, on the other hand, implies that LTP can be induced in a limited set of input fibers only when they are stimulated in conjunction with another, larger set of input fibers. Although associativity implies that cooperativity also exists, the converse is not true: Cooperativity does not necessarily imply associativity.

This concept can be illustrated in the following hypothetical example (figure 15.8). In part A the I/O curves for this fictitious experiment are shown before and after high-frequency stimulation (HFS) and the induction of LTP. HFS was given at the intensity indicated (I_1), and a larger-amplitude pEPSP would be observed following HFS when tested at this same stimulus intensity. Suppose, however, that a lower stimulus intensity (I_2) was used to test for LTP (after HFS was given at the higher stimulus intensity, I_1). In this case one would conclude that LTP had not been induced, because the pEPSP amplitude would be the same as before HFS. This is an example of an experiment in which cooperativity might exist (there could be a stimulus intensity threshold for the induction of LTP) but not associativity. LTP was not expressed in the smaller subset of fibers even though they were stimulated in conjunction with the rest of the fibers during HFS. The result of this fictitious experiment implies that, in addition to a requirement that a minimum number of fibers be stimulated for the induction of LTP, there could also be a requirement that a minimum number of fibers be activated for the expression of the LTP. Therefore, at least in theory, the presence of cooperativity does not necessitate the presence of associativity.

An experiment in which both cooperativity and associativity exist is illustrated in figure 15.8B. In this experiment LTP would be observed after HFS at virtually any intensity of stimulation throughout the I/O curve. The extreme case would be the observation of LTP in a single input fiber. For example, according to the scheme shown in figure 15.8A, if one tested for the expression of LTP at a *single* input fiber after inducing LTP by HFS at

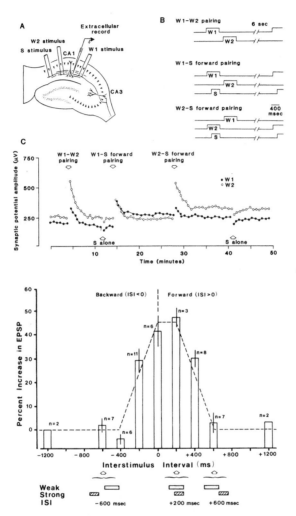

Figure 15.7 Temporal contiguity requirements for associative LTP. The locations of three stimulating electrodes and one recording electrode are shown in (A). S represents a strong stimulus-intensity input, and W1 and W2 are weak inputs. The timing of the three pairing protocols are indicated in (B). The square pulse represents a 100-Hz train of that duration. The results of the three stimulus patterns are shown in (C). Note that only the forward pairing patterns produce LTP. (From Kelso and Brown 1986.) The plot at the bottom illustrates another experiment in which LTP is plotted as a function of the time between the *onset* of the train given to W and the *onset* of the train given to S (i.e., interstimulus interval or ISI). (From Brown et al. 1989.)

a high stimulus intensity, one would conclude that no LTP had been induced. On the other hand, in the scheme shown in figure 15.8B, LTP would be measured in even a single fiber. Some forms of LTP indeed appear to be expressed in single input pathways. Whether or not LTP is expressed in single fibers, such as illustrated by the hypothetical experiments of figure 15.8, would suggest different underlying mechanisms for the LTP.

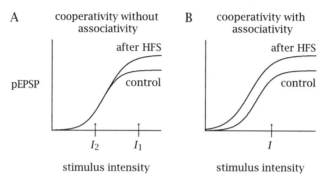

Figure 15.8   Hypothetical I/O curves for experiments in which LTP is cooperative but not associative (A) and both cooperative and associative (B) (see text for explanation).

15.2.1.8   The Hebb rule for synaptic plasticity   In 1949 Donald Hebb proposed what has proven to be an extremely important idea for the study of learning. He suggested that the efficacy or "efficiency" of a synapse would increase when there was concurrent activity in the pre- and postsynaptic elements. Specifically, he suggested that (Hebb, 1949, p. 62)

> When an axon of cell A is near enough to excite cell B or repeatedly or persistently takes part in firing it, some growth process or metabolic change takes place in one or both cells such that A's efficiency, as one of the cells firing B, is increased.

and that these changes in synaptic efficacy could form the basis of learning. This deceptively simple idea has far-reaching implications for our understanding of the cellular mechanisms of learning. It has been quantified and put into the form of a learning "rule" and then incorporated into numerous computer models for learning and memory. It has also guided numerous investigations into the mechanisms underlying synaptic plasticity.

A relatively simple quantitation of the Hebb rule would be the following. Suppose we let synaptic efficacy be defined as a synaptic *weight* or W_{AB}, where W_{AB} represents the weight or strength of the synaptic connection from neuron A to neuron B, and let V_A and V_B be the frequency of firing of neurons A and B, respectively. By the Hebb rule, the change in synaptic weight (ΔW_{AB}) would be

$$\Delta W_{AB} = F(V_A, V_B),$$

where $F(V_A, V_B)$ is some function of the firing rates of A and B. One type of function could be simply the product of the individual firing rates, or

$$\Delta W_{AB} = k(V_A \cdot V_B), \tag{15.2.2}$$

where k is just a numerical constant whose value is small. There are many other possible functions for specifying changes in synaptic weight according to the Hebb rule, and each has important functional implications. The reader interested in further discussion of this topic should consult the suggested reading list for this chapter.

From an input-output point of view, there are several ways of implementing the Hebb learning rule. This was nicely put forth by Sejnowski and Tesauro (1989) and is illustrated in figure 15.9. In (a) there are just two neurons and one synapse. When neuron B is fired repeatedly by the activity of neuron A, some change takes place such that the strength of the synaptic connection is increased. In this example there must be some mechanism for sensing firing rate in neuron B before enhancing the synaptic weight. The simplest way for this to occur would be for the sensing of the association and the mechanism for the change in efficacy to be located in neuron B.

(a)

(b)

(c)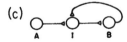

Figure 15.9   Three different implementations of the Hebb learning rule (see text for explanation). (From Sejnowski and Tesauro 1989.)

In (b), neuron B feeds back and makes a synaptic connection onto the synapse of neuron A. In this case the presynaptic element senses the firing rate of neuron B along with its own firing, and some change takes place to increase synaptic efficacy. The feedback from neuron B could be an actual synaptic connection or some type of retrograde message or signal (i.e., chemical or electrical).

A final scheme for implementing the Hebb rule is depicted in (c). In this case there is an interneuron (I) that senses the firing rates of neurons A and B. A change in the excitability of I would result in an overall increase in the efficiency of neuron A in firing neuron B, as specified by the Hebb rule. Although the cellular mechanisms would differ for each of these implementations, they are all functionally equivalent forms of the Hebb rule.

The first experimental support for the existence of a Hebb-like learning rule came more or less simultaneously from a number of groups working on induction mechanisms for LTP (reviewed in Brown et al. 1990). They found that if the postsynaptic neuron was hyperpolarized to prevent action potentials during the HFS, LTP was not induced, whereas if the neuron was depolarized during HFS, LTP was induced. In contrast, if the postsynaptic neuron was depolarized to fire action potentials in the absence of any synaptic stimulation, then LTP was not induced. There did not appear to be any specific requirement for action potentials, as proposed by Hebb, but there was a requirement for *both* depolarization in the postsynaptic neuron *and* firing of the presynaptic axons. Although this work was first done using Schaffer collateral synapses, Hebbian-type learning rules have also been described for LTP at perforant path and mossy fiber synapses in hippocampus. An experiment testing for a Hebbian rule is illustrated in figure 15.10.

From the results of such experiments, it was proposed that the strong pathway in the associative LTP experiments might simply be providing the postsynaptic depolarization that appeared necessary for the induction of LTP. An important experiment to test this idea was to pair postsynaptic depolarization with a single (weak) stimulus to the afferent fibers. At some synapses—for example, Schaffer collaterals and perforant path (but not mossy fibers)—LTP is induced with this paradigm (figure 15.11).

The cellular mechanisms underlying this Hebb rule, and also the mechanisms for associative LTP, will be discussed in the next section.

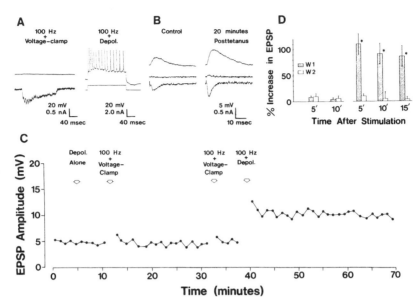

Figure 15.10 Experiment that tested the Hebbian rule for synaptic plasticity. (A) HFS (100 Hz) given under voltage clamp (left) and HFS given with a depolarizing current injection (right). (B) The EPSPs and EPSCs before and 20 min post tetanus are indicated. LTP was elicited when HFS was given in conjunction with postsynaptic depolarization. (C) EPSP amplitude is plotted as a function of time for the different stimulus protocols. (D) Summary data are shown from two pathways (W1 and W2) at different times following HFS given separately to each pathway. On the left is the mean increase in the EPSP when a voltage clamp was applied to the postsynaptic cell during HFS is indicated. On the right is the same experiment except that a depolarizing current pulse was given to the cell during the HFS to W1. (From Kelso et al. 1986.)

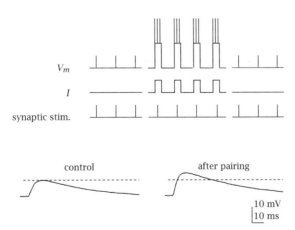

Figure 15.11 Pairing postsynaptic depolarization with a single weak stimulus (with the pairing repeated every 20–30 sec) induces LTP at some synapses.

15.2.1.9 Cellular mechanisms for long-term potentiation Currently, LTP can be broadly categorized into two forms, those that require the activation of NMDA receptors for their induction and those that do not. These have been called NMDA-receptor-dependent (NMDA) and NMDA-receptor-independent (non-NMDA) forms of LTP. There may be other distinct forms of LTP, for example, those that might require activation of metabotropic glutamate receptors or a particular second messenger system, but this is an active area of research and it will take time for these questions to be completely sorted out. There are also many controversial aspects to the mechanisms for both the induction and expression of LTP, and we will try and confine ourselves here to discussing what is most generally agreed upon by current researchers in the field.

NMDA-receptor–dependent vs. NMDA-receptor-independent LTP The common mechanism that seems to underlie all forms of LTP is that induction depends on a rise in intracellular Ca^{2+} in the postsynaptic neuron. The induction of both NMDA LTP and non-NMDA LTP can be blocked by the postsynaptic injection of Ca^{2+} chelators. Chelating Ca^{2+} *after* the induction phase, however, has no effect on expression. The rise in intracellular Ca^{2+} during the induction phase may occur from an influx through NMDA receptors, an influx through voltage-gated Ca^{2+} channels, or release from intracellular stores. The two most common mechanisms studied so far are those involving Ca^{2+} influx through NMDA receptors or through voltage-gated Ca^{2+} channels. Not surprisingly, Ca^{2+} influx through NMDA receptors is believed to be the initial trigger for NMDA LTP, while Ca^{2+} influx through voltage-gated Ca^{2+} channels may be the trigger for non-NMDA LTP. NMDA LTP has been studied extensively at the Schaffer collateral and perforant path synapses in the hippocampus but has also been observed in neocortical neurons and elsewhere. Non-NMDA LTP, on the other hand, has been studied less extensively but has been observed in some invertebrate preparations, in peripheral ganglia, in dorsolateral septal nucleus, in neocortex, in the amygdala, and in the hippocampus at Schaffer collateral, commissural/associational, perforant path, and mossy fiber synapses.

For NMDA LTP the general scheme for LTP induction involves depolarization of the postsynaptic neuron to relieve the Mg^{2+} block of NMDA receptors and influx of Ca^{2+}. The requirement for postsynaptic activity results from the need for sufficient depolarization to unblock the NMDA receptors, whereas that for presynaptic activity results from the need to have glutamate released from the presynaptic fiber to activate the NMDA (and non-NMDA) receptors. It is assumed that non-NMDA LTP can occur at many of these same synapses when there is sufficient depolarization to allow Ca^{2+} influx through voltage-gated Ca^{2+} channels (see figure 15.12). There also is a requirement for presynaptic activity under these conditions, but what that requirement might be is not entirely clear. Under certain conditions both NMDA and non-NMDA LTP can show input specificity, cooperativity, associativity, and Hebbian rules for induction.

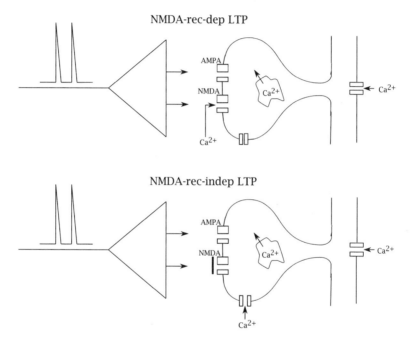

Figure 15.12 Induction mechanisms for NMDA and non-NMDA LTP. For NMDA-rec-dep LTP, Ca^{2+} entry through NMDA receptors during the HFS is believed to initiate LTP. For NMDA-rec-indep LTP, Ca^{2+} may enter through voltage-gated Ca^{2+} channels to initiate LTP. Ca^{2+} may also be released from intracellular stores.

There are many proposed biochemical mechanisms for the induction and expression of LTP. Most prominent among them is that there is a persistent activation of one or more protein kinases associated with the expression of LTP. Such a mechanism could lead to the persistent phosphorylation of one or more proteins that are involved in synaptic transmission or in initiating gene expression. It is worth noting that many of the mechanisms associated with LTP may also be important during development whereby new synapses are made and existing synapses are either strengthened or weakened. Also suggested are changes in second messenger systems during the induction phase that may lead to longer-term changes during maintenance. A considerable amount of work is needed in this area before a coherent picture can be drawn for the molecular mechanisms of LTP induction and expression. As mentioned previously, it is likely that different molecular mechanisms may be involved during different phases of LTP.

Another important question is whether the changes associated with the expression of LTP are located at the pre- or postsynaptic side of the synapse. This was also discussed in chapter 11. Various methods of quantal analysis have been used to test for changes in transmitter release or for changes in postsynaptic responsiveness to transmitter. Given the difficulty of performing a proper quantal analysis in a cortical neuron, it is not surprising that different groups have reached somewhat different conclusions. The bulk of the evidence, however, appears to favor increases in transmitter release accounting for at least part of the increase in synaptic efficacy during LTP. These conclusions are based on work in *Aplysia*, crayfish, and hippocampus. In hippocampus it has also been suggested that there are changes in both quantal content (m) and quantal size (q). Given that both the pre- and postsynaptic structural elements appear to develop in concert (Lisman and Harris 1993), it would again not be surprising if changes occur on both sides of the synapse during LTP.

In chapter 11 we discussed facilitation of transmitter release during paired stimuli and reviewed the evidence that facilitation resulted from changes in m. One method of testing for pre- vs. postsynaptic changes during LTP is to test for changes in paired-pulse facilitation (PPF) during LTP. The assumption would be that if both facilitation and LTP result from changes in transmitter release, m, then they might interfere with one another and lead to decreases in PPF during the expression of LTP. For example, the number of quanta released can be described by the equation $m = np$, where n is the number of release sites and p is the average probability of release at each site (chapter 11). Presumably, PPF results from a transient increase in p following the first stimulus. If LTP is also due to an increase in p, then one would predict a decrease in PPF during LTP. In other words, as $p \rightarrow 1$ during LTP, there would be no further increase in p during PPF. It is also possible, however, that LTP could result from increases in n. Increases in n with LTP and increases in p with PPF would cause multiplicative increases in m and not be expected to interfere with one another. Changes in PPF with LTP would therefore be suggestive of some type of presynaptic mechanisms for LTP, but the lack of change in PPF would have no similar predictive value.

PPF has been measured many times before and during LTP, and, on average, there appears to be no change. In some experiments, however, there are decreases while in others there are increases during LTP, and the direction of this change appears to depend on the initial magnitude of PPF. The results therefore do not support a simple scheme whereby LTP results in increases in p, but instead a more complicated set of mechanisms with perhaps changes in both n and p so that the changes in n are accompanied by different values of p. Nevertheless, these experiments as well as the results of quantal analysis (chapter 11) support the hypothesis that at least part of the changes associated with the expression of LTP occurs presynaptically. The results of quantal analysis, however, also support a partial postsynaptic mechanism and thus again changes on both sides of the synapse are most likely involved in the expression of LTP.

If Ca^{2+} influx into the postsynaptic neuron is an initial trigger for LTP and if at least part of the change in synaptic efficacy during LTP is due to increases in transmitter release, then it follows that there must be some signal or message relayed from the postsynaptic neuron to the presynaptic terminal for the expression of LTP. This has been suggested to be some type of retrograde messenger or chemical signal that is released from the postsynaptic neuron following Ca^{2+} influx, diffuses to the presynaptic terminal, and initiates some biochemical change leading to an increase in transmitter release. Nitric oxide, arachidonic acid, and carbon monoxide are among the proposed candidates for such a retrograde messenger for which some evidence exists. Retrograde communication could also take place at the active zone through proteins that span the cleft from pre- to postsynaptic sides of the synapse. Regardless of the mechanism, what seems clear is that in addition to some type of retrograde signal, there must also be presynaptic activity. The retrograde signal is probably not sufficient to confer the synapse specificity discussed in previous sections for both NMDA and non-NMDA LTP.

15.2.2 Long-term depression

Most of the above discussion has focused on LTP because more is known about LTP than other forms of plasticity. Another type of long-term plasticity that has been observed, however, is long-term depression, or LTD. LTD has been described in the cerebellum, where it may play a role in motor learning, in neocortex, and in hippocampus. Under certain condi-

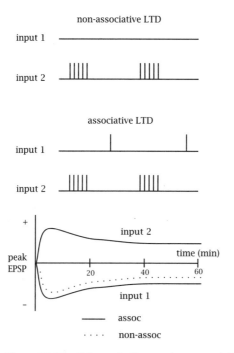

Figure 15.13   Schematic diagram (not to scale) for the induction of associative and nonassociative LTD. In nonassociative LTD, trains of stimuli (input 2) lead to a depression in the nonstimulated input (1). In associative LTD, stimuli to input 1, which are out of phase with the trains in input 2, also lead to a depression of the synaptic response. The plot at the bottom shows the change in EPSP amplitude (for single stimuli) seen under the indicated conditions.

tions stimulation of afferent fibers can lead to an enduring depression of synaptic efficacy (see figure 15.13). LTD, at least in the hippocampus, can have either associative or nonassociative properties, and can be either homosynaptic or heterosynaptic, depending on the conditions. There is also good evidence that a rise in postsynaptic Ca^{2+} is again an initial trigger for LTD. One possibility is that smaller increases in Ca^{2+} lead to LTD while larger increases lead to LTP.

Some examples of LTD have been called "anti-Hebb" because they are induced under non-Hebbian conditions. For example, such non-Hebbian conditions might exist when there is activity in the presynaptic fibers but not in the postsynaptic neuron, or when there is activity in the postsynaptic neuron but not in the presynaptic fibers. It has long been believed that for learning to occur there must be some type of LTD-like mechanism for reducing synaptic strength. Otherwise (if there was nothing but LTP), there would be a continual increase in synaptic strength until saturation occurred. At the point of saturation, no further learning would occur. Moreover, on theoretical grounds it has been demonstrated that more information can be stored in a system in which there are both increases and decreases in synaptic strength.

If correlated activity in the pre- and postsynaptic elements leads to increases in synaptic efficacy as discussed above, while uncorrelated activity leads to decreases in efficacy, then the simple algorithm for synaptic weight changes presented by equation 15.2.2 should be modified to account for both increases and decreases in synaptic efficacy. There have been many suggested algorithms, but the two that have received the most attention are the covariance rule of Sejnowski (1981, 1989) and the sliding modification threshold rule of Bienenstock et al. (1982). Which of these (or any of the other) algorithms best mimics real synapses remains to be determined, but this is currently an active area of research.

The covariance rule can be stated fairly simply. If the activity in the pre- and postsynaptic elements is correlated, synaptic weight is increased, while negatively correlated activity leads to decreases in synaptic weight. Uncorrelated activity has no effect on synaptic weight. This covariance rule can be expressed as follows:

$$\Delta W_{AB} = k[(V_A - \overline{V}_A)(V_B - \overline{V}_B)], \tag{15.2.3}$$

where ΔW_{AB} represents the change in synaptic weight, V_A and V_B represent the activities of the pre- and postsynaptic neurons, respectively, and $\overline{V}_A$ and $\overline{V}_B$ represent the average values of the activities or firing rates of neurons A and B. From equation 15.2.3, if the activities of A and B are correlated in time, then the change in weight will be positive, while if they are negatively correlated, the weight change will be negative. If there is no correlation, then there will be no change. The relationship between synaptic weight changes and activity for the covariance rule is illustrated in figure 15.14.

The sliding modification threshold rule or BCM rule (for Bienenstock, Cooper, and Munro 1982) can also be stated rather simply. This rule is similar in certain respects to the covariance rule in that correlated activity in the pre- and postsynaptic neurons leads to enhancement in efficacy, while negative correlation leads to depression. It differs, however, in several important respects. First, the relationship is nonlinear. Second, there can be a threshold for depression as well as for enhancement. And third, the threshold for LTP vs. LTD varies as a function of the average activity of the postsynaptic neuron. In other words, the higher the average

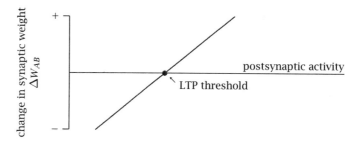

Figure 15.14   Plot of covariance rule. The abscissa represents the postsynaptic activity that is correlated with, or in response to, presynaptic activity.

background activity of the postsynaptic neuron, the higher the threshold for enhancing synaptic efficacy. Conversely, the lower the average background activity, the lower the threshold for an LTP-like process. This is illustrated in figure 15.15. The variation in threshold (θ_M) varies nonlinearly with postsynaptic activity, as given below.

$$\Delta W_{AB} = \phi(c_B, \overline{c}_B)d_B \tag{15.2.4}$$

and

$$\theta_M = \overline{c}^2 \tag{15.2.5}$$

where ϕ is the nonlinear function relating changes in synaptic weight resulting from the correlated activity of neurons A and B (c_B) and the average correlated activity ($\overline{c}_B$); d_B is a local variable for synapse *AB*; and θ_M is the modification threshold for LTP vs. LTD (see figure 15.15).

Because of the importance of postsynaptic Ca^{2+} in initiating LTP as well as LTD, one attractive idea is that if such a sliding modification threshold for LTP vs. LTD does exist in neurons (and there is some experimental support for it), the changing threshold may be related to changes in Ca^{2+} levels or to changes in a neuron's ability to buffer intracellular Ca^{2+}. There are certainly many other possible mechanisms including changes in the sensitivity of NMDA receptors.

15.3   Associative and nonassociative forms of learning

Any discussion of the cellular neurophysiology of learning and memory should include at least a brief review of some of the simple forms of learning that have been described at the behavioral level. It is for these forms of

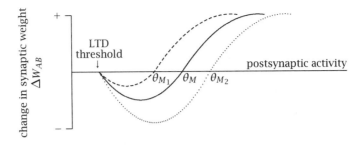

Figure 15.15 Sliding modification threshold rule for synaptic plasticity. The higher the average activity levels in the postsynaptic neuron, the more activity that is necessary to induce LTP (i.e., the larger the θ_M).

learning that the cellular analogs and mechanisms are being investigated. Certainly, the various forms of short- and long-term synaptic plasticity discussed in this book, as well as changes in many cellular processes, are likely to be involved at some level in these forms of learning. It is the challenge of neurophysiology to determine what the relationships are between neuronal events and behavior. Only brief definitions of a few of the more important simple forms of learning will be provided here, and the interested reader is referred to several excellent reviews listed in the bibliography for this chapter.

15.3.1 Habituation

Habituation refers to a decrease in some response (e.g., a behavioral response) following repeated elicitation of the response. At the cellular level, repeated stimulation of a synaptic input can lead to a decrease in the amplitude of the EPSP (figure 15.16). Habituation is not simply synaptic depression as discussed in chapter 11. There are several unique features that define habituation (Kandel and Spencer 1968). For example, a weaker stimulus habituates faster whereas a strong stimulus may not produce any habituation at all. Also, presentation of a strong stimulus during habituation can restore the original response amplitude (this is called *dishabituation*). The duration of habituation can be from minutes to weeks.

15.3.2 Classical conditioning

Classical or Pavlovian conditioning refers to a type of learning in which a stimulus (conditioned stimulus, or CS) that does not normally produce a

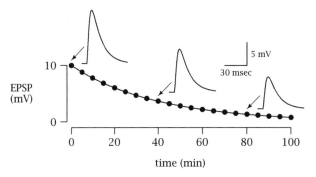

Figure 15.16 Hypothetical habituation experiment showing response decrement of EPSP with repeated stimulation.

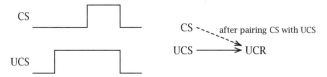

Figure 15.17 Diagram of classical conditioning. Repeated pairings of the CS with the UCS produces a response from the CS (dashed line) that was not present before the pairings.

response is paired with a stimulus (the unconditioned stimulus, or UCS) that does reliably produce a response (the UCR). Following repeated pairings of the CS followed by the UCS (each pairing is called a *reinforcement*), the CS begins to elicit a response (the conditioned response, or CR) that mimics the UCR (figure 15.17). In the original experiments of Pavlov, the CS was a bell and the UCS was the presentation of meat; the UCR was salivation. After the bell was repeatedly paired with the presentation of meat, the bell began to elicit salivation in the absence of meat.

Alpha conditioning refers to a similar paradigm except that the CS elicits a response before conditioning, and the pairing of the CS with the UCS serves to strengthen the response. *Pseudoconditioning* refers to a CR that develops (or is strengthened) by repeated presentations of the UCS or repeated unpaired presentations of the CS and UCS. Pseudo-alpha conditioning is more commonly called *sensitization* and is simply the strengthening of a preexisting CR by repeated presentation of the UCS or unpaired presentations of the CS and UCS.

An example of sensitization at the cellular level made famous by Dr. Eric Kandel, a pioneer in this field, is from the marine mollusk *Aplysia*. Dr. Kandel and his colleagues demonstrated that repeated stimulation of a sensory neuron results in habituation (or homosynaptic depression) of the resulting monosynaptic EPSP in the motor neuron. After habituating the EPSP, they found that stimulation of an entirely different set of sensory neurons resulted in an increase in the amplitude of the habituated sensory-to-motor neuron connection. The increase in the EPSP amplitude will occur even in the absence of prior habituation. This increase in response amplitude due to a completely different stimulus was considered a cellular analog to behavioral sensitization. In more cellular terms it has also been called heterosynaptic facilitation. Both habituation and sensitization are considered nonassociative forms of learning, whereas classical, alpha, and operant conditioning (see the following) are all associative forms of learning.

15.3.3 Instrumental or operant conditioning

Operant conditioning is a kind of trial-and-error learning in which the organism is rewarded, or receives a reinforcement, for a particular behavioral response. The reward could also involve the avoidance of punishment. Cellular analogs of operant conditioning are much more difficult to study because the behavioral responses involved are not usually under experimental control.

15.4 Role of hippocampus in learning and memory

The hippocampus is thought to play a fundamental role in certain forms of learning and memory. This notion derives primarily from the study of surgical cases in which bilateral removal of the hippocampus for intractable epilepsy led to profound and selective loss of memory. The most famous of these cases is the patient called H. M. Following removal of both hippocampi (and some surrounding temporal cortex), H. M. lost the ability to learn new information, although memories from up to a year before surgery and short-term memory are relatively unimpaired. H. M. can also learn new motor skills.

Subsequent to these surgical cases, the hypothesis for a role of the hippocampus and medial temporal lobe in memory has been strengthened by studies of human amnestics, who have selective damage to this area of the brain from a variety of causes, and from patients with Alzheimer's disease. Because LTP was first described in the hippocampus and because LTP has been extensively studied in the hippocampus, an intense effort

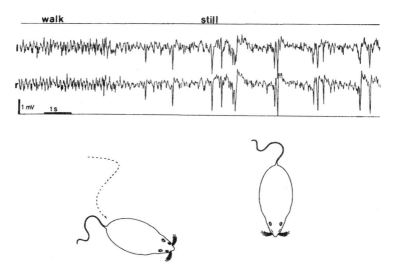

Figure 15.18   Theta rhythm during walking. Note that when the animal is still the EEG changes and shows sharp waves (from Churchland and Sejnowski 1992).

is being made to understand the relationship between the hippocampus and memory and, in particular, the relationship between LTP and memory in the hippocampus and elsewhere.

The fact that H. M. (and other patients with similar lesions) are able to learn certain motor skills (without even remembering that they have done so) has led to a division of human memory into two broad categories, one involving memories for events, facts, and people, and the other involving procedures and skills. These have been called *declarative* and *procedural* memory, respectively. The hippocampus and the medial temporal lobe are associated only with declarative memory. These structures are not considered to be involved in the permanent storage of memory but only in the facilitation of memory storage elsewhere.

The *theta rhythm* (figure 15.18) is a 4–8 Hz, rhythmic field potential that can be recorded from the rat hippocampus during exploratory behavior. A number of theories have been proposed for how the theta rhythm (also called *rhythmic slow activity*) can be involved in memory storage (Buzsáki 1989; see also Buzsáki and Vanderwolf 1985). As mentioned previously, brief bursts of synaptic inputs at theta frequencies (i.e., 5 Hz) are very effective in eliciting LTP.

Other more direct studies that have attempted to link LTP with memory have used antagonists for the NMDA receptor and shown that both LTP and learning are impaired. The learning paradigms included the eight arm radial maze, the Morris swim test, the delayed match-to-sample task, or various types of odor discrimination, which are all thought to require the hippocampus. Also, a correlation has been suggested between the amount of LTP at perforant path synapses and the ability of an animal to perform on some of these learning tasks. More recent experiments, which represent an exciting new approach for the study of molecular mechanisms of learning, have involved so-called genetic "knockout" experiments. In these experiments transgenic mice have been made that lack a gene which encodes a particular protein thought to be critical for the induction or maintenance of LTP. These animals have been shown to have learning difficulties in addition to impaired LTP.

The hippocampus is considered the highest level of association cortex. All sensory modalities have projections to the entorhinal cortex, which then project to the hippocampus through the perforant path. The rather brief and simplistic view of the hippocampus and memory outlined in this section, along with the role of the hippocampus as an associator, leads to what will be presented in the final section of this chapter. We will discuss one highly simplified and theoretical model as an example of how a neural network might learn and store information. We will also show how various forms of this model bear some rudimentary resemblance to different regions of the hippocampus.

15.5 Computational model for learning and memory

The intent of this section is not to present a unified theory for how memory is processed or stored in the hippocampus; nor is it our intent to foster this model as a realistic representation of hippocampal function over other excellent models of the hippocampus available in the literature (e.g., Rolls 1989a, b; Treves and Rolls 1994). Our purpose is instead to discuss a single example from a general class of models for information storage. We choose this particular model for discussion because it can be easily understood by all readers and because it attempts to embody some of the known hippocampal structure and physiology, including a fundamental role for synaptic inhibition. The model assumes that memory is stored as changes in synaptic weights throughout the network, and derives originally from Marr (1971) with recent refinements by McNaughton and colleagues (1987, 1989).

15.5.1 Correlation matrix

To understand fully the principle of memory storage in this model, it is necessary first to examine what is called a *correlation matrix*. This is illustrated in figure 15.19. There are two sets of binary inputs, the X inputs and the Y inputs, mapping onto a 6×6 square matrix. The value for each element of the input *vectors* is either 1 or 0 (on or off, active or inactive). Multiplying the X and Y vectors gives the values for the matrix elements that are shown in figure 15.19A. (The elements in each row of the matrix are obtained by multiplying the corresponding X element by each element in the Y vector. A 1 is obtained only if both the X and Y elements are 1.)

Once the elements of the matrix have been established, the X vector is multiplied by this matrix (the corresponding X element multiplies each row element of the matrix) to obtain a new matrix. If each column of this new matrix is summed, an output vector with value of 330300 is obtained. A critical feature of this correlation matrix, and for the model as a whole, is that if this output vector is divided by the sum of the X input vector (i.e., 3 in this case), then a vector of 110100 is obtained, which is identical to the input vector Y. In other words, having established the correlation matrix by multiplying X and Y, one can "recall" the Y vector by multiplying the matrix by the input vector X.

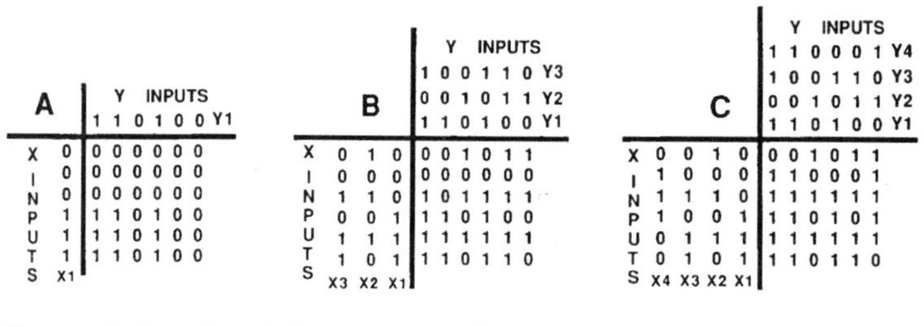

Figure 15.19 Correlation matrices after storing 1 (A), 3 (B), and 4 (C) pairs of inputs. Multiplying the X and Y vectors establishes each matrix according to the rules: A matrix element becomes 1 if the corresponding X and Y elements are 1. If the matrix element is already 1, it remains 1 regardless of the corresponding X and Y values. Once the matrix is established, each Y vector can be recalled by multiplying the matrix by the paired X vector, summing each column of product values, and then dividing the total by the sum of the X vector elements. For example, in (A) multiplying $X1$ times the matrix and then adding together the resulting elements in each column yields the product vector 330300. The sum of the $X1$ vector elements is 3, so dividing the product vector by 3 gives 110100, the original $Y1$ vector. The division is integer division so that any quotient less than 1 becomes 0. In (C) the problem of saturation is illustrated in that storage of the $X4$, $Y4$ pair causes errors in the recall of $Y3$ given $X3$. See text for further explanation. (From McNaughton 1989.)

In figure 15.19B there are 3 sets of X inputs and three sets of Y inputs. If the $X1$ vector is multiplied by the $Y1$ vector, the $X2$ vector by the $Y2$ vector, and so forth, in an identical fashion to that discussed above, the matrix shown in figure 15.19B is obtained (note that the maximum value of each of the matrix elements is 1). This matrix now has "stored" information about each of the input vectors Y. For example, if $X3$ is multiplied by the matrix and the columns are summed as before, an output vector of 322332 is obtained. Dividing this vector by the sum of the elements in $X3$ gives 100110, which is the same as $Y3$ (note that each element of the output vector can be either 1 or 0 and a quotient less than 1 is set equal to 0). Similarly, multiplying the matrix by $X2$ and dividing the output vector by the sum of $X2$ yields $Y2$, and multiplying the matrix by $X1$ and dividing the output vector by the sum of $X1$ yields $Y1$. Again, we were able to recall each of the Y input vectors by multiplying the correlation matrix by the corresponding X input vector.

We can also illustrate the idea of saturation by adding another set of inputs $X4$ and $Y4$. Multiplying these inputs together alters the matrix in such a way that errors are now introduced. This is illustrated in figure 15.19C. Multiplying the matrix by one of the X input vectors, as we did before, will lead to an output vector that is slightly different from the corresponding Y vector. The matrix has thus exceeded its maximum storage capability.

15.5.2 Neurophysiological implementation of the correlation matrix

The correlation matrix obviously has some interesting features. It is also easy to imagine how the input vectors could be groups of axons and the matrix itself a set of postsynaptic neurons. If the group of axons representing the Y vector strongly excites the postsynaptic neurons such that when a Y axon is active it will always fire the neuron, then a simple Hebbian learning rule is implemented whereby the synaptic "weight" of an X input onto any neuron will be changed from 0 to 1 (i.e., it will now fire that particular neuron) if and only if the X axon is active along with the corresponding Y axon. Extending the analogy, the Y axons would represent the UCS and the X axons the CS in a classical conditioning paradigm. This is illustrated in figure 15.20.

Recall that one of the key features of the correlation matrix is the division or normalization of the output vector. This can be plausibly implemented in a simple neural-like circuit by adding inhibition. Inhibitory interneurons generally are excited by large numbers of inputs and project

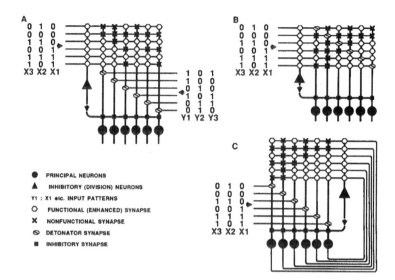

Figure 15.20   Neurophysiological implementation of a correlation matrix. In (A), a heteroassociative network is depicted in which the X vectors are represented by weak synaptic inputs and the Y vectors by strong synaptic inputs ("detonator synapses") to an array of 6 neurons. The Y synapses fire the neuron when they are active, while the X synapses are strengthened when they are active in conjunction with an active Y synapse. A strengthened X synapse will then fire the neuron the next time it is active and recall the corresponding Y in the same way as in the correlation matrix illustrated in figure 15.19. The division function of the correlation matrix is implemented by inhibitory interneurons that receive feedforward connections and inhibit each of the neurons by an amount that is dependent on the total activity in the input fibers. In (B) and (C) two different implementations of an autoassociative network are illustrated. In an autoassociative network there are only X input patterns, which get stored via the interspersed strong synapses (B) or through feedback synapses (C). Once stored, the patterns can be recalled even if some elements of the input pattern are missing. This is called pattern completion (see text for further explanation). (From McNaughton 1989.)

diffusely to large populations of target neurons. If we assume that an inhibitory interneuron receives excitation from each of the X inputs and inhibits each of the postsynaptic output neurons, then the division or normalization function can be readily realized. The inhibitory neuron will thus provide information to each postsynaptic neuron that is dependent on the total number of active inputs (i.e., the sum of the X inputs). This simple neural network, which functions as the correlation matrix discussed in the previous section, is called a *heteroassociative network*.

An important feature of this network is something called *pattern completion*. If, after this network has been "conditioning" with pairs of inputs, an X input vector is presented that only partially represents one of the original X vectors, an output vector will be obtained that completely rep-

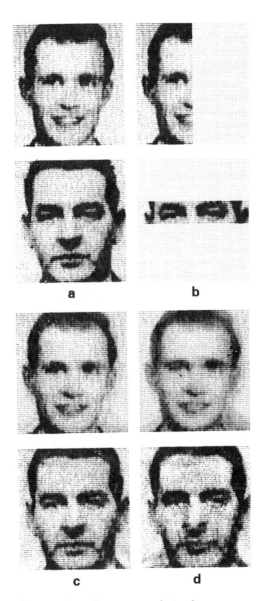

Figure 15.21 Pattern completion by an autoassociative network. In (a) the two faces used for training the network are illustrated. In (b) the two patterns used to elicit an output are illustrated. The outputs are shown in (c) and (d) after different numbers of patterns were stored by the network. Recall in (d) is poorer because of saturation. (From Kohonen 1978.)

resents the appropriate Y vector. In other words, the matrix was able to "recall" the entire pattern of Y inputs even though it was given only partial information. This feature of the network is illustrated in figure 15.21.

The features of the heteroassociative network can also be demonstrated with only one set of inputs if the outputs from the neurons are allowed to feed back onto themselves in a reentrant or recurrent excitatory circuit. This is also illustrated in figure 15.20. In this case the network forms a kind of reverberatory loop in which the feedback fibers carry information about the inputs back onto the same set of neurons. This is called an *autoassociative* network and is important theoretically because it does not require any additional inputs other that those that are to be stored by the circuit.

In figures 15.22 and 15.23 the hippocampus is illustrated in simplified anatomical detail as well as in a kind of neural network arrangement. The dentate and CA1 regions bear remarkable similarity to the heteroassociative network, whereas the CA3 region, with its extensive recurrent collaterals, incorporates features of the autoassociative network. Whether the hippocampus functions in this way, or even whether the associative networks can perform in the ways illustrated above if implemented with more realistic neurons and synapses, remains to be determined. Nevertheless, the discussion of this model is a useful exercise for appreciating how theoretical principles of general network function can help stimulate neurophysiological thinking at the cellular level.

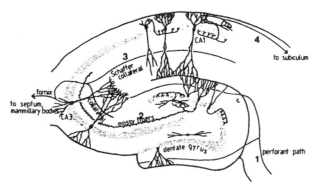

Figure 15.22 Schematic diagram of a cross section of the hippocampus. The dentate and CA1 areas appear to approximate heteroassociative networks. Because of the recurrent excitation among CA3 neurons, CA3 has been compared to an autoassociative network. Note the inhibitory interneurons depicted in dentate and CA1 making divergent connetions with the principal neurons. Similar inhibitory interneurons, which are activated by both feedforward and feedback connections, also exist in CA3. (From Rolls 1989a.)

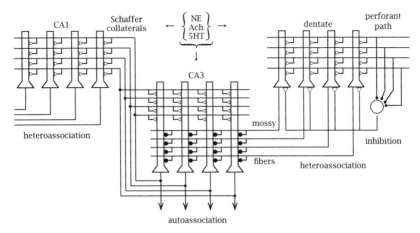

Figure 15.23 Memory circuits in the hippocampus. The anatomical arrangement of neurons and synaptic connections illustrated for a slice of the hippocampus in figure 15.22 is redrawn. NE, 5-HT, and ACh are potential memory modulators. An inhibitory neuron that could function as a divisor or for normalizing inputs is depicted only for the dentate. Similar inhibitory interneurons, however, exist in both CA1 and CA3 regions.

15.6 Summary of important concepts

1. Effects of dendritic spines on synaptic signals.

2. EPSP vs. E-S potentiation.

3. Heterosynaptic vs. homosynaptic.

4. Cooperativity vs. associativity.

5. Hebb rule for learning.

6. NMDA-receptor–dependent and NMDA-receptor–independent LTP.

7. LTD.

8. Habituation, sensitization, and classical conditioning.

9. Memory storage by a correlation matrix.

15.7 Homework problems

1. Your colleague recently demonstrated that during LTP of a glutamatergic synapse (the Schaffer collaterals in the hippocampus) there

was an increase in the non-NMDA component of the synaptic response while the NMDA mediated component of the response was unchanged. The assumption was that the NMDA and non-NMDA receptors are adjacent to each other on the spine head (as shown in the figure below) and that the only explanation for the results is that the number of activated non-NMDA receptors must increase during LTP. (If the number of both types of receptors increased, then there would be an increase in both components of the response.) The explanation for LTP, then, is that there is an increase in G_s for the non-NMDA response with no change in G_s for the NMDA response.

Show that there may be an alternative explanation for these results and that a change in spine neck resistance could differentially enhance two separate components of a synaptic input to the same spine head (i.e., non-NMDA and NMDA components). Assume that G_s for the non-NMDA component is 10×10^{-9} S, G_s for the NMDA component is 1×10^{-9} S, E_s for both components is 0 mV, the spine neck resistance is 100×10^6 Ω before LTP and 50×10^6 Ω during LTP, all synaptic conductance changes are steady-state, and you can voltage clamp to the resting potential (-70 mV) the dendritic shaft at the base of the spine.

(a) Calculate the non-NMDA synaptic current that reaches the dendritic shaft before and during LTP.

(b) Calculate the NMDA synaptic current that reaches the dendritic shaft before and during LTP.

(c) Explain why this alternative mechanism may or may not account for your colleague's results.

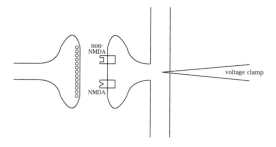

2. You wish to study an excitatory synapse for which no previous information is available. You are able to record intracellularly from the soma of the postsynaptic cell, stimulate action potentials in the

presynaptic nerve, and freely vary the composition of the extracellular medium. In your answers to the following questions, be very specific in describing your experiments, and generate hypothetical results where necessary.

(a) Test for the chemical vs. electrical nature of the synapse. Your results should be consistent with a chemical mode for transmission.

(b) Test the hypothesis that neurotransmitter quanta are released independently of each other.

(c) Test the hypothesis that evoked release follows a Poisson process. Determine the appropriate parameters for the Poisson equation.

(d) Calculate from a series of 100 trials the expected number of single quantal releases, and compare this value to your experimental data.

(e) Test for the relative permeability of the postsynaptic channels to Na^+, K^+, Mg^{2+}, Cs^+, and $Tris^+$. All these ions are partially permeable.

(f) Test for the degree of nonisopotentiality of the synapse from the recording site.

(g) Measure the peak conductance and kinetic properties of the synaptic response.

3. LTP is normally measured as an increase in an EPSP following brief, high-frequency stimulation. An observation of a sustained increase in intracellular Ca^{2+} levels following application of excitatory amino acids has been made, and this observation suggests that a long-term change in voltage-gated Ca^{2+} channels can occur. The investigators proposed a hypothesis that LTP might be expressed through a long-lasting change in postsynaptic voltage-gated Ca^{2+} conductances. If voltage-gated Ca^{2+} currents make a substantial contribution to EPSP amplitude, increased Ca^{2+} channel activity after tetanus might mediate the expression of LTP.

To test the hypothesis that a change in Ca^{2+} channel activity is solely responsible for the expression of LTP, you use the technique of whole-cell patch clamping in thick hippocampal slices to measure both EPSPs under current clamp and EPSCs under voltage clamp from CA3 cells. You stimulate afferent fibers that you believe synapse electrotonically close to the soma for voltage clamping. In a typical cell, you measure a control EPSP amplitude of 10 mV, and after tetanus the EPSP amplitude is 20 mV. Under voltage clamp and before tetanus, you measure an excitatory synaptic conductance of 10 nS with a reversal potential of 0 mV.

(a) Assuming the above hypothesis is correct (i.e., that a change in Ca^{2+} channel activity is solely responsible for expression of LTP), will EPSCs measured under voltage clamp be potentiated as were the EPSPs following tetanus?

(b) Draw the *I-V* curve of EPSCs at various potentials (e.g., −90 to +20 mV) before and after tetanus.

You further test this hypothesis by making cell-attached patch-clamp recordings of single Ca^{2+} channels in CA3 neurons before and after a tetanic stimulation that produces LTP of EPSPs. You are able to measure ensemble averages of single Ca^{2+} currents activated by voltage steps from a holding potential of −80 mV. Results of one experiment are shown in the following table.

Command Potential (mV)	Unitary Current (pA) Control	Unitary Current (pA) Tetanus	Peak Ensemble Current (pA) Control	Peak Ensemble Current (pA) Tetanus
−60	−1.5	−1.6	—	−0.1
−50	—	—	−0.04	−0.2
−40	−1.2	−1.1	−0.12	−0.39
−30	—	—	−0.39	−0.75
−20	−0.9	−0.88	−0.75	−0.93
−10	—	—	−0.93	−0.88
0	−0.7	−0.72	−0.9	−0.72
10	—	—	−0.7	−0.55
20	—	—	−0.5	−0.35
30	—	—	−0.25	−0.15

(c) What is the conductance of this calcium channel?

(d) Draw the activation curves for this channel, before and after tetanus. Assume a reversal potential of +40 mV and that the calcium current can be approximated by $I_{Ca} = g_{Ca}(V_m - E_{Ca})$.

(e) Is the effect of tetanus on the characteristics of this channel consistent with the LTP hypothesis?

(f) Under current clamp, at what holding potentials would LTP not be observable?

4. The following correlation matrix has stored 3 pairs of X, Y inputs. What Y vector would be obtained by presenting the given X vector?

							Y
1	0	0	1	0	1	1	
0	1	1	0	0	0	1	
1	1	1	1	1	1	1	
0	1	1	0	1	0	1	
1	1	1	1	1	1	1	
0	1	1	0	1	1	0	
X							

5. What types of LTP are being illustrated in A and B of the figure below? Explain your answer.

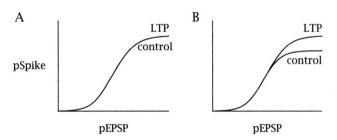

A   Basic Electricity and Electrical Circuits

A.1   Introduction

As do most fields of science, neurophysiology involves the art of measuring things. The things that neurophysiologists typically want to measure are electrical signals such as action potentials and synaptic potentials, or the membrane currents responsible for these potentials. Under ideal circumstances, the physical act of measuring a neurophysiological event would have no effect on the electrical signal of interest. Unfortunately, this is seldom the case in neurophysiology. One of the purposes of this appendix is to provide the theoretical and practical framework with which to determine how intrusive the measurement one wants to make really is. Also, this section should provide some understanding of electrical circuits for those readers who have a poor background in this area. An intuitive grasp of concepts of electricity and electrical circuits is helpful for understanding some of the basic theory in cellular neurophysiology. This appendix should also serve as a useful reference. The reader, however, is encouraged to review the more thorough treatments of the subject material in the references listed for this appendix.

The goal of this appendix is twofold. The first goal is to provide the reader with enough basic understanding to evaluate some of the methodology and techniques associated with neurophysiology. The second goal is to help the reader become facile at analyzing the electrical circuit models that are used throughout this book. The appendix is divided into four main parts. After the introduction, various terms that are frequently used with measuring systems are listed and defined. The third section defines the most useful terms, concepts, and laws for dealing with electricity and electrical circuits. The final section describes some of the basic circuits used for neurophysiological recordings.

A.2   Definitions related to measuring systems

Measurement The assignment of numerals to objects or events according to rules.[1]

Instruments All devices whose primary task is measurement or that employ a measurement component to carry out a task.

Variables Manifest themselves as forms of energy (e.g., voltage, amperage, temperature).

Properties Characteristics of things that manipulate energy (e.g., resistance, capacitance, thermal capacity).

Value Is used when relating magnitude to true zero.

Amplitude Is used when relating magnitude to the average value. The average value is also called the DC, mean, or baseline value. It is often assumed to be, or set equal to, zero.

Peak value Greatest positive or negative deviation from zero. *Maximum* value is used for positive peak value; *minimum* value for negative peak value. These same two points when measured from the average value are called the positive and negative peak amplitudes, respectively, and the measure between the two is the peak-to-peak amplitude. (Note: A distinction is rarely made anymore between *value* and *amplitude*, and the two are usually used interchangably. A good example of an instance where the distinction should be made is when one measures the membrane potental before and during an action potential (see figure A.1). The peak value of membrane potential during the action potential might be +30 mV while its peak amplitude might be +100 mV, assuming a resting potential, or average value, of −70 mV.)

Average amplitude The average deviation (without respect to sign) from the average value. It is calculated by averaging the absolute magnitude of the signal during the entire period of observation, or

$$V_{avg} = \frac{1}{T} \int_0^T |V(t)| dt.$$

---

[1] By convention, capital letters are usually used to represent constant parameters (e.g., an equilibrium potential such as E_{Na}) and peak values (e.g., G_s); small letters are used to represent parameters that are functions of other parameters (e.g., a sodium conductance that is a function of time and voltage such as g_{Na}). For simplicity, however, voltage will always be represented by V. Other exceptions to this rule are discussed in chapters 3 and 4.

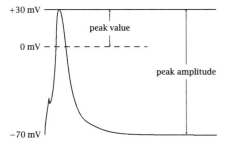

Figure A.1 Action potential measurements illustrating the difference between value and amplitude (see text for further explanation).

As an example, for a sine wave,

$$V_{avg} = \frac{1}{\pi} \int_0^{\pi} V_{\text{peak}} (\sin \omega t) d\omega t$$

$$= \frac{2V_p}{\pi}$$

$$= 0.637 \, V_{\text{peak}}.$$

Root mean square (RMS) amplitude A value found by taking the square root of the average of the squares of the deviation from the average value over the chosen time interval (refer to figure A.2), or

$$V_{\text{RMS}} = \left[\frac{(V_{avg} - V_1)^2 + (V_{avg} - V_2)^2 + \cdots + (V_{avg} - V_n)^2}{n} \right]^{1/2}.$$

RMS amplitude is also the effective value used in power calculations. The RMS amplitude of a waveform will produce the same heating in a resistor as an identical DC current (or voltage), or

$$P_{avg} = \frac{1}{T} \int_0^T p(t) dt$$
$$\text{Power} = i^2 R \text{ or } V^2/R, \text{ so}$$
$$P_{avg} = \frac{1}{T} \int_0^T \frac{V^2(t) dt}{R}$$
$$= V_{eff}^2 / R.$$
$$V_{eff} = V_{\text{RMS}} = \sqrt{\frac{1}{T} \int_0^T V^2(t) dt}.$$

The average power is proportional to the average of the square of the current (or voltage) over a given interval. The current (or voltage)

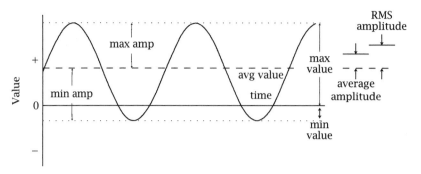

Figure A.2 Diagram of different types of electrical measurements (see text for further explanation). (Adapted from lectures by J. W. Moore.)

equivalent is the square root of the average of the squared amplitude.

For a sine wave,

$$V_{\text{RMS}} = \left(\frac{1}{\pi}\int_0^{\pi} V_{\text{peak}}{}^2 (\sin \omega t)^2 d\omega t\right)^{\frac{1}{2}}$$
$$= \frac{V_{\text{peak}}}{\sqrt{2}}$$
$$= 0.707\, V_{\text{peak}}.$$

Accuracy The degree to which a measurement indicates the true magnitude of a measurable quantity.

Precision The resolution and reproducibility of a measurement; implies nothing about accuracy. A measurement can be precise without being accurate. The reverse, however, is usually not true. Accuracy generally implies precision. Ideally one would want a measuring device to be accurate and precise. Unfortunately, most measuring devices in neurophysiology are precise without being accurate. For example, a recording amplifier may make extremely precise measures of the membrane potential but be consistently in error by many mV due to an unknown junction potential at the ground electrode, a poorly adjusted capacitance neutralization circuit, or a poorly adjusted DC offset.

Linearity Refers to the constancy by which a measuring system treats all inputs with respect to their magnitudes (refer to figure A.3). Ideally one wants the output of a measuring device to be equal to some constant times its input (output = $K \times$ input).

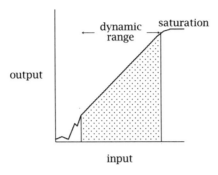

Figure A.3 Linearity of measuring systems. The amplitude of the output measure (for example, current, voltage, or power) as a function of the input amplitude.

Dynamic range The range of input signals between saturation of output and the minimum acceptable signal level. The latter is usually determined by noise (refer to figure A.3). Dynamic range is often a function of the gain of the system (among other things). One wants to make sure that a measurement is made within an adequate dynamic range of the measuring device. The concept of dynamic range is easily explained by using the familiar example of the speakers attached to your stereo. With a relatively high setting on your volume control (the "gain" of the system), the speakers will not reproduce sounds very well that are of low amplitude, and you will often hear hiss or noise mixed in with your music. In contrast, extremely high-amplitude sounds can reach saturation of the response of your speakers and produce garbled noises and poor reproduction of the recorded music. For optimal performance you set the gain or volume control to a middle range that is adequate for most soft sounds while being below saturation for the loudest parts of the input signal. Such a setting will utilize the maximum dynamic range of your system.

Fidelity The degree to which the ratio of output to input (the "gain") is constant for all frequencies of input. *Frequency response* is a measure of the fidelity of an instrument or system (refer to figure A.4). For example, suppose the fidelity or frequency response of an FM tape recorder used for recording intracellular signals is in the range of 5 kHz. This means that the tape recorder will reproduce accurately the resting potential and synaptic potentials but will likely attenuate particularly fast action potentials that have frequency components greater than 5 kHz.

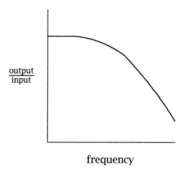

Figure A.4 Frequency response of measuring systems. The gain (output/input) is plotted as a function of frequency.

A.3 Definitions and units for electrical circuits

Much of what is in this section is presented in far greater detail in most introductory physics texts. One particularly good source (Halliday and Resnick) is listed in the references for this appendix.

A.3.1 Electromotive force (EMF, unit = volt)

One volt (V) is the potential difference between two points that requires the expenditure of 1 joule of work to move one coulomb of charge between the two points. We will use the symbol E for EMF sources.

A.3.2 Coulomb

A coulomb (C) is the quantity of charge (q) that repels an identical charge, 1 meter away, with a force of 9×10^9 newtons (nt); or is the charge experiencing a force of 1 nt in an electric field of 1 V/m. The elementary charge ($\pm e$) is

$e = 1.602 \times 10^{-19}$ C.

From Coulomb's law,

$$F_e = \frac{1}{4\pi\epsilon_0} \frac{q_1 q_2}{r^2},$$

where $\frac{1}{4\pi\epsilon_0} = 9 \times 10^9$ nt-m^2/C^2 and ϵ_0 = the permittivity constant (8.85×10^{-12} F/m).

A.3.3 Faraday

Faraday's constant is the magnitude of charge on one mole of electrons, or

$$F = N_A e,$$

where N_A = Avogadro's number (6.023×10^{23} molecules/mol). Therefore, $F = 96,495$ C.

A.3.4 Current

Current is defined as the rate of flow of electrical charges, or

$$i = \frac{dq}{dt} .$$

One ampere (A) is the flow of one C/sec. Positive current is defined as current flowing in the direction outward from the positive pole of a battery toward the negative pole (even though *electrons* are flowing in the opposite direction).

A.3.5 Ohm's law

Ohm's law states that the ratio of voltage to current is a constant:

$$R = V/i, \tag{A.3.1}$$

where 1 ohm = 1 Ω = 1 $\frac{V}{A}$. Also,

$$G = i/V, \tag{A.3.2}$$

where 1 mho or 1 siemen (S) = 1 $\frac{A}{V}$. (Note: Resistance is usually defined as the product of the resistivity of a conductor times its length divided by the cross-sectional area, or $R = R_i l / A$, where R_i is given in Ω-cm (see also chapter 4). Ohm's law is one of the more important concepts in electrophysiology. Any conductor that has a linear current-voltage (*I-V*) curve is said to be *ohmic*. Not all conductors have linear *I-V* curves. Most neurons, for example, have nonlinear or *nonohmic I-V* relations.

In describing the resistance (or Ohm's law) properties of neurons, one either injects current and measures the resulting change in voltage (a current clamp) or changes the voltage to different values and measures the current necessary to hold (or clamp) the voltage to these values (a voltage clamp). With a current clamp, voltage is the dependent variable and current is the independent variable, $V = f(I)$; with a voltage clamp the current is the dependent variable and voltage is the independent variable

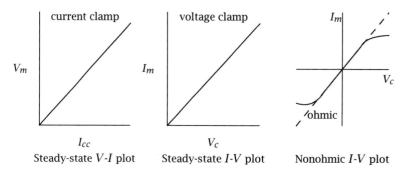

<table>
<tr><td>current clamp</td><td>voltage clamp</td><td>I_m</td></tr>
</table>

V_m I_m V_c

ohmic

I_{cc} V_c

Steady-state V-I plot   Steady-state I-V plot   Nonohmic I-V plot

Figure A.5 Schematized steady-state V-I and I-V curves for a typical neuron (see text for further explanation).

or $I = f(V)$. The resulting V-I or I-V curves should be plotted accordingly, and examples are given in figure A.5. Only a narrow region of the V-I or I-V curves of a neuron can usually be considered ohmic.

A.3.6 Capacitance

Capacitance refers to the ability to store charge, or

$$C = q/V,\tag{A.3.3}$$

where 1 Farad (F) = 1 $\frac{C}{V}$. Some important equations for the relationships among capacitance, charge, current, and voltage are

$$q \;=\; C \cdot V, \text{ and}\tag{A.3.4}$$

$$i \;=\; \frac{dq}{dt}, \text{ so}\tag{A.3.5}$$

$$i \;=\; C\frac{dV}{dt}, \text{ and}\tag{A.3.6}$$

$$V \;=\; \frac{1}{C}\int i\,dt.\tag{A.3.7}$$

Capacitance is also defined as the quantity of charge required to create a given potential difference between two conductors (or parallel plates), or

$$C = \epsilon_0 (K_d \cdot a/d),\tag{A.3.8}$$

where ϵ_0 is the permittivity constant, K_d is the dielectric constant, a is the surface area of parallel plates, and d is the distance between plates.

For a neuron, the lipid membrane acts as an insulator between two conductors—the intra- and extracellular solutions. From equation A.3.8

we can see that as the surface area of the parallel conductors (plates) increases, the capacitance increases. For a neuron, this means that as the surface area of the membrane increases so will the surface area of the conductors, and, therefore, larger neurons have larger capacitances. Also, from equation A.3.8 we can see that the capacitance decreases as the distance between the conductors (plates) increases. The lipid membrane has a fairly uniform thickness so this is unlikely to vary much in different parts of a neuron or among neurons. The myelin wrapping around an axon, however, will greatly increase the distance between conductors and thus decrease the capacitance. The insulation between the nodes of Ranvier in myelinated axons will therefore not only increase the effective transmembrane resistance in the internodal region but also decrease the capacitance. A decrease in the myelin sheath, such as occurs in demyelinating diseases, results in an increase in capacitance. This can lead to a further reduction in the conduction of an action potential from node to node (refer to chapter 7 for a description of saltatory conduction in myelinated nerves).

Another concept that is sometimes difficult to grasp is that there is never any actual movement of charges directly across the insulator between the two plates of a capacitor. For this reason there is no current flow across a capacitor for an applied DC voltage even though a separation of charge is induced by the voltage ($q = CV$). When the voltage changes with time, however, charge increases and decreases (positive on one side and then the other) with time on each side of the conductor *as if* current were actually flowing across the capacitor ($i = C \cdot dV/dt$). There is, however, no actual loss of charge across the capacitor—the charge merely shifts from one side to the other by way of the rest of the circuit.

A.3.7 Kirchhoff's laws

An understanding of Kirchhoff's laws will be useful in analyzing circuit models for neurons as well as for understanding the circuits used for measuring electrophysiological signals.

1. **Current law**: The algebraic sum of all current flowing toward a junction is zero (i.e., charge can neither be created nor destroyed). By convention, positive current flows into a node or intersection and negative current flows away from a node (see figure A.7).

2. **Voltage law**: The algebraic sum of all potential sources and voltage drops across passive elements around a closed conducting path or

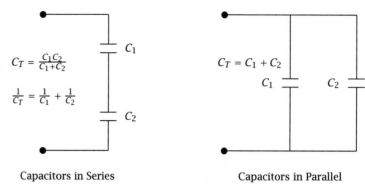

Capacitors in Series   Capacitors in Parallel

Figure A.6 Adding capacitors in series and in parallel. (The black dots are frequently used to represent points of connection to other circuits.)

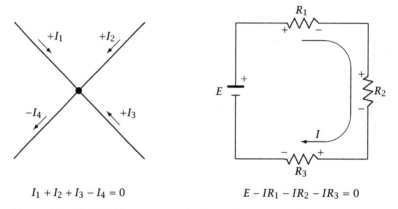

$$I_1 + I_2 + I_3 - I_4 = 0$$ $$E - IR_1 - IR_2 - IR_3 = 0$$

Figure A.7 Kirchhoff's current and voltage laws (see text for further explanation).

"loop" is zero (conservation of electrical energy, refer to figures A.7–A.9). This is just another form of the first law of thermodynamics. By convention, positive current produces a voltage drop across a resistor with polarity (+ to −) in the direction of current flow (see figure A.7). The relevant equations for figure A.8 are

$$E_1 - I_1R_1 - (I_1 + I_2)R_3 = 0,$$

$$E_2 - I_2R_2 - (I_1 + I_2)R_3 = 0,$$

and

$$V_3 = (I_1 + I_2)R_3.$$

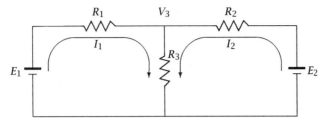

Figure A.8 Kirchhoff's laws in a multi-loop circuit.

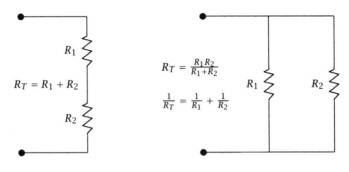

Resistances in Series   Resistances in Parallel

Figure A.9 Adding resistors in series and in parallel.

A.3.8 Voltage source

An ideal voltage (potential) source is one that will maintain an absolutely unchanging value despite any value of load resistance across its output terminals (i.e., one that has zero internal or source resistance; refer to figure A.11). A battery is the simplest example of a voltage source. Another (not so simple) example is a voltage clamp.

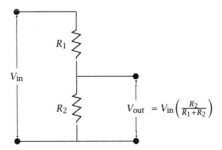

$$V_{\text{out}} = V_{\text{in}} \left(\frac{R_2}{R_1 + R_2} \right)$$

Figure A.10 Voltage divider circuit.

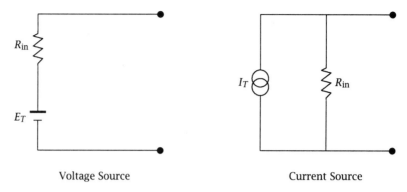

Voltage Source   Current Source

Figure A.11   Current and voltage sources. For a voltage source, $R_{in} \simeq 0$. For a current source, $R_{in} \simeq \infty$. E_T is the total (internal) EMF of the voltage source while I_T is the total (internal) current of the current source. The intersecting circles on the right are often used to represent a current source.

A.3.9   Current source

An ideal current source is one that maintains an absolutely unchanging value of current despite changes in the load resistance across its output terminals (i.e., it has an infinite internal or source resistance, refer to figure A.11). When one injects current into a neuron, one wants the current to be constant regardless of the input resistance (or changes in input resistance) of the neuron. This is achieved via a circuit that acts as a current source. A *current clamp* is simply a current source at which one can vary the amplitude of the injected current.

Other potentially useful theories of electrical circuits include *Thévenin's theorem* and the *superposition principle.* Thévenin's theorem simply states that any circuit of linear elements (e.g., resistors, capacitors, voltage and current sources) can be replaced by a series combination of an ideal voltage source and a linear impedance. One practical application of Thévenin's theorem in neurophysiology is in the use of an input resistance to represent a neuron. No matter how complex the neuron, we can let it (at least the linear portion) be represented by a single resistor (R_N) in series with the resting potential (E_r).

The superposition principle states that in any circuit of linear elements with more than one source of voltage or current, the current in any branch of the circuit is simply the linear sum of the currents due to each of the sources being treated individually. The superposition principle also has many practial applications in neurophysiology, including the summation of synaptic potentials, the analysis of field potentials (chapter 14), and the "two-port" network analysis of dendrites (chapters 4 and 13). The reader is referred to the bibliography for additional information about these theorems.

A.3.10 Voltage and current measurement

An ideal *voltmeter* is one with an infinite internal resistance. The measurement is made in parallel with the circuit, and the voltmeter will ideally draw zero current from the circuit. The measurement of voltage should therefore not alter the voltage or current flow in the circuit.

An ideal *ammeter* is one with zero internal resistance. The measurement of current is made in series with the circuit, and the ammeter will ideally add zero potential difference to the circuit (refer to figure A.12). Again, the measurement of current should not alter the voltage or current flow in the circuit.

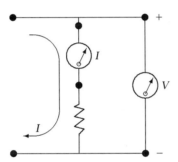

Figure A.12 Current and voltage measurement. Ideally, the current through the voltmeter on the right will be ≃ 0 while the voltage across the ammeter on the left will also be ≃ 0. (A circle with an arrow inside is frequently used to represent a meter or measuring device.)

A.3.11 Impedance

Impedance is the ratio of forcing quantity to resulting moving quantity. For electricity,

$$Z = E/i,$$

where E and i can be functions of time or frequency and Z is the *impedance*. This is the more general form of Ohm's law where E and i can be any waveform that can be represented by exponentials. (Note: A detailed discussion of impedance and AC circuits is beyond the scope of this book, and the reader should consult the references given for this appendix for

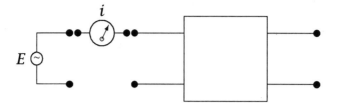

Figure A.13   Measurement of input impedance. A sinusoidal voltage source (the circle with the ~ inside) is applied to a circuit and the input current is measured. The input impedance is the ratio E/i (see text).

further reading.) In AC circuits Z will have real and imaginary components, but $Z = R$ if the forcing function (E) is not a function of time or frequency (i.e., DC). Also,

$1/Z = Y = $ Admittance.

A.3.12 Input impedance and input resistance

The terms input impedance and input resistance refer to the ratio of the input forcing function to the resulting flow of energy through the system (refer to figure A.13).

$Z_N = E/i.$

The concept of input impedance and input resistance follows directly from Thévenin's theorem. Assume that the box in figure A.13 represents a neuron. No matter how complicated the neuron, a single parameter (input impedance) can be measured by applying a voltage and measuring the resulting current or by injecting a current and measuring the resulting voltage. Note that the input impedance of a neuron is not equal to its input resistance. Input resistance is the steady-state measure of input impedance. The input impedance is a function of the frequency of the applied current or voltage. For example, the input impedance of a neuron for a 1 kHz sine wave of injected current will be much less than for a 1 Hz sine wave. The input impedance measured after the voltage has reached a steady state following a step change in injected current is defined as the input resistance.

A.3.13 Source impedance

The source impedance of a device is also called the output impedance. The value of the source or output impedance is important when connecting one device to another for the purpose of making a measurement. Typically, the source or output impedance should be much lower than the input impedance of the measuring device (see section A.3.16).

A.3.14 Filters

Electrical filters are frequently used in neurophysiology to remove unwanted high-frequency noise. When computers are used to digitize electrical signals one must ensure that the signal is filtered at 1/2 (or less) of the maximum rate of digitization to minimize aliasing noise (Nyquist frequency). These would be applications of *low-pass* filters. In measurements of extracellular electrical signals, it is also sometimes useful to filter out low-frequency changes in potential. This would be an example of a *high-pass* filter. Filters that pass low frequencies, high frequencies, and a *band* (or specific range) of frequencies can be made from simple RC circuits (see figure A.14). The nerve membrane acts as a *low-pass* filter, because it behaves as a parallel resistor and capacitor (figure A.15).

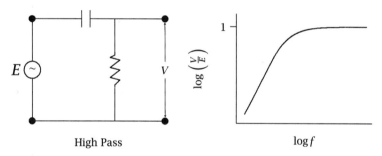

Figure A.14 Circuit for high-pass filtering.

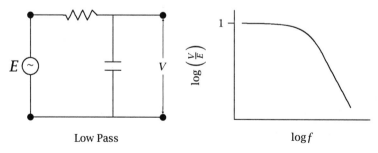

Figure A.15 Circuit for low-pass filtering.

A plot of gain (output/input) of a system vs. log(frequency) is called a *Bode plot* (see figure A.16). It is a common way of representing the frequency response of a system. The gain in a Bode plot is usually given in decibels (db), where gain in db = $20 \log(V_{out}/V_{in})$. Using the gain at zero frequency (DC) as the reference or 0 db point, the gain of the system will decrease at higher frequencies for a low-pass response. The steepness of the falloff in gain with higher frequencies is often given in "poles" so that an 8-pole filter has a steeper slope than a 4-pole filter and is thus a better filter. The frequency response of a system is usually given as the frequency at which the gain has declined by 3 db or by $1/\sqrt{2}$. This is also sometimes called the *corner* frequency.

A high-pass filter in series with a low-pass filter yields a *band-pass* filter (shaded area in figure A.16). A *notch filter* passes all frequencies except for a very narrow range (often near 60 Hz).

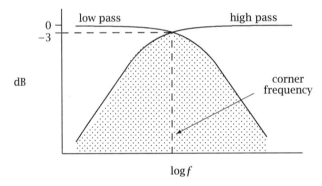

Figure A.16   Bode plot for low, high, and band pass filters.

A.3.15   Response of resistors and capacitors to different forcing functions

The relevant equations are

$$V = IR,$$
$$I = V/R,$$
$$i = CdV/dt,$$
and
$$V = \tfrac{1}{C}\int i\,dt.$$

With the above equations one can derive the response of resistors (R) and capacitors (C) to different forcing functions. For example, what is the response of a capacitor to a step of current, a step of voltage, or a ramp of voltage? In figure A.17 we illustrate the responses of a resistor and a capacitor to the forcing functions one is most likely to encounter in neurophysiology. When resistors and capacitors are combined in electrical circuits, the overall circuit response depends on the responses of each of the individual elements.

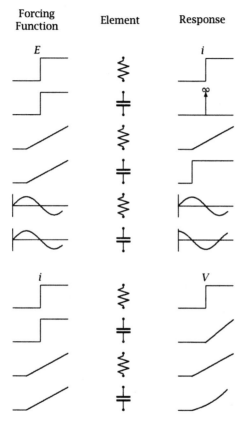

Figure A.17   Forcing functions and their responses for different circuit elements. (Adapted from lecture notes by J. W. Moore.)

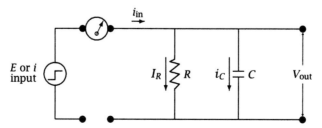

Figure A.18   Cell membrane with a voltage or current step input.

For example, in the parallel RC circuit of figure A.18, if the forcing function is a voltage step (E), then $I_R = E/R$, $i_c = CdE/dt$, and $I_R + i_C = i_{in}$.

The response of a nerve membrane to a voltage step, such as that produced by a voltage clamp, is relatively simple. The capacitance current is ideally an impulse response, and the current flow across the resistor mirrors the voltage step. (In reality, of course, the impulse response is slowed by the finite frequency responses of the voltage generating and current measurement devices as well as the series resistance associated with the measurement electrodes (see next section) and ends up being a sum of exponentials.) The total current flow is just the sum of the two currents (see figure A.19).

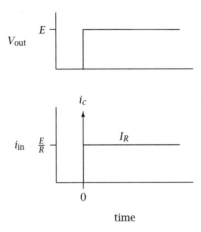

Figure A.19   Response of cell to a voltage step.

If instead the forcing function is a current step (i), then the response is a bit more complicated and requires the solution of a simple first-order differential equation. This is shown below:

$$I_R = V_{out}/R,$$

and

$$i_C = CdV_{out}/dt;$$

$$i = I_R + i_C = V_{out}/R + CdV_{out}/dt;$$

$$i \cdot R = V_{out} + RC \frac{dV_{out}}{dt};$$

$$i \cdot R - V_{\text{out}} = RC \, \frac{dV_{\text{out}}}{dt};$$

$$\int_0^{V_{\text{out}}} \frac{dV_{\text{out}}}{i \cdot R - V_{\text{out}}} = \int_0^t \frac{dt}{RC};$$

$$- \left[\ln(i \cdot R - V_{\text{out}}) \right]_0^{V_{\text{out}}} = \left[\frac{t}{RC} \right]_0^t;$$

$$\frac{i \cdot R - V_{\text{out}}}{i \cdot R} = e^{-t/RC};$$

and finally,

$$V_{\text{out}} = i \cdot R \left(1 - e^{-t/RC} \right).$$

This is the so-called first-order response of a nerve membrane to a step of current (figure A.20). The voltage response to a step of current is a single exponential for isopotential cells and multiexponential for nonisopotential neurons (see chapter 4), and is also sometimes called the *charging curve*.

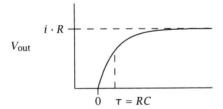

Figure A.20 Voltage response of cell to a current step. This is sometimes called a charging curve (see text for further explanation).

A.3.16 Measuring biological signals

The measurement of the electrical signals from biological tissue involves three main stages. The first involves the interface or junction between current flow via ions in solution and current flow via electrons in wires. This first stage usually involves some type of electrode. The types of electrodes are extracellular (glass or metal), intracellular (glass), and patch (glass). The second stage of the measurement involves the preamplifier. This is a device used to match the *source* impedance of the electrode to the *input* impedance of the measuring device or amplifier (see section A.4.2.3).

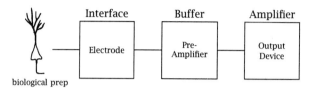

Figure A.21   Stages of a measuring system for biological signals.

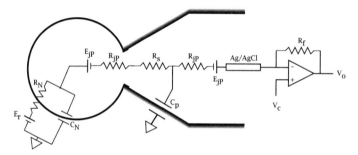

Figure A.22   Schematic diagram of a glass microelectrode in a cell or a patch electrode in the whole-cell mode. E_r, R_N, and C_N are the resting potential, input resistance, and input capacitance of the cell, respectively. E_{jP} and R_{jP} are the potentials and resistances of the liquid-liquid junction and the liquid-silver wire junction. C_p is the pipette capacitance. R_s is the resistance of the electrode. It is usually called the series resistance (R_s) for patch electrodes, and tip resistance (R_e) for intracellular electrodes. The amplifier represents a patch-clamp amplifier (see section A.4.4). (The inverted triangles at the bottom of the circuit elements are frequently used to represent connections to ground.)

The third stage involves the final amplifier and output device where the actual measurement of the signal takes place.

A great deal has been written about techniques for constructing electrodes used in neurophysiology, and the reader is referred to several references given at the end of this appendix. There are several practical considerations, however, that are important to mention here because they are relevant to the discussion of circuits that will follow (refer to figure A.22). The actual point of junction between the solution and the wire in the circuit (usually a silver/silver chloride (Ag/AgCl) wire) has resistance to current flow and a potential across the interface (liquid-junction potential). A well-chlorided silver wire will minimize this resistance and potential but not eliminate them. Some circuitry for neutralizing the junction potential (DC offset) will therefore be necessary.

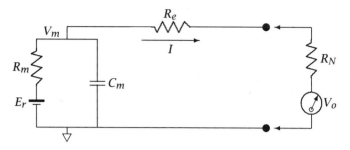

Figure A.23 Measuring the membrane potential (V_m) with a low input resistance device.

Glass microelectrodes also have a resistance and a potential associated with their tips, and these are usually much more of a concern than those associated with the silver/silver chloride wire. The resistance of a microelectrode can be very high. As an example, we can make a rough calculation of the tip resistance of a microelectrode filled with 3 M KCl, using the equation $R_e = R_i l / A$. Let $R_i = 5$ Ω-cm, $l = 0.2$ cm (the length of the tip), and $A = 1$ μm^2 (the opening of tip). Putting these numbers into the equation gives a tip resistance of 100 MΩ, a typical value for an intracellular microelectrode. This tip resistance is also a nonlinear function of current passed through the electrode. The resistance can get very large depending on the amplitude and polarity of the current being passed through the electrode. The tip potential can also vary considerably with time or with the amount of current passed. The electrode tip in solution also has a capacitance resulting from the two solutions (conductors) on either side of the glass wall. These electrical characteristics of microelectrodes necessitate the use of special circuitry. The need for a preamplifier is explained in the following, and a brief description of bridge balance and negative capacitance circuitry is given.

Figure A.23 illustrates the equivalent circuit associated with measuring the membrane potential from a neuron. R_m, C_m, and E_r are the usual membrane parameters described first in chapter 3. R_e is the electrode resistance, and R_N and V_o are the input resistance and meter associated with a voltage measurement device. Let's assume that this voltage measurement is being done by an oscilloscope. Some typical values for R_e and R_N are $R_e = 10^8$ Ω and $R_N = 10^6$ Ω.

Using these values, V_o can be written in terms of the circuit parameters:

$$V_o = \frac{R_N V_m}{R_e + R_N} = \frac{10^6}{1.01 \times 10^8} V_m \approx 0.01 \, V_m.$$

It can be seen then that the *measured* membrane potential will be only 1/100th of the actual value because of the relatively large source resistance of the microelectrode compared to the input resistance of the oscilloscope. The optimal situation for this type of measurement is for the source or output resistance to be very low and the input resistance to be very high. This can be achieved with a preamplifier as shown in figure A.24. Here the preamplifier is actually a *follower* or *buffer* amplifier, which will be described in more detail in the next section. The preamplifier (at least the first stage) has a high input resistance, low output resistance, and usually a gain of 1. Higher gain is achieved by adding additional amplifier stages in series with the first stage.

With an input resistance of 10^{12} Ω and an output resistance of 100 Ω for the preamplifier and with it placed between the electrode and the oscilloscope, the equation for the measurement of membrane potential can now be rewritten as

$$V_o = \frac{R_N \, V_m}{R_e + R_N} = \frac{10^{12}}{1.0001 \times 10^{12}} \, V_m \approx V_m.$$

With the high input resistance of the preamplifier or *headstage*, the measured V_m will now be very close to the actual V_m across the cell membrane.

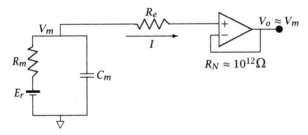

Figure A.24   Measuring V_m with high-impedance buffer amplifier.

The other problems with microelectrodes mentioned above, namely the tip resistance and tip capacitance, can be addressed with special circuitry, as illustrated in figure A.25. When current is passed through a microelectrode, a large potential develops across the electrode tip. This potential is in series with the membrane potential so that the measured potential is the sum of the membrane potential V_m and the electrode potential V_e. With an electrode of 50 MΩ and a current of 1 nA, V_e will be 50 mV, which would obviously be significant when measuring a membrane potential on the order of 50–100 mV. The voltage drop across the electrode can be subtracted from the total signal using a so-called bridge circuit as shown in figure A.25. This circuit merely subtracts a signal proportional to the applied current from the total output voltage of the preamplifier. The adjustment is made manually and is possible because the time constant of the microelectrode is usually much smaller than that of the cell.

The capacitance of the electrode is in parallel with the electrode resistance and causes the membrane potential to be low-pass filtered by the electrode. This can cause significant loss of fidelity of the measurement, especially if one is measuring fast changes in membrane potential such as during the action potential. One way to compensate for this electrode capacitance is to feed back into the input a signal that represents the loss associated with the electrode capacitance. This is done by taking the output of the preamplifier and feeding it through a capacitor and adding it back to the input. This is called negative capacitance or capacitance compensation and is also illustrated in figure A.25.

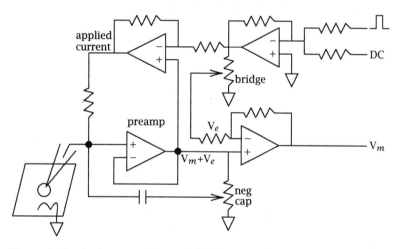

Figure A.25   Basic preamplifier with bridge and negative capacitance.

A.4   Amplifiers and voltage-clamp circuits

A.4.1   Ideal amplifier

The intent of this section is not to describe in detail the design and use of operational amplifiers, but to provide some minimal level of analysis and understanding of several very common circuits in neurophysiology: a) the *inverter*, b) the *noninverter*, and c) the *follower*. With a knowledge

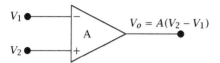
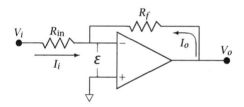

of these simple amplifier circuits, the means by which voltage and patch clamping are achieved can more readily be appreciated.

A schematic diagram of an ideal amplifier is shown in figure A.26. The amplifier has two inputs and an output that is equal to the gain (A) times the difference between the two inputs.

$V_o = A(V_2 - V_1)$

Figure A.26   Ideal amplifier.

A is called the *open loop gain* of the amplifier. It is very large and is independent of frequency over a given range. The input impedance of the amplifier is considered infinite, and the output impedance zero, for most applications. If indeed these ideal characteristics are met by such an amplifier, then it can be used for a number of extremely useful circuits.

A.4.2   Practical circuits

Figure A.27   Inverting amplifier. ε is the small "error" voltage between the + and − inputs to the amplifier. The high gain of the amplifier makes this very small and negligible for most applications.

A.4.2.1   Inverter   The first step for analyzing the circuit shown in figure A.27 is to write the equations for voltage and current at the input and output terminals using Ohm's and Kirchhoff's laws.

At the input terminal

$$V_i - I_i R_{in} = \varepsilon. \tag{A.4.9}$$

ε is usually very small so that the negative input to the amplifier is at essentially the same potential as the positive input. This configuration is thus sometimes called a *virtual ground*. If, as stated above, the input impedance of the amplifier is very large (i.e., $R_N \cong \infty$), then

$$I_i = -I_o. \tag{A.4.10}$$

At the output terminal,

$$V_o - I_o R_f = \varepsilon, \tag{A.4.11}$$

and

$$V_o = -A\varepsilon. \tag{A.4.12}$$

These equations can then be combined to solve for the output (V_o) in terms of the input signal (V_i). First, we will combine equations A.4.9, A.4.10, and A.4.11, and then rearrange to obtain

$$-I_i = I_o = \frac{V_o - \varepsilon}{R_f} = -\left(\frac{V_i - \varepsilon}{R_{\text{in}}}\right).$$

Then we will eliminate ε by using equation A.4.12:

$$\frac{V_o - (-V_o/A)}{R_f} = -\frac{V_i - (-V_o/A)}{R_{\text{in}}}.$$

What follows is just a step-by-step rearrangement of the above equations.

$$\frac{(V_o + V_o/A)}{R_f} = -\frac{(V_i + V_o/A)}{R_{\text{in}}},$$

$$\frac{AV_o + V_o}{AR_f} = -\left(\frac{AV_i + V_o}{AR_{\text{in}}}\right),$$

$$\frac{AV_o}{R_f} + \frac{V_o}{R_f} = -\frac{AV_i}{R_{\text{in}}} - \frac{V_o}{R_{\text{in}}},$$

$$\frac{AV_o}{R_f} + \frac{V_o}{R_f} + \frac{V_o}{R_{\text{in}}} = \frac{-AV_i}{R_{\text{in}}},$$

$$V_o\left(\frac{1+A}{R_f} + \frac{1}{R_{\text{in}}}\right) = -V_i\frac{A}{R_{\text{in}}},$$

$$\frac{V_o}{V_i}\left[\frac{(1+A)R_{\text{in}} + R_f}{R_f R_{\text{in}}}\right] = \frac{-A}{R_{\text{in}}},$$

$$\frac{V_o}{V_i}\left[\frac{(1+A)R_{\text{in}} + R_f}{AR_f}\right] = -1,$$

and

$$\frac{V_o}{V_i} \left[\frac{(1 + A)R_{in}}{AR_f} + \frac{R_f}{AR_f} \right] = -1.$$

As mentioned above, we will assume that the internal gain (A) of this ideal amplifier is very large. We can therefore consider the above equation in the limit as $A \to \infty$, or

as $A \to \infty$

$$\frac{V_o}{V_i} \left[\frac{R_{in}}{R_f} \right] = -1,$$

and the overall gain of this inverter is

$$V_o = -\frac{R_f}{R_{in}} V_i. \tag{A.4.13}$$

This is an extremely important result. It says that the output of this amplifier configuration is just the negative of the input times the ratio of the feedback resistor to the input resistor. This proves to be a very useful and easily implemented circuit, although for best results the gain is usually limited to the range of $1 < \text{gain} < 50$.

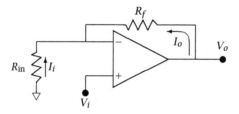

Figure A.28   Noninverting amplifier.

A.4.2.2 Noninverter   The noninverting amplifier (shown in figure A.28) is also a useful circuit. The noninverting amplifier, however, is often more difficult to build and design than the inverter for reasons that are beyond the scope of this book. (To achieve a noninverting amplifier it is often easier to use two inverters in series.) The overall gain for the noninverting amplifier (equation A.4.14) is given below; its derivation is left as a homework exercise.

$$V_o = \frac{R_f + R_{in}}{R_{in}} V_i. \tag{A.4.14}$$

A.4.2.3 Follower The follower amplifier is one of the most useful and versatile circuits one will encounter, second only perhaps to the inverter (figure A.29). It is also by far the easiest to implement. The equations for the follower can be derived as follows:

$$V_o = A(V_i - V_o),$$

$$V_o = AV_i - AV_o,$$

$$V_o(1 + A) = AV_i,$$

and

$$V_o \frac{(1 + A)}{A} = V_i.$$

As before, we can consider the above equation in the limit, or as $A \to \infty$,

$$V_o = V_i. \tag{A.4.15}$$

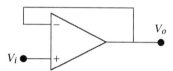

Figure A.29   Follower amplifier.

The above result appears trivial and raises the question of the purpose of this circuit. The main purpose of this configuration is as an impedance buffer (it is thus also called a *buffer* amplifier). Because of its high input impedance and low output impedance, the follower can be inserted into a circuit without drawing any current. Its main use is to buffer (or match) high-impedance sources (microelectrodes) with low-impedance measuring devices or recorders (see figure A.24).

A.4.3 Voltage-clamp circuit

In figure A.30 a simplified circuit diagram of a two-electrode voltage clamp is illustrated using the ideal amplifiers described in the previous sections. The equivalent circuit of the cell is given with two electrodes attached. One electrode (R_e) is connected directly to a follower (the preamp), which in turn is connected to the negative input of the ideal amplifier. The output of this amplifier is connected to the cell through the second electrode, which is represented by R_a, and is called the *access* resistance to the cell.

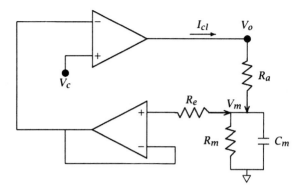

Figure A.30   Simplified two-electrode voltage-clamp circuit. The arrows attached to R_a and R_e represent electrodes in the cell. The resistance of the R_a electrode is often called an access resistance (see text for further explanation).

At the positive input, we apply the potential we want to "clamp" the cell to, or the clamp potential, V_c. The circuit equations can be written for this voltage clamp in a similar manner as in the previous sections:

$$V_o = A(V_c - V_m),$$

$$V_m = V_o - I_{cl}R_a,$$

$$V_m = A(V_c - V_m) - I_{cl}R_a,$$

$$V_m = AV_c - AV_m - I_{cl}R_a,$$

$$V_m + AV_m = AV_c - I_{cl}R_a,$$

and

$$V_m = \left(\frac{A}{1+A}\right)V_c - \frac{I_{cl}R_a}{1+A}. \tag{A.4.16}$$

As before, we consider this equation in the limit as A gets very large, that is, if A is large,

$$V_m \approx V_c.$$

This is a very important result. It says that if A is large compared to $I_{cl}R_a$, then the membrane potential will be clamped at whatever potential (V_c) is applied to the positive input of the clamp amplifier. There are many practical considerations in the design of a voltage clamp having to do with frequency response, or with how fast the membrane potential will be clamped to the applied potential, that are beyond the scope of this book. The basic theory of the voltage clamp, however, is quite simple given a little knowledge of the basic building blocks of amplifier circuits.

One practical consideration worth mentioning is something called *series resistance*. In the case of the two-electrode voltage clamp shown in figure A.30, series resistance would represent the resistance of the extracellular fluid between the membrane and the ground wire in the bath. This would be added to figure A.30 by putting a resistor (R_s) between the ground triangle and the membrane. The clamp current I_{cl} flows across the membrane and then across this resistor so that the error in V_m caused by this series resistance is dependent on the product of $R_s \cdot I_{cl}$. Equation A.4.16 now becomes

$$V_m = \left(\frac{A}{1 + A}\right) V_c - \frac{I_{cl}R_a}{1 + A} - I_{cl}R_s, \tag{A.4.17}$$

or, in the limit as A gets very large,

$$V_m = V_c - I_{cl}R_s. \tag{A.4.18}$$

R_s is said to be "outside the feedback loop" and thus the error term is not affected by the gain of the voltage clamp.

Note that the circuit in figure A.30 requires two electrodes. A modification of the basic voltage clamp that requires only a single electrode is called a switched or discontinuous voltage clamp. This type of voltage clamp uses a single electrode for both current passing and voltage recording, but switches or time shares the electrode between these two modes (Smith et al. 1985). Another method for doing voltage clamping with a single electrode is described in the next section.

A.4.4 Patch-clamp circuit (current-to-voltage converter)

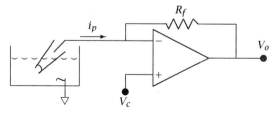

Figure A.31 Patch clamp circuit (see text).

The concept behind the patch-clamp amplifier is actually derived from a very early circuit used in voltage-clamp systems, called the virtual ground circuit. This circuit converted the total current flow to ground into a voltage signal that could be measured and was thus called a *current-to-voltage converter*. The basic principle of such a circuit (but with the ground connection replaced with V_c) is illustrated in figure A.31 and can be derived as follows:

$$V_o - V_c = -i_p R_f, \tag{A.4.19}$$

and

$$V_p = V_c, \tag{A.4.20}$$

where V_p is the pipette potential (again, assuming that A is large).

The output voltage of the amplifier is simply the product of the current through the patch (or whole cell) and the value of the feedback resistor. It is obviously important to choose the proper size of feedback resistor depending on the amplitude of the current one wishes to measure. For example, single-channel currents of a few pA would require a feedback resistor of at least 10 GΩ to produce measurable signals in the tens of mV range. On the other hand, whole-cell currents of around 0.1 nA would require a feedback resistor of about 100 MΩ.

From the characteristics of the ideal amplifier described in the previous sections, the potential at the electrode, or the patch (V_p), will be approximately equal to the potential applied to the positive input of the amplifier (V_c). Changing the membrane potential across the patch or across the whole cell is therefore simply a matter of changing the potential applied to the amplifier. The polarity of the potential applied to the electrode, however, will depend on the configuration of the membrane under the electrode. For example, in the whole-cell mode one would obviously apply −70 mV to the patch electrode to clamp the cell to near the resting potential. On the other hand, one would apply 0 mV to the patch electrode to clamp the membrane to near rest if one were recording in the so-called *cell-attached patch configuration* (refer to chapter 8).

Furthermore, in the cell-attached patch mode one must consider that the resting potential of the cell is in series with the patch potential. Excising the patch from the cell or using high K^+ in the bath to zero the membrane potential is usually necessary to control the patch potential accurately in such cases. Different configurations of the patch clamp (cell-attached, excised-patch, inside-out, outside-out, whole-cell, and perforated-patch) are illustrated in chapter 8, and the reader is referred to Sakmann and Neher (1983) for further technical details of patch clamping.

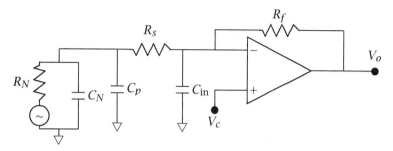

Figure A.32   Circuit diagram for a patch-clamp electrode in the whole-cell mode.

There are a number of sources of error in patch clamping (especially in the whole-cell mode) that should be mentioned. Some of these can be best explained by referring to figures A.22 and A.32. The pipette resistance (R_s in figure A.22) is a series resistance that is "outside the feedback loop." As such, it will produce an error in the membrane potential in a similar way to that described above for series resistance of the extracellular space. In other words, the actual potential across the membrane will differ from the clamp potential V_c by a factor $I_{cl}R_s$. For example, if R_s = 20 MΩ and I_{cl} = 1 nA, the error will be 20 mV. Obviously, one cannot clamp very well using a patch clamp when large currents are involved. There are two additional sources of error, however, associated with the presence of this series resistance.

1. When a step change in potential is applied to V_c, the potential across the membrane will be low-pass filtered by the R_s and C_N. For example, if R_s is as above and C_N is 100 pF, then the time constant of this filter will be 2 msec. To reach a steady-state membrane potential after a step command to V_c would require at least 10 msec (5 times the time constant).

2. The effect of R_s and C_N is also to filter membrane signals. All membrane signals, such as synaptic currents, will be low-pass filtered by R_s and C_N (also by C_p and C_{in}). The corner frequency or –3db point of this filter will be equal to

$$-3 \text{ db} = \frac{1}{2\pi R_s C_N} = 80 \text{ Hz}.$$

This would represent a significant amount of filtering for many fast synaptic signals.

A.5   Summary of important concepts

1. Ohm's law.
2. Definitions for and understanding of capacitance.
3. Kirchhoff's laws.
4. Voltage-divider circuit.
5. Input impedance (and resistance).
6. Measurement of membrane potential with high-impedance electrode.
7. Analysis of simple two-electrode voltage-clamp circuit.
8. Analysis of simple patch-clamp circuit.

A.6   Homework problems

1. Using Kirchhoff's laws and the relations $V = \frac{1}{C} \int i\, dt$ and $i = C\frac{dV}{dt}$, show that parallel capacitors sum and series capacitors sum as reciprocals.

2. Using figure A.8 and Kirchhoff's laws, derive V_3 in terms of E_1, E_2, R_1, R_2, and R_3.

3. Using Kirchhoff's laws, derive the equation for a voltage divider in figure A.10.

4. Derive the equation for the gain of the following circuit (i.e., V_2/V_1) in terms of R_1, R_2, R_3, R_4.

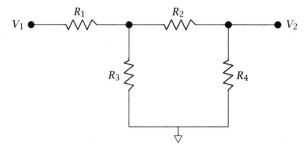

5. Derive the gain equation (V_o/V_i) for the noninverting amplifier in figure A.27.

6. For the voltage-clamp circuit

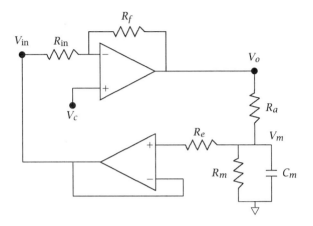

(a) Derive the equation for V_m as a function of V_c. (*Hint:* You can assume that the overall gain of the upper differential amplifier is $\frac{V_o}{V_c - V_{in}} = R_f / R_{in}$.)

(b) If you want V_m to equal 50 mV, determine what V_c must be with the following values: $R_{in} = 10^3$ Ω, $R_f = 10^5$ Ω, $R_a = 10^7$ Ω, $R_e = 10^7$ Ω, $R_m = 10^7$ Ω, and $C_m = 10^{-9}$ F.

7. For the circuit below, the applied signal is a sawtooth voltage and A is a current measuring device. If the sawtooth voltage goes from 0–10 mV in 100 msec, what is the waveform of the measured current (both amplitude and time course for 2 full cycles)? Let $C = 10^{-6}$ F and remember that $i = C dV/dt$ and $V = 1/C \int i\, dt$.

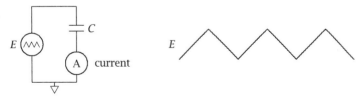

8. For the patch-clamp circuit below, what potential must be given as V_c to clamp the membrane (across C_N) in the steady state to -10 mV? Assume that E_i (resting potential) = 0 mV, R_s (series resistance) = 30 MΩ, R_f (feedback resistor) = 1 GΩ, and $R_N = 100$ MΩ. At steady state, what will V_o be?

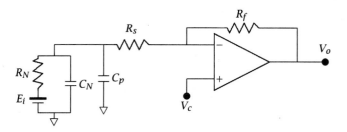

9. Using the figure below,

   (a) Derive an equation for V_m (the actual transmembrane potential) in terms of V_c, A, R_a, R_s, and I_{cl}. Assume that the output of amplifier A is just A times the difference between its inputs and that $R_1 = R_2 = R_3 = R_4$.

   (b) If V_c = 100 mV, A = 100, R_s = 1 MΩ, R_a =100 MΩ, and I_{cl} =1 nA, what is V_m in mV? What i R_s is 20 MΩ?

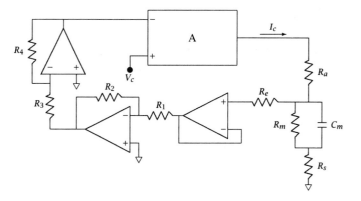

B   Optical Methods in Cellular Neurophysiology

B.1   Introduction

The purpose of this appendix is to present the basic concepts underlying some of the optical methods commonly used in cellular neurophysiology. This appendix is not intended to substitute for a physics course in optics or for a more in-depth study of microscopy that may eventually be undertaken by the serious student. The authors thought it would be useful for the student to have access to a beginning reference source for some of the optical methods he or she is likely to encounter in neurophysiology.

B.2   Definitions

B.2.1   Optics

Light Visible light composes a small part of the electromagnetic spectrum (figure B.1). All electromagnetic waves in a vacuum travel at the same velocity ($c = 2.998 \times 10^8$ m/sec) and differ only in their wavelength (λ) and frequency (f), that is,

$$c = \lambda \cdot f.$$

The velocity of light in a material substance depends on the wavelength of the light and the nature of the material. Visible light is typically considered in the range of 400–700 nm (figure B.1). Light is characterized by its wavelength, amplitude, phase, and plane of polarization.

Polarization Light is a transverse electromagnetic wave. (A transverse wave is one in which the displacement of the medium by the wave

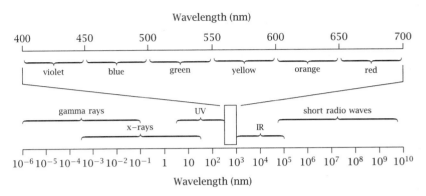

Figure B.1   The electromagnetic spectrum with the visible region expanded.

is perpendicular to the direction of propagation. Only transverse waves can be polarized.) There are three perpendicular vectors that describe light: the electrical and magnetic vectors, **E** and **B**, and the Poynting vector, **S**. **S** is in the direction of propagation while **E** and **B** oscillate at right angles to **S** (and to each other). Light is said to be polarized if **E** oscillates in certain ways and is *linearly polarized* if **E** oscillates in a single plane. Typically, a source such as a lamp will emit light at all angles of polarization from 0° to 180°. Polarized filters will pass light predominantly at a particular angle of polarization.

Phase   If one views light as a sine wave, then it also has a certain phase associated with it. For example, two sine waves that are 180° out of phase will cancel each other. Similarly, two light rays that are in phase will sum. Light normally contains waves of all phase relationships. When light from one source is split and then recombined, *interference* may result such that local maxima and minima occur in the combined light.

Reflection   Light will reflect from a surface at the same angle as the incident light.

Refraction   The bending of light when traveling from one medium into another (e.g., from air into water). The light bends because the velocity of light differs in the two media. The relationship between the angle (all angles are measured from a line perpendicular to the surface) of incident light in the first medium (I) and the angle of refracted light in the second medium (R) is given by *Snell's law* (figure B.2):

$$\frac{\sin I}{\sin R} = \frac{n_2}{n_1},$$

where n_2 is the refractive index of medium 2 and n_1 is the refractive index of medium 1. The refractive index of one medium with respect to another varies with wavelength.

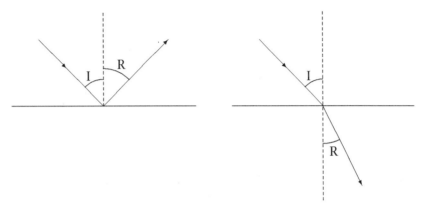

Figure B.2 Reflection (left) and refraction (right) of light. I = incident beams, and R = reflected or refracted beams.

Snell's law is thus derived from the definition of n or

$$\frac{v_1}{v_2} = \frac{n_2}{n_1},$$

and

$$n_1 = \frac{c}{v_1} \quad \text{and} \quad n_2 = \frac{c}{v_2},$$

where v_1 and v_2 are the velocities of light in the two media.

The refractive index for some common media are as follows:

Medium	Refractive index, n
air	1.00
glass	1.52
water	1.33
immersion oil	1.51

In general, materials show different refractive indexes at different orientations and hence are "nonisotropic." Also, the refractive index of materials is wavelength dependent.

Refraction of light at spherical surfaces (*thin lens*)   Parallel light incident to a lens will refract at both surfaces, forming a focal point (figure B.3). The focal length (f) of a lens in air is

$$\frac{1}{f} = (n-1)\left(\frac{1}{r_1} - \frac{1}{r_2}\right), \qquad\qquad (B.2.1)$$

where n is the refractive index of the lens and r_1 and r_2 are the radii of the two surfaces of the lens. The focal length of a lens is also defined by the distance of the object from the lens (o) and the distance of the image from the lens (i) by

$$\frac{1}{f} = \frac{1}{o} + \frac{1}{i}.$$

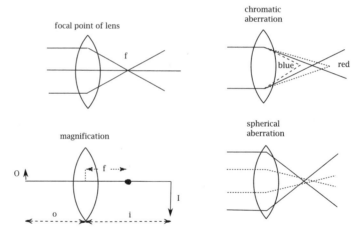

Figure B.3   Focal point of a lens, chromatic and spherical aberrations, and magnification.

Chromatic aberration Because the refractive index of a material varies with wavelength, the focal point of a lens will be different for different wavelengths (see equation B.2.1). This will cause image distortion when light of more than one wavelength is used to illuminate an object (see figure B.3).

Spherical aberration Light rays at the periphery of the lens will be refracted more than rays in the center (i.e., the amount of refraction depends on the angle of the incident light). This will prevent the focal point from being sharp and will produce distortion of an image (see figure B.3).

Resolution The least distance between two points that can be distinguished. A common definition for resolution is from Sir George Airy
and states that two point light sources are just resolved when the
bright center of the diffraction pattern of one source just touches
the first dark ring of the diffraction pattern of the other.

Magnification The ratio of the image size (I) to the object size (O). This
is also equal to the ratio of the distance of the image from the lens
(i) to the distance of the object from the lens (o) or the ratio of the
distance of the image from the focal point ($i - f$) to the focal length
(f) of the lens, or (see figure B.3)

$$M = \frac{I}{O} = \frac{i}{o} = \frac{i-f}{f}. \tag{B.2.2}$$

B.2.2 Microscopy

Stereo microscope A microscope with two separate image-forming paths,
one for each eye. The stereo microscope is characterized by a long
working distance, good depth perception, no lateral inversion of the
image, and (usually) low magnification.

Compound microscope The compound microscope uses two lenses for
image forming. The first lens (*objective*) forms an intermediate image of the object that is then enlarged further by the second lens
(*eyepiece*) to form a virtual image for observation (see figure B.4).
The *optical tube length* of a compound microscope is the distance of
the primary image formed by the objective to the focal point of the
objective ($i - f$ in figure B.3 and equation B.2.2). The primary image
is arranged so that it occurs at the focal point of the eyepiece. The
final image has the same orientation as the primary image (inverted
from the object) and is enlarged. The position of the final image can
be considered at infinity so that it can be viewed at any position. The
total magnification of the microscope (final image size/object size)
is therefore simply the product of the magnification of the objective
and the magnification of the eyepiece, or

$$
\begin{aligned}
M_T &= \frac{\text{final image size}}{\text{object size}} & \text{(B.2.3)}\\[2mm]
&= \frac{\text{primary image size}}{\text{object size}} \times \frac{\text{final image size}}{\text{primary image size}}\\[2mm]
&= M_{obj} \times M_{eyep}. & \text{(B.2.4)}
\end{aligned}
$$

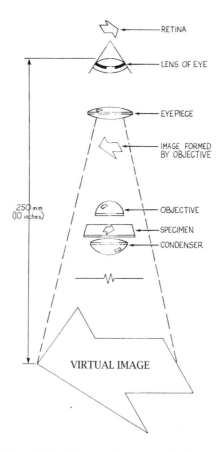

RETINA

LENS OF EYE

EYEPIECE

IMAGE FORMED
BY OBJECTIVE

250 mm
(10 inches)

OBJECTIVE

SPECIMEN

CONDENSER

VIRTUAL IMAGE

Figure B.4   Diagram of compound microscope. (From Delly 1980.)

- **Upright compound microscope** An ordinary compound microscope in which the objectives and eyepieces are above the specimen and the light source is below the specimen (see figure B.4). For focusing, either the microscope stage (i.e., the specimen) moves up and down or the objectives and eyepieces move up and down. These are called *stage-focusing* and *fixed-stage* microscopes, respectively.

- **Inverted compound microscope** A compound microscope in which the light source is above the specimen and the objectives are below the specimen. These are usually fixed-stage microscopes with movable objectives.

Objectives These are the most important parts of a compound microscope. They are differentiated on the basis of

- **Magnification** (See equation B.2.2.)

- **Numerical aperture (*NA*)** The numerical aperture is one of the most important characteristics of an objective lens. It determines the maximal resolution of the microscope. The *NA* of an objective is also a measure of the light-gathering ability of the lens. It is defined by the angular aperture (the angle of the cone of light coming into the objective) and the refractive index of the medium in which the lens operates (e.g., air, water, or immersion oil), that is,

$$NA = n \sin \alpha/2, \tag{B.2.5}$$

where n is the refractive index and α is the angular aperture (see figure B.5). The maximum possible angular aperture would be 180°, so the maximum possible *NA* would be 1 in air, 1.33 in water, and 1.51 in oil. One cannot achieve these values in practice, but they illustrate the importance of the working medium for obtaining high *NA*s. The highest *NA*s are normally achieved with oil immersion objectives.

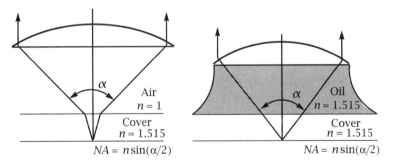

Figure B.5 Numerical aperture of an objective lens in air (left) or in oil (right). The cover slip is glass. (Adapted from Bradbury 1989 by permission of Oxford University Press, copyright © 1991 Oxford University Press.)

- **Working distance** The distance from the focal plane in the specimen to the surface of the objective for immersion objectives with no coverslip. Otherwise, the working distance depends on the refractive index of the coverslip and the immersion and mounting media.

- **Degree of optical correction** Most objectives are corrected to some degree for chromatic and spherical aberrations. Objectives are usually classified by the number of wavelengths for which the corrections have been made. There are *achromat* objectives, which have chromatic corrections at 2 wavelengths and spherical corrections at 1 wavelength; *semiapochromatic* or *fluorite* objectives, which have chromatic and spherical corrections at 2 wavelengths; and *apochromat* objectives, which have chromatic corrections at 3 wavelengths and spherical corrections at 2 wavelengths.

- **Spectral filtering properties** Specify which wavelengths are passed through the lens.

Resolution The resolving power of a microscope in the x, y dimensions (i.e., within the plane of focus) is diffraction limited (see previous discussion of resolution) and is determined by the NA of the objective and the wavelength of light, or

$$R(x, y) = 1.22 \frac{\lambda}{2NA}, \tag{B.2.6}$$

where R is the distance between two just resolvable points in the specimen. The above equation assumes that the numerical aperture of the light condenser matches that of the objective. Otherwise the equation for resolving power is

$$R(x, y) = 1.22 \frac{\lambda}{NA_{obj} + NA_{cond}}.$$

Illumination methods One of the more important and least understood aspects of microscopy is specimen illumination. Improper illumination will result in a reduction in resolution, poor image quality due to scattered light, and poor color reproduction. The essential components of a typical light source (see figure B.6) are the lamp, a field diaphragm, an aperture diaphragm, and one or two lenses for focusing the light. The goal is to provide a cone of light that fills the aperture of the objective. A cone of light larger than the aperture of the objective results in excessive scattered light, whereas a cone smaller than the aperture of the objective reduces the resolution of the microscope. The two most common methods of illumination are *source focus* and *Köhler*.

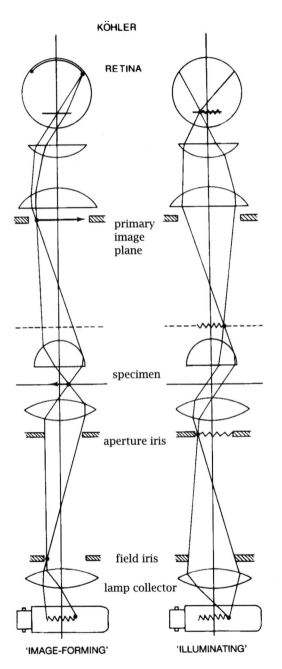

KÖHLER

RETINA

primary
image
plane

specimen

aperture iris

field iris

lamp collector

'IMAGE-FORMING'   'ILLUMINATING'

Figure B.6   Köhler illumination for a compound microscope. (From Bradbury 1989 by permission of Oxford University Press, copyright © 1991 Oxford University Press.)

In *source focus* illumination, one lens (substage condenser) is used to focus the light source directly on the specimen. The lamp itself must have a uniform surface so that the specimen is properly illuminated. In *Köhler* illumination the lamp is at the focal plane of the lamp condenser, which then projects a uniform beam of parallel light. The iris or field diaphragm, placed on the other side of the lamp condenser from the lamp, is uniformly illuminated and becomes a virtual light source. The aperture diaphragm is at the focal plane of the substage condenser so that an image of the lamp is at the plane of the aperture diaphragm and at the back focal plane of the objective. The specimen, however, is uniformly illuminated because the virtual light source of the field diaphragm is focused on the specimen by the substage condenser. Details of how to adjust for Köhler illumination should be obtained from the particular microscope manufacturer.

Depth of field The distance from the nearest to the farthest focal planes considered in focus. It can also be considered the resolution of the microscope in the z or depth dimension and is given by

$$R(z) \propto \frac{\lambda}{NA^2}.$$

Depth of focus Depth of focus refers to the distance around the image on a film plane that is considered in focus. Depth of focus (d_{focus}) is related to depth of field (d_{field}) by

$$d_{\text{focus}} = M_T^2 \cdot d_{\text{field}}.$$

Brightness Brightness is dependent on the magnification and NA by

$$\text{brightness} \propto \frac{NA^2}{M_{obj}^2}. \tag{B.2.7}$$

Empty magnification From equation B.2.6, the maximum resolution of a microscope is dependent on the NA of the objective and not on the total magnification. Magnification beyond a certain value (approximately $1000 \times NA_{obj}$) increases the size of the image but adds nothing to the resolution. One simply gets a larger, fuzzier image. This is called *empty magnification*.

Contrast enhancement techniques In addition to having adequate resolution, there must also be sufficient contrast of a specimen for proper viewing. A number of techniques are commonly in use for enhancing the contrast of specimens viewed through a compound microscope. These are briefly described below.

- **Dark field** The central portion of the light beam is blacked out so that only light scattered by the specimen enters the aperture of the objective. The specimen appears as a bright object against a dark background. A modification of this method in which different colors are substituted for the dark and light fields is called *Rheinberg differential color* illumination.

- **Phase contrast** The principle behind phase contrast enhancement is that the incident light beam is split into two (deviated and primary) beams with the deviated beam passing through a transparent medium to change its phase relationship with the primary beam. The two beams then pass through the specimen and recombine to produce an image. The specimen itself further changes the phase relationships between the two beams so that when they are recombined the edges and fine details produce interference (local maxima and minima) that enhance the contrast of the specimen.

- **Interference contrast (Nomarski)** In the interference contrast method, there are again two beams of light (object and reference). The object and reference beams are recombined before they reach the eyepiece. The specimen alters the phase of the object beam so that when the two beams are recombined interference occurs, which in turn enhances the contrast of the specimen. The *Nomarski* system merely refers to the method of separating the two beams. The Nomarski method of interference contrast gives an excellent (but artificial) three-dimensional appearance to the specimen.

- **Modulation contrast (Hoffman)** The Hoffman modulation contrast system provides similar contrast-enhancing capabilities to Nomarski, but is simpler to use and less expensive. The Hoffman system also has a longer depth of field and can be used with relatively thick specimens as compared to Nomarski. Modification to the light condenser and objective are necessary along with the use of a polarizing filter.

Fluorescence microscopy Fluorescence is the ability of some materials to emit light of a longer wavelength when excited by light of a shorter wavelength. The wavelength of the emitted light is normally longer than that of the exciting light (*Stokes law*). Fluorescence microscopy refers to the use of a compound microscope to excite and visualize fluorescent materials. All of the principles of microscopy discussed thus far hold for fluorescence microscopy except that there are a number of additional components to the fluorescence microscope that need to be considered.

The principal parts of a fluorescence microscope are the *light source*, which supplies sufficient light at the proper wavelength, the *excitation* or *primary* filter, which transmits light at only the wavelengths needed to excite the fluorescent material, and the *barrier* or *secondary* filter, which transmits light at the emitting wavelengths of the fluorescent material. The arrangement of these components in a transmission fluorescence microscope is illustrated in figure B.7.

It is usually desirable to have the excitation light incident to the specimen and the emitted light reflected from the specimen so that the illumination and observation are made from the same side of the specimen. This is called *epifluorescence* and is also illustrated in figure B.7. For epifluorescence there is an additional key component called the *epi-illumination mirror, chromatic beamsplitter,* or *dichroic mirror.* The dichroic mirror reflects the short-wavelength light from the excitation filter and transmits the longer-wavelength light to the barrier filter for subsequent observation. The brightness of the image for epifluorescence varies with NA^4 of the objective rather than with NA^2 (compare with equation B.2.7).

Video microscopy The use of a video camera to visualize, record, and analyze images from a light microscope is becoming quite common in cellular neurophysiology. Depending on the application, the video camera can range from a relatively inexpensive black-and-white camera to a very expensive, cooled, charge-coupled device (CCD) camera. The camera is usually mounted on the microscope with a trinocular head, tetraocular head (upright microscopes), or from the side (inverted microscope). In these cases, one can view the specimen through the eyepieces or through the camera, depending on the position of a mirror. The output of the camera can be viewed with a standard video monitor or on a computer screen after the video image is converted to digital form. A few of the most common terms associated with video microscopy are described below (see also Inoué 1986).

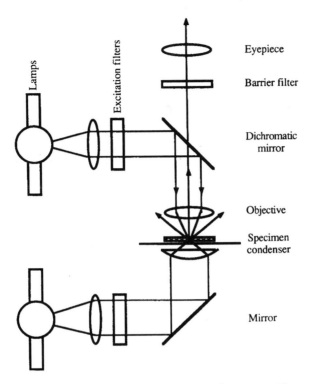

Figure B.7 Diagram for measuring simultaneous epifluorescence and transmitted fluorescence. (From Rost 1992 by permission of Cambridge University Press, copyright © 1992 Cambridge University Press.)

- **NTSC** The video standard adopted by the National Television Systems Committee. The standard is 525 horizontal scan lines and 60 fields/sec, with each field consisting of 262.5 lines. A *frame* consists of two interlaced fields.

- **Frame rate** A frame consisting of 525 lines (2 fields) will be repeated 30 times/second. The maximum *time resolution* of video microscopy is therefore 1/30 sec or 33 msec. There are also two other common standards (PAL and SECAM) that use 625 lines-50 fields/sec. Unfortunately, all are incompatible with each other.

- **Frame grabber** The hardware used for acquisition and storage of a video frame in digital form. This normally consists of analog-to-digital (A/D) conversion hardware and digital memory for storing the data.

- **Pixels** Picture elements in a digitized video frame. During the A/D conversion, each horizontal scan line is converted into a series of discrete sample values, which are called pixels. The number of total pixels in the image determines its *spatial resolution.*

- **Gray levels** The brightness levels of the pixels. If each pixel carries 2 bits of brightness information, then there will be 4 (or 2^2) gray levels. If there are 8 bits of information for each pixel, then there will be 256 (or 2^8) gray levels, and so forth. Similarly for color, if each pixel is 8 bits, then there can be 256 different colors. There is no reason to display more than about 256 gray levels because normally that is all the human eye can distinguish.

- **Contrast (image) enhancement** In addition to the various methods of contrast enhancement for the microscope discussed earlier, there are many methods for enhancing the contrast of a video image. These include simple analog methods as well as sophisticated digital image enhancing techniques. One of the simplest methods for image enhancement is to decrease the overall brightness levels in the image. For example, if most of the differences in contrast within a specimen occur in a narrow range of gray levels, then the rest of the brightness levels from the image can be subtracted out so that the entire dynamic range of the recording system can be used to explore the relatively narrow range of brightness differences within the image.

- **Nyquist frequency** When analog signals are digitized, the rate of digitization must be at least twice that of the highest-frequency components in the analog signal. Otherwise, spurious noise will appear in the digitized signals. For example, if the analog signals are low-pass filtered at 10 kHz, then the A/D sampling rate must be at least 20,000/sec. Also, in digitized images the sampling rate (or pixelization) should be 2.3 times denser than the resolution of the optics. Otherwise, high spatial frequencies in the original image will appear at lower frequencies in the digitized image. This is called *aliasing*. It is interesting to note that the digitization done by the photoreceptors in the eye exactly matches the Nyquist theorem, that is, the density of photoreceptors is 2.3 times larger than the resolving power of the lens.

- **Photodiode** A special diode that converts light energy into an electrical signal. The output from a photodiode array would be equivalent to a digitized video image (although at reduced spatial resolution) with each photodiode representing one pixel.

- **Photomultiplier** A device that is extremely sensitive to low levels of light. A photomultiplier will generate electrical signals for brief flashes of light consisting of only a few photons.

Confocal microscopy A microscope system in which illuminating light is focused through a pinhole and then on the specimen, and the transmitted (or reflected) light from that point in the specimen is focused to a point in the detector (see figure B.8). Such a microscope is thus "confocal." The confocal microscope can be explained in the following way:[1] a light source emits light that is parallelized by means of a collector lens. A system of two achromatic lenses and a pinhole, which is positioned in the focal point of both achromats, form a point light source. The second achromat parallelizes the light coming from the point source and this light is directed onto the specimen after it has passed a beamsplitter (for reflected light) or a dichroic mirror (for fluorescence). Light reflected or emitted (in the case of fluorescence) by the specimen passes the beamsplitter or dichroic mirror and is detected by a photomultiplier tube after it has been imaged via the achromatic lens and a second pinhole. It is important that in the objectives both pinholes overlap and are "confocal." This ensures that only light from a very small region around the illumination focal point is detected and all other light is blocked by the pinhole. An image of a specimen is detected pointwise by scanning the stage under the objective. The data are stored in a computer and afterwards images are reconstructed. The confocal microscope has a lateral resolution that is slightly higher than a conventional microscope (by a factor of 1.414) while the depth (z) resolution can be an order of magnitude higher. Modern confocal microscopes use lasers for illumination sources. One reason is that it is easier to scan the laser beam than the stage. Also, lasers produce intense monochromatic light, which eliminates chromatic aberration.

Light sources The most common types of light sources are tungsten-filament lamps, gas-discharge lamps, and lasers. The considerations for choosing a light source include intensity, spectral density, and

---

[1]Description from Johannes Helm

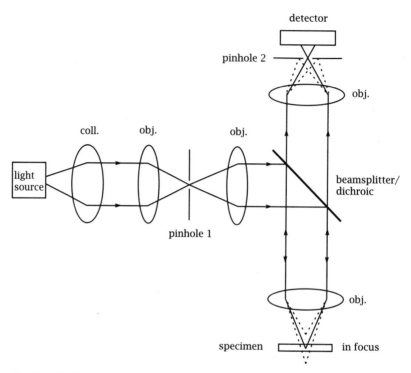

Figure B.8 Diagram of a confocal microscope (see text for description).

stability. Tungsten-filament lamps are the most common method of illumination for most forms of microscopy. Gas-discharge lamps and lasers are used for fluorescence microscopy. The two most common gas-discharge lamps are mercury and xenon arc lamps. The mercury arc lamp has much higher emission intensity than the xenon arc, but the highest emission is concentrated in a few narrow bands of wavelengths. These wavelengths may or may not correspond to the *fluorophore* (fluorescent molecule or dye) being used. The xenon arc lamp has lower intensity but broader band emission than the mercury arc lamp. Lasers emit pure monochromatic light, which may be ideal for some fluorescence microscopy applications, in particular, for confocal microscopy.

B.3 Optical probes

Optical techniques have been used for many years to monitor different aspects of neuronal function. They have a number of advantages: They are relatively noninvasive, are fast responding, can be "imaged," giving information about more than one location and can be used to supplement electrical measurements. As early as 1968 several groups reported that when an action potential was generated in a large axon (typically squid or crab) there were intrinsic changes in light scattering and birefringence of the axon. Furthermore, when the axon was stained with a dye there were extrinsic changes in fluorescence or light absorption associated with changes in membrane potential.[2] The intrinsic light signals are often small, whereas the fluorescence and absorbance associated with the use of dyes are generally larger and easier to measure. Much effort has thus been spent on the design of these dyes, which are also called *optical* or *molecular probes*.

In addition to those that measure membrane potential, there are also optical probes available that bind to specific ions or molecules and change the intensity of their absorption or fluorescence emission as a function of the concentration of the particular ion (e.g., Ca^{2+}). Variations in the design of some of these molecular probes have also led to so-called caged compounds, in which previously bound ions or molecules are released from these probes upon illumination of light at particular wavelengths.

B.3.1 Potential-dependent light signals

Birefringence refers to the ability of an object to have different values of refractive index for polarized light at different planes of polarization. Birefringence is defined as the difference between the refractive index with the plane of polarization parallel to the optic axis and the refractive index with the plane of polarization at 90° to the optic axis. If an axon is placed at an angle of 45° to the plane of polarization of incident light, there will be light transmitted with a plane of polarization at 90° to that of the incident light. The intensity of this transmitted light is taken as a measure of the birefringence of the preparation, and has been found to decrease transiently during neural activity. The birefringence signal is usually indicated as a change

---

[2] There are also intrinsic light scattering and reflected signals associated with neural activity that can be recorded from the surface of a mammalian brain (see Lieke et al. 1989).

in the intensity of this transmitted light during activity divided by the resting intensity, or $\Delta I/I$. The measurement of birefringence is most useful for linear structures like axons. For spherical cells the birefringent signals at different angles tend to cancel each other.

Light scattering (also called *turbidity*) refers to the ability of a neuronal preparation to scatter light at different angles from the incident light. Depending on the angle of measurement, the intensity of the scattered light ($\Delta I/I$) can either increase or decrease during nerve activity.

Fluorescence refers to the change in fluorescence emission intensity as a function of the membrane potential of a dye-stained neuron (usually indicated as $\Delta F/F$).

Absorbance refers to the change in the absorption of incident light at a particular wavelength as a function of the membrane potential of a dye-stained neuron (usually indicated as $\Delta I/I$).

Thousands of potentiometric dyes have been tested for monitoring neuronal activity via changes in fluorescence or absorbance. The useful dyes have been classified into two groups. Those of one group, called *permeant dyes*, cross the membrane as a function of the transmembrane potential; those classified as *impermeant dyes* are bound to the membrane and sense the potential either directly or indirectly (Cohen et al. 1989; Salzberg 1989; Tsien 1989). Permeant dyes produce relatively large changes in optical signals but are slow to respond (high msec range) while impermeant dyes yield smaller signals but respond much faster to changes in membrane potential (μsec range).

Some of the problems associated with the use of any dyes on live tissue are possible pharmacological effects, dye bleaching, and photodynamic damage. Photodynamic damage results from the generation of toxic free radicals when the dye is illuminated in the presence of oxygen. In some cases this damage can be quite severe and limit the usefulness of the dye.

B.3.2   Concentration-sensitive fluorescent probes

By far the most common use of optical probes in cellular neurophysiology is for measuring the concentration of molecules (usually ions) inside neurons. Monitoring changes in ion concentrations can be useful for determining physiological functions and for providing information about neural activity. There are fluorescent indicators available for Ca^{2+}, Na^+, pH, K^+, Cl^-, and Mg^{2+} as well as more exotic indicators for cAMP and protein

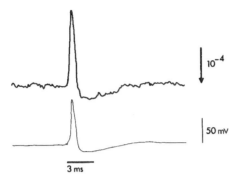

Figure B.9   Absorption change (top) for a dyed axon during action potential (bottom). (From Ross et al. 1974 by permission of the Biophysical Society, copyright © 1974 Biophysical Society.)

kinase C (see the chapter by Poenie and Chen in Herman and Lemasters 1993).

The use of such indicators involves either injecting a membrane-impermeant dye inside the cell or applying a membrane-permeant form of the dye (if available) to the outside of the cell. The latter is usually the acetoxymethyl (AM) ester form of the dye molecule. This ester form is membrane permeant, but the ester gets removed by cytosolic esterases, leaving the membrane-impermeant form of the dye "trapped" inside the cell. The cell is then illuminated at the appropriate excitation wavelength for the particular dye, and the fluorescence emission intensity is measured (see *fluorescence microcopy* described in section B.2.2).

Although changes in concentrations of ions can be readily monitored by this method (i.e., as $\Delta F/F$), it is difficult to calibrate the magnitude of the change unless the dye can be used for *ratioing*. The ratiometric method for measuring ion concentration is best explained by illustrating the use of ratioing for the dye fura-2. In figure B.10 the fluorescence emission intensity for fura-2 as a function of excitation wavelength and Ca^{2+} concentration is illustrated. At 360 nm excitation fura-2 emission is insensitive to changes in Ca^{2+} concentration. At 340 nm excitation, however, the emission intensity increases with increasing Ca^{2+} concentration, and at about 380 nm excitation the emission decreases with Ca^{2+} concentration.

In principle, it is possible to determine the concentration of Ca^{2+}, independent of dye concentration, optical pathlength, and instrumentation parameters, by determining the ratio (R) of emission intensities at the two wavelengths (e.g., 340 and 380 nm) or

$R = F_{340}/F_{380}.$

This leads to the equation (Grynkiewicz et al. 1985)

$$[Ca^{2+}]_{in} = K_d \frac{R - R_{min}}{R_{max} - R} \cdot \frac{S_{f2}}{S_{b2}}, \tag{B.3.8}$$

where R_{min} is the ratio at 0 $[Ca^{2+}]_{in}$, R_{max} is the ratio at saturating concentrations of Ca^{2+}, K_d is the dissociation constant for fura-2, and S_{f2}/S_{b2} is the ratio of emission intensities at 380 nm for fura-2 in 0 $[Ca^{2+}]$ to fura-2 in saturating $[Ca^{2+}]$. When fura-2 is used intracellularly, there is also a viscosity correction factor that is multiplied by R in equation B.3.8 because its spectra depend on the presence of ions and proteins in the solution. (Ratioing of fura-2 with 360 and 380 nm (or even 390 nm) wavelengths is also quite common, because objectives that pass 340 nm are usually quite expensive.)

It is possible to determine the magnitude of changes in $[Ca^{2+}]_{in}$ using single-wavelength excitation (usually 380 nm) if the initial or resting concentration of Ca^{2+} is known (usually through the use of ratioing as just described). The equation for determining $[Ca^{2+}]_{in}$ based on single-wavelength excitation is given by (Lev-Ram et al. 1992)

$$[Ca^{2+}]_{in} = \frac{[Ca^{2+}]_{rest} + K_d(\Delta F/F/(\Delta F/F)_{max})}{(1 - (\Delta F/F/(\Delta F/F)_{max}))}, \tag{B.3.9}$$

where $(\Delta F/F)_{max}$ is determined from the calibration of the microscope and imaging setup.

B.4 Photoactivated "caged" compounds

In neurophysiology it is often useful to induce a sudden change in the concentration of an ion or other biologically important molecule. For example, a sudden increase in $[Ca^{2+}]_{in}$ could be useful for studying neurotransmitter release, or a sudden decrease in $[Ca^{2+}]_{in}$ might allow one to determine the timing requirements for a rise in $[Ca^{2+}]_{in}$ during the induction phase of LTP. Also, sudden increases in second-messenger molecules like cAMP, cGMP, or GTP would permit high time resolution of certain chemical processes. "Caged" compounds were developed for just such purposes and represent molecules that contain a photosensitive masking or caged group that inhibits the activity of the biologically relevant

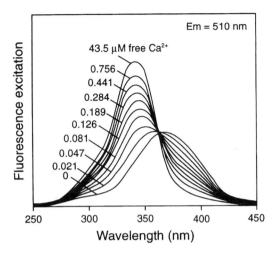

Figure B.10 Excitation spectra (emission at 510 nm) for fura-2 in different concentrations of free Ca^{2+}. (From Molecular Probes 1993.)

molecule. With the photolytic removal of the masking group, the biologically active molecule becomes released and available for action.

Caged compounds have been developed for most nucleotides, amino acid neurotransmitters, and Ca^{2+} buffers. In the case of Ca^{2+} buffers, compounds have been developed in which, upon photolysis, Ca^{2+} can be released from the caged buffer molecule or the buffer itself can become active, bind Ca^{2+}, and reduce the free Ca^{2+} concentration.

B.5 Summary of important concepts

1. Snell's law.

2. Magnification.

3. Numerical aperture and relationship to resolution.

4. Fluorescence microscopy.

5. Pixels.

6. Gray levels.

7. Confocal microscopy.

8. Optical probes.

C Short Answers to Homework Problems

Chapter 2

2. (a) 10^{-8}

    (b) 3.1×10^{-5}

    (c) 2.02×10^{-4}

3. (a) The final concentrations (mM) are:

	a	b
R^+	150	–
K^+	193	257
Cl^-	343	257

    (b) -7.2 mV

    (c) Yes, H_2O will flow from b to a.

4. (a) $P_K = 5.9 \times 10^{-7}$ cm/sec.

    $P_{Na} = 7.99 \times 10^{-9}$ cm/sec.

    (b) $V_{rest} = -89$ mV.

    (c) If $P_K/P_{Na} \to 0$, $V_{rest} = +64.7$ mV.

    If $P_K/P_{Na} = 1$, $V_{rest} = -4.96$ mV.

5. (a) $P_K > P_{Ca}$.

    (b) $P_K < P_{Ca}$.

    (c) $[K^+]_{in} > [K^+]_{out}$.

    (d) $[Ca^{2+}]_{in} < [Ca^{2+}]_{out}$.

6. (c) $\dfrac{P_{Na}}{P_K} = 0.03$.

7. (a) Outward rectified

   (b) -59.5 mV

8. (a) The neuron is *not* at ECE; the principle of space-charge neutrality is obeyed.

   (b)

	inside (mM)	outside (mM)
K^+	200	100
Na^+	10	100
Cl^-	100	200
A^-	110	0

   (c) $E_K = -17.5$ mV, $E_{Na} = 0$, $E_{Cl} = -17.5$ mV, and $V_{rest} = -17.5$ mV.

9. (a)

	Inside	Outside
K^+	200	100
Na^+	10	90
Cl^-	100	200
A^-	110	0
Ca^{2+}	10^{-4}	5

   (b) -3.5 mV

Chapter 3

2. (a) In darkness, $V_{rest} = -86$ mV. Under constant light, $V_{rest} = 43$ mV.

   (b) g_{Cl} increases 5.14-fold. $\Delta g_{Cl} = 0.37 \times 10^{-8}$ S.

   (c) 2 mV hyperpolarization

3. (a) At rest, $g_a = 0.029$ S and $g_b = 0.0077$ S. During excitation, $g_a = 0.029$ S and $g_b = 0.4$ S.

   (b) $V_{rest} = -42.7$ mV.

   (c) $V_p = +51.2$ mV.

Chapter 4

1. (b) $R_N = 3 \times 10^5$ Ω.

(c) $V(\infty, 0) = 3$ mV.

$V(\infty, \lambda) = 1.1$ mV.

$V(\infty, 10\,\lambda) = 0.00014$ mV.

(d) $V\left(\dfrac{\tau_m}{2}, 0\right) = 2.04$ mV.

3. (a) $G_N = 2.56$ nS.

   (b) $G_1 = 1.57$ pS.

   (c) $B_1 = 6 \times 10^{-4}$ (open circuit).

5. (a) 225 Ω; 450 Ω; 591 Ω.

   (b) 3.7 mV; 3.7 mV; 6.5 mV.

6. (a) No

   (b) Yes

   (c) Yes

7. $V_0 = +84$ mV.

8. (a) 1.414

   (b) 0.354

   (c) 1.414

   (d) 1

9. (a) $R_N \simeq 2 \times 10^9$ Ω.

   (b) $C_N = 12.6$ pF; $\tau = 25$ msec.

   (c) i. $\rho = 5.6$.

       ii. $G_N = 3.3 \times 10^{-9}$ S; $V_{ss} = 3$ mV.

   (d) $R_N = 2.6 \times 10^8$ Ω; $\rho = 2.8$.

10. (a) True

    (b) False

    (c) False

    (d) False

    (e) False

11. (a) $L = 2$.

    (b) $X = 1$.

(c) $V_L/V_0 = 0.266$.

(d) $\lambda_{10\text{Hz}} = 0.26$ cm.

12. (a) $L_1 = 0.85$.

$L_2 = 0.65$.

(c) $G_N = 9 \times 10^{-9}$ S.

(d) $\rho = 0.8$.

(e) $V_m = -181$ mV with bias current and +41 mV with 2 nA current.

13. (a) $G_S = 56.5 \times 10^{-9}$ S.

(b) $5\, G_D = 13.5 \times 10^{-9}$ S.

(c) $G_\infty = 1.26 \times 10^{-8}$ S.

(d) $G_N = 82.6 \times 10^{-9}$ S.

(e) $V_m = 51$ mV.

14. (a) $G_N = 20$ nS.

(b) $G_{den} = 10$ nS; $\rho = 1$.

(c) $\lambda = 0.01$ cm.

$r_i \cong 8 \times 10^9\ \Omega/\text{cm}$.

$r_m = 8 \times 10^5\ \Omega\text{-cm}$.

$R_m = 2510\ \Omega\text{-cm}^2$.

$r \cong 14\ \mu\text{m}$.

$\tau_m = 2.5$ msec.

(d) $V(\infty, L) = 32.4$ mV.

(e) $V(\infty, \lambda) = 18.4$ mV.

18. (a) $G_N = 4.48 \times 10^{-3}$ S.

(b) At a, $V = 223.4$ mV; at b, $V = 78.9$ mV; at c, $V = 78.9$ mV; and at d, $V = 113.1$ mV.

19. $\tau = -\dfrac{0.434(t_2 - t_1)}{\log(V_2/V_1)}$.

20. (a) $R_N(\text{sphere}) = 9$ mV; $R_N(\text{inf}) = 11.3$ mV; $R_N(\text{semi}) = 22.6$ mV; $R_N(\text{fin}) = 29$ mV.

(b) $\theta = 14$ cm/sec; $t = 7.5$ msec.

21. (a) True, false, false, true.

(b) True, false, true, true.

22. (a) $d_{new} = 6.6\ \mu m$.

   (b) $L = 0.9$.

   (c) 30%

   (d) Greater than

23. (a) $d_A = 1.6 \times 10^{-4}$ cm; $L_A = 0.7$; $L_B = 3.5$.

   (b) $r_i(A) = 10^{10}\ \Omega$; $G_N(A) = 1.2 \times 10^{-9}$ S; $r_i(B) = 2.6 \times 10^{10}\ \Omega$; $G_N(B) = 0.96 \times 10^{-9}$ S; $G_N = 3.2 \times 10^{-9}$ S.

   (c) $V_{rev} = +283$ mV.

Chapter 5

3. (a) Cl^-

   (b) $a_A = 0$; $a_C = 0.042$; $a_F = 0.137$; $a_J = 0.672$; and $a_K = 1$.

   (c) Constant field model

4. (a) A = single energy-barrier model; B = either; C = constant field model; D = either; and E = single energy-barrier model.

   (b) 24

   (c) +2

   (d) -34.8 mV

5. (a) Activated by depolarization

   (b) $\overline{g}_y = 571$ pS; $\tau_y = 80$ msec.

   (c) $\tau_m = 1.256$ msec; $\Delta V = 40$ mV.

7.

V_p (mV)	I_∞ (mA/cm^2)	τ (msec)	y_∞ (unitless)	α (msec^{-1})	β (msec^{-1})
-20	0.6	1	1	1	0
-40	0.3	1.5	0.75	0.5	0.17
-60	0.1	2	0.5	0.25	0.25
-80	0	1.8*	0.35*	0.19*	0.37*
-100	-0.05	1.5	0.25	0.17	0.5

* Extrapolated values from the graph.

8. (a) $V_{rest} = -60$ mV; $V^*_{stimulus} = +16$ mV.

   (b) $V_{rest} = -60$ mV; $V^*_{stimulus} = +16$ mV; $V^{(\infty)}_{stimulus} = 55.4$ mV.

Chapter 6

1. (b) For condition 1, $g_{Na}^{peak} = \frac{1}{90}$ S/cm^2. For condition 2, $I_{Na}^{peak} = -0.55$ mA/cm^2. For condition 3, $I_{Na}^{peak} = -0.25$ mA/cm^2, and for condition 4, $I_{Na}^{peak} = +0.39$ mA/cm^2.

4. 1.554 mA/cm^2

5. (a) K$^+$

    (b) Cl$^-$

    (c) No contribution

    (d) At $t = 1$ msec, $V = -50$ mV, $g_{Cl} = 6.8$ nS and $g_{Na} = 0$. $V = +150$ mV, $g_{Cl} = 16.5$ nS.

    At $t = 7$ msec, $V = -50$ mV, $g_{Cl} = 0$ and $g_{Na} = 2.3$ nS. $V = +150$ mV, $g_{Cl} = 0$ and $g_{nA} = 5.6$ nS.

6. (a) Leakage conductance

9. (a) At rest, $x = 12$; at threshold, $x = 5.5$; and at AP peak, $x = 0.18$.

    (b) $V_{rest} = -77.8$ mV.

10. (a) If $V_P = 100$ mV, $g_{K\infty} = 83.3$ mS/cm^2.

    If $V_P = 85$ mV, $g_{K\infty} = 82.4$ mS/cm^2.

    If $V_P = 60$ mV, $g_{K\infty} = 71.4$ mS/cm^2.

    If $V_P = 25$ mV, $g_{K\infty} = 47.6$ mS/cm^2.

    (b)

V_P(mV)	α_n(sec^{-1})	β_n(sec^{-1})
100	833	0
85	714	0
60	600	25
25	435	66

11. (b) $[Ca^{2+}]_{in} = 1.78$ μM; $[Cl^-]_{in} = 2.087$ mM.

12. (a) 1.87×10^{10} channels/cm^2

    (b) 1.07×10^{-11} S/channel

    (c) Activated by hyperpolarization

13. (a) 1.02×10^{10} channels/cm^2

    (b) 100 pS

14. (b) g_{Na} is not ohmic.

Chapter 9

1.

$$F(X) = \begin{cases} 0 & \text{if } x < 2 \\ \frac{1}{4}(x-2) & \text{if } 2 \le x \le 6 \\ 1 & \text{if } x > 6 \end{cases}$$

2.

$$F(x) = \begin{cases} 0 & \text{if } x < 0 \\ \frac{1}{3}x^2 - \frac{2}{27}x^3 & \text{if } 0 \le x \le 3 \\ 1 & \text{if } x > 3 \end{cases}$$

3. (a)

$$F(x) = \begin{cases} 0 & \text{if } x < 0 \\ \frac{3}{5}x^2 & \text{if } 0 \le x \le 1 \\ 1 - \frac{2}{5x^3} & \text{if } x > 1 \end{cases}$$

   (b) $\frac{4}{5}$

   (c) The median is 0.9129; for the upper quartile, $F(x) = \frac{3}{4}$; and for the lower quartile, $F(x) = \frac{1}{4}$. The semi-interquartile range is 0.2621.

4. (b) $y = 6.25$ nS, and $N \approx 42$ (channels)

5. (a) At $V = -100$ mV, 0.025%; at $V = -50$ mV, 0.5%; and at $V = 0$ mV, 9.1%.

   (b) 8,000,000

   (c) $-7,500$ pA, -9.1×10^5 pA, -22.75 nA

   (d) 167 pA, 498 pA, 1017 pA

Chapter 10

1. (a)

$$Q_{-80} = \begin{pmatrix} -2956 & 2956 \\ 0.73 & -0.73 \end{pmatrix}$$

$$Q_{+20} = \begin{pmatrix} -243 & 243 \\ 109 & -109 \end{pmatrix}$$

(b) At $V = -80$ mV, $\mu_I = -65$ pA and $\tau = 0.34$ ms. At $V = +20$ mV, $\mu_I = -18.6$ nA and $\tau = 2.8$ ms.

(c) $\tau(-80$ mV$) = 0.34$ msec. $\tau(+20$ mV$) = 2.8$ msec

2. (a) $Q = \begin{bmatrix} -100 & 100 \\ 2 & -2 \end{bmatrix}$ det $Q = 0$

(b) $\overline{\tau}_{open} = 0.01$ sec.

$\overline{\tau}_{closed} = 0.5$ sec. $P_1(\infty) = 1.96\%$, and $P_2(\infty) = 98.04\%$.

(c) -2×10^{-10} A

(d) 1.4×10^{-11} A

(f) $S(0) = 3.84 \times 10^{-21}$ A$^2 \cdot$ sec, and $f_c = \frac{1}{2\pi\tau} = \frac{1}{2\pi(0.01)S} = 16$ Hz

3. (a) $Q = \begin{pmatrix} -2.71 & 2.71 \\ 0.0736 & -0.0736 \end{pmatrix}$ (sec^{-1})

(b) $P_1(\infty) = 2.6\%$, and $P_2(\infty) = 97.4\%$.

(c) $N = 2500$.

4. (b) 2nd power function of a single exponential.

6. (a) $\lambda_1 = 0$, $\lambda_2 = -6.63$ sec^{-1}, and $\lambda_3 = -158$ sec^{-1}.

(b) Mean open lifetime $= 6.7$ msec; mean blocked lifetime $= 200$ msec; and mean closed lifetime $= 100$ msec.

(c) $f_{c_1} = 1.06$ Hz, and $f_{c_2} = 25$ Hz.

7. (a)

$$Q = \begin{bmatrix} -k_{-3} & k_{-3} & 0 & 0 \\ k_3 & -(k_3 + k_{-2}) & k_{-2} & 0 \\ 0 & k_2 & -(k_2 + k_{-1}) & k_{-1} \\ 0 & 0 & k_1 & -k_1 \end{bmatrix}$$

(b) $\overline{\tau}_1 = \frac{1}{k_{-3}}$; $\overline{\tau}_2 = \frac{1}{k_3 + k_{-2}}$; $\overline{\tau}_3 = \frac{1}{k_2 + k_{-1}}$; and $\overline{\tau}_4 = \frac{1}{k_1}$.

8. (a) $\lambda_1 = 0$, $\lambda_2 = -55$ sec^{-1}, and $\lambda_3 = -445$ sec^{-1}.

(b) $f_{c_2} = 8.75$ Hz. $f_{c_3} = \frac{1}{2\pi\tau_3} = 70.9$ Hz.

9. (a) 0.4 nS

(b) 20 pA

(c) $\alpha = 0.0044$ sec^{-1}; $\beta = 0.0029$ sec^{-1}; $k_B X_B = 0.022$ sec^{-1}; and $k_{-B} = 0.02$ sec^{-1}.

10. (a) Macroscopic current: $\tau = \frac{1}{\alpha+\beta}$. Sum of five channels: $\tau = \frac{1}{\alpha}$.

11. (d) $+70 : \mu_I = 5.03$ nA; $+50 : \mu_I = 1.88$ nA; $0 : \mu_I = -0.25$ nA; $-50 : \mu_I = -0.91$ nA, and $-70 : \mu_I = -0.63$ nA.

12. (a)

$$\mathbf{Q} = \begin{pmatrix} -150 & 100 & 50 \\ 10 & -10 & 0 \\ 5 & 0 & -5 \end{pmatrix}$$

$\lambda_1 = 0$; $\lambda_2 = -6.5$ sec^{-1}; and $\lambda_3 = -158$ sec^{-1}.

13. (a) $k_1 = 100$ sec^{-1}; $k_2 = 20$ sec^{-1}.

(b)

$$\mathbf{Q} = \begin{pmatrix} -150 & 100 & 0 & 0 & 0 & 0 & 50 \\ 10 & -45 & 10 & 0 & 0 & 25 & 0 \\ 0 & 20 & -80 & 40 & 0 & 20 & 0 \\ 0 & 0 & 30 & -50 & 20 & 0 & 0 \\ 0 & 0 & 0 & 40 & -40 & 0 & 0 \\ 0 & 50 & 20 & 0 & 0 & -120 & 50 \\ 10 & 0 & 0 & 0 & 0 & 50 & -60 \end{pmatrix}$$

(c) 6.67 msec

(d) 10 msec

14. (a) -24 nA

(b) $\tau_1 = 15.9$ msec; $\tau_2 = 0.8$ msec.

(c) $\alpha = 290$ sec^{-1}, $\beta = 29$ sec.

$$\mathbf{Q} = \begin{pmatrix} -290 & 290 & 0 \\ 29 & -1029 & 100 \\ 0 & 100 & -100 \end{pmatrix}$$

15. (b) 0.32 msec

(c) $P = 0.04$

Chapter 11

2. (a) 3 or 4 failures

    (b) 42 times

4. (a) Mean amplitude = 1.0 mV.

    (b) Independent

    (c) $r = 0.27/\text{sec}$.

5. (b) Direct: $m_d = 1.21$; failures: $m_f = 1.27$.

    (c) 18 times

6. (a) 10.6 nS

    (b) 5.8

    (c) 3

7. (b) Before LTP: 3.33; during LTP: 5.0

    (c) Before: $p = 0.0067$; after: $p = 0.01$.

    (d) Poisson: 29 before LTP and 149 during LTP; binomial: 29 before LTP and 153 during LTP. Yes.

9. (a) mean mEPSP = 0.6 mV; $\sigma = 0.18$.

    (b) mean EPSP = 1.46 mV; $\sigma = 0.93$.

    (c) $m_f = 2.6$; $m_d = 2.4$; $m_{cv} = 2.5$.

    (d) 1 or 2 failures

10. $\frac{\text{EPSP}(t) - \text{EPSP}_0}{\text{EPSP}_0} = A \exp(-t/10) + P \exp(-t/60)$.

Chapter 13

1. (b) $G_T = 120$ nS; $E_T = -58$ mV; inhibitory.

2. (a) $\tau_{\text{decay}} = 17.4$ min.

    (b) G_s for control and MTP is 37.5 nS.
    E_s for control is -22 mV.
    E_s for MTP is $+20$ mV.

    (c) $G_K = 18.8$ nS, $G_K/G_{Na} = 1$ (control); $G_K = 8.6$ nS, $G_K/G_{Na} \approx 0.3$ (during MTP).

3. (a) $V_{rest} = -64.3$ mV.

   (b) i. $\Delta I_A = -2.3$ mA/cm^2.
       ii. $\Delta I_B = -0.7$ mA/cm^2.
      iii. $\Delta I = -3.0$ mA/cm^2.

   (c) (i) = +50 mV; (ii) = -100 mV; (iii) does not exist.

4. (a) Reversal potentials: A = -5 mV; B = -40 mV; C = -70 mV; and D = -60 mV.

   (b) A: conductance increase; B: conductance decrease; C: conductance decrease; D: conductance increase.

5. (a) $G_S = 100$ nS; $E_{rev} = -20$ mV from rest.

   (b) Inhibitory

   (c) $R_N \cong 50$ MΩ.

   (d) $\tau_m \cong 44$ msec.

   (e) No.

6. (a) *A*: $G_A = 50$ nS; $V_{rev} = 0$ mV. *B*: $G_B = 75$ nS; $V_{rev} = -60$ mV.

   (b) A is excitatory, B is inhibitory.

   (c) No

   (d) $V_{rev} = -36$ mV; $G_{A+B} = 125$ nS; excitatory.

   (e) Conductance increase.

   (f) Excitatory.

7. (a) ΔG_{Na}; +50 mV.

   (b) EPSP.

   (c) IPSP.

8. (a) $V_m = -33$ mV; $I_{cl} = 2$ nA.

   (b) $V_m = -20$ mV; $I_{cl} = 4$ nA.

9. (a)

	Before	During
$G_s =$	2.5 nS	3.2 nS
$V_{rev} =$	91 mV	76 mV
$R_N =$	226 MΩ	636 MΩ

   (d) $E_s = -70$ mV before, +70 mV after.

12. (b) $G_{s_{A+B}} = 30$ nS, $E_{s_{A+B}} = -67$mV; $R_N = 25$ MΩ.

13. (a) $G_{\text{slope}} = 2.2 \times 10^{-8}$ S.
14. (a) $V_{den} = -66$ mV; $V_{sh} = -44$ mV.
    (b) $V_{den} = -68.2$ mV; $V_{sh} = -51.1$ mV.

Chapter 15

1. (a) and (b)
     Control: $I_{\text{non-NMDA}} = -0.35$ nA; $I_{\text{NMDA}} = -0.064$ nA.
     LTP: $I_{\text{non-NMDA}} = -0.47$ nA; $I_{\text{NMDA}} = -0.067$ nA.
4. $Y = 001011$.

Appendix A

2. $V_3 = \dfrac{R_3(E_1 R_2 + E_2 R_1)}{R_1 R_2 + R_1 R_3 + R_2 R_3}.$

4. $V_2/V_1 = \dfrac{R_3 R_4}{R_4(R_1 + R_3) + R_3(R_1 + R_2) + R_1 R_2}.$

6. (a) $V_m = \dfrac{\frac{R_f}{R_{\text{in}}} V_c}{1 + \frac{R_f}{R_{\text{in}}}} - \dfrac{I_{cl} R_a}{1 + \frac{R_f}{R_{\text{in}}}}.$
   (b) $V_c = 51$ mV.

8. $V_c = -13$ mV; $V_o = -113$ mV.
9. (b) $V_m = 98$ mV; $V_m = 79$ mV.

Complete Solutions to Problems

Chapter 2

1.

$$\frac{RT}{F} \ln \frac{[C]_{out}}{[C]_{in}} = \frac{1.98\ \text{cal}}{{}^\circ\text{K mol}} 293^\circ\text{K} \frac{\text{mol}}{96,000\ \text{C}} \frac{4.2\ \text{joules}}{\text{cal}} 2.3\ \log_{10} \frac{[C]_{out}}{[C]_{in}}$$

$$= 5.83 \times 10^{-2} \frac{\text{joules}}{\text{C}} \log \frac{[C]_{out}}{[C]_{in}} = 58\ (\text{mV})\log_{10} \frac{[C]_{out}}{[C]_{in}}.$$

2. Membrane capacitance $= C = \frac{q}{V} = 1\ \mu\text{F}/\text{cm}^2 = 10^{-6}\ \frac{\text{C}}{\text{V}\cdot\text{cm}^2}$.

   Charge carried by each ion $(z = 1) = e = 1.6 \times 10^{-19}$ C

   $= (96,000\ \text{C}/\text{mol})/(6.02 \times 10^{23}/\text{mol})$.

   (a) The number of ions needed to charge up 1 cm^2 membrane by 100 mV

   $$n = \frac{q \times 1\ \text{cm}^2}{e} = \frac{C \cdot V \times 1\ \text{cm}^2}{e} = \frac{10^{-6}\ \text{C}}{1.6 \times 10^{-19}\ \text{C}} = 0.6 \times 10^{12}\ (\text{ions})$$

   $$= 10^{-12}\ \text{mol}.$$

   $$\text{Fraction of uncompensated ions} = \frac{10^{-12}\ \text{mol}}{(0.1\ \text{mol}/\text{L}) \times 10^{-3}\ \text{L}/\text{cm}^3}$$

   $$= 10^{-8}.$$

   (b) Sphere of radius 10 μm

   $$\text{Surface area} = 4\pi a^2 = 4(3.14)(0.001\ \text{cm})^2$$

   $$= 1.26 \times 10^{-5}\ \text{cm}^2.$$

$$\text{Total volume} = \frac{4}{3}\pi a^3 = \frac{4}{3}(3.14)(0.001 \text{ cm})^3$$

$$= 4.19 \times 10^{-9} \text{ cm}^3$$

Moles of uncompensated ions for the whole spherical surface
$= 10^{-12} \text{ mol/cm}^2 \times 1.26 \times 10^{-5} \text{ cm}^2 = 1.3 \times 10^{-17} \text{ mol}.$

Total moles of ions in the sphere $= 10^{-4} \text{ mol/cm}^3 \times 4.19 \times 10^{-9} \text{ cm}^3 = 4.19 \times 10^{-13} \text{ mol}.$

$$\text{Fraction uncompensated} = \frac{1.3 \times 10^{-17}}{4.19 \times 10^{-13}} = 3.1 \times 10^{-5}.$$

(c) Cylindrical surface area $= 2\pi a^2 + 2\pi a l$

$$= 2(3.14)(0.0001 \text{ cm})(0.0001 + 0.01) \text{ cm} = 6.34 \times 10^{-6} \text{ cm}^2.$$

Cylindrical volume $= \pi a^2 l = 3.14 \times 10^{-10} \text{ cm}^3.$

Moles of uncompensated ions on the cylindrical surface

$$= 10^{-12} \text{ mol/cm}^2 \times 6.34 \times 10^{-6} \text{ cm}^2 = 6.34 \times 10^{-18} \text{ mol}.$$

Total moles of ions in the cylinder

$$= 10^{-4} \text{ mol/cm}^3 \times 3.14 \times 10^{-10} \text{ cm}^3 = 3.14 \times 10^{-14} \text{ mol}$$

$$\text{Fraction uncompensated} = \frac{6.34 \times 10^{-18}}{3.14 \times 10^{-14}} = 2.02 \times 10^{-4}.$$

3. Let X be the concentrations of K^+ and Cl^- moving from b to a.

(a) $\dfrac{K_a^+ + X}{K_b^+ - X} = \dfrac{Cl_b^- - X}{Cl_a^- + X} \rightarrow \dfrac{150 + X}{300 - X} = \dfrac{300 - X}{300 + X}.$

Solve for X, $45,000 + 450X + X^2 = 90,000 - 600X + X^2.$

$X = 45,000/1,050 = 42.85 \simeq 43 \text{ mM}.$

The final concentrations (mM)	a	b
R^+	150	–
K^+	193	257
Cl^-	343	257

(b) $V_m = V_a - V_b = 58 \log_{10} \dfrac{193}{257} = -58 \log_{10} \dfrac{343}{257} = -7.2 \text{ mV}.$

(c) Yes, total ions in $a = 150 + 193 + 343 = 686$, and in $b = 257 + 257 = 514$. Thus, H_2O will flow from b to a.

4. Using the constant field model, the unidirectional fluxes and membrane permeabilities are related by the following expressions:

(a) $J_{efflux} = P\dfrac{\xi[C]_{in}}{1 - e^{-\xi}}$.

$J_{influx} = P\dfrac{\xi e^{-\xi}[C]_{out}}{1 - e^{-\xi}}$.

At $V_m = -90$ mV,

$$\frac{\xi}{1 - e^{-\xi}} = \frac{\frac{V_m F}{RT}}{1 - \frac{-V_m F}{eRT}} = \frac{-3.55}{1 - e^{3.55}}$$

$$= \frac{-3.55}{-33.8} = 0.105.$$

$$\frac{\xi e^{-\xi}}{1 - e^{-\xi}} = 0.105 \times e^{3.55} = 3.65.$$

$$P_K = \frac{J_{efflux}}{[K^+]_{in} \times 1.05} = \frac{8.8 \times 10^{-12}\ \text{mol/cm}^2\text{-sec}}{140\ \text{mM} \times 1.05}$$

$$= 5.9 \times 10^{-7}\ \text{cm/sec}.$$

$$P_K = \frac{J_{influx}}{[K^+]_{out} \times 3.65} = \frac{5.4 \times 10^{-12}\ \text{mol/cm}^2\text{-sec}}{2.5 \times 10^{-6}\ \text{mol/cm}^3 \times 3.65}$$

$$= 5.9 \times 10^{-7}\ \text{cm/sec}.$$

So P_K obtained from efflux agrees with that obtained from influx.

$$P_{Na} = \frac{J_{influx}}{[Na^+]_{out} \times 3.65} = \frac{3.5 \times 10^{-12}\ \text{mol/cm}^2\text{-sec}}{120 \times 10^{-6}\ \text{mol/cm}^3 \times 3.65}$$

$$= 7.99 \times 10^{-9}\ \text{cm/sec}.$$

(b) If Cl^- are passively distributed across the membrane, then

$$V_{rest} = \frac{RT}{F} \ln \frac{P_K[K^+]_{out} + P_{Na}[Na^+]_{out}}{P_K[K^+]_{in} + P_{Na}[Na^+]_{in}},$$

$$\frac{P_{Na}}{P_K} = \frac{7.99 \times 10^{-9}}{5.9 \times 10^{-7}} = 0.013$$

$$V_{rest} = 58 \log \frac{2.5 + 0.013 \times 120}{140 + 0.013 \times 9.2}$$

$$= -89\ \text{mV}.$$

(c) If $P_K/P_{Na} \to 0$,

$$V_{rest} = \frac{RT}{F} \ln \frac{[\text{Na}^+]_{out}}{[\text{Na}^+]_{in}} = 58 \log \frac{120}{9.2} = +64.7 \text{ mV}.$$

If $P_K/P_{Na} = 1$,

$$V_{rest} = \frac{RT}{F} \ln \frac{[\text{K}^+]_{out} + [\text{Na}^+]_{out}}{[\text{K}^+]_{in} + [\text{Na}^+]_{in}} = 58 \log \frac{2.5 + 120}{140 + 9.2} = -4.96 \text{ mV}.$$

5. (a) At the resting state, $P_K > P_{Ca}$ because RMP depends more on $[\text{K}^+]_{out}$ than on $[\text{Ca}^{2+}]_{out}$.

   (b) During the peak of the AP, $P_K < P_{Ca}$ because AP overshoot depends more on $[\text{Ca}^{2+}]_{out}$ than on $[\text{K}^+]_{out}$.

   (c) $[\text{K}^+]_{in} > [\text{K}^+]_{out}$, because at rest, $P_K > P_{Ca}$ and $V_{rest} = -30$ mV, which should be close to E_K, thus $58 \log \frac{\text{K}^+_{out}}{\text{K}^+_{in}}$ is negative; therefore, $[\text{K}^+]_{in} > [\text{K}^+]_{out}$.

   (d) $[\text{Ca}^{2+}]_{in} < [\text{Ca}^{2+}]_{out}$, because $P_K < P_{Ca}$ during AP and AP $\approx$ +20 mV. Thus,

   $$E_{Ca} = 29 \log \frac{[\text{Ca}^{2+}]_{out}}{[\text{Ca}^{2+}]_{in}} \text{ is positive} \Rightarrow [\text{Ca}^{2+}]_{in} < [\text{Ca}^{2+}]_{out}.$$

   (e) The hyperpolarization is most likely caused by increasing P_K because V_{rest} is between E_K and E_{Ca}. E_K is more negative than V_{rest}. $P_K \uparrow$ will result in more hyperpolarization. It is also possible that the hyperpolarization is caused by decreasing P_{Ca}, but this is less likely because $P_K > P_{Ca}$ at rest. There isn't much room for P_{Ca} to decrease.

6. (a) NPE applied within membrane; ionic movements independent of one another; $\vec{E}$ is constant and so V is linear in the membrane.

   (b) Curves were calculated with $[\text{C}]_{in}/[\text{C}]_{out}$ values of 0/100, 10/100, 100/100, 100/3.333, and 100/0.

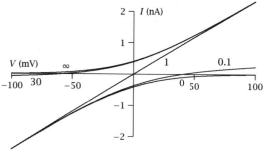

(c) From the data,

$V_m = -68$ mV when $[K^+]_{out} = 5$ mM, thus,

$$-68 = 58 \log \frac{5 + 430 \frac{P_{Na}}{P_K}}{270 + 12 \frac{P_{Na}}{P_K}}$$

$$\Rightarrow \frac{P_{Na}}{P_K} = 0.03.$$

7. (a) $E_{Cl} = -58 \log \frac{340}{41}$

    $\qquad = -53$ mV.

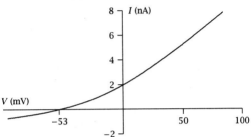

The *I-V* relation is outward rectified. Because $[Cl^-]_{out} > [Cl^-]_{in}$, it is easier for Cl^- ions to flow from outside to inside than from in to out. Hence it is easier for I_{Cl} to flow from inside to outside because $Z_{Cl} = -1$.

(b) $V_m = 58 \log \dfrac{P_K[K^+]_{out} + P_{Na}[Na^+]_{out} + P_{Cl}[Cl^-]_{in}}{P_K[K^+]_{in} + P_{Na}[Na^+]_{in} + P_{Cl}[Cl^-]_{out}}$

$\qquad = 58 \log \dfrac{1(6) + (0.019)(337) + (0.381)(41)}{1(168) + (0.019)(50) + (0.381)(340)}$

$\qquad = -59.5$ mV.

8. (a) The neuron is *not* at ECE because

    $$\frac{[K^+]_{in}}{[K^+]_{out}} = \frac{150}{150} \neq \frac{[Cl^-]_{out}}{[Cl^-]_{in}} = \frac{250}{50}.$$

    The principle of space-charge neutrality is obeyed because:

    Inside $(+)150 + 10 = 160 = (-)50 + 110$;

    outside $(+)150 + 100 = 250 = (-)250$.

(b) Cl^- will flow from outside to inside down its concentration gradient, and K^+ will flow with Cl^- (outside $\rightarrow$ inside) to maintain space-charge neutrality.

The final equilibrium concentrations of K^+ and Cl^- can be calculated by Donnan's rule:

$$\frac{[K^+]_{in}}{[K^+]_{out}} = \frac{[Cl^-]_{out}}{[Cl^-]_{in}}, \text{ thus,}$$

$$\frac{150 + X}{150 - X} = \frac{250 - X}{50 + X}.$$

X mM of KCl flows from outside to inside.

Therefore, X = 50 mM.

Therefore, the final equilibrium concentrations are

	Inside (mM)	Outside (mM)
K^+	200	100
Na^+	10	100
Cl^-	100	200
A^-	110	0

(c) The final equilibrium potentials:

$$E_K = 58 \text{ (mV)} \log \frac{100}{200} = -17.5 \text{ mV}.$$

$E_{Na} = 0$ because the membrane is impermeable to Na^+.

$$E_{Cl} = -58 \text{ (mV)} \log \frac{200}{100} = -17.5 \text{ mV}.$$

$V_{rest} = E_K = E_{Cl} = -17.5 \text{ mV}.$

9. (a) At rest, $P_K : P_{Na} : P_{CL} : P_A : P_{Ca} \equiv P_{K-Ca} = 1 : 0 : 1 : 0 : 0$. The membrane is permeable only to K^+ and Cl^-, the neuron is *not* in ECE because $\frac{[K^+]_{out}}{[K^+]_{in}} = \frac{150}{150} \neq \frac{[Cl^-]_{in}}{[Cl^-]_{out}} = \frac{50}{250}$ although space-charge neutrality is satisfied.

To reach equilibrium, $\frac{[K^+]_{out}}{[K^+]_{in}} = \frac{[Cl^-]_{in}}{[Cl^-]_{out}}$, X mM KCl must flow into the neuron.

$$\frac{150 - X}{150 + X} = \frac{50 + X}{250 - X} \rightarrow X = 50 \text{ mM}.$$

Thus, the final concentrations at equilibrium are as follows:

	Inside	Outside
K^+	200	100
Na^+	10	90
Cl^-	100	200
A^-	110	0
Ca^{2+}	10^{-4}	5

ECE is satisfied because $\left(\frac{[K^+]_{out}}{[K^+]_{in}} = \frac{100}{200}\right) = \left(\frac{[Cl^-]_{in}}{[Cl^-]_{out}} = \frac{100}{200}\right)$ and space-charge neutrality is satisfied.

Inside: $200 + 10 + 2 \times 10^{-4} = 100 + 110$.

Outside: $100 + 90 + 2 \times 5 = 200$.

V_{rest} (at equilibrium) $= E_K = E_{Cl}$ (because membrane is permeable only to K^+ and Cl^- and $P_K : P_{Cl} = 1$).

Therefore, V_{rest} (at equilibrium) $= 58 \log \frac{100}{200} = -17.5$ mV.

(b) At the onset of stimulus, $P_{K-Ca} = 1 : 10 : 1 : 0 : 0$.

$$V_m = 58 \log \frac{1(100) + 10(90) + 1(100)}{1(200) + 10(10) + 10(200)} = -3.5 \text{ mV}.$$

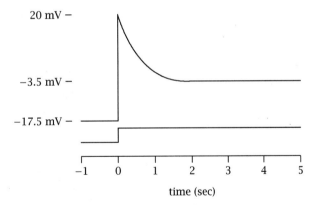

Chapter 3

1. $I = PZF\xi \dfrac{[C]_{in}e^\xi - [C]_{out}}{e^\xi - 1}$, the slope conductance at rest $= \left(\dfrac{dI}{dV}\right)_{I\to 0}$.

$$\left(\frac{dI}{dV}\right)_{I\to 0} = PZF\frac{ZF}{RT}\frac{d}{dV}\left(V\frac{[C]_{in}\exp\left(\frac{RFV}{RT}\right) - [C]_{out}}{\exp\left(\frac{ZFV}{RT}\right) - 1}\right)$$

$$= \frac{PZ^2F^2}{RT}\left[V\frac{d}{dV}\left(\frac{[C]_{in}\exp\left(\frac{ZFY}{RT}\right) - [C]_{out}}{\exp\left(\frac{ZFV}{RT}\right) - 1}\right)\right.$$
$$\left.+ \left(\frac{[C]_{in}\exp\left(\frac{ZFY}{RT}\right) - [C]_{out}}{\exp\left(\frac{ZFV}{RT}\right) - 1}\right)\frac{dV}{dV}\right]$$

$$= \frac{PZ^2F^2}{RT}V\left[\left([C]_{in}\exp\left(\frac{ZFV}{RT}\right) - [C]_{out}\right)\right.$$
$$(-1)\left(\exp\left(\frac{ZFV}{RT}\right) - 1\right)^{-2}\exp\left(\frac{ZFV}{RT}\right)\frac{ZF}{RT}\frac{dV}{dV}$$
$$\left.+ \left(\exp\left(\frac{ZFV}{RT}\right) - 1\right)^{-1}[C]_{in}\exp\left(\frac{ZFV}{RT}\right)\frac{ZF}{RT}\frac{dV}{dV}\right]$$

$$= \frac{PZ^3F^3V}{R^2F^2}\exp\left(\frac{ZFV}{RT}\right)\left(\exp\left(\frac{ZFV}{RT}\right) - 1\right)^{-1}$$
$$\left[-\left(\frac{[C]_{in}\exp\left(\frac{ZFV}{RT}\right) - [C]_{out}}{\exp\left(\frac{ZFV}{RT}\right) - 1}\right) + [C]_{in}\right]$$

$$= \frac{PZ^3F^3V}{R^2F^2}\frac{\exp\left(\frac{ZFV}{RT}\right)[C]_{in}}{\exp\left(\frac{ZFV}{RT}\right) - 1}$$

$$= \frac{PZ^2F^2\xi}{RT}\frac{[C]_{out}}{(\exp^\xi - 1)}.$$

As $I \to 0$, $V = \dfrac{RT}{ZF}\ln\dfrac{[C]_{out}}{[C]_{in}}$.

$$[C]_{out} = \exp\left(\frac{ZFV}{RT}\right) - [C]_{in}.$$

2. (a) In darkness,

$$V_{rest} = \frac{g_K E_K + g_{Na} E_{Na} + g_{Cl} E_{Cl}}{g_K + g_{Na} + g_{Cl}}$$

$$= \frac{(1)(-90) + (0.005)(50) + (0.1)(-50)}{1 + 0.005 + 0.1}$$

$$= -86 \text{ mV}.$$

Under constant light,

$$V_{rest} = \frac{(1)(-90) + (20)(50) + (0.1)(-50)}{1 + 20 + 0.1}$$

$$= 43 \text{ mV}.$$

(b) When the interneuron is stimulated in darkness, the photoreceptor is depolarized by 10 mV. Thus,

$$V_m = -86 + 10 = -76 \text{ mV}$$

$$= \frac{(1)(-90) + (0.005)(50) + \left(\frac{g_{Cl}}{g_K}\right)(-50)}{1 + 0.005 + \left(\frac{g_{Cl}}{g_K}\right)};$$

$$\frac{g_{Cl}}{g_K} = 0.514.$$

Therefore, g_{Cl} *increases* 5.14-fold ($\frac{g_{Cl}}{g_K}$: 0.1 → 0.514). Qualitatively, it is obvious that g_{Cl} must increase because V_{rest}(dark) = −86 mV, E_{Cl} = −50 mV, and a depolarization (pushing V_m from −86 mV toward E_{Cl} = −50 mV) must be accompanied by g_{Cl} *increase* (Ohm's law).

In darkness, R_m is ohmic and equals 10^8 Ω, thus

$$g_K + g_{Na} + g_{Cl} = \frac{1}{R_m} = 10^{-8} \text{ S}$$

and $g_K : g_{Na} : g_{Cl} = 1 : 0.005 : 0.1$.

11.05 $g_{Cl} = 10^{-8}$ S, therefore, $g_{Cl} = 0.09 \times 10^{-8}$ S.

When the interneuron is stimulated,

$$\Delta g_{Cl} = (5.14 - 1)0.09 \times 10^{-8} \text{ S}$$

$$= 0.37 \times 10^{-8} \text{ S}.$$

(c) g_{Cl} is constant between -100 and $+50$ mV, and the interneuron stimulation gives the same synaptic Δg_{Cl}. Thus,

$$\frac{g_{Cl}}{g_K} = 0.514.$$

$$V_m = \frac{(1)(-90) + (20)(+50) + (0.514)(-50)}{1 + 20 + 0.514}$$

$$= 41 \text{ mV}.$$

The voltage response is a 2 mV hyperpolarization.

If you interpret the question as g_{Cl} is constant and the interneuron has no effect (i.e., $\frac{g_{Cl}}{g_K}$ is always 0.1), then $V_m = 43$ mV, and thus interneurons cause *no* voltage response.

3. (a) From the figure, $g_a = $ slope of A, $g_b = $ slope of B.

    At rest:

    $$g_a = \frac{2 \text{ mA}}{70 \text{ mV}} = 0.029 \text{ S},$$

    $$g_b = \frac{1 \text{ mA}}{130 \text{ mV}} = 0.0077 \text{ S}.$$

    During excitation:

    $$g_a = \frac{2 \text{ mA}}{70 \text{ mV}} = 0.029 \text{ S},$$

    $$g_b = \frac{4 \text{ mA}}{10 \text{ mV}} = 0.4 \text{ S}.$$

    (b) The resting potential of the cell:

    $$\begin{aligned} V_{\text{rest}} &= \frac{g_a E_a + g_b E_b}{g_a + g_b} \\ &= \frac{(0.029)(-70 \text{ mV}) + (0.0077)(+60 \text{ mV})}{0.029 + 0.0077} \\ &= \frac{-2.03 + 0.462}{0.0367} = \frac{-1.568}{0.0367} \\ &= -42.7 \text{ mV}. \end{aligned}$$

    (c) The peak membrane potential during excitation:

    $$\begin{aligned} V_p &= \frac{g_a E_a + g_b E_b}{g_a + g_b} = \frac{(0.029)(-70 \text{ mV}) + (0.4)(60 \text{ mV})}{0.029 + 0.4} \\ &= \frac{-2.03 + 24}{0.429} = +51.2 \text{ mV}. \end{aligned}$$

4. (a) See graph below.

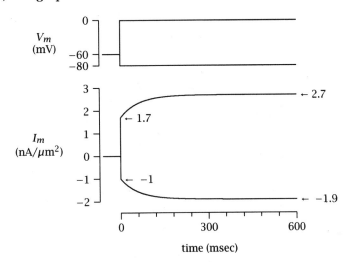

(b) Instantaneous current equals 0 because there is no *change* in $I_{\text{instantaneous}}$ when V is stepped from $+50$ to $+100$ mV.

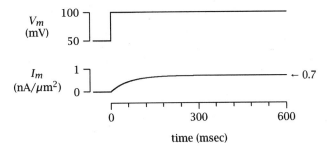

Chapter 4

1. (a)

$$
\begin{aligned}
\text{erfc}(\infty) &= 1 - \text{erf}(\infty) \\
&= 1 - \frac{2}{\sqrt{\pi}} \int_0^\infty e^{-y^2} dy \\
&= 1 - \frac{2}{\sqrt{\pi}} \left(\frac{\sqrt{\pi}}{2} \right) \quad \text{(from table of integrals)} \\
&= 1 - 1 = 0.
\end{aligned}
$$

$$
\begin{aligned}
\text{erfc}(0) &= 1 - \text{erf}(0) \\
&= 1 - \frac{2}{\sqrt{\pi}} \int_0^0 e^{-y^2}\, dy = 1 - 0 = 1.
\end{aligned}
$$

$$
\begin{aligned}
\text{erfc}(-\infty) &= 1 - \text{erf}(-\infty) \\
&= 1 - \frac{2}{\sqrt{\pi}} \int_0^{-\infty} e^{-y^2}\, dy \\
&= 1 - \frac{2}{\sqrt{\pi}} \left[-\int_{-\infty}^0 e^{-y^2}\, dy \right] \\
&= 1 + \frac{2}{\sqrt{\pi}} \left(\frac{\sqrt{\pi}}{2} \right) = 2.
\end{aligned}
$$

$$
\begin{aligned}
\text{erf}(-x) &= \frac{2}{\sqrt{\pi}} \int_0^{-x} e^{-y^2}\, dy \\
&= -\frac{2}{\sqrt{\pi}} \int_{-x}^0 e^{-y^2}\, dy \\
&= -\frac{2}{\sqrt{\pi}} \int_0^x e^{-y^2}\, dy = -\text{erf}(x).
\end{aligned}
$$

(b)

$$
\begin{aligned}
R_N &= V(\infty, 0)/I_0 \\
&= \frac{\sqrt{r_m r_i}}{2} = \sqrt{\frac{R_m R_i}{2}} \left(\frac{1}{2\pi a^{3/2}} \right) \\
&= \sqrt{\frac{2000\ \Omega\text{-cm}^2 \cdot 60\ \Omega\text{-cm}}{2}} \cdot \frac{1}{2\pi \cdot (25 \times 10^{-4}\ \text{cm})^{3/2}} \\
&= 245\ \Omega\text{-cm}^{3/2}/785 \times 10^{-6}\ \text{cm}^{3/2} \\
&= 3 \times 10^5\ \Omega.
\end{aligned}
$$

(c) Steady-state voltage response

$$
\begin{aligned}
V(\infty, x) &= \frac{r_i I_0 \lambda}{2} e^{-x/\lambda} = I_0 R_N e^{-x/\lambda} \\
&= 10 \times 10^{-9}\ \text{A} \cdot 3 \times 10^5\ \Omega \cdot e^{-x/\lambda} \\
&= 3 \times 10^{-3}\ \text{V} \cdot e^{-x/\lambda} = 3\ \text{mV}\, e^{-x/\lambda}.
\end{aligned}
$$

Therefore,

$$
\begin{aligned}
V(\infty, 0) &= 3\ \text{mV}, \\
V(\infty, \lambda) &= 3e^{-\lambda/\lambda} = 3 \times 0.3678 = 1.1\ \text{mV}, \\
V(\infty, 10\lambda) &= 3 \cdot e^{-10} = 3 \times 0.0000454 = 0.00014\ \text{mV}.
\end{aligned}
$$

(d) Transient-state voltage response at $x = 0$,

$$
\begin{aligned}
V(t, 0) &= \frac{r_i I_0 \lambda}{2} \mathrm{erf}(t/\tau_m)^{1/2} \\
&= I_0 R_N \cdot \mathrm{erf}(t/\tau_m)^{1/2} = 3 \text{ mV} \cdot \mathrm{erf}(t/\tau_m)^{1/2}.
\end{aligned}
$$

Therefore,

$$
\begin{aligned}
V\left(\frac{\tau_m}{2}, 0\right) &= 3 \text{ mV} \cdot \mathrm{erf}\left(\frac{1}{2}\right)^{1/2} \\
&= 3 \, \mathrm{erf}(0.707) = 3 \times 0.68 = 2.04 \text{ mV}.
\end{aligned}
$$

2.

$$
\begin{aligned}
V &= A_1 e^X + A_2 e^{-X}, \\
\frac{dV}{dX} &= A_1 e^X - A_2 e^{-X}, \\
\frac{d^2 V}{dX^2} &= A_1 e^X - (-) A_2 e^{-X} = V. \\
V &= A_1 \cosh(X) + A_2 \sinh(X), \\
\frac{dV}{dX} &= A_1 \sinh(X) + A_2 \cosh(X), \\
\frac{d^2 V}{dX^2} &= A_1 \cosh(X) + A_2 \sinh(X) = V. \\
V &= A_1 \cosh(L - X) + A_2 \sinh(L - X) \\
&= \frac{A_1}{2} \left(e^{L-X} + e^{X-L} \right) + \frac{A_2}{2} \left(e^{L-X} - e^{X-L} \right). \\
\frac{dV}{dX} &= \frac{A_1}{2} \left(-e^{L-X} + e^{X-L} \right) + \frac{A_2}{2} \left(-e^{L-X} - e^{X-L} \right), \\
\frac{d^2 V}{dX^2} &= \frac{A_1}{2} \left(e^{L-X} + e^{X-L} \right) + \frac{A_2}{2} \left(e^{L-X} - e^{X-L} \right) = V.
\end{aligned}
$$

3. (a)

$$
\begin{aligned}
G_N &= G_\infty = (\pi/2)(R_m R_i)^{-1/2} d^{3/2} \\
&= (\pi/2)(5000 \ \Omega\text{-cm}^2 \cdot 75 \ \Omega\text{-cm})^{-1/2}(1 \times 10^{-4} \text{ cm})^{3/2} \\
&= \frac{\pi}{2} 1.63 \times 10^{-9} \text{ S} \\
&= 2.56 \text{ nS}.
\end{aligned}
$$

(b)

$$
\begin{aligned}
G_L &= \pi d^2 / 4 R_m \\
&= \frac{\pi}{4} \cdot \frac{1 \times 10^{-8} \text{ cm}^2}{5000 \ \Omega\text{-cm}^2} = \frac{\pi}{2} \times 10^{-12} \text{ S} \\
&= 1.57 \text{ pS}.
\end{aligned}
$$

(c) B_L = G_L/G_∞

$\quad\quad$ = $1 \times 10^{-12}/1.63 \times 10^{-9} = 6 \times 10^{-4}$ (open circuit).

4. $\dfrac{\partial V_m(x,t)}{2}$ = $-r_i i_i$,

$\quad\quad \dfrac{\partial i_i}{\partial x}$ = $-i_m$,

$\quad\quad \dfrac{\partial^2 V_m}{\partial x^2}$ = $-r_i \dfrac{\partial i_i}{\partial x} = r_i i_m$,

$\quad\quad i_m$ = $i_c + i_i = c_m \dfrac{\partial V_m}{\partial t} + \dfrac{V_m}{r_m}$.

$\dfrac{1}{r_i} \dfrac{\partial^2 V_m}{\partial x^2}$ = $c_m \dfrac{\partial V_m}{\partial t} + \dfrac{V_m}{r_m}$ $\lambda^2 \dfrac{\partial^2 V_m}{\partial x^2} = \tau_m \dfrac{\partial V_m}{\partial t} + V_m$. Let $\lambda = \sqrt{\dfrac{r_m}{r_i}}$ and $\tau_m = r_m c_m$, then $\dfrac{\partial^2 V_m}{\partial X^2} - \dfrac{\partial V_m}{\partial T} - V_m = 0$, where $X = x/\lambda, T = t/\tau_m$.

5. (a) R_N = $\dfrac{1}{\pi}(R_m R_i)^{1/2} d^{-3/2}$

$\quad\quad$ = $\dfrac{1}{\pi}(707)1^{-3/2}$

$\quad\quad$ = $225\ \Omega$ (infinite).

$\quad R_N$ = $2(225) = 450\ \Omega$ (semi-infinite).

$\quad G_N$ = $\left(\dfrac{1}{\lambda r_i}\right) \tanh(L)$,

$\quad \lambda$ = $\left(\dfrac{(0.5)5000}{2(100)}\right)^{1/2} = 3.53$ cm.

$\quad r_i$ = $\dfrac{R_i}{\pi a^2} = \dfrac{100}{\pi(0.5)^2} = 127.3\ \Omega/\text{cm}$.

$\quad G_N$ = $\dfrac{1}{(3.53)(127.3)} \tanh(1)$

$\quad\quad$ = $0.00223(0.76)$.

$\quad R_N$ = $591\ \Omega$ (finite).

(b) $e^{-\lambda/\lambda} \cdot 10$ mV = 3.7 mV (infinite).

$\quad\quad\quad\quad\quad\quad$ = 3.7 mV (semi-infinite).

$\quad\quad\quad\quad\quad V$ = $\dfrac{10\text{ mV}}{\cosh(1)} = 6.5$ mV (finite).

6. (a) $(100\text{ mV})e^{-2\lambda/\lambda} = 13.5$ mV

$\quad\quad$ or -70 mV $+ 13.5 = -56.5$ mV.   No.

(b)
$$\lambda(\text{new}) = 2\lambda \ (\text{old}),$$
$$(100 \text{ mV})e^{2\lambda \text{old}/\lambda \text{new}} = (100 \text{ mV})e^{-2\lambda \text{old}/2\lambda \text{old}}$$
$$= 36.7 \text{ mV},$$
$$\text{or} -70 \text{ mV} + 36.7 = -33.3 \text{ mV}. \quad \text{Yes.}$$

(c) For A,
$$2 \times \text{diameter} \rightarrow \sqrt{2}\lambda_{\text{old}} = \lambda_{\text{new}},$$
$$(100)e^{-2\lambda_{\text{old}}/1.41\lambda_{\text{old}}} = 24.2 \text{ mV},$$
$$\text{or} -70 \text{ mV} + 24.2 = -45.8 \text{ mV} \quad \text{Yes.}$$

For B, $\lambda_{\text{new}} = 2\sqrt{2}\lambda_{\text{old}}$, still Yes.

7. $L = 1$ for all 3 cables. Assume $V_{\text{rest}} = 0$ mV, then $E_{Na} = +100$ mV.

$$V(\infty, L) = 100 \text{ mV} = V_0 \frac{\cosh(L - X)}{\cosh(L)} = V_0 \frac{\cosh(0)}{\cosh(1)} = V_0 \frac{1}{1.54}.$$
$$V_0 = 100(1.54)$$
$$= 154 \text{ mV above rest or} +84 \text{ mV above zero.}$$

8. $a_1 = 5 \ \mu\text{m}; a_2 = 10 \ \mu\text{m}.$

(a) $\theta \propto \sqrt{a} = K\sqrt{a},$

$$\frac{\theta_2}{\theta_1} = \frac{\sqrt{10/2}}{\sqrt{5/2}} = \sqrt{2} = 1.414/1.$$

(b) $R_N = \frac{(R_m R_i/2)^{1/2}}{2\pi a^{3/2}},$

$$\frac{R_{N_2}}{R_{N_1}} = \frac{\frac{1}{a_2^{3/2}}}{\frac{1}{a_1^{3/2}}} = \frac{a_1^{3/2}}{a_2^{3/2}}$$

$$= \left(\frac{10}{5}\right)^{3/2} = 2^{-3/2} = \frac{0.354}{1}.$$

(c) $\lambda = \sqrt{\frac{aR_m}{2R_i}},$

$$\frac{\lambda_2}{\lambda_1} = \frac{\sqrt{\frac{a_2}{2}}}{\sqrt{\frac{a_1}{2}}} = \sqrt{\frac{a_2}{a_1}} = \sqrt{2} = \frac{1.414}{1}.$$

(d) $\dfrac{1}{1}$.

9. (a) Surface area of sphere, $S = 4\pi a^2 = 12.6 \times 10^{-6}$ cm^2, so

$$R_N = \frac{25,000 \ \Omega\text{-cm}^2}{12.6 \times 10^{-6} \ \text{cm}^2} \simeq 2 \times 10^9 \ \Omega.$$

(b) For C_N,

$$
\begin{aligned}
C_N &= C_m \times S \\
&= (1 \times 10^{-6} \ \text{F/cm}^2)(12.6 \times 10^{-6} \ \text{cm}^2) \\
&= 12.6 \ \text{pF}.
\end{aligned}
$$

$\tau = R_N C_N = R_m C_m = 25$ msec.

Calculate and plot single exponential response with the steady-state value

$$V_{ss} = (10^{-11})(2 \times 10^9) = 20 \ \text{mV}.$$

(c)

$$
\begin{aligned}
R_N(\text{cable}) &= \frac{(R_m R_i / 2)^{1/2}}{\pi a^{3/2}} \\
&= 0.36 \times 10^9 \ \Omega.
\end{aligned}
$$

   i. $\rho = G_D/G_S = R_S/R_D = 2/0.36 = 5.6$.
   ii. $G_N = G_S(1 + \rho) = (0.5 \times 10^{-9})(6.6) = 3.3 \times 10^{-9}$ S.
      $R_N = 3.0 \times 10^8 \ \Omega$,
      so
      $V_{ss} = 3.0$ mV.
      Because $\rho > 5$, the response is close to that for a semi-infinite cable (error function). Plot an error function with $V_{ss} = 3.0$ mV along with the exponential curve with $V_{ss} = 20$ mV. Also, plot both again but on same graph with normalized amplitudes to illustrate different time courses.

(d) Reduce R_m of soma by $1/2$, so $R_S = 1 \times 10^9 \ \Omega$ and

$$R_N = \frac{R_S R_{\text{cable}}}{R_S + R_{\text{cable}}} = 2.6 \times 10^8 \ \Omega.$$

$\rho = G_S/G_D = 2.8$.

10. True, false, false, false, false.

11. (a) $L = 1$ cm$/0.5$ cm $= 2$.

(b) $X = 0.5$ cm/0.5 cm = 1.

(c) $V_L/V_0 = \frac{\cosh(0)}{\cosh(2)} = 0.266$.

(d)
$$\lambda_{10Hz} = \lambda_{DC} \sqrt{\frac{2}{1 + \sqrt{1 + (2\pi f \tau_m)^2}}}$$
$$= (0.5 \text{ cm})0.52 = 0.26 \text{ cm}.$$

AC signals attenuate more with distance than DC (i.e., **smaller** λ).

12. (a) $L_1 = \frac{l_1}{\lambda_1} + \frac{l_{11}}{\lambda_{11}} = \frac{300}{500} + \frac{100}{400} = 0.85$.

$L_2 = \frac{l_2}{\lambda_2} + \frac{l_{21}}{\lambda_{21}} = \frac{200}{500} + \frac{100}{400} = 0.65$.

(b) It can be reduced to a soma and 2 equivalent finite-length cables.

(c) $G = \frac{1}{\lambda r_i} \tanh(L)$.

$$G_1 = \frac{1}{\lambda R_i/(\pi a^2)} \tanh(L_1)$$
$$= \frac{1}{(0.05)(200)/(\pi \times 10^{-8})} \tanh(L_1)$$
$$= \frac{\pi \times 10^{-8}}{10} \tanh(L_1).$$
$$G_1 = 2.2 \times 10^{-9} \text{ S}.$$
$$G_2 = 1.8 \times 10^{-9} \text{ S}.$$

$G_{\text{dendrites}} = G_1 + G_2 = 4.0 \times 10^{-9}$ S.

$$G_{\text{soma}} = \frac{4\pi a^2}{10,000} = 5.0 \times 10^{-9} \text{ S}.$$

$G_N = 4.0 + 5.0 = 9 \times 10^{-9}$ S or $R_N = 111$ MΩ.

(d) $\rho = 4.0/5.0 = 0.8$.

(e) $R_N = 1.1 \times 10^8$ Ω.

$V_m = V_{\text{rest}} \pm (I \cdot R_N)$.

$I \cdot R_N = 111$ mV.

$V_m = -181$ mV with bias current and 41 mV with 2 nA current.

13. (a) Soma conductance $G_S = \dfrac{S}{R_m} = \dfrac{4\pi r^2}{R_m}$.

$$G_S = \frac{4 \times 3.14 \times (30 \times 10^{-4})^2 \text{ cm}^2}{2000 \ \Omega\text{cm}^2} = 56.5 \times 10^{-9} \text{ S}.$$

(b) From example 4.3 in text, $G_D = 2.7 \times 10^{-9}$ S;

$$5G_D = 5 \times 2.7 \times 10^{-9} \text{ S} = 13.5 \times 10^{-9} \text{ S}.$$

(c) Input conductance of an infinitely long axon connected to soma

$$\begin{aligned}
G_\infty &= \frac{1}{\lambda_0 r_i} = \frac{\pi}{2}\sqrt{\frac{d_0^{\,3}}{R_m R_i}} = 1.59\sqrt{\frac{(2 \times 10^{-4})^3 \text{ cm}^3}{2000 \ \Omega\text{-cm}^2 \cdot 60 \ \Omega\text{-cm}}} \\
&= 1.54 \times 0.8 \times 10^{-8} \text{ S} = 1.26 \times 10^{-8} \text{ S}.
\end{aligned}$$

(d) $$\begin{aligned}
G_N &= \frac{S}{R_m} + 5G_D + G_\infty \\
&= 56.5 \times 10^{-9} \text{ S} + 13.5 \times 10^{-9} \text{ S} + 12.6 \times 10^{-9} \text{ S} \\
&= 82.6 \times 10^{-9} \text{ S}.
\end{aligned}$$

(e) When 10 nA is applied to the cell, the steady-state voltage response

$$V(\infty) = I_0/G_N = \frac{10 \times 10^{-9} \text{ A}}{82.6 \times 10^{-9} \text{ S}} = 0.121 \text{ V} = 121 \text{ mV}.$$

Therefore, the steady-state membrane potential

$$\begin{aligned}
V_m &= V(\infty) + V_{\text{rest}} \\
&= (121 \text{ mV}) + (-70 \text{ mV}) \\
&= 51 \text{ mV}.
\end{aligned}$$

14. From graph, $I = 0.5$ nA and $V = 25$ mV.

(a) $R_N = \dfrac{25}{0.5} = 50 \text{ M}\Omega$.

$G_N = 1/R_N = 20$ nS.

(b) $R_{\text{soma}} = 100 \text{ M}\Omega$; $G_{\text{soma}} = 10$ nS.

$G_N = G_{\text{soma}} + G_{den}$. So, $G_{den} = 10$ nS.

$\rho = \dfrac{G_{den}}{G_{\text{soma}}} = \dfrac{10}{10} = 1$.

(c) G_N (finite cable) $= \dfrac{1}{\lambda r_i} \tanh(L) = G_{den}$

$\lambda = \dfrac{l}{L} = \dfrac{0.01}{1} = 0.01$ cm.

$$r_i = \frac{1}{G_{den}\lambda} \tanh(1) = \frac{0.76}{(10 \times 10^{-9})(10^{-2})} \ \Omega/\text{cm}$$
$$\cong \ 8 \times 10^9 \ \Omega/\text{cm}.$$

$\lambda = \sqrt{r_m/r_i}$, so
$$r_m = \lambda^2 r_i = (10^{-4} \text{ cm}^2)(8 \times 10^9 \ \Omega/\text{cm})$$
$$= \ 8 \times 10^5 \ \Omega\text{-cm}.$$

$$R_m = 2\pi a r_m = 2\pi (0.0005 \text{ cm})(8 \times 10^5 \ \Omega\text{-cm})$$
$$= \ 2510 \ \Omega\text{-cm}^2.$$

$R_{soma} = \dfrac{R_m}{4\pi r^2}$, so
$$r = \sqrt{\frac{R_m}{4\pi R_{soma}}} = 1.41 \times 10^{-3} \text{ cm}$$
$$\cong \ 14 \ \mu\text{m}.$$

$\tau_m = c_m r_m = C_m R_m = 2.5$ msec.

(d) $\begin{aligned} V(\infty, X) &= V_o \frac{\cosh(L - X)}{\cosh(L)} = 50 \text{ mV} \frac{\cosh(0)}{\cosh(1)} \\ &= 32.4 \text{ mV}. \end{aligned}$

(e) For semi-infinite cable at $x = \lambda$,

$$V(\infty, x) = V_o e^{-x/\lambda}.$$

$$V(\infty, \lambda) = 50 \text{ mV} e^{-1} = 18.4 \text{ mV}.$$

15. Assumptions:

(a) Uniform R_m

(b) Uniform C_m

(c) Passive (linear)

(d) Same L at ends

(e) Sealed end or open end

(f) Uniform V_m for initial condition.

(g) $R_o = 0$.

(h) radial $I = 0$.

(i) 3/2 power rule applies.

(j) Soma isopotential

16.
$$L_{da} = L_{11} = l_{11}/\lambda_{11} = 10 \text{ cm}/5 \text{ cm} = 2.$$
$$L_{db} = L_{12} = 1.$$
$$L_{dc} = L_{13} = 0.67.$$

$$G_{11} = \frac{1}{\lambda r_i} \tanh(L) = \frac{1}{(5)(42)}(.964)$$
$$= 4.6 \times 10^{-3} \text{ S}.$$
$$G_{12} = \frac{.762}{(2)1660} = 2.3 \times 10^{-4} \text{ S}.$$
$$G_{13} = \frac{.58}{(3)(327)} = 5.9 \times 10^{-4} \text{ S}.$$
$$G_N = G_{11} + G_{12} + G_{13} = 5.42 \times 10^{-3} \text{ S}.$$

$$V_d = I/G_N = \frac{10^{-3}}{5.4 \times 10^{-3}}$$
$$= 0.185 \text{ V}.$$

$$V_a = V_d \frac{\cosh(L - X)}{\cosh(L)} = \frac{0.185}{\cosh(2)} = 49 \text{ mV}.$$
$$V_b = \frac{0.185}{\cosh(1)} = 120 \text{ mV}.$$
$$V_c = \frac{0.185}{\cosh(.67)} = 150 \text{ mV}.$$

17. (a) See answer to problem 15.

    (b) See figure 4.15.

18. (a) $(a_{ad})^{3/2} = (a_{db})^{3/2} + (a_{dc})^{3/2}$,

    $(0.75)^{3/2} = (0.39)^{3/2} + (0.55)^{3/2}$,

    $0.65 = 0.24 + 0.41 = 0.65$.

Therefore, 3/2 power rule applies.

$$L_{ac} = \frac{l_{ad}}{\lambda_{ad}} + \frac{l_{dc}}{\lambda_{dc}} = \frac{4}{5} + \frac{4}{4.3} = 1.7.$$

$$L_{ab} = \frac{l_{ad}}{\lambda_{ad}} + \frac{l_{db}}{\lambda_{db}} = \frac{4}{5} + \frac{3.3}{3.6} = 1.7.$$

Therefore, both ends are at the same L, so

$$G_N(\text{at } a) = \frac{1}{(\lambda_{ad})(r_i)} \tanh(L) = \frac{\tanh(1.7)}{(5 \text{ cm})(42 \text{ }\Omega/\text{cm})}$$

$$= 4.48 \times 10^{-3} \text{ S}, \quad \text{or } R_N = 223.4 \text{ }\Omega.$$

(b) 1×10^{-3} A; $V_{in} = IR$.

At a, $V_{in} = (10^{-3})(223.4) = 223.4$ mV.

At b,

$$V_{in} = V_0 \frac{\cosh(L - X)}{\cosh(L)} = 223.4 \frac{\cosh(0)}{\cosh(1.7)} \text{ mV} = 78.9 \text{ mV}.$$

At c, $V_{in} = 78.9$ mV, because c is at the same L.

At d, $X = \frac{x}{\lambda} = \frac{4 \text{ cm}}{5 \text{ cm}} = 0.8$, so

$$V_{in} = 223.4 \frac{\cosh(1.7 - 0.8)}{\cosh(1.7)} = 113.1 \text{ mV}.$$

19. From the exponential equation, $V_m = V_0 e^{-t/\tau}$, and $\ln V_m = \ln V_0 - t/\tau$.

Convert to common log $\log V = \log V_0 - \frac{1}{2.3}\frac{1}{\tau}t$, so the slope of the graph on common log coordinates is

$$\text{slope} = \frac{\log V_2 - \log V_1}{t_2 - t_1}$$

$$= \frac{\log(V_2/V_1)}{t_2 - t_1}.$$

From (1) above,

$$\text{slope} = -\frac{1}{2.3} \cdot \frac{1}{\tau}, \quad \text{or}$$

$$\tau = -\frac{0.434}{\text{slope}} = -\frac{0.434(t_2 - t_1)}{\log(V_2/V_1)}.$$

20. (a) $R_N(\text{sphere}) = R_m/4\pi a^2 = 0.9 \times 10^8 \ \Omega$, so $V_m(ss) = 9$ mV.

$$R_N(\text{inf}) = \frac{(R_m R_i/2)^{1/2}}{2\pi a^{3/2}} = 1.13 \times 10^8 \ \Omega; \ V_m(ss) = 11.3 \text{ mV}.$$

$$R_N(\text{semi}) = 2R_N(\text{inf}) = 2.26 \times 10^8 \ \Omega; \ V_m(ss) = 22.6 \text{ mV}.$$

$$
\begin{aligned}
R_N(\text{fin}) &= \lambda r_i \coth(L) \\
&= \frac{2}{\pi}(R_m R_i)^{1/2}(d)^{-3/2} \coth(L) \\
&= \frac{2}{\pi}(10^4 10^2)^{1/2}(2 \times 10^{-4})^{-3/2} \coth(L) \\
&= 2.9 \times 10^8 \ \Omega; \ V_m(ss) = 29 \text{ mV}.
\end{aligned}
$$

Refer to figure 4.16 for shapes of curves.

(b) $\lambda = 0.0707$ cm; $\tau_m = 10$ msec.

$$\theta = \frac{2\lambda}{\tau_m} = 14 \text{ cm/sec},$$

$X = 2T - 0.5.$

At $X = 1$, $T = 0.75$, so

$t = (0.75)(10 \text{ msec}) = 7.5 \text{ msec}.$

(c)

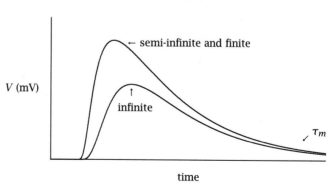

21. (a) True, false, false, true.

(b) True, false, true, true.

22. (a) $d_{\text{new}}^{3/2} = d_1^{3/2} + d_2^{3/2} + d_3^{3/2} + d_4^{3/2}$,

$$d_{\text{new}} = (1^{3/2} + 2^{3/2} + 3^{3/2} + 4^{3/2})^{2/3}$$

$$= (1 + 2.83 + 5.2 + 8)^{2/3}$$

$$= 6.6 \ \mu\text{m}.$$

(b) $V_m(\infty, X) = V_0 \dfrac{\cosh(L - X)}{\cosh(L)}$,

$$\frac{V_m(\infty, L)}{V_0} = 0.7 = \frac{\cosh(0)}{\cosh(L)},$$

$$\cosh(L) = \frac{1}{0.7} = 1.43,$$

$$L = 0.9.$$

(c) 30%

(d) Greater than

23. (a) $d_A = (1^{3/2} + 1^{3/2})^{2/3} = 1.6 \times 10^{-4}$ cm.

$$L_A = \frac{l_A}{\lambda_A} + \frac{l_{A_1}}{\lambda_{A_1}} = 0.5 + 0.2 = 0.7.$$

$$L_B = \frac{l_B}{\lambda_B} + \frac{l_{B_1}}{\lambda_{B_1}} = \frac{600}{400} + \frac{600}{300} = 3.5.$$

(b) $r_i(A) = \dfrac{200}{\pi(0.8 \times 10^{-4})^2} = 10^{10} \ \Omega.$

$$G_N(A) = \frac{1}{0.05 \times 10^{10}} \tanh(0.7) = 1.2 \times 10^{-9} \ \text{S}.$$

$$r_i(B) = \frac{200}{\pi(0.5 \times 10^{-4})^2} = 2.6 \times 10^{10} \ \Omega.$$

$$G_N(B) = \frac{1}{0.04 \times 2.6 \times 10^{10}} \tanh(3.5) = 0.96 \times 10^{-9} \ \text{S}.$$

$$G_N = (1 + 1 + 1.2) \times 10^{-9} = 3.2 \times 10^{-9} \ \text{S}.$$

$$R_N = 1/G_N = 310 \ \text{M}\Omega.$$

(c) For A ($X = L$),

$$(V_{rev} - E_r) \frac{\cosh(0)}{\cosh(0.7)} = E_s - E_r,$$

$$V_{rev} - E_r = 1.26(E_s - E_r),$$

$V_{rev} = +31$ mV.

For B ($X=1.5$)

$$(V_{rev} - E_r)\frac{\cosh(2)}{\cosh(3.5)} = E_s - E_r,$$

$V_{rev} = +283$ mV.

Because $L_B > 2$, it acts like a semi-infinite cable, so

$V_X = V_0 e^{-X/L}$, and

$(E_s - E_r) = (V_{rev} - E_r)e^{-1.5};$

$V_{rev} = +288$ mV.

(d) *A* larger than *B*; *A* faster rise than *B*; *A* shorter half width than *B*; *A* same decay time constant as *B*.

Chapter 5

1. (a) The driving forces of ion flux:

    i. Concentration gradients of ions

    ii. Electric field or electrical potentials: (1) Ions always flow from regions of high concentration to regions of low concentrations; and (2) cations flow from regions of high electrical potential to regions of low electrical potential—anions flow in the opposite direction.

   (b) $E_K = 58(\text{mV}) \log \dfrac{[K^+]_{out}}{[K^+]_{in}} = -58$ mV.

   $E_{Na} = 58(\text{mV}) \log \dfrac{[Na^+]_{out}}{[Na^+]_{in}} = +58$ mV.

   I_K is outward rectified because $[K^+]_{in} > [K^+]_{out}$, so it is easier for K^+ to flow outward (down the concentration gradient) than inward. I_{Na} is inward rectified because $[Na^+]_{out} > [Na^+]_{in}$, thus it is easier for Na^+ to flow inward (down the concentration gradient) than outward.

2. (a) $\dfrac{[C]_{out}}{[C]_{in}} = 1$ because all curves go through the origin. i.e.,

$$E_i = \frac{RT}{zF} \ln \frac{[C]_{out}}{[C]_{in}} = 0 \text{ when } I_i = 0 \text{ for all } \delta.$$

(b)
$$\left.\begin{aligned}\delta_a &= 0.1 \\ \delta_b &= 0.2 \\ \delta_c &= 0.3 \\ \delta_d &= 0.4\end{aligned}\right\} \text{ energy barrier closer to the outside of the membrane, therefore outward rectified}$$

$$\left.\begin{aligned}\delta_e &= 0.5 \\ \delta_f &= 0.6 \\ \delta_g &= 0.7 \\ \delta_h &= 0.8 \\ \delta_i &= 0.9\end{aligned}\right\} \text{ energy barrier closer to the inside, thus inward rectified}$$

(c) $I = zF\beta k_0[C]_{out}\left(e^{\delta zFV/RT} - e^{-z(1-\delta)FV/RT}\right)$, since $\dfrac{[C]_{out}}{[C]_{in}} = 1$

$$= zF\beta k_0[C]_{out}e^{z\delta FV/RT}\left(e^{-zFV/RT} - 1\right).$$

3. (a) Given $\dfrac{[C]_{in}}{[C]_{out}} = a$, $a_B = 0.028$.

$E_B = -90$ mV (from figure), then

$$-90 \text{ mV} = \frac{58}{z} \log \frac{1}{0.028} \Rightarrow z = -1.$$

Therefore, the most likely ion is Cl^-, since it is the only permeable biological ion having $z = -1$.

(b) $E_A = 58 \log a_A$, thus

$$a_A = \log^{-1}\left(\frac{E_A}{58}\right) = \text{antilog}\left(\frac{-\infty}{58}\right) = 0,$$

$$a_C = \log^{-1}\left(\frac{-80}{58}\right) = 0.042,$$

$$a_F = \log^{-1}\left(\frac{-50}{58}\right) = 0.137,$$

$$a_J = \log^{-1}\left(\frac{-10}{58}\right) = 0.672,$$

$$a_K = \log^{-1}\left(\frac{0}{58}\right) = 1.$$

(c) Constant field model. Energy barrier model *cannot* because it does *not* predict a straight line for K^+ (when $\frac{[C]_{out}}{[C]_{in}} = a_K = 1$). When $[C]_{out} > [C]_{in}$ ($a < 1$), it is easier for Cl^- to flow in ($I_{out} \rightarrow$ outward rectified). When $[C]_{out} < [C]_{in}$ ($a > 1$), it is easier for Cl^- to flow out ($I_{in} \rightarrow$ inward rectified).

4. (a) A = single energy-barrier model; B = either; C = constant field model; D = either; and E = single energy-barrier model.

(b) For ion a, $E_a = -80 \text{ mV} = \underbrace{\frac{RT}{-F} \ln \frac{[C_a]_{out}}{[C_a]_{in}}}_{= -58 \text{ mV}} \log \frac{[C_a]_{out}}{[C_a]_{in}}$.

Therefore, $\dfrac{[C_a]_{out}}{[C_a]_{in}} = \log^{-1} \dfrac{-80}{-58} = 24$.

(c) $E_e = +40 \text{ mV} = \dfrac{58}{z} \text{ mV} \log 24$,

$z = \dfrac{58}{40} \times 1.38 = 2 \qquad z = +2.$

(d) From part a, $[C_a]_{out} = 24[C_a]_{in} = 24 \times 10 \text{ mM} = 240 \text{ mM}$.

$$E_b = -40 \text{ mV} = 58 \log \frac{[C_b]_{out}}{[C_b]_{in}},$$

$$[C_b]_{out} = \left(\log^{-1} \frac{-40}{58} \right) [C_b]_{in}$$

$$= 0.204 \times 100 \text{ mM} = 20.4 \text{ mM}.$$

$$E_d = +20 \text{ mV} = 58 \log \frac{[C_d]_{out}}{[C_d]_{in}},$$

$$[C_d]_{out} = \left(\log^{-1} \frac{20}{58} \right) [C_d]_{in}$$

$$= 2.21 \times 100 \text{ mM} = 221 \text{ mM}.$$

$$V_{rest} = 58 \log \frac{P_a[C_a]_i + P_b[C_b]_{out} + P_d[C_d]_{out}}{P_a[C_a]_{out} + P_b[C_b]_i + P_d[C_d]_i}$$

since $z_a = -1, z_b = z_d = +1.$

$$V_{rest} = 58 \log \frac{1 \times 10 + 10 \times 20.4 + 0.5 \times 221}{1 \times 240 + 10 \times 100 + 0.5 \times 100}$$

$$= 58 \log \frac{324}{1290} = -34.8 \text{ mV}.$$

5. (a) $I_y(V, t)$ is activated by depolarization because $y_\infty(V)$ is larger at depolarized voltages. $I_y(V, t)$ is an *outward* current because depolarization (activation) from -60 mV to -20 mV (step offset) causes a time-dependent outward current.

(b) The value of $I_{y\infty}(V)$ at the step offset is about 20 pA. Since $I_{y\infty}(V) = y_\infty(V) \cdot \overline{g}_y(V - E)$, thus

$$
\begin{aligned}
20 \text{ pA} &= I_\infty(-20) - I_\infty(-60) \\
&= y_\infty(-20)\overline{g}_y[-20 + 70] - y_\infty(-60)\overline{g}_y[-60 + 70] \\
&= \overline{g}_y[1 \times 50 \text{ mV} - 0.3 \times 10 \text{ mV}] \\
&= \overline{g}_y \times 47 \text{ mV}
\end{aligned}
$$

Therefore, $\overline{g}_y = 20$ pA/47 mV= 426 pS.

τ_y can be measured from figure A as the time to reach 63% of the steady-state $I_y(V, t)$. From $I_y(V, t)$ at step offset,

$$
\begin{aligned}
\tau_y &= 0.08 \text{ sec} \\
&= 80 \text{ msec}
\end{aligned}
$$

(c) For spherical neuron

$$
\begin{aligned}
\tau_m &= R_m C_m = 4\pi a^2 R_N \cdot C_m \\
&= 12.56(10^{-3} \text{ cm})^2 \cdot 10^8 \ \Omega \times 10^{-6} I/\text{cm}^2 \\
&= 1.256 \text{ msec}
\end{aligned}
$$

Steady-state voltage response elicited by current step

$$
\begin{aligned}
\Delta V &= \Delta I \cdot R_N = 0.4 \times 10^{-9} \text{ A} \times 10^8 \ \Omega \\
&= 40 \text{ mV}.
\end{aligned}
$$

The neuron depolarized from -60 mV to -20 mV. From (b), approximately 20 pA of $I_y(V, t)$ will be activated by this depolarization, and since this 20 pA current is outward, it will hyperpolarize the neuron about 2×10^{-11} A $\times 10^8 \ \Omega = 2$ mV with a time course $\tau = 80$ msec.

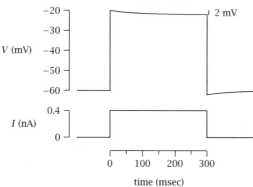

6. See text.

7. (a) The values of τ and I_∞ at each V_c can be measured from the current traces:

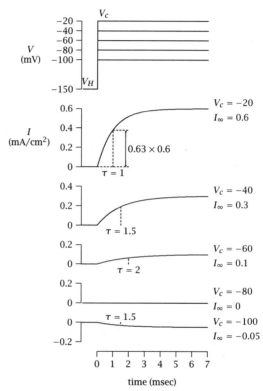

(b) $y_\infty = \dfrac{I_\infty}{\bar{g}_y(V - E_y)}.$ $\bar{g}_y = 10 \text{ mS/cm}^2, \ E_y = -80 \text{ mV}$

$\alpha = y_\infty/\tau, \qquad \beta = \dfrac{1 - y_\infty}{\tau}.$ These equations yield

V_c (mV)	I_∞ (mA/cm^2)	τ (msec)	y_∞ (unitless)	α (msec^{-1})	β (msec^{-1})
-20	0.6	1	1	1	0
-40	0.3	1.5	0.75	0.5	0.17
-60	0.1	2	0.5	0.25	0.25
-80	0	1.8*	0.35*	0.19*	0.37*
-100	-0.05	1.5	0.25	0.17	0.5

* Extrapolated values from the graph.

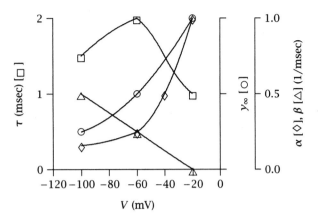

8. (a) From the slopes of the *I-V* relations in part A of the figure, at rest, $g_K/g_{Na} = 5$. During stimulus, $g_K/g_{Na} = 0.25$, $E_K = -80$ mV, $E_{Na} = +40$ mV.

Therefore,

$$V_{rest} = \frac{g_K E_K + g_{Na} E_{Na}}{g_K + g_{Na}} = \frac{5(-80) + 1(40)}{6} = -60 \text{ mV};$$

$$V_{stimulus} = \frac{0.25(-80) + (40)}{1.25} = +16 \text{ mV}.$$

The voltage response is shown below.

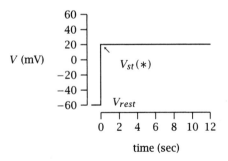

(b) At rest, $g_K/g_{Na} = 5$. Immediately after stimulus onset ($*$), $g_K/g_{Na} = 0.25$. 10 seconds after stimulus onset (∞), $g_K/g_{Na} = 1 : 4 : 8$.

Therefore,

$$V_{rest} = -60 \text{ mV};$$

$$V_{stimulus}^{(*)} = +16 \text{ mV};$$

$$V_{\text{stimulus}}^{(\infty)} = \frac{g_K E_K + g_{Na} E_{Na} + g_{Ca} E_{Ca}}{g_K + g_{Na} + g_{Ca}}$$

$$= \frac{(-80) + 4(40) + 8(80)}{13} = 55.4 \text{ mV}.$$

Immediately after the stimulus onset, V_m moves from $V_{\text{rest}} = -60$ mV to $V_{\text{stimulus}}(*) = +16$ mV. This depolarization fully activates Y_{Ca} (i.e., y_∞ moves from 0 to 1) and depolarizes the S_2 neuron to $V_{\text{stimulus}}(\infty) = +55.4$ mV with exponential time course of $\tau = 1$ second. Voltage response is shown in the plot below.

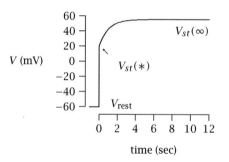

Chapter 6

1. (a)

	$[Na^+]_{\text{out}}$	E_{Na}	$I_{Na} \propto (-50 - E_{Na})$
1	150 mM	+40 mV > −50	inward
2	30	0 > −50	inward
3	10	−27 > −50	inward
4	1	−85 < −50	outward

(b) $I_{Na}^{\text{peak}} = -1$ mA/cm^2 normal : $[Na^+]_{\text{out}} = 150$ mM

$$-1 \text{ mA/cm}^2 = g_{Na}^{\text{peak}}(-50 - 40) \Rightarrow g_{Na}^{\text{peak}} = \frac{1}{90} \text{ S/cm}^2.$$

For the other conditions:

2. $I_{Na}^{\text{peak}} = \frac{1}{90}(-50 - 0) = -.55 \text{ mA/cm}^2.$

3. $I_{Na}^{\text{peak}} = \frac{1}{90}(-50 - (-27)) = -0.25 \text{ mA/cm}^2.$

4. $I_{Na}^{peak} = \dfrac{1}{90}(-50 - (-85)) = +0.39 \text{ mA/cm}^2$.

(c) $[Na^+]_{out}$ is adjusted so that $E_{Na} = -50$ mV. Now if V_m is clamped to -20 mV, I_{Na} will flow *out*. Thus the observed inward current *cannot* be I_{Na}. $E_K \approx -90$ mV and $E_{Cl} \approx -80$ mV for frog muscle $\Rightarrow$ thus I_K, I_{Cl} are outward too. The observed inward current *cannot* be I_K or I_{Cl}.

(c) $E_{Ca} = 29 \log \dfrac{2.5}{10^{-2}} = +70$ mV. Thus, when $V_c = -20$ mV, I_{Ca} flows *inward*. Thus it is possible that the observed inward current is I_{Ca}.

2. $g_K = \overline{g}_K n^4 = (24.3 \text{ mS/cm}^2)(0.891 - 0.376 e^{-t/1.7 \text{ msec}})^4$.

t (msec)	g_K (mS/cm^2)
1	2.59
2	6.29
3	9.58
4	11.88
5	13.33
6	14.19
7	14.68
8	14.96
9	15.12
10	15.20
∞	15.315
0	0.239

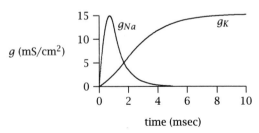

3. $g_{Na} = \overline{g}_{Na} m^3 h$.

t	$m = 0.963\,(1 - e^{-t/0.252})$	$h = 0.605\,e^{-t/0.84}$	$\overline{g}_{Na}m^3h$
0.5	0.8306	0.336	13.52
1	0.9448	0.1839	12.75
1.5	0.9605	0.1014	6.526
2	0.9626	0.0559	3.525
2.5	0.963	0.0308	1.945
3	0.9629	0.01701	1.074
3.5	0.9629	0.0094	0.5935
4	0.9629	0.00517	0.3264
4.5	0.963	0.00285	0.1799
5	0.963	0.00157	0.0991

$$
\begin{aligned}
g_{Na} &= \overline{g}_{Na}m^3h \\
&= \left(70.7 \text{ mS/cm}^2\right)\left(0.963(1 - e^{-t/0.252})\right)^3 \left(0.605\,e^{-t/0.84}\right)
\end{aligned}
$$

For the largest value of g_{Na}, $\dfrac{dg_{Na}}{dt} = 0$.

Thus, $\dfrac{dg_{Na}}{dt} = \dfrac{d}{dt}\left[38.2\left(1 - e^{-t/0.252}\right)^3\left(e^{-t/0.84}\right)\right] = 0$

yields g_{max} occurs at 0.6 msec and $g_{max} = 13.4 \text{ mS/cm}^2$.

4. According to figure 5 of Hodgkin's and Huxley's paper (1952d), $n_\infty(-60 \text{ mV}) \approx 0.88$.

$$
\begin{aligned}
I_{K\infty} &= \overline{g}_{K\infty}n_\infty^4(V - E) \\
&= \left(36 \text{ mS/cm}^2\right)(0.88)^4(0 - (-71)) \\
&= 1.554 \text{ mA/cm}^2.
\end{aligned}
$$

5. $E_{Na} = 58\log\frac{0.031}{57} = -189.34 \text{ mV}$,

   $E_K = 58\log 0.046/65 = -182.7 \text{ mV}$,

   $E_{Cl} = -58\log 0.04/112 = +200 \text{ mV}$.

   (a) At rest, $V_m = -182 \text{ mV}$. Thus the cell is permeable primarily to K^+ because V_{rest} is very closed to E_K (slightly permeable to Na^+ and a tiny bit permeable to Cl^-).

   (b) During the peak of action potential, $V_{peak} = 198 \text{ mV}$, thus the cell is permeable primarily to Cl^- because V_{peak} is very closed to E_{Cl} (slightly permeable to K^+).

(c) Since the pump ratio is 1 and z (Na$^+$) = $-z$ (Cl$^-$), it is a neutral pump → no contribution to the resting potential of the cell.

(d) At t = 1 msec,

$V = -50$ mV $g_{Cl} = \frac{-1.7 \text{ nA}}{-250 \text{ mV}} = 6.8$ nS.

$g_{Na} = 0$ because I_{Na} (1 msec) = 0.

$V = +150$ mV $g_{Cl} = \frac{-0.825 \text{ nA}}{-50 \text{ mV}} = 16.5$ nS.

$g_{Na} = 0$ because I_{Na} (1 msec) = 0.

At t = 7 msec,

$V = -50$ mV $g_{Cl} = 0$ because I_{Cl} (7 msec) = 0.

$g_{Na} = \frac{0.3235 \text{ nA}}{-50-(-189)} = \frac{0.325 \text{ nA}}{139 \text{ mV}} = 2.3$ nS.

$V = +150$ mV $g_{Cl} = 0$ because I_{Cl} (7 msec) = 0.

$$g_{Na} = \frac{1.9 \text{ nA}}{150-(-189)} = \frac{1.9 \text{ nA}}{339 \text{ mV}} = 5.6 \text{ nS}.$$

6. (a) The small sustained current is due to leakage conductance g_L.

(b) Between -30 and -5 mV,

$V_c \uparrow \rightarrow g_{Na} \uparrow$ (slope of B > slope A) → $I_{Na} \uparrow$.

(c) Between -5 and $+57$ mV, $V_c \uparrow \rightarrow g_{Na}$ does not change too much (slope $B \approx$ slope C almost constant), but $(V - E_{Na}) \downarrow$ as $V_c \uparrow$, thus $I_{Na} \downarrow$.

(d) V_{th} is between -40 and -30 mV because no I_{Na}, I_K at $V_c = -40$ mV, but they appear at -30 mV

(e) $E_{Na} \approx +57$ mV because $I_{Na} =$
$$\begin{matrix} \text{inward} & V_c < +57 \text{ mV} \\ 0 & V_c = 57 \text{ mV} \\ \text{outward} & V_c > +57 \text{ mV} \end{matrix}$$

(f) No reversal of the late outward current when $V_c = -65$ mV because it can only be activated above -30 mV.

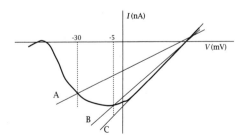

7.

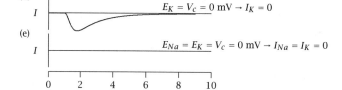

V_c 0
-70

(a)

I TTX blocks I_{Na}

(b)

I TEA blocks I_K

(c)

I $E_{Na} = V_c = 0$ mV $\rightarrow I_{Na} = 0$

(d)

I $E_K = V_c = 0$ mV $\rightarrow I_K = 0$

(e)

I $E_{Na} = E_K = V_c = 0$ mV $\rightarrow I_{Na} = I_K = 0$

     0   2   4   6   8   10

time (msec)

8.

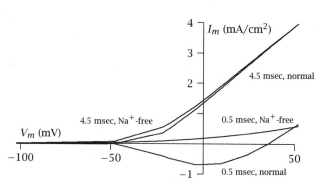

I_m (mA/cm^2)

4.5 msec, normal

4.5 msec, Na$^+$-free   0.5 msec, Na$^+$-free

V_m (mV)

-100 -50 50

0.5 msec, normal

t	I_{Na}	I_K	g_{Na}	g_K
0	0	0	0	0
0.25	-0.38	0.11	5.9	2
0.5	-0.85	0.16	13.3	2.4
0.75	-1.03	0.22	16.1	3.3
1	-1.01	0.29	15.8	4.4
1.25	-0.9	0.36	14.2	5.5
.	.	.	.	.
.	.	.	.	.
.	.	.	.	.
5	-0.07	1.33	1.2	20

$$g_{Na} = \frac{I_{Na}}{V - 60}.$$

$$g_K = \frac{I_K}{V + 70}.$$

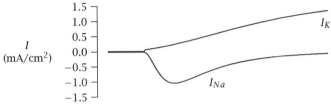

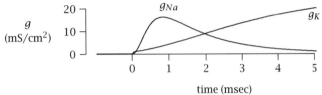

9. (a) Let $g_K/g_{Na} = x$.

At rest, $V = -70$ mV $= \dfrac{(x)(E_K) + (E_{Na})}{1 + x}$

$$= \frac{(-80)x + (+50)}{1 + x},$$

$-70 - 70x = -80x + 50,$

$x = 12.$

At threshold, $-60 = \dfrac{(-80)x + 50}{1 + x}$,

$-60 - 60x = -80x + 50,$

$x = 5.5.$

At AP peak, $+30 = \dfrac{-80x + 50}{1 + x}$,

$30 + 30x = -80x + 50,$

$x = \dfrac{20}{110} = 0.18.$

(b) At time T,

$$V_{rest} = \frac{5x(-80) + 50}{5x + 1} = \frac{(5 \times 12)(-80) + 50}{61} = -77.8 \text{ mV}.$$

Threshold and V_{peak} do *not* exist. The axon is in refractory period.

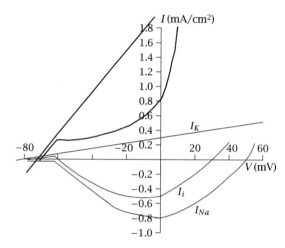

10. (a) $V_c = 100$ mV,

$$I_{K\infty} = 15\text{mA/cm}^2 = g_{K\infty}(V - E_K),$$

$$\begin{aligned} g_{K\infty} &= (15\text{mA/cm}^2)/[100 - (-80)] \text{ mV} \\ &= 0.0833 \times 10^{-3} \times 10^3 \text{ S/cm}^2 \\ &= 83.3 \text{ mS/cm}^2. \end{aligned}$$

$V_c = 85$ mV,

$$g_{K\infty} = \frac{13.6 \text{ mA/cm}^2}{85 + 80 \text{ mV}} = 82.4 \text{ mS/cm}^2.$$

$V_c = 60$ mV,

$$g_{K\infty} = \frac{10 \text{ mA/cm}^2}{60 + 80 \text{ mV}} = 71.4 \text{ mS/cm}^2.$$

$V_c = 25$ mV,

$$g_{K\infty} = \frac{5 \text{ mA/cm}^2}{25 + 80} = 47.6 \text{ mS/cm}^2.$$

(b) Since $g_{K\infty} = n_\infty^4 \bar{g}_K$,

V_c	$n_\infty = \left[\frac{g_{K\infty}}{\bar{g}_K}\right]^{1/4}$
100	1
85	1
60	$\left[\frac{71.4}{83}\right]^{1/4} = 0.96$
25	$\left[\frac{47.6}{83}\right]^{1/4} = 0.87$

Since $n^4(t) = [n_\infty - [(n_\infty - n_0)e^{-t/\tau}]]^4$,

τ can be estimated by measuring time to reach $[0.63]^4$ of the $I_K\infty$.

$(0.63)^4 = 0.16$, thus

τ = time to reach 0.16 of $I_K\infty$.

V_c	τ_n
100 mV	1.2 msec
85 mV	1.4 msec
60 mV	1.6 msec
25 mV	2.0 msec

Since $\alpha_n = n\infty/\tau_n, \quad \beta_n = (1 - n_\infty)/\tau_n$.

Therefore,

$V_c = 100$ mV; $\quad \alpha_n = \frac{1}{1.2\times10^{-3}} = 833$ sec^{-1}
$\quad\quad\quad\quad\quad\quad \beta_n = 0.$

$V_c = 85$ mV; $\quad \alpha_n = \frac{1}{1.4\times10^{-3}} = 714$ sec^{-1}
$\quad\quad\quad\quad\quad\quad \beta_n = 0$

$V_c = 60$ mV; $\quad \alpha_n = \frac{0.96}{1.6\times10^{-3}} = 600$ sec^{-1}
$\quad\quad\quad\quad\quad\quad \beta_n = \frac{0.04}{1.6\times10^{-3}} = 25$ sec^{-1}

$V_c = 25$ mV; $\quad \alpha_n = \frac{0.87}{2.0\times10^{-3}} = 435$ sec^{-1}
$\quad\quad\quad\quad\quad\quad \beta_n = \frac{0.13}{2\times10^{-3}} = 66$ sec^{-1}.

11. (a)

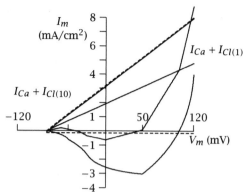

(b) From the right figure, $E_{Ca} = +100$ mV and $E_{Cl} = -80$ mV. Thus,

$$\frac{[\text{Ca}^{2+}]_{\text{out}}}{[\text{Ca}^{2+}]_{\text{in}}} = \text{antilog}\frac{100}{29} = 2807.2 \text{ because } z = +2.$$

$$\begin{aligned}
[\text{Ca}^{2+}]_{\text{in}} &= [\text{Ca}^{2+}]_{\text{out}}/2807.2 = \frac{5 \text{ mM}}{2807.2}\\
&= 1.78 \times 10^{-3} \text{ mM} = 1.78 \ \mu\text{M}.
\end{aligned}$$

$$\frac{[\text{Cl}]_{\text{in}}}{[\text{Cl}^-]_{\text{out}}} = \text{antilog}\frac{-80}{58} = 0.042.$$

$$[\text{Cl}^-]_{\text{in}} = [\text{Cl}^-]_{\text{out}} \times 0.042 = 50 \text{ mM} \times 0.042 = 2.087 \text{ mM}.$$

12. (a) $\begin{aligned}[t]
Q = \int_0^b I\,dt &= \tfrac{1}{2}(10 \ \mu\text{A/cm}^2)(3 \text{ msec})\\
&= 15\frac{\mu\text{A}}{\text{cm}^2} \cdot \text{msec}\\
&= 15 \times 10^{-6} \times 10^{-3} \text{ A} \cdot \text{sec/cm}^2\\
&= 15 \times 10^{-9} \text{ C/cm}^2\\
&= 9.37 \times 10^{10} e/\text{cm}^2.
\end{aligned}$

Five gating changes are required to open one K^+ channel.

$$D = \frac{Q}{ze} = \frac{9.37 \times 10^{10} \ e/\text{cm}^2}{5 \times e} = 1.87 \times 10^{10} \text{ channels/cm}^2.$$

(b) $\bar{g}_K = 2 \times 10^{-1}$ S/cm^2

$$\begin{aligned}
\gamma &= \frac{\bar{g}_K}{D} = \frac{2 \times 10^{-1} \text{ S/cm}^2}{1.87 \times 10^{10} \text{ channels/cm}^2}\\
&= 1.07 \times 10^{-11} \text{ S/channel}\\
&= 10.7 \text{ pS/channel}.
\end{aligned}$$

(c) The K$^+$ current is activated by hyperpolarization because I_g is inward and only hyperpolarization can result in a net inward electric field.

$$+ + + + + + + + + + +$$ out

$V_{hyperpol.} \longrightarrow \vec{E} \downarrow \longrightarrow I_g \downarrow$ inward

$$- - - - - - - - - - - - - - - -$$ in

13. (a) The total amount of charge movement carried by the gating current equals to the area underneath the I_g trace.

$$\begin{aligned} Q &= 2.625 \ \mu A \cdot msec/cm^2 \\ &= 2.625 \times 10^{-6} \times 10^{-3} \ A \cdot sec/cm^2 \\ &= 2.625 \times 10^{-9} \ C/cm^2 \end{aligned}$$

Number of charges $= \dfrac{Q}{e} = \dfrac{2.625 \times 10^{-9} \ C/cm^2}{1.6 \times 10^{-19} \ C}$.

$$= 1.64 \times 10^{10} \ particles/cm^2.$$

Two gating particles are required to open a channel.

$$CD = \text{channel density} = \frac{1.64 \times 10^{10}}{2 \times 80\%} = \frac{0.82 \times 10^{10}}{0.8}$$

$$= 1.02 \times 10^{10} \ channels/cm^2$$

(because from -70 mV to -10 mV only opens 80% of the channels).

(b) From part B of the figure, $\bar{g}_{Na}$ is 1 S/cm^2.

Thus, $\gamma = \dfrac{\bar{g}_{Na}}{CD} = \dfrac{0.8 \ S/cm^2}{1.02 \times 10^{10}/cm^2} = 10^{-10} \ S = 100 \ pS.$

(c)

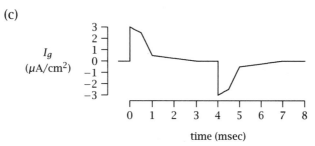

14. (a)

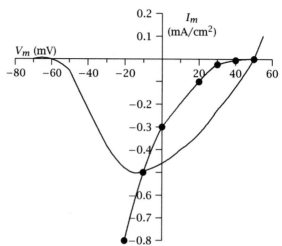

(b) g_{Na} is *not* ohmic because the instantaneous *I-V* relation is not linear. This nonlinearity can be explained by both the constant field model (*I-V* is inward rectified with positive equilibrium potential) and the energy barrier model.

(c) See dashed curves in the figure below. Data points for dashed curves are obtained by adding I_K to I_{Na}. I_K is linear and its slope is determined by $g_K(0.5 \text{ msec}) = 1 \text{ mS/cm}^2$, which gives a straight line from -80 mV (E_K) to $I_K(0 \text{ mV}) = (1 \text{ ms/cm}^2)(0 - (-80)) = 0.08 \text{ mA/cm}^2$.

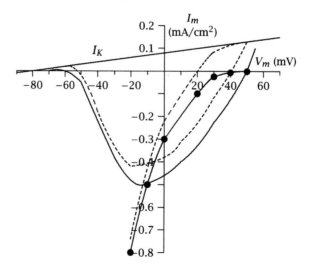

15. (a)

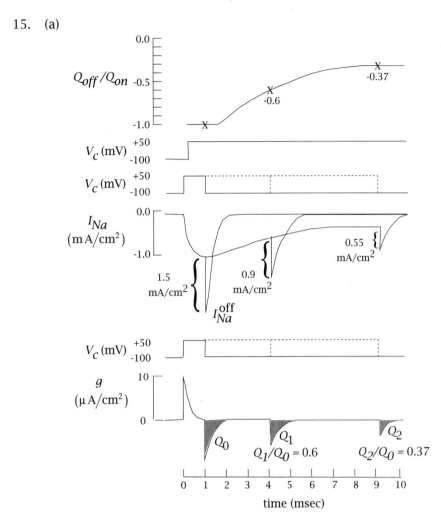

(b) The major difference is that the Hodgkin and Huxley model predicts that the relaxation of I_{Na}^{off} is 3 times faster than that of the I_g^{off}, because only one m particle needs to be moved from a permissive to a nonpermissive state to close a channel. Experimental results show that the relaxation time courses of I_{Na}^{off} and I_g^{off} are about the same. This indicates that the Hodgkin and Huxley model is not completely correct in describing detailed mechanisms of Na$^+$ channel gating. The movements of gating particles, for example, may not be totally independent from each other.

Chapter 7

1. (a) See solid line in plot below.

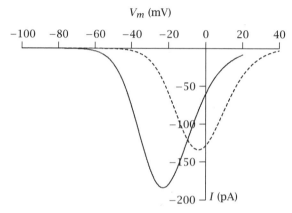

   (b) Give step depolarizations from different holding potentials to a fixed voltage that activates I_{Na} maximally (i.e., $m_\infty = 1$). Then,

   $$h_\infty = \frac{I_{peak}}{I_{peak_{(max)}}} \quad \text{for different holding potentials.}$$

   Give step depolarizations to different values from a very negative holding potential (i.e., $h_\infty = 1$).

   $$m_\infty = \frac{I_{peak}}{I_{peak_{(max)}}} \cdot \frac{(V_{m_{(max)}} - E_{Na})}{(V_m - E_{Na})} \quad \text{for different step potentials.}$$

   (c) Slow depolarization in voltage range of -80 to 0 mV. Depending on kinetics, could cause oscillations or new resting potential.

   (d) See dashed line in above plot.

2. See graph below. Solid line is V_m under control conditions; dashed line is after addition of 4-AP.

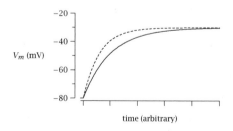

3. See figure below.

4. See figure below.

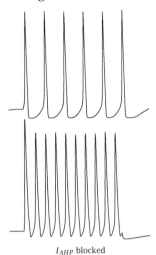

I_{AHP} blocked

5. (a) To distinguish a low-threshold (T) from a high-threshold (L) calcium current, at least five of the following points should be made in some descriptive detail:

   i. Hold at hyperpolarized potentials ($\sim$ –100 mV) and step positive (for T); measure activation (and inactivation) properties.

   ii. Hold at depolarized potentials ($\sim$ –40 mV) and step positive (for L). Subtract these currents from those above.

   iii. Peak vs. steady-state. Discuss inactivation properties.

   iv. Dihydropyridines (L), low concentration of Ni^{2+} (T), low concentration of Cd^{2+} (L).

   v. cAMP increases L.

   vi. ACh increases T, decreases L.

   vii. Single-channel conductances (8 vs. 25 pS in Ba^{2+}); above properties using single-channel measurements.

   (b) i. Chord conductance = 875 nS.

    ii. Yes

    iii. *I-V* curve nonohmic at depolarized potentials; conductance decreases at depolarized potentials; better fit by GHK current equation (see text).

6. (a) Solid line in left graph below for a); dashed line for b); and right graph for c).

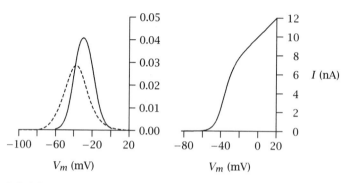

    (b) Solid lines (superimposed) for a) and c); dashed line for b).

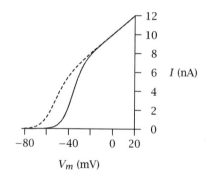

    (c) a) = $I_{K(A)}$; b) = $I_{K(D)}$; c) = $I_{K(M)}$.

    (d) See graph.

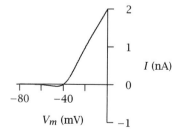

7. (a)   i. g_L is a leakage conductance; g_x is potentially a function of
            voltage and time (has m and h in equation). Get g_L and E_L
            from measuring I-V curve with respect to the zero current
            level from a hyperpolarized holding potential.

      ii. Get $\overline{g}_x$ from peak current responses in I-V curve taken from
           hyperpolarized potentials. Tail currents from maximum
           response will yield $\overline{g}_x$. Slope of total I-V curve near reversal
           will also yield $\overline{g}_x$. E_x can be taken from total I-V curve.

      iii. Determine m_∞ by giving depolarizing steps from negative
            holding potential and plotting $g_x(V)/\overline{g}_x$, as described also
            in problem 1. Determine h_∞, by giving a depolarizing step
            to a fixed positive potential from different holding poten-
            tials and plotting $I_{\text{peak}}(V)/I_{\text{peak}}(max)$. This is assuming
            that m and h have different rates of activation.

    (b) See solid curves on graph below.

    (c) See dashed curves on graph below.

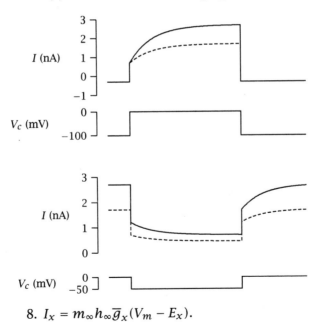

8. $I_x = m_\infty h_\infty \overline{g}_x (V_m - E_x)$.

   $I_y = m_\infty h_\infty \overline{g}_y (V_m - E_y)$.

V_m (mV)	m_∞	h_∞	I_x	I_y ($\times10^{-12}$)
−80	0	0.95	0	0
−70	0.025	0.825	−124	+12.4
−60	0.1	0.7	−385	+56
−50	0.18	0.5	−450	+90
−40	0.295	0.295	−392	+104
−30	0.5	0.15	−300	+105
−20	0.7	0.1	−245	+112
−10	0.83	0.03	−75	+45
0	0.93	0	0	0

Make a plot from values in table (see graph).

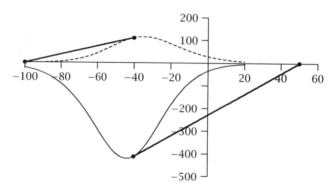

9. From notes.

10. From notes.

11. Hyperpolarization activated current. Could be $I_{K(IR)}$, I_Q, or $I_{Cl(V)}$.

12. In sequence: $I_{K(A)}$, $I_{Ca(T)}$, $I_{Na(slow)}$, $I_{Na(fast)}$, $I_{Ca(N,L)}$ for depolarizing phases, and $I_{K(A)}$, $I_{K(DR)}$, $I_{K(C)}$, $I_{K(M)}$, $I_{K(AHP)}$ for hyperpolarizing phases.

Chapter 9

1. The pdf is $f(x) = \frac{1}{4}$ (for $2 \le x \le 6$.)

   If $x < 2$, $P(X \le x) = 0$,

   and if $x > 6$, $P(X \le x) = 1$.

   If $2 \le x \le 6$, $P(X \le x)$ is the shaded area, which is $\frac{1}{4}(x - 2)$.

Hence the cdf is

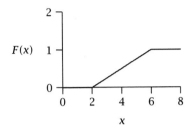

$$F(X) = \begin{cases} 0 & \text{if } x < 2 \\ \frac{1}{4}(x - 2) & \text{if } 2 \le x \le 6 \\ 1 & \text{if } x > 6 \end{cases}$$

2. If $x < 0$, then $F(x) = 0$; and if $x > 3$, $F(x) = 1$.

If $0 \le x \le 3$,

$$\begin{aligned} F(x) &= \int f(x)dx = \int \frac{2}{9}(3x - x^2)dx \\ &= \frac{1}{3}x^2 - \frac{2}{27}x^3 + C. \end{aligned}$$

Since $F(0) = P(\alpha \le 0) = 0$, $C = 0$. Thus $F(x) = \frac{1}{3}x^2 - \frac{2}{27}x^3$. Note that $F(3) = 3 - 2 = 1$.

The cdf is

$$F(x) = \begin{cases} 0 & \text{if } x < 0 \\ \frac{1}{3}x^2 - \frac{2}{27}x^3 & \text{if } 0 \le x \le 3 \\ 1 & \text{if } x > 3 \end{cases}$$

Then $P(X > 2) = 1 - F(2) = \dfrac{7}{27}$.

3. (a) If $0 \le x \le 1$, $F(x) = \int \frac{6}{5}x\,dx = \frac{3}{5}x^2 + C$.

    $F(0) = 0$, and so $C = 0$,

    $$F(x) = \frac{3}{5}x^2; \qquad F(1) = \frac{3}{5}.$$

    If $x > 1$, $\qquad F(x) = \int \frac{6}{5x^4}dx = -\frac{2}{5x^3} + D$.

    $F(1) = \frac{3}{5}$, and so $D = 1$.

    Hence the cdf is

    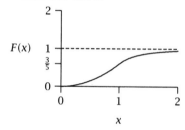

    $$F(x) = \begin{cases} 0 & \text{if } x < 0 \\ \frac{3}{5}x^2 & \text{if } 0 \le x \le 1 \\ 1 - \frac{2}{5x^3} & \text{if } x > 1 \end{cases}$$

    As $x \to \infty$, $F(x) \to 1 - 0 = 1$.

    (b) $P\left(\frac{1}{2} < X < 2\right) = F(2) - F\left(\frac{1}{2}\right) = \left(1 - \frac{2}{5 \times 8}\right) - \frac{3}{5} \times \left(\frac{1}{2}\right)^2$

    $\qquad\qquad\qquad\quad = \frac{4}{5}$.

    (c) For the median, $F(x) = \frac{1}{2}$, i.e., $\frac{3}{5}x^2 = \frac{1}{2}$, i.e., $x^2 = \frac{5}{6}$.

    The median is $\sqrt{\frac{5}{6}} \approx 0.9129$.

    For the upper quartile, $F(x) = \frac{3}{4}$, i.e., $1 - \frac{2}{5x^3} = \frac{3}{4}$,

    so $x^3 = \frac{8}{5}$, i.e., $x \approx 1.1696$

    For the lower quartile, $F(x) = \frac{1}{4}$, i.e., $\frac{3}{5}x^2 = \frac{1}{4}$,

    so $x^2 = \frac{5}{12}$, i.e., $x \approx 0.6455$.

    The semi-interquartile range is $\frac{1}{2}(1.1696 - 0.6455) = 0.2621$.

4. (a)

From data	σ_{I_N}	$\sigma_{I_N}{}^2$	μ_I
0 μM Glu	0	0	0
1 μM Glu	±0.7 nA	0.49 $(nA)^2$	2 nA
2 μM Glu	±1.0 nA	1.00 $(nA)^2$	5 nA
3 μM Glu	±0.7 nA	0.49 $(nA)^2$	8 nA

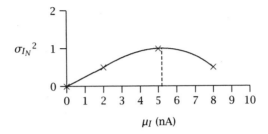

(b) $\sigma_{I_N}{}^2 = I_1\mu_I - \mu_I{}^2/N.$

$$\frac{d}{d\mu_I}(\sigma_{I_N}{}^2) = I_1 - 2\mu_I/N.$$

At $\mu_I = 0$ (origin in the above plot), $\frac{d}{d\mu_I}(\sigma_{I_N}{}^2) = I_1.$

$I_1 \quad = \quad$ (slope of the curve at $\mu_I = 0$)

$$= \quad \frac{0.5\ (nA)^2}{2\ nA} = 0.25\ nA.$$

$$\gamma = \frac{I_1}{V - E_{Na}} = \frac{0.25\ nA}{(10-50)\ mV} = 6.25\ nS.$$

At $\frac{d}{d\mu_I}(\sigma_{I_N}{}^2) = 0_1;$ $\quad N = \frac{2\mu_I{}^*}{I_1}.$

From the graph, $\mu_I{}^*$ (maximum $\sigma_{I_N}{}^2$ or when $\frac{d}{d\mu_I}(\sigma_{I_N}{}^2) = 0$) is approximately 5.2 nA.

$$N = \frac{2 \times 5.2\ nA}{0.25\ nA} \approx 42\ \text{channels}.$$

There are 42 glutamate-gated channels in the neuron.

5. $P_\infty(V) = \dfrac{\beta(V)}{\alpha(V) + \beta(V)}.$

(a) At $V = -100$ mV,

$\alpha = 500\ e^{+100/50} = 3694\ \text{sec}^{-1},$

$$\beta = 50\ e^{-100/25} = 0.9\ \text{sec}^{-1}.$$

$$P_\infty(-100\ \text{mV}) = \frac{0.9}{3694 + 0.9} = 0.025\%.$$

At $V = -50$ mV,

$$\alpha = 500\ e^{+50/50} = 1359\ \text{sec}^{-1}.$$

$$\beta = 50\ e^{-50/25} = 6.77\ \text{sec}^{-1}.$$

$$P_\infty(-50\ \text{mV}) = \frac{6.77}{1359 + 6.77} = 0.5\%.$$

At $V = 0$ mV,

$$\alpha = 500\ e^{-0} = 500\ \text{sec}^{-1},$$

$$\beta = 50\ e^{0} = 50\ \text{sec}^{-1},$$

$$P_\infty(0\ \text{mV}) = \frac{50}{500 + 50} = 0.091 = 9.1\%.$$

(b) The mean whole-cell current at steady state

$$\mu_I = NP_\infty^{-50}I_1,$$

$$-100\ \text{nA} = N(0.005)(25\ \text{pS})(-50 - 50)\ \text{mV},$$

$$\begin{aligned}
N &= \frac{-100 \times 10^{-9}\text{A}}{(0.005 \times 25) \times 10^{-12}\ \text{S} \times (-100 \times 10^{-3})\ \text{V}} \\
&= 8,000,000.
\end{aligned}$$

(c)
$$\begin{aligned}
\mu_I(-100\ \text{mV}) &= NP_\infty^{(-100)}I_1 \\
&= (8 \times 10^6)(0.00025)(25 \times 10^{-12}) \\
&\quad (-100 - 50)10^{-3} \\
&= -7500\ \text{pA}. \\
\mu_I(0\ \text{mV}) &= +8 \times 10^6(0.091)(25 \times 10^{-12}) \\
&\quad (-50) \times 10^{-3} \\
&= 9.1 \times 10^5\ \text{pA}.
\end{aligned}$$

(d) $\sigma^2 = \gamma^2(V - E_i)^2 NP_\infty(1 - P_\infty).$

$V = -100$ mV.

$$
\begin{aligned}
\sigma^2(-100) &= (25 \times 10^{-12})^2(-100 - 50)^2 10^{-6} \times 8 \\
&\quad \times 10^6 (0.00025)(0.99975) \\
&= 28,118 \times 10^{-24}. \\
|\sigma|(-100) &= 167 \times 10^{-12} \text{ A} = 167 \text{ pA}. \\
\sigma^2(-50) &= (25 \times 10^{-12})^2(-50 - 50)^2 10^{-6} \times 8 \\
&\quad \times 10^6 \times (0.005)(0.995) \\
&= 248750 \times 10^{-24} \text{ A}. \\
\sigma(-500) &= 498 \times 10^{-12} \text{ A} = 498 \text{ pA}. \\
\sigma^2(0) &= (25 \times 10^{-12})^2(0 - 50)^2 10^{-6} \times 8 \\
&\quad \times 10^6 (0.091)(0.909) \\
&= 1,033,987.5 \times 10^{-24}. \\
\sigma(0) &= 1017 \times 10^{-12} \text{ A} = 1.017 \text{ pA}.
\end{aligned}
$$

6. Single-channel gating transition is a "memoryless" Markov process because the lifetime of the channel in each state, once entered, is *exponentially distributed.* For an n-state channel, the pdf for the k-th state is $P_k(t) = |q_{kk}|e^{q_{kk}t}$. For the two-state scheme, $P_1(t) = \alpha e^{-\alpha t}$, $P_2(t) = \beta e^{-\beta t}$.

Mathematically, a "memoryless" random process is defined as:

Prob[lifetime at k state $> t + t_1$| lifetime at k state $> t_1$].

= Prob [lifetime at k state $> t$].

It can be shown that exponential pdf is the only lifetime distribution that satisfies this "memoryless" definition. Since channel lifetime in each state follows exponential distribution, channel gating transition is a "memoryless" process, often called the Markov process.

Chapter 10

1. (a) At $V = -80$ mV,

$$
\alpha = 2956 \text{ sec}^{-1}, \qquad \beta = 0.73 \text{ sec}^{-1}.
$$

$$
Q_{-80} = \begin{pmatrix} -\alpha & \alpha \\ \beta & -\beta \end{pmatrix} = \begin{pmatrix} -2956 & 2956 \\ 0.73 & -0.73 \end{pmatrix}
$$

At $V = +20$ mV, $\alpha = 243$ sec^{-1}, $\beta = 109$ sec^{-1}.

$$Q_{+20} = \begin{pmatrix} -243 & 243 \\ 109 & -109 \end{pmatrix}$$

(b) $V = -80$ mV, $P_1(\infty) = \dfrac{\beta}{\alpha + \beta} = \dfrac{0.73}{2956 + 0.73} = 0.00025.$

$$\begin{aligned} \mu_I &= N\gamma(V - E)P_1(\infty) \\ &= 10^5 \cdot 20 \times 10^{-12}\,(\text{S})(-80 - 50)10^{-3}\,(\text{V})(0.00025) \\ &= -65 \text{ pA}. \end{aligned}$$

At $V = +20$ mV, $p_1(\infty) = \dfrac{108}{243 + 109} = 0.31.$

$$\begin{aligned} \mu_I &= 10^5 \cdot 20 \times 10^{-12}\,(\text{S})(20 - 50)10^{-3}\,(\text{V})(0.31) \\ &= -18.6 \text{ nA}. \end{aligned}$$

(c)

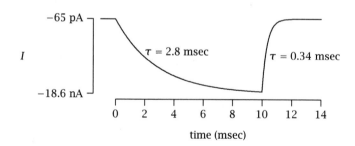

time (msec)

$$\begin{aligned} \tau(-80 \text{ mV}) &= \frac{1}{\alpha + \beta} \\ &= \frac{1}{2956 + 0.73} \\ &= 0.34 \text{ msec}. \\ \tau(+20 \text{ mV}) &= \frac{1}{243 + 109} \\ &= 2.8 \text{ msec}. \end{aligned}$$

2.   (a) $\mathbf{Q} = \begin{bmatrix} -k_2 & k_2 \\ k_1 x_A & -k_1 x_A \end{bmatrix} = \begin{bmatrix} -100 & 100 \\ 2 & -2 \end{bmatrix}$

each element is in $\sec^{-1}$

$\det \mathbf{Q} = k_2 k_1 x_A - k_1 x_A k_2 = 0.$

(b) $\overline{\tau}_{\text{open}} = \dfrac{1}{k_2} = \dfrac{1}{100 \text{ sec}^{-1}} = 0.01 \text{ sec.}$

$\overline{\tau}_{\text{closed}} = \dfrac{1}{k_1 x_A} = \dfrac{1}{2 \text{ sec}^{-1}} = 0.5 \text{ sec.}$

At steady state, the probability of the channel in open state:

$P_1(\infty) = \dfrac{k_1 x_A}{k_1 x_A + k_2} = \dfrac{2}{2 + 100} = 0.0196 = 1.96\%;$

in closed state:

$P_2(\infty) = \dfrac{k_2}{k_1 x_A + k_2} = \dfrac{100}{2 + 100} = 0.9804 = 98.04\%.$

Therefore,   the open time is about 2%;

the closed time is about 98%.

(c)

$$\begin{aligned} \mu_I &= \gamma(V - E_{Na})P_1(\infty) \cdot N \\ &= 10 \times 10^{-12} \text{ S}(-50 - 50) \text{ mV}(0.02)10,000 \\ &= -2 \times 10^{-10} \text{A.} \end{aligned}$$

(d)

$$\begin{aligned} \sigma_N{}^2 &= \gamma^2 (V - E_{Na})^2 N P_1(\infty) P_2(\infty) \\ &= (10 \times 10^{-12} \text{ S})^2 \left[(-50 - 50) \text{ mV} \right]^2 \\ &\qquad (10,000)(0.02)(0.98) \\ &= 1.96 \times 10^{-22} \text{ A}^2. \end{aligned}$$

$|\sigma_N| = 1.4 \times 10^{-11}$ A noise.

(e)

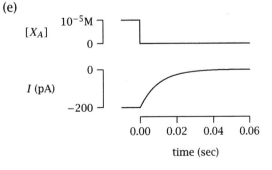

$$\tau = \frac{1}{k_1 x_A + k_2}$$

$$= \frac{1}{0 + 100}$$

$$= 0.01 \text{ sec.}$$

(f)

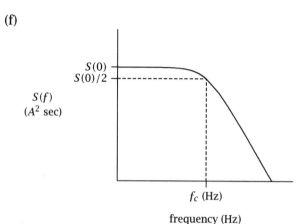

$$S(0) = \frac{2\mu_I \gamma (V - E_i)\alpha}{(\alpha + \beta)^2} = \frac{2\mu_I \gamma (V - E_i)k_2}{(k_1 x_A + k_2)^2}$$

$$= \frac{2(-2 \times 10^{-10})(10 \times 10^{-12})(-50 - 50) \cdot 100}{(2 + 100)^2}$$

$$= 3.84 \times 10^{-21} \text{ A}^2 \text{ sec.}$$

$$f_c = \frac{1}{2\pi\tau} = \frac{1}{2\pi(0.01) \text{ sec}} = 16 \text{ Hz.}$$

3. (a) In the presence of 10^{-4} M glutamate, $V_m = -40$ mV,

$$k_1(-40) = 2000 \ e^{-1} = 736 \text{ sec}^{-1} \text{ M}^{-1}, (k_1 x_A = 0.0736 \text{ sec}^{-1}).$$

$$k_2(-40) = e^1 = 2.71 \text{ sec}^{-1}.$$

$$\mathbf{Q} = \begin{pmatrix} -k_2 & k_2 \\ k_1 X_A & -k_1 X_A \end{pmatrix} = \begin{pmatrix} -2.71 & 2.71 \\ 0.0736 & -0.0736 \end{pmatrix} (\text{sec}^{-1})$$

   (b) $P_1(\infty) = \dfrac{k_1 x_A}{k_1 x_A + k_2} = \dfrac{0.0736}{2.71 + 0.0736} = 0.026 = 2.6\%$ (open)

   $P_2(\infty) = \dfrac{k_2}{k_1 x_A + k_2} = \dfrac{2.71}{2.71 + 0.0736} = 0.974 = 97.4\%$ (closed)

(c) At rest, the neuron is only permeable to K^+, thus $V_m = E_K = -80$ mV, $g_K = \frac{1}{R_{in}} = 10^{-7}$ S. In the presence of 10^{-4} M glutamate, $V_m = -40$ mV. Therefore,

$$V_m = -40 \text{ mV} = \frac{g_K E_K + g_{Na} E_{Na}}{g_K + g_{Na}}$$

$$-40 \text{ mV} = \frac{(g_K/g_{Na})(-80 \text{ mV}) + (+40 \text{ mV})}{(g_K/g_{Na}) + 1}$$

$$g_K/g_{Na} = 2.$$

Therefore, $g_{Na} = \frac{1}{2} 10^{-7}$ S $= 5 \times 10^{-8}$ S

$$N = \frac{g_{Na}}{\gamma_{Na}} = \frac{5 \times 10^{-8} \text{ S}}{20 \times 10^{-12} \text{ S}} = \frac{50,000}{20} = 2500$$

4. (a) See text.

   (b) The whole-cell I_K and its activation curve are given below.

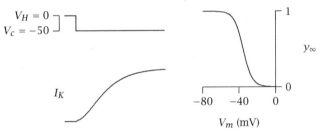

   The time course of the onset of I_K follows a 2nd power function of a single exponential.

   $$I_K(t) \propto \left(1 - e^{-t/\tau}\right)^2.$$

5. $E_{Na} = 58 \log \frac{[Na^+]_{out}}{[Na^+]_{in}} = 58 \log 2 = 17.5$ mV.

   Since k_{+1} and k_{-1} are much faster than α and β, the channel can be approximately described with a two-state transition scheme.

(a)

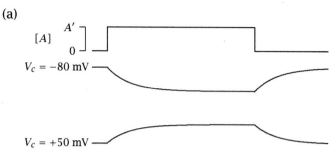

(b) The approximate relaxation time constant $\tau = \frac{1}{\alpha+\beta}$

(c)

$S(f)$
$(A^2 \text{ sec})$

frequency (Hz)

6. (a) $Q = \begin{pmatrix} -(100+50) & 50 & 100 \\ 5 & -5 & 0 \\ 10 & 0 & -10 \end{pmatrix}$

$$\det Q - \lambda I = \begin{vmatrix} -150 - \lambda & 50 & 100 \\ 5 & -5 - \lambda & 0 \\ 10 & 0 & -10 - \lambda \end{vmatrix} = 0$$

$$-(150 + \lambda)(5 + \lambda)(10 + \lambda) + 1000(5 + \lambda) + 250(10 + \lambda) = 0.$$

$$\lambda(\lambda^2 + 165\lambda + 1050) = 0.$$

$$\lambda_1 = 0.$$

$$\lambda_2 = \frac{-165 + \sqrt{(165)^2 - 4 \cdot 1050}}{2} = -6.63 \ \text{sec}^{-1}.$$

$$\lambda_3 = \frac{-165 - \sqrt{(165)^2 - 4 \cdot 1050}}{2} = -158 \ \text{sec}^{-1}.$$

(b) Mean open lifetime $= -\frac{1}{q_{11}} = \frac{1}{150} = 6.7$ msec,

mean blocked lifetime $= \frac{-1}{q_{22}} = \frac{1}{5} = 200$ msec,

mean closed lifetime $= \frac{-1}{q_{33}} = \frac{1}{10} = 100$ msec.

(c) $$S(f) = \frac{S(0)}{\left[1 + \left(\frac{2\pi f}{\lambda_2}\right)^2\right]\left[1 + \left(\frac{2\pi f}{\lambda_3}\right)^2\right]},$$

$$f_{c_1} = \frac{-\lambda_2}{2\pi} = \frac{6.63}{2\pi} = 1.06 \ \text{Hz},$$

$$f_{c_2} = \frac{-\lambda_3}{2\pi} = \frac{158}{2\pi} = 25 \ \text{Hz}.$$

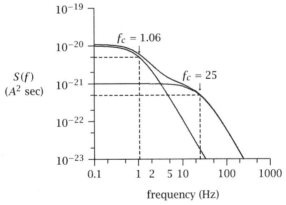

7. (a)
$$\mathbf{Q} = \begin{bmatrix} -k_{-3} & k_{-3} & 0 & 0 \\ k_3 & -(k_3 + k_{-2}) & k_{-2} & 0 \\ 0 & k_2 & -(k_2 + k_{-1}) & k_{-1} \\ 0 & 0 & k_1 & -k_1 \end{bmatrix}$$

(b) Mean lifetimes of the channel in each state:

$$\overline{\tau}_1 = \frac{1}{k_{-3}},$$

$$\overline{\tau}_2 = \frac{1}{k_3 + k_{-2}},$$

$$\overline{\tau}_3 = \frac{1}{k_2 + k_{-1}},$$

$$\overline{\tau}_4 = \frac{1}{k_1}.$$

8.   (a) $\mathbf{Q} = \begin{pmatrix} -(300+50) & 50 & 300 \\ 50 & -50 & 0 \\ 100 & 0 & -100 \end{pmatrix} = \begin{pmatrix} -350 & 50 & 300 \\ 50 & -50 & 0 \\ 100 & 0 & -100 \end{pmatrix}$

$$\det |\mathbf{Q} - \lambda\mathbf{I}| = \begin{vmatrix} -350 - \lambda & 50 & 300 \\ 50 & -50 - \lambda & 0 \\ 100 & 0 & -100 - \lambda \end{vmatrix} = 0$$

$$-(350 + \lambda)(50 + \lambda)(100 + \lambda) + (100)(50 + \lambda)(300)$$
$$+(100 + \lambda)(50)(50) = 0.$$

$$-(350 \times 50 \times 100 + 57,500\lambda + 500\lambda^2 + \lambda^3)$$
$$+1,500,000 + 30,000\lambda + 250,000 + 2,500\lambda = 0.$$

$$\lambda^3 + 500\lambda^2 + 25,000\lambda = 0.$$

$$\lambda(\lambda^2 + 500\lambda + 25,000) = 0.$$

$$\lambda = 0.$$

$$\lambda = \frac{-500 \pm \sqrt{(500)^2 - 4 \times 25,000}}{2} = \frac{-500 \pm 390}{2} = \begin{matrix} -55 \\ -445 \end{matrix}$$

Eigenvalues: $\lambda_1 = 0$, $\lambda_2 = -55 \text{ sec}^{-1}$, $\lambda_3 = -445 \text{ sec}^{-1}$

(b) $\tau_1 = -\dfrac{1}{\lambda_1} \rightarrow -\infty$ does not exist.

$$\tau_2 = -\frac{1}{55} = 0.0182 \text{ sec} = 18.2 \text{ msec}.$$

$$\tau_3 = -\frac{1}{445} = 0.0022 \text{ sec} = 2.2 \text{ msec}.$$

$$f_{c_2} = \frac{1}{2\pi\tau_2} = 8.75 \text{ Hz}.$$

$$f_{c_3} = \frac{1}{2\pi\tau_3} = 70.9 \text{ Hz}.$$

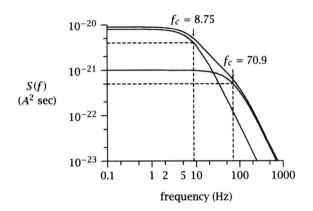

9. (a)

$$
\begin{aligned}
\gamma &= \frac{I_1}{(V - E_i)} = \frac{-20 \text{ pA}}{(-40 - 10) \text{ mV}} \\
&= 0.4 \text{ nS.}
\end{aligned}
$$

(b) At $V_m = +60$,

$$
\begin{aligned}
I_1 &= \gamma(V - E_i) = (0.4 \text{ nS})[(60 - 10) \text{ mV}] \\
&= 0.4 \times 50(10^{-9} \cdot 10^{-3} \text{ S} \cdot \text{V}) \\
&= 20 \text{ pA.}
\end{aligned}
$$

(c) Mean open lifetime

$$
\begin{aligned}
&= \frac{1}{\alpha + k_B x_B} \\
&= \frac{1}{18}(3 + 2 + 5 + 2 + 5 + 6 + 3 + 2 + 5 \\
&\qquad + 8 + 3 + 5 + 2 + 2 + 3 + 4 + 5 + 3)(100 \text{ msec}) \\
&= 3.78 \times 100 \text{ msec.}
\end{aligned}
$$

Mean blocked lifetime (gap with a burst)

$$
\begin{aligned}
&= \frac{1}{k_{-B}} = \frac{1}{15}(4 + 8 + 5 + 5 + 3 + 3 + 7 + 2 + 4 + 5 \\
&\qquad + 3 + 5 + 5 + 11 + 5)(100 \text{ msec}) \\
&= 5 \times 100 \text{ msec.}
\end{aligned}
$$

Mean closed lifetime (gap between bursts)

$$
= \frac{1}{\beta}
$$

$$= \frac{1}{2}(35 + 35)$$
$$= 35 \times 100 \text{ msec.}$$

Mean open time per burst

$$= \frac{1}{\alpha}$$
$$= \frac{1}{3}(23 + 21 + 24)(100 \text{ msec}) = 22.7 \times 100 \text{ msec.}$$

$$\alpha = 0.0044 \text{ sec}^{-1}.$$
$$\beta = 0.0029 \text{ sec}^{-1}.$$
$$k_B X_B = 0.022 \text{ sec}^{-1}.$$
$$k_{-B} = 0.02 \text{ sec}^{-1}.$$

10. (a) Macroscopic current: relaxation time constant $\tau = \frac{1}{\alpha+\beta}$. Sum of five channels: $\tau = \frac{1}{\alpha}$.

   (b) The macroscopic current relaxation involves N channels opened and closed at random intervals with *decreasing* open probabilities with respect to time, i.e.,

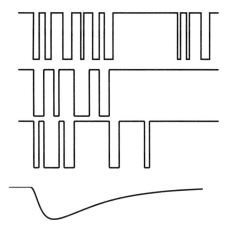

Since individual channels undergo both opening and closing transitions (α and β), according to rule 6, $\tau = \frac{1}{\alpha+\beta}$.

The five channels in figure 10.4C are in *special* situations: (1) they all are in open state at $t = 0$ (synchronized), and (2) once they are closed, they never open again. Therefore, the relaxation of the ensemble of these channels involves transition from open to closed (α) but not vice versa (β). Thus $\tau = \frac{1}{\alpha}$, which is equivalent to the time constant of the open time distribution function.

11. See graph for answers to (a), (b), and (c).

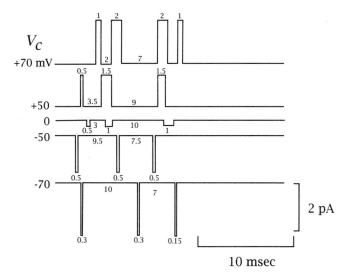

$1/\alpha$	$1/\beta$	α	β	$\frac{\beta}{\alpha+\beta}$
1.5	3.7	666	272	0.29
1.17	6.25	857	160	0.16
0.83	6.5	1200	154	0.11
0.5	8.5	2000	118	0.056
0.25	8.5	400	118	0.029

$P_1(\infty)$ at $-30\,\mathrm{mV} = 0.08$

b)

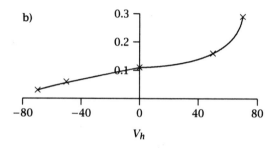

c)

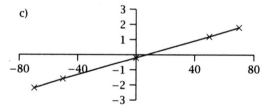

$$\gamma = \frac{1.2\,\mathrm{pA}}{(50-8)\,\mathrm{mV}} = 28\,\mathrm{pS}$$

(d) $\mu_I = NP_1I_1 = NP_1\gamma(V-E)$ $E = 8\,\mathrm{mV}$, $\tau = \dfrac{1}{\alpha+\beta}$.

$+70$	$\mu_I =$	$10^4(0.29)28(62)$	$=$	$5.03\,\mathrm{nA}$,
$+50$	$\mu_I =$	$10^4(0.16)28(42)$	$=$	$1.88\,\mathrm{nA}$,
0	$\mu_I =$	$10^4(0.11)28(-8)$	$=$	$-0.25\,\mathrm{nA}$,
-50	$\mu_I =$	$10^4(0.056)28(-58)$	$=$	$-0.91\,\mathrm{nA}$,
-70	$\mu_I =$	$10^4(0.029)28(-78)$	$=$	$-0.63\,\mathrm{nA}$.

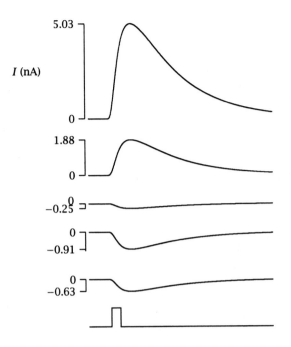

12. (a)

$$\mathbf{Q} = \begin{pmatrix} -150 & 100 & 50 \\ 10 & -10 & 0 \\ 5 & 0 & -5 \end{pmatrix}$$

$$\det(Q - \lambda I) = 0$$

$$\begin{vmatrix} -150 - \lambda & 100 & 50 \\ 10 & -10 - \lambda & 0 \\ 5 & 0 & -5 - \lambda \end{vmatrix}$$

$$= -(150 + \lambda)(10 + \lambda)(5 + \lambda) + (5)(50)(10 + \lambda) + 1000(5 + \lambda) = 0.$$

$$\lambda_1 = 0,$$
$$\lambda_2 = -6.5 \text{ sec}^{-1},$$
$$\lambda_3 = -158 \text{ sec}^{-1}.$$

(b) Mean open lifetime should follow exponential pdf.

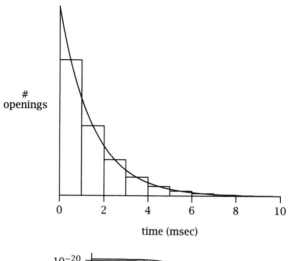

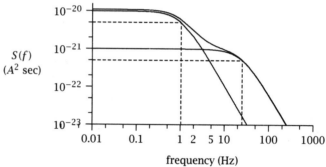

13. (a)

OC$_4$I$_1$I$_2$ loop : $k_1 \times 25 \times 50 \times 10 = 50 \times 50 \times 50 \times 10.$
                           $k_1 = 100 \ (\text{sec}^{-1}).$

OC$_4$C$_3$I$_1$I$_2$ loop : $100 \times 10 \times 20 \times 50 \times 10 = 50 \times 50$
                              $\times k_2 \times 20 \times 10.$
                           $k_2 = 20 \ (\text{sec}^{-1}).$

(b)

$$\mathbf{Q} = \begin{bmatrix} -150 & 100 & 0 & 0 & 0 & 0 & 50 \\ 10 & -45 & 10 & 0 & 0 & 25 & 0 \\ 0 & 20 & -80 & 40 & 0 & 20 & 0 \\ 0 & 0 & 30 & -50 & 20 & 0 & 0 \\ 0 & 0 & 0 & 40 & -40 & 0 & 0 \\ 0 & 50 & 20 & 0 & 0 & -120 & 50 \\ 10 & 0 & 0 & 0 & 0 & 50 & -60 \end{bmatrix}$$

(c) $\bar{\tau} = \dfrac{-1}{q_{11}} = \dfrac{1}{150} = 6.67$ msec.

(d) In pronase, I_1 and I_2 disappear. $q_{11} = k_1 = 100$ sec^{-1}. $\bar{\tau} = \dfrac{1}{100} = 10$ msec.

14.   (a)

$$\begin{aligned} \mu_I &= Np\gamma(V - E) = 10^5(0.2 - 0)(20\text{ pS})(-40 - 20)\text{ mV} \\ &= -24\text{ nA} \end{aligned}$$

(b) From the figure:

$fc_1 = 10$ sec$^{-1} \rightarrow \lambda_1 = -62.8$ sec^{-1},

$fc_2 = 200$ sec$^{-1} \rightarrow \lambda_2 = -1257$ sec^{-1},

$\tau_1 = 15.9$ msec,

$\tau_2 = 0.8$ msec.

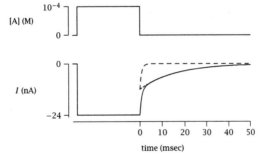

(c) $k_{-1} = 1000$ sec^{-1}, $k_1 = 100$ sec^{-1}.

$|\lambda_1 + \lambda_2| = |63 + 1257| = 1320$ sec$^{-1} = \alpha + \beta + k_{-1}$.

$\alpha = 10\,\beta$; therefore, $\alpha = 290$ sec^{-1}; $\beta = 29$ sec.

$$\mathbf{Q} = \begin{bmatrix} -\alpha & \alpha & 0 \\ \beta & -(\beta + k_{-1}) & k_{-1} \\ 0 & k_1 & -k_1 \end{bmatrix} = \begin{bmatrix} -290 & 290 & 0 \\ 29 & -1029 & 100 \\ 0 & 100 & -100 \end{bmatrix}$$

15.   (a)

$$\mathbf{Q} = \begin{array}{c} \\ 1 \\ 2 \\ 3 \\ 4 \\ 5 \end{array} \begin{array}{ccccc} 1 & 2 & 3 & 4 & 5 \\ -(3\beta_m + \beta_h) & 3\beta_m & 0 & 0 & \beta_h \\ \alpha_m & -(\alpha_m + 2\beta_m + 2\beta_h) & 2\beta_m & 0 & \beta_h \\ 0 & 2\alpha_m & -(2\alpha_m + \beta_m) & \beta_m & 0 \\ 0 & 0 & 3\alpha_m & -3\alpha_m & 0 \\ \alpha_h & \alpha_h & 0 & 0 & -2\alpha_h \end{array}$$

(b) Mean open lifetime $= \frac{-1}{q_{11}} = \frac{1}{3\beta_m + \beta_h} = \frac{1}{(3 \times 1000 + 100) \text{ sec}}$

$= 0.00032 \text{ sec} = 0.32 \text{ msec}.$

(c) $I_N = NPI_1 = NP\gamma(V - E_{Na})$,

$-2 \text{ nA} = 10,000 \, P \quad (100 \text{ pS}) \quad (0 - 50) \text{ mV}.$

$$P = \frac{-2 \times 10^{-9} \text{A}}{10,000 \times 100 \times 10^{-12} \text{ S}(-50) \times 10^{-3} \text{ V}}$$

$$= \frac{-2 \times 10^{-9} \text{ A}}{-5 \times 10^{-8} \text{ A}}$$

$$= 0.04.$$

Chapter 11

1. For binomial model,

$$N_x = \frac{Nn!}{x!(n - x)!} p^x (1 - p)^{n-x}.$$

For failures, $N_0 = N(1 - p)^n$, or $\frac{N_0}{N} = (1 - p)^n$, and

$\ln(N_0/N) = n \cdot \ln(1 - p).$

Remember $m = np$, so $\ln(N_0/N) = (m/p) \ln(1 - p)$, and

$$m = \ln(N_0/N) \cdot \frac{p}{\ln(1 - p)}.$$

For variance, $\sigma^2 = m(1 - p)$,

$$CV = \sigma/m \text{ or } CV^2 = \sigma^2/m^2 = \frac{1 - p}{m}, \text{ and}$$

$$m = (1 - p)/CV^2.$$

2. (a) $N_0 \ = \ Ne^{-m}$

$= \ 500 \, e^{-5}$

$= \ 3.37 \ (3 \text{ or } 4 \text{ failures}).$

(b) $N_2 = \dfrac{Nm^2 e^{-m}}{2!}$

$\quad\quad\quad = \dfrac{(500)(25)e^{-5}}{2}$

$\quad\quad\quad = 42.1$ (approximately 42 observations).

3. $\displaystyle\lim_{n\to\infty} P_n(k) = \lim_{n\to\infty} \frac{n!}{k!(n-k)!}\, p^k(1-p)^{n-k}$

$\quad\quad\quad\quad\quad\quad = \displaystyle\lim_{n\to\infty} \frac{n(n-1)(n-2)\ldots(n-k)!}{k!(n-k)!}\, p^k(1-p)^{n-k}$

$\quad\quad\quad\quad\quad\quad = \displaystyle\lim_{n\to\infty} \frac{n^k}{k!}\, p^k(1-p)^{n-k} = \lim_{n\to\infty} \frac{m^k}{k!}\,(1-m/n)^n.$

Expand $(1-m/n)^n$ in a series:

$$\left(1-\frac{m}{n}\right)^n = 1 - \frac{n}{1!}\left(\frac{m}{n}\right) + \frac{n(n-1)}{2!}\left(\frac{m}{n}\right)^2$$
$$-\frac{n(n-1)(n-2)}{3!}\left(\frac{m}{n}\right)^3 + \cdots.$$

As $n \to \infty$,

$$\left(1-\frac{m}{n}\right)^n \to 1 - m + \frac{m^2}{2!} - \frac{m^3}{3!} = e^{-m}.$$

Substitute e^{-m} for $(1-m/n)^n$ in above, and

$\lim_{n\to\infty}$ binomial $= \frac{m^k}{k!} e^{-m} =$ Poisson.

4. (a) From table, mean amplitude = 1.0 mV.

(b) From table,
$\quad$ 9 intervals at 0.5 ms
$\quad$ 8 intervals at 1.0 ms
$\quad$ 7 intervals at 1.5 ms
$\quad$ 6 intervals at 2.0 ms
$\quad$ 5 intervals at 2.5 ms
$\quad$ 5 intervals at 3.0 ms
$\quad$ 4 intervals at 3.5 ms
$\quad$ 2 intervals at 6.0 ms.

$\quad$ Construct a histogram from the above table. A plot of log(histogram) vs. t results in a straight line so mEPPS are *independent*.

(c) From semilog plot in (b), calculate the slope of the straight line. From text $f_1(t) = re^{-rt}$, so

$$r = -\frac{\log(y_2/y_1)}{0.434(t_2 - t_1)} = 0.27/\sec.$$

5. (a) See figure below.

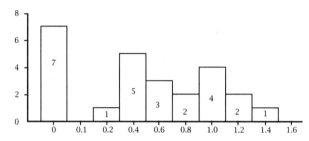

(b) Direct method: $m = \frac{0.608}{0.5} = 1.21$.

Failures method: $m = \ln(N/N_0) = \ln(25/7) = 1.27$.

(c) $N_3 = \dfrac{200(1.24)^3 e^{-1.24}}{3 \cdot 2 \cdot 1} = 18.4$ (or 18 times).

6. From table, mean EPC = 0.84 nA.

(a) $G_S = \dfrac{0.84}{80 \text{ mV}} = 10.6$ nS.

(b) From text, $CV = \sigma/\text{mean}$. From list of EPCs,

$$\sigma = \left(\frac{\Sigma(\text{epp}-\text{mean epp})^2}{17} \right)^{1/2}$$

$$= 0.35$$

$$CV = 0.35/0.84 = 0.42.$$

$$m = \frac{1}{CV^2} = 5.8.$$

(c) $N_0 = Ne^{-m}$

$\quad\quad = 1000\, e^{-5.8} \cong 3.$

(d) (1) Noise level too high.
(2) Release not Poisson.
(3) Too few trials.

7. (a) Presynaptic because (1) no change in amplitude, and (2) there was a doubling in the frequency of mEPSCs.

(b) m (before) $= \dfrac{2.0}{0.6} = 3.33$.

$$m \text{ (LTP)} = \frac{3.0}{0.6} = 5.0.$$

(c) $m = np$.

Before: $3.33 = 500(p)$ so $p = 0.0067$.

During LTP: $p = 5.0/500 = 0.01$.

(d) For Poisson:

$$m = \ln(N/N_0) \quad \text{or} \quad e^{-m} = \frac{N_0}{N} \quad \text{and} \quad N_0 = Ne^{-m}.$$

$$1 = Ne^{-3.33}, \quad \text{and}$$

$$N = 28.02 \quad \text{(or 29 before)}.$$

$$N = e^{+5.0} = 148.4 \quad \text{(or 149 during LTP)}.$$

For binomial:

$$m = \frac{p}{\ln(1-p)} \ln(N_0/N).$$

$$\frac{\ln(N_0/N)}{\ln(1-p)} = \frac{p}{m}.$$

$$\begin{aligned} N &= \frac{N_0}{(1-p)^{m/p}} \\ &= 28.34 \quad \text{(or 29 before)} \\ &= 152.2 \quad \text{(or 153 during LTP)}. \end{aligned}$$

Yes, depends on statistic, although both are close in this case. Binomial is better at higher release rates (i.e., during LTP m is larger so binomial should be used).

8. $$\begin{aligned} P(1, t + \Delta t) &= P(0,t) \cdot P(1, \Delta t) \\ &= e^{-rt} \cdot r\Delta t. \end{aligned}$$

$$\begin{aligned} \text{pdf} = \lim_{\Delta t \to 0} \frac{P(1, t + \Delta t)}{\Delta t} = \frac{dP(1,t)}{dt} &= \frac{e^{-rt} \cdot r\Delta t}{\Delta t} \\ &= re^{-rt}. \end{aligned}$$

9. (a) Mean mEPSP = 0.6 mV; $\sigma = 0.18$.

(b) Mean EPSP = 1.46 mV; $\sigma = 0.93$.

(c) m_f = $\ln(14/1) = 2.6$.

   m_d = $1.46/0.6 = 2.4$.

   m_{cv} = $\dfrac{1}{(\sigma/\text{mean})^2} = 2.5$.

   Nonlinear summation a factor for m_d and m_{cv}.

(d) Use $m_d = 2.4 = np$.

   $p = 2.4/10 = 0.24$.

   $m_f = \dfrac{0.24}{\ln(1 - 0.24)} \ln(N_0/N)$.

   $N_0 = 1.55$ (or 1 to 2 failures).

10. The equation is

$$\frac{\text{EPSP}(t) - \text{EPSP}_0}{\text{EPSP}_0} = A \exp(-t/10) + P \exp(-t/60),$$

where $A = P = 1$. This is obtained by plotting $\frac{\text{EPSP}(t) - \text{EPSP}_0}{\text{EPSP}_0}$ vs. time on semilog paper and "peeling" exponentials to obtain time constants of 10 and 60 sec and the coefficients A and P.

Chapter 12

1. Refer to text.

2. Refer to text.

3. (a) $I_m \cong V_1 - V_2$, refer to text. I_{clamp} includes I_m but also the current flow in the other direction from the synapse and the transmembrane current between I_{clamp} and V_2.

   (b) Slow onset to I_{Ca}; nonlinear relationship between I_{Ca} and EPSP; delay in EPSC during I_{Ca}; *off* response bigger than *on* response; short delay from tail current to EPSC.

Chapter 13

1.   (a) See the figure below.

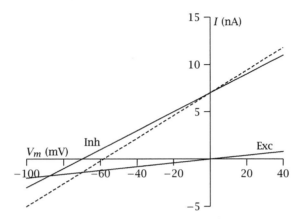

   (b) Add graphically from the figure above, or

$$G_T = G_I + G_E = 120 \text{ nS}.$$

$$E_T = \frac{G_I E_I + G_E E_E}{G_T} = -58 \text{ mV}.$$

Because E_T is negative to threshold the response is inhibitory.

2.   (a) Decay is exponential with time constant of 17.4 min.

   (b) G_s for control and MTP is 37.5 nS.

E_s for control is -22 mV.

E_s for MTP is $+20$ mV.

   (c) Slope $= \frac{G_K}{G_K + G_{Na}}$. Control slope $= 0.5$.

$G_K + G_{Na} = 37.5 \text{ nS}$ (from (b)).

$G_K = (0.5)(37.5) = 18.8 \text{ nS}.$

$G_{Na} = 18.8 \text{ nS}.$

$G_K/G_{Na} = 1$ during control.

MTP slope $= 0.23$.

$G_K = (0.23)(37.5) = 8.6 \text{ nS}.$

$G_{Na} = 37.5 - 8.6 = 28.9 \text{ nS}.$

$G_K/G_{Na} = \frac{8.6}{28.9} \cong 0.3.$

(d) MTP is caused by a change in the reversal potential in depolarizing direction due to a decrease in the permeability of the channel to K^+ ions and an increase in permeability to Na^+ ions. Mechanism is postsynaptic.

3. (a) $E_{Cl} = -58 \log 15 = -68$ mV.

$$
\begin{aligned}
V_{\text{rest}} &= \frac{G_K E_K + G_{Na} E_{Na} + G_{Cl} E_{Cl}}{G_K + G_{Na} + G_{Cl}} \\
&= \frac{(-100)(1) + (50)(0.8) + (-68)(15)}{1 + 0.8 + 15} \\
&= \frac{-100 + 40 - 1020}{16.8} = -64.3 \text{ mV}.
\end{aligned}
$$

(b) i. Synaptic current evoked by transmitter A:

$$
\begin{aligned}
\Delta I_A &= \Delta G_{Na}(V_m - E_{Na}) \\
&= (0.02 \text{ S/cm}^2)(-64.3 - 50) \text{ mV} \\
&= -2.3 \text{ mA/cm}^2.
\end{aligned}
$$

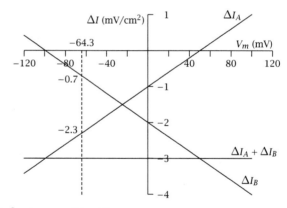

ii. by transmitter B:

$$
\begin{aligned}
\Delta I_B &= \Delta G_K(V_m - E_K) \\
&= (-0.02 \text{ S/cm}^2)(-64.3 + 100) \text{ mV} \\
&= -0.7 \text{ mA/cm}^2.
\end{aligned}
$$

iii. by transmitters $A + B$:

$$\Delta I = \Delta I_A + \Delta I_B = (-2.3 \text{ mA} - 0.7 \text{ mA})/\text{cm}^2 = -3.0 \text{ mA/cm}^2.$$

(c) Reversal potential for stimuli:

i. $V_{rev} = E_{Na} = +50$ mV.

ii. $V_{rev} = E_K = -100$ mV.

iii. V_{rev} does not exist.

4. (a) Reversal potentials: A = -5 mV; B = -40 mV; C = -70 mV; D = -60 mV.

    (b) A: Conductance increase because when $V_m > V_{rev}$, ΔV is hyperpolarizing, which leads to $\Delta I_{\text{outward}}$ or ΔI is positive. $\Delta G = \frac{\Delta I}{V_m - E_{rev}}$ is therefore also positive.
    B: Conductance decrease, similar argument as A.
    C: Conductance decrease
    D: Conductance increase

5. (a) $G_s = 100$ nS. $V_{rev} = -20$ mV from resting. (Obtained from plot of peak synaptic current vs. holding potential.)

    (b) Inhibitory

    (c) $R_N \cong 50$ MΩ. (Obtained from plot of holding current vs. holding potential.)

    (d) $\tau_m \cong 44$ msec. (Obtained from the time to 37% of I peak.)

    (e) No.

    (f) The synapse is electrotonically remote from recording site.

6. (a) A: $G_A = \frac{1 \text{ nA}}{20 \text{ mV}} = 50$ nS; $V_{rev} = 0$ mV.
    B: $G_B = \frac{3 \text{ nA}}{40 \text{ mV}} = 75$ nS, $V_{rev} = -60$ mV.

    (b) A = excitatory; B = inhibitory.

    (c) No.

    (d) $V_{rev} = -36$ mV; $G_{A+B} = 125$ nS; excitatory.

    (e) Conductance increase.

    (f) Excitatory.

7. (a) Slope = ΔG_{Na}. $V_{rev} = +50$ mV.

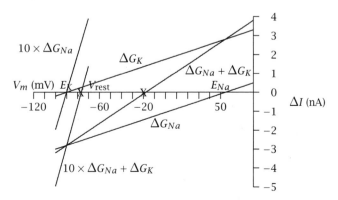

(b) $$E_s = \frac{\Delta G_K E_K + \Delta G_{Na} E_{Na}}{\Delta G_K + \Delta G_{Na}}$$

$$= \frac{\Delta G_{Na}(E_K + E_{Na})}{2\Delta G_{Na}}$$

$$= \frac{E_K + E_{Na}}{2} = \frac{-90 + 50}{2} = -20 \text{ mV}.$$

It is therefore an EPSP.

(c) $$E_s = \frac{\Delta G_{Na}(10E_K + E_{Na})}{11\Delta G_{Na}}$$

$$= \frac{-900 + 50}{11} = -77.3 \text{ mV}.$$

It is therefore an IPSP.

8. (a) $$G_{s_{X,Y}} = \frac{2 \text{ nA}}{100 \text{ mV}} = 0.02 \times 10^{-6} = 20 \text{ nS}.$$

$$E_{sum} = \frac{G_s}{G_s + G_r} E_s \quad \text{from rest).}$$

$$= \frac{20}{20 + 10} E_s = 67 \text{ mV} \quad \text{(from rest)}$$

or -33 mV from zero. If X and Y use same channels), E_{sum} is the same for X alone, Y alone, or with X and Y together. Also, $I_{cl} = 2$ nA for X alone, Y alone, or with X and Y together.

(b) If X and Y use different channels, then for X alone, Y alone, same as above. But for X + Y,

$$E_{sum} = \frac{20 + 20}{40 + 10} E_s = 80 \text{ mV} \quad \text{(from rest)},$$

or -20 mV from zero. Also,

$$I_{cl_{X+Y}} = 2 + 2 = 4 \text{ nA}.$$

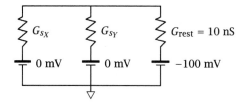

Alternate method:

(a) $\quad E_{sum} \quad = \quad \dfrac{G_X E_X + G_r E_r}{G_X + G_r}$

$\quad\quad\quad\quad = \quad \dfrac{20(0) + 10(-100)}{30} = -33 \text{ mV.}$

(b) $\quad E_{sum} \quad = \quad \dfrac{G_X E_X + G_Y E_Y + G_r E_r}{G_X + G_Y + G_r}$

$\quad\quad\quad\quad = \quad \dfrac{20(0) + 20(20) + 10(-100)}{50} = -20 \text{ mV.}$

9. (a) From graphs,

	Before	During Drug X
G_s =	2.5 nS	3.2 nS
V_{rev} =	91 mV	76 mV
R_N =	226 MΩ	636 MΩ.

(b) $\lambda = \sqrt{\dfrac{a R_m}{2 R_i}}$; obtain R_m from τ_m assuming $C_m = 1 \; \mu\text{F/cm}^2$.

	Before	During
λ_1 =	0.12 cm	0.23 cm
λ_2 =	0.1	0.18
λ_3 =	0.08	0.14

(c) R_m, R_i, and C_m are constants (and uniform), soma isopotential, $I_{\text{radial}} = I_{\text{external}} = 0$, passive, nontapering, all terminate same (open or sealed), all at same L, 3/2 power rule, V_m constant at $t = 0$.

$$L = \frac{l_1}{\lambda_1} + \frac{l_2}{\lambda_2} + \frac{l_3}{\lambda_3}.$$

Before: $L = 0.8$; after: $L = 0.43$.

$X = \frac{l_1}{\lambda_1} + \frac{l_2}{\lambda_2}$.

Before: $X = 0.65$; after: $X = 0.35$.

(d) $\quad E_s \quad = \quad V_{rev} \dfrac{\cosh(L - X)}{\cosh L}$

$\quad\quad\quad\quad = \quad 90 \text{ mV} \dfrac{\cosh(0.8 - 0.65)}{\cosh(0.8)} = 70 \text{ mV} \quad \text{(before)}$

$\quad\quad\quad\quad = \quad 76 \dfrac{\cosh(0.65 - 0.35)}{\cosh(0.65)} = 70 \text{ mV} \quad \text{(after).}$

Drug X does not appear to change synaptic equilibrium potential. It alters the apparent reversal potential by changing the electrotonic distance of the synapse from recording site.

(e) Drux X can alter the *apparent* G_s by changing the electrotonic distance of synapse to recording site. One would measure more (or less) synaptic current for a given voltage change depending on the "remoteness" of synapse.

10. See figure below. It is likely to be excitatory.

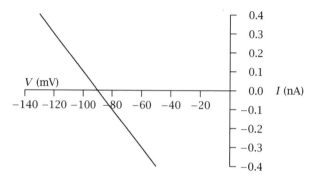

11. Two possibilities are:

(a) Part of the EPSP is due to activation of NMDA receptors. Hyperpolarization reduces the NMDA response, making the EPSP smaller. This can be tested by using APV, 0 Mg^{2+}, or, under voltage clamp, measuring the chord conductance as a function of membrane potential.

(b) The input resistance of the neuron decreases with hyperpolarization, making the EPSP smaller. This could be due to some type of anomalous rectification. Test by trying to block AR with external Cs^+ or Ba^{2+} or by measuring the *I-V* curve of the synaptic response under voltage clamp.

(c) At the resting potential the EPSP may be activating a voltage-gated inward current (e.g., $I_{Na(slow)}$ or $I_{Ca(T)}$) that contributes to its amplitude. When the neuron is hyperpolarized, the EPSP may be below the activation range for these inward currents, and this would reduce its apparent amplitude.

12. (a) See the figure below.

(b) $G_{s_{A+B}} = 10 + 20 = 30$ nS; $E_{s_{A+B}} = -67$ mV; $R_N = 25$ MΩ.

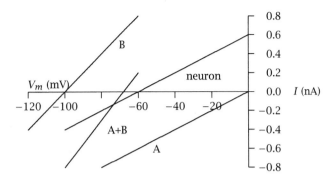

(c) At 1 msec: G_A = (peak G_A) while G_B = $(1/3)\cdot$ (peak G_B); at 3 msec: G_A = $(e^{-2/4})\cdot$ (peak G_A)= 0.6 (peak G_A) while G_B = (peak G_B); at 6 msec: G_A = $(e^{-5/4})\cdot$ (peak G_A) = 0.3 (peak G_A) while G_B = $(e^{3/10})\cdot$ (peak G_B) = 0.6 (peak G_B), so

$$I_s(1\ \text{msec}) = G_A(V_m - E_{s_A}) + G_B/3(V_m - E_{s_B}),$$
$$I_s(3\ \text{msec}) = 0.6\,G_A(V_m - E_{s_A}) + G_B(V_m - E_{s_B}),$$
$$I_s(6\ \text{msec}) = 0.3\,G_A(V_m - E_{s_A}) + 0.6G_B(V_m - E_{s_B}).$$

Calculate and plot *I-V* curves according to above equations.

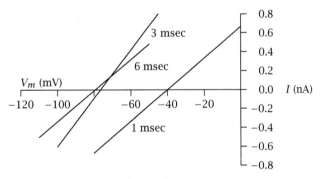

13. (a) Slope resistance at rest = $\frac{10\ \text{mV}}{0.1\ \text{nA}}$ = 100 MΩ. Slope conductance = 1×10^{-8} S. During transmitter X,

$$R_{\text{slope}} = \frac{10\ \text{mV}}{0.22\ \text{nA}} = 45\ \text{MΩ}.$$

$$G_{\text{slope}} = 2.2 \times 10^{-8}\ \text{S}.$$

(b) Conductance increase response.

(c) Inward current at -70 mV so V_{rev} is depolarized from rest. Whether excitatory or inhibitory depends on V_{th}.

14. (a) V_{den} $= \dfrac{R_{den}E_s}{R_{den} + R_{sp} + R_s} = \dfrac{100}{1600}E_s$

$= 4.4$ mV from rest or -66 mV

V_{sh} $= \dfrac{R_{den} + R_{sp}}{1600}E_s = \dfrac{600}{1600}E_s$

$= 26$ mV from rest or -44 mV.

(b) See figure below.

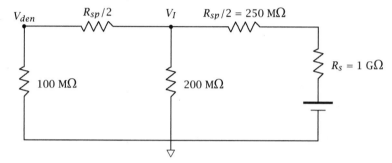

Combine R_{den} (i.e., 100 MΩ) in series with the $R_{sp}/2$ on the left and then combine this equivalent resistance (i.e., 350 MΩ) in parallel with 200 MΩ. This leads to an equivalent resistance of 120 MΩ in series with $R_{sp}/2$ on the right.

V_{sh} $= \dfrac{120 + 250}{1000 + 120 + 250}E_s$

$= \dfrac{370}{1370}E_s = 18.9$ mV from rest or -51.1 mV.

For V_{den} use Kirchhoff's current law and then voltage divider equation.

$$\dfrac{V_I}{200} + \dfrac{V_I - V_{den}}{250} = \dfrac{70 - V_I}{1250},$$

$$\dfrac{V_I - V_{den}}{250}100 = V_{den},$$

$$V_I = 3.5\,V_{den},$$

$$\dfrac{3.5\,V_{den}}{200} + \dfrac{2.5\,V_{den}}{250} = \dfrac{70 - 3.5\,V_{den}}{1250},$$

$V_{den} = 1.8$ mV from rest or -68.2 mV.

Chapter 14

1. (a) Sink; active; negative.

    (b) Membrane current; slope is less contaminated by other events than peak; peak current.

    (c) Make measurements at fixed distances along a laminar profile and take second spatial derivative. See text for details.

2. (a) A = active source with positive-going waveform. B = passive sink with negative-going waveform. Intracellular response is hyper-polarizing. Extracellular response is positive-going with faster time course (faster rise and decay) that the intracellular signal, because field potential is proportional to membrane current.

    (b) A = active sink. B = passive source.

Chapter 15

1. See figure below.

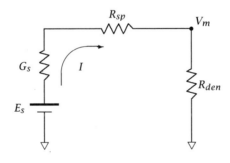

(a) and (b)

From diagram, $I = \frac{V_m - E_s}{\frac{1}{G_s} + R_{sp}}$. During control:

G_s (non-NMDA) $= 10 \times 10^{-9}$ S; $\frac{1}{G_s} = 10^8$ Ω; and $R_{sp} = 100 \times 10^6$ Ω.

So, $I_{\text{non-NMDA}} = \frac{(-70-0) \text{ mV}}{10^8 + 10^8} = -0.35$ nA   (i.e. an inward current).

Also, G_s (NMDA) $= 10^{-9}$ S; $\frac{1}{G_s} = 10^9$ Ω.

So,

$$I_{NMDA} = \frac{-70\text{ mV}}{10^8 + 10^9} = -0.064 \text{ nA}.$$

During LTP:

$$R_{sp} = 50 \times 10^6 \ \Omega.$$

So,

$$I_{non\text{-}NMDA} = \frac{-70\text{ mV}}{10^8 + 0.5 \times 10^8} = -0.47 \text{ nA};$$

and

$$I_{NMDA} = \frac{-70\text{ mV}}{10^9 + 0.5 \times 10^8} = -0.067 \text{ nA}.$$

(c) NMDA shows very little change during LTP (0.063 to 0.067 = 6% increase), while the non-NMDA shows a big increase (0.35 to 0.47 = 34% increase). Changes in spine neck resistance can therefore differentially increase the component with a conductance that closely matches the spine neck conductance, even though there is no change in synaptic conductance.

2. (a) through (g) Devise experiments and make up results. All directly from text.

3. (a) There should be no change in EPSC amplitude; under voltage clamp, there should be no change in potential and, therefore, no activation of voltage-gated Ca^{2+} channels.

    (b) There should be no change in the *I-V* curve before and after tetanus.

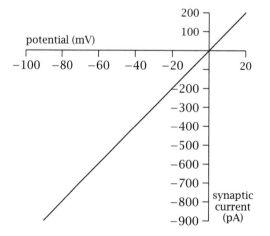

(c) The single-channel conductance is 14 pS before and after tetanus.

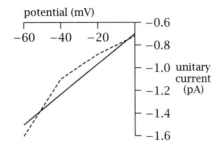

(d) See figure below.

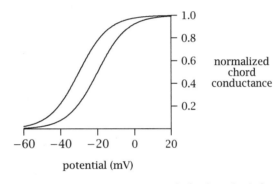

(e) Yes, the activation curve is shifted to the left, such that a smaller depolarization will cause greater influx of Ca^{2+} and produce a greater depolarization leading to potentiation of the postsynaptic response.

(f) LTP will be observed only between membrane potentials of -70 mV and -20 mV. At membrane potentials less than -70 mV, a 10–15 mV EPSP will not activate a signficant Ca^{2+} current. If the membrane potential is above -20 mV, almost all Ca^{2+} channels will have been activated and, with time, will inactive to some extent such that LTP will not be observed.

4. Y vector is 001011.

5. (a) EPSP LTP.

   (b) E-S potentiation.

Appendix A

1. (a) Capacitors in series (refer to figure below)

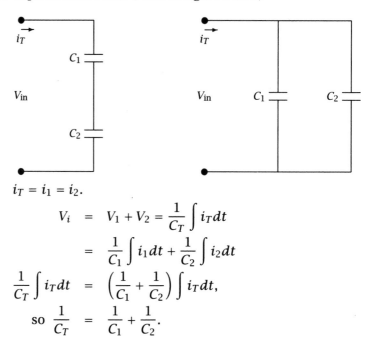

$$i_T = i_1 = i_2.$$

$$
\begin{aligned}
V_i &= V_1 + V_2 = \frac{1}{C_T} \int i_T dt \\
&= \frac{1}{C_1} \int i_1 dt + \frac{1}{C_2} \int i_2 dt
\end{aligned}
$$

$$\frac{1}{C_T} \int i_T dt = \left(\frac{1}{C_1} + \frac{1}{C_2} \right) \int i_T dt,$$

$$\text{so } \frac{1}{C_T} = \frac{1}{C_1} + \frac{1}{C_2}.$$

(b) Capacitors in parallel

$$V_i = V_1 = V_2.$$

$$
\begin{aligned}
i_T &= i_1 + i_2 = C_T \frac{dV_i}{dt} \\
&= C_1 \frac{dV_1}{dt} + C_2 \frac{dV_2}{dt}
\end{aligned}
$$

$$C_T \frac{dV_i}{dt} = C_1 \frac{dV_i}{dt} + C_2 \frac{dV_i}{dt}$$

$$\text{so } C_T = C_1 + C_2.$$

2. Given

$$E_1 - I_1 R_1 - (I_1 + I_2) R_3 = 0,$$

$$E_2 - I_2 R_2 - (I_1 + I_2) R_3 = 0,$$

and

$$V_3 = (I_1 + I_2) R_3,$$

a brute-force method of solution is:

$$E_1 = I_1(R_1 + R_3) + I_2(R_3),$$

$$E_2 = I_1(R_3) + I_2(R_2 + R_3),$$

$$V_3 = I_1(R_3) + I_2(R_3),$$

then $I_2 = \frac{E_1 - I_1(R_1 + R_3)}{R_3}$,

and $I_1 = \frac{E_2 - I_2(R_2 + R_3)}{R_3}$.

$$
\begin{aligned}
V_3 &= I_1 R_3 + E_1 - I_1(R_1 + R_3) \\
&= I_1(R_3 - R_1 - R_3) + E_1 \\
&= E_1 - I_1 R_1.
\end{aligned}
$$

$$I_1 = \frac{E_2 - \frac{[E_1 - I_1(R_1 + R_3)]}{R_3}(R_2 + R_3)}{R_3},$$

$$I_1 \left(1 - \frac{(R_1 + R_3)(R_2 + R_3)}{R_3{}^2} \right) = \frac{E_2}{R_3} - \frac{(R_2 + R_3)}{R_3{}^2} E_1,$$

$$I_1 \left(\frac{R_3{}^2 - R_1 R_2 - R_2 R_3 - R_1 R_3 - R_3{}^2}{R_3{}^2} \right) = \frac{R_3 E_2 - (R_2 + R_3)E_1}{R_3{}^2},$$

$$I_1 = \frac{(R_2 + R_3)E_1 - R_3 E_2}{R_1 R_2 + R_2 R_3 + R_1 R_3}.$$

$$
\begin{aligned}
V_3 &= E_1 - \left[\frac{R_1(R_2 + R_3)E_1 - R_1 R_3 E_2}{R_1 R_2 + R_2 R_3 + R_1 R_3} \right] \\
&= \frac{E_1 R_1 R_2 + E_1 R_2 R_3 + E_1 R_1 R_3 - R_1 R_2 E_1 - R_1 R_3 E_1 + R_1 R_3 E_2}{R_1 R_2 + R_2 R_3 + R_1 R_3} \\
&= \frac{E_1(R_2 R_3) + E_2(R_1 R_3)}{R_1 R_2 + R_2 R_3 + R_1 R_3}.
\end{aligned}
$$

A simpler method is:

$$\frac{E_1 - V_3}{R_1} + \frac{E_2 - V_3}{R_2} = \frac{V_3}{R_3},$$

$$\frac{E_1}{R_1} + \frac{E_2}{R_2} = V_3 \left(\frac{1}{R_1} + \frac{1}{R_2} + \frac{1}{R_3} \right),$$

$$V_3 = \frac{\frac{E_1}{R_1} + \frac{E_2}{R_2}}{\frac{1}{R_1} + \frac{1}{R_2} + \frac{1}{R_3}} = \frac{R_3(E_1 R_2 + E_2 R_1)}{R_1 R_2 + R_1 R_3 + R_2 R_3}.$$

3. $V_i = I(R_1 + R_2)$ or $I = \dfrac{V_i}{(R_1 + R_2)}$.

$$V_o = IR_2 = \frac{V_i}{R_1 + R_2} \cdot R_2.$$

4. From circuit diagram:

$$\left.\begin{aligned}
&I_1 = I_2 + I_3 \\
&I_2 = I_4 \\
&V_1 - I_1 R_1 - I_3 R_3 = 0 \\
&V_2 - I_2 R_4 = 0 \\
&I_2 R_2 + I_2 R_4 - I_3 R_3 = 0
\end{aligned}\right\} \text{Solve}$$

$$\begin{aligned}
V_1 &= (I_2 + I_3) R_1 + I_3 R_3 \\
&= \left(I_2 + \frac{I_2 R_2 + I_2 R_4}{R_3}\right) R_1 + I_2 R_2 + I_2 R_4 \\
&= I_2 \left(R_1 + \frac{R_2 R_1}{R_3} + \frac{R_4 R_1}{R_3} + R_2 + R_4\right).
\end{aligned}$$

$$V_2 = I_2 R_4.$$

$$V_2/V_1 = \frac{R_3 R_4}{R_4(R_1 + R_3) + R_3(R_1 + R_2) + R_1 R_2}.$$

5. Derivation for noninverter:

$V_o = i(R_f + R_{in}) = A(V_i - V_1)$, and $V_i = iR_{in}$, where V_1 is the voltage at the negative input.

(a) Easy derivation:

A is large, so $\dfrac{V_o}{A} = V_i - V_1 \sim 0$,

so $V_i \sim V_i$.

$$V_o = i(R_f + R_{in}) = \frac{V_1}{R_{in}}(R_f + R_{in}) = V_i \frac{(R_f + R_{in})}{R_{in}}.$$

(b) Harder derivation (but not much):

$$V_o = i(R_f + R_{\text{in}}) = \frac{V_1}{R_{\text{in}}}(R_f + R_{\text{in}})$$

$$= \frac{V_i - \frac{V_o}{A}}{R_{\text{in}}}(R_f + R_{\text{in}})$$

$$= \frac{AV_i - V_o}{A}\frac{(R_f + R_{\text{in}})}{R_{\text{in}}}.$$

$$\lim_{A \to \infty} V_o = V_i \frac{(R_f + R_{\text{in}})}{R_{\text{in}}}.$$

6. (a) $V_o = (V_c - V_m)R_f/R_{\text{in}}$,

$$V_m = V_o - I_{cl}R_a;$$

$$V_m = R_f/R_{\text{in}}(V_c - V_m) - I_{cl}R_a,$$

$$V_m + \frac{R_f}{R_{\text{in}}}V_m = \frac{R_f}{R_{\text{in}}}V_c - I_{cl}R_a, \quad \text{and}$$

$$V_m = \frac{\frac{R_f}{R_{\text{in}}}V_c}{1 + \frac{R_f}{R_{\text{in}}}} - \frac{I_{cl}R_a}{1 + \frac{R_f}{R_{\text{in}}}}.$$

(b) With values given:

$$R_f/R_{\text{in}} = 100,$$

$$I_{cl} = I_m = V_m/R_m = \frac{50\text{ mV}}{10^7\ \Omega} = 5 \times 10^{-9}\text{ A},$$

$$\text{so } 50\text{ mV} = \frac{100}{101}V_c - \frac{(5 \times 10^{-9}\text{ A})(10^7\ \Omega)}{101},$$

$$50.5\text{ mV} = 100V_c - 5 \times 10^{-2}\text{ V}, \quad \text{and}$$

$$V_c = 51\text{ mV}.$$

7. Sawtooth changes 10 mV from 0 to 100 msec;

$$i = CdV/dt = (10^{-6})(10\text{ mV}/100\text{ msec})$$

$$= 10^{-7}\text{ A for } 0 < t < 100\text{ msec}.$$

Sawtooth changes −10 mV from 100 msec to 200 msec,

$$i = CdV/dt = (10^{-6})(-10\text{ mV}/100\text{ msec})$$

$$= -10^{-7}\text{ A for } 100 < t < 200\text{ msec}.$$

Measured current will be a square wave from 0 to 10^{-7} A for 100 msec and then to -10^{-7} A for the next 100 msec, and so forth.

8. At steady-state:

$$V_c - I_{cl}R_s - I_{cl}R_N = E_i = 0,$$

so

$$I_{cl} = \frac{V_c}{R_s + R_N}$$

$$V_m = I_{cl}R_N + E_i$$

$$= \frac{V_c}{R_s + R_N} \cdot R_N.$$

$$-10 \text{ mV} = V_c \cdot \frac{100}{130}, \text{ or}$$

$$V_c = -13 \text{ mV}.$$

At steady state,

$$V_o - V_c = I_{cl}R_f = \frac{-13 \times 10^{-3}}{130 \times 10^6} \times 10^9, \text{ and}$$

$$V_o = -113 \text{ mV}.$$

9. (a) $V_m = \dfrac{AV_c}{1 + A} - \dfrac{I_{cl}R_a}{1 + A} - I_{cl}R_s.$

    (b) For $R_s = 10^6 \ \Omega$,

$$V_m = \frac{100}{101}10^{-1} - \frac{(10^{-9})(10^8)}{101} - (10^{-9})(10^6)$$

$$= 0.099 - 0.00099 - 0.001$$

$$= 98 \text{ mV}.$$

    For $R_s = 20 \times 10^6 \ \Omega$,

$$V_m = 0.099 - 0.00099 - 0.02$$

$$= 79 \text{ mV}.$$

Suggested Readings

Chapter 1

Books and Reviews

1. Churchland, P. S. and Sejnowski, T. J. *The Computational Brain*. Cambridge, MA: MIT Press, 1992.

2. Hall, Z. W. *Introduction to Molecular Neurobiology*. Sunderland, MA: Sinauer, 1992.

3. Kandel, E. R., Schwartz, J. H., and Jessell, T. M. *Principles of Neural Science*, 3rd edition. Norwalk, CT: Appleton and Lange, 1991.

4. Nicholls, J. G., Martin, A. R., and Wallace, B. G. *From Neuron to Brain*, 3rd edition. Sunderland, MA: Sinauer, 1992.

Original Articles

1. Fisher, S. K. and Boycott, B. B. Synaptic connections made by horizontal cells within the outer plexiform layer of the retina of the cat and rabbit. *Proc. R. Soc. Lond. [Biol.]* 186: 317-331, 1974.

2. Iversen, L. L. The chemistry of the brain. *Scientific American* 241: 134-149, 1979.

3. Stevens, C. F. How cortical interconnectedness varies with network size. *Neural Computation* 1: 473-479, 1989.

Chapter 2

Books and Reviews

1. Aidley, D. J. *The Physiology of Excitable Cells*, 3rd edition. Cambridge: Cambridge University Press, 1989.

2. Andreoli, T. E., Hoffman, J. F., and Fanestil, D. D. *Membrane Physiology*. New York: Plenum Press, 1980.

3. Bahill, A. T. *Bioengineering: Biomedical, Medical and Clinical Engineering*. Englewood Cliffs, NJ: Prentice-Hall, 1981.

4. Bull, H. B. *An Introduction to Physical Biochemistry*, 2nd edition. Philadelphia: Davis, 1971.

5. Carslaw, H. S. and Jaeger, J. C. *Conduction of Heat in Solids*. Oxford: Clarendon Press, 1959.

6. Conway, E. J. Nature and significance of concentration relations of potassium and sodium ions in skeletal muscle. *Physiol. Rev.* 37: 84-132, 1957.

7. Crank, J. *The Mathematics of Diffusion*, 2nd edition. Oxford: Clarendon Press, 1975.

8. Einstein, A. *Investigations on the Theory of Brownian Movement*, edited by R. Furth, translated by A.D. Cowper. New York: Dover Publications, Inc., 1956, p. 75.

9. Ferreira, H. G. and Marshall, M. W. *The Biophysical Basis of Excitability*. Cambridge: Cambridge University Press, 1985.

10. Hille, B. *Ion Channels of Excitable Membranes*, 2nd edition. Sunderland, MA: Sinauer, 1992.

11. Jack, J. J. B., Noble, D., and Tsien, R. W. *Electric Current Flow in Excitable Cells*. Oxford: Clarendon Press, 1975.

12. Jackson, J. D. *Classical Electrodynamics*. New York: Wiley, 1962.

13. Jewett, D. L. and Rayner, M. D. *Basic Concepts of Neuronal Function*. Boston, MA: Little, Brown, 1984.

14. Junge, D. *Nerve and Muscle Excitation*, 3rd edition. Sunderland, MA: Sinauer, 1992.

15. MacGregor, R. J. and Lewis, E. R. *Neural Modeling: Electrical Signal Processing in the Nervous System*. New York: Plenum Press, 1977.

16. Moore, W. J. *Physical Chemistry*, 4th edition. Englewood Cliffs, NJ: Prentice-Hall, 1972.

17. Nicholls, J. G., Martin, A. R., and Wallace, B. G. *From Neuron to Brain*, 3rd edition. Sunderland, MA: Sinauer, 1992.

18. Purcell, E. M. *Electricity and Magnetism: Berkeley Physics Course*, vol. 2. New York: McGraw-Hill, 1965.

19. Tuckwell, H. C. *Introduction to Theoretical Neurobiology.* Vol. 1, *Linear Cable Theory and Dendritic Structure.* Cambridge: Cambridge University Press, 1988.

20. Williamson, R. E., Crowell, R. H., and Trotter, H. F. *Calculus of Vector Functions*, 2nd edition. Englewood Cliffs, NJ: Prentice-Hall, 1968.

21. Wylie, C. R. *Advanced Engineering Mathematics.* New York: McGraw-Hill, 1975.

Original Articles

1. Goldman, D. E. Potential, impedance, and rectification in membranes. *J. Gen. Physiol.* 27: 37–60, 1943.

2. Hodgkin, A. L. and Horowicz, P. The influence of potassium and chloride ions on the membrane potential of single muscle fibres. *J. Physiol. (Lond.)* 148: 127–160, 1959.

3. Hodgkin, A. L. and Horowicz, P. The effect of sudden changes in ionic concentrations on the membrane potential of single muscle fibres. *J. Physiol. (Lond.)* 153: 370–385, 1960a.

4. Hodgkin, A. L. and Horowicz, P. Potassium contractures in single muscle fibres. *J. Physiol. (Lond.)* 153: 386–403, 1960b.

5. Hodgkin, A. L. and Katz, B. The effect of sodium ions on the electrical activity of the giant axon of the squid. *J. Physiol. (Lond.)* 108: 37–77, 1949.

6. Hodgkin, A. L. and Keynes, R. D. Active transport of cations in giant axons from *Sepia* and *Loligo. J. Physiol. (Lond.)* 128: 28–60, 1955a.

7. Hodgkin, A. L. and Keynes, R. D. The potassium permeability of a giant nerve fibre. *J. Physiol. (Lond.)* 128: 61–88, 1955b.

8. Hodgkin, A. L. and Keynes, R. D. Movements of labelled calcium in squid giant axons. *J. Physiol. (Lond.)* 138: 253–281, 1957.

9. Hodgkin, A. L. and Rushton, W. A. H. The electrical constants of a crustacean nerve fibre. *Proc. R. Soc. Lond. [Biol.]* 133: 444–479, 1946.

Chapter 3

Books and Reviews

1. Aidley, D. J. *The Physiology of Excitable Cells*, 3rd edition. Cambridge: Cambridge University Press, 1989.

2. Bull, H. B. *An Introduction to Physical Biochemistry*, 2nd edition. Philadelphia: Davis, 1971.

3. Cole, K. S. *Membranes, Ions and Impulses: A Chapter of Classical Biophysics.* Berkeley: University of California Press, 1968.

4. Ferreira, H. G. and Marshall, M. W. *The Biophysical Basis of Excitability.* Cambridge: Cambridge University Press, 1985.

5. Hille, B. *Ion Channels of Excitable Membranes*, 2nd edition. Sunderland, MA: Sinauer, 1992.

6. Jack, J. J. B., Noble, D., and Tsien, R. W. *Electric Current Flow in Excitable Cells*. Oxford: Clarendon Press, 1975.

7. Junge, D. *Nerve and Muscle Excitation*, 3rd edition. Sunderland, MA: Sinauer, 1992.

8. MacGregor, R. J. and Lewis, E. R. *Neural Modeling: Electrical Signal Processing in the Nervous System*. New York: Plenum Press, 1977.

9. Nicholls, J. G., Martin, A. R., and Wallace, B. G. *From Neuron to Brain*, 3rd edition. Sunderland, MA: Sinauer, 1992.

10. Tuckwell, H. C. *Introduction to Theoretical Neurobiology*. Vol. 1, *Linear Cable Theory and Dendritic Structure*. Cambridge: Cambridge University Press, 1988a.

11. Tuckwell, H. C. *Introduction to Theoretical Neurobiology*. Vol. 2, *Nonlinear and Stochastic Theories*. Cambridge: Cambridge University Press, 1988b.

Original Articles

1. Hodgkin, A. L. and Rushton, W. A. H. The electrical constants of a crustacean nerve fibre. *Proc. R. Soc. Lond. [Biol.]* 133: 444–479, 1946.

Chapter 4

Books and Reviews

1. Brown, T. H., Zador, A. M., Mainen, Z. F., and Claiborne, B. J. Hebbian computations in hippocampal dendrites and spines. In *Single Neuron Computation*, edited by T. McKenna, J. Davis, and S. F. Zornetzer. San Diego: Academic Press, 1992, pp. 81–116.

2. Claiborne, B. J., Zador, A. M., Mainen, Z. F., and Brown, T. H. Computational models of hippocampal neurons. In *Single Neuron Computation*, edited by T. McKenna, J. Davis, and S. F. Zornetzer. San Diego: Academic Press, 1992, pp. 61–80.

3. Ferreira, H. G. and Marshall, M. W. *The Biophysical Basis of Excitability*. Cambridge: Cambridge University Press, 1985.

4. Jack, J. J. B., Noble, D., and Tsien, R. W. *Electric Current Flow in Excitable Cells*. Oxford: Clarendon Press, 1975, chapters 3, 4, and 7.

5. Major, G. The physiology, morphology, and modelling of cortical pyramidal neurones. Ph.D. dissertation, University of Oxford, 1992.

6. Rall, W. Core conductor theory and cable properties of neurons. In *Handbook of Physiology*. Sec. 1, *The Nervous System*, vol. 1. Bethesda, MD: Am. Physiol. Soc., 1977, pp. 39–97.

7. Rall, W., Burke, R. E., Holmes, W. R., Jack, J. J. B., Redman, S. J., and Segev, I. Matching dendritic neuron models to experimental data. *Physiol. Rev.* 72: S159-S186, 1992.

8. Tuckwell, H. C. *Introduction to Theoretical Neurobiology.* Vol. 1, *Linear cable theory and dendritic structure.* Cambridge: Cambridge University Press, 1988.

9. Zador, A. M. Biophysics of computation in single hippocampal neurons. Ph.D. dissertation, Yale University, 1992.

Original Articles

1. Brown, T. H., Fricke, R. A., and Perkel, D. H. Passive electrical constants in three classes of hippocampal neurons. *J. Neurophysiol.* 46: 812-827, 1981.

2. Carnevale, N. T. and Johnston, D. Electrophysiological characterization of remote chemical synapses. *J. Neurophysiol.* 47: 606-621, 1982.

3. Durand, D., Carlen, P. L., Gurevich, N., Ho, A., and Kunov, H. Electrotonic parameters of rat dentate granule cells measured using short current pulses and HRP staining. *J. Neurophysiol.* 50: 1080-1097, 1983.

4. Hodgkin, A. L. and Rushton, W. A. H. The electrical constants of a crustacean nerve fibre. *Proc. R. Soc. Lond. [Biol.]* 133: 444-479, 1946.

5. Holmes, W. R. and Rall, W. Electrotonic length estimates in neurons with dendritic tapering or somatic shunt. *J. Neurophysiol.* 68: 1421-1437, 1992a.

6. Holmes, W. R. and Rall, W. Estimating the electrotonic structure of neurons with compartmental models. *J. Neurophysiol.* 68: 1438-1452, 1992b.

7. Holmes, W. R., Segev, I., and Rall, W. Interpretation of time constant and electrotonic length estimates in multicylinder or branched neuronal structures. *J. Neurophysiol.* 68: 1401-1420, 1992.

8. Iversen, L. L. The chemistry of the brain. *Scientific American* 241:134-149, 1979.

9. Jack, J. J. B. and Redman, S. J. An electrical description of the motoneurone, and its application to the analysis of synaptic potentials. *J. Physiol. (Lond.)* 215: 321-352, 1971.

10. Johnston, D. and Brown, T. H. Interpretation of voltage-clamp measurements in hippocampal neurons. *J. Neurophysiol.* 50: 464-486, 1983.

11. Rall, W. Branching dendritic trees and motoneuron membrane resistivity. *Exp. Neurol.* 1: 491-527, 1959.

12. Rall, W. Time constants and electrotonic length of membrane cylinders and neurons. *Biophys. J.* 9: 1483-1508, 1969.

13. Rihn, L. L. and Claiborne, B. J. Dendritic growth and regression in rat dentate granule cells during late postnatal development. *Dev. Brain Res.* 54: 115-124, 1990.

14. Spruston, N., Jaffe, D. B., Williams, S. H., and Johnston, D. Voltage- and space-clamp errors associated with the measurement of electronically remote synaptic events. *J. Neurophysiol.* 70: 781–802, 1993.

15. Spruston, N. and Johnston, D. Perforated patch-clamp analysis of the passive membrane properties of three classes of hippocampal neurons. *J. Neurophysiol.* 67: 508–529, 1992.

16. Tsai, K. Y., Carnevale, N. T., Claiborne, B. J., and Brown, T. H. Efficient mapping from neuroanatomical to electronic space. *Network* 5: 21–46, 1994.

Chapter 5

Books and Reviews

1. Andreoli, T. E., Hoffman, J. F., and Fanestil, D. D. *Membrane Physiology.* New York: Plenum Press, 1980.

2. Bahill, A. T. *Bioengineering: Biomedical, Medical and Clinical Engineering.* Englewood Cliffs, NJ: Prentice-Hall, 1981.

3. Cole, K. S. *Membranes, Ions and Impulses: A Chapter of Classical Biophysics.* Berkeley: University of California Press, 1968.

4. Eggers, D. F., Jr., Gregory, N. W., Halsey, G. D., Jr., and Rabinovitch, B. S. *Physical Chemistry.* New York: Wiley, 1964.

5. Ferreira, H. G. and Marshall, M. W. *The Biophysical Basis of Excitability.* Cambridge: Cambridge University Press, 1985.

6. Hille, B. *Ion Channels of Excitable Membranes*, 2nd edition. Sunderland, MA: Sinauer, 1992.

7. Jack, J. J. B., Noble, D., and Tsien, R. W. *Electric Current Flow in Excitable Cells.* Oxford: Clarendon Press, 1975.

8. Junge, D. *Nerve and Muscle Excitation*, 3rd edition. Sunderland, MA: Sinauer, 1992.

9. Kittel, C. *Thermal Physics.* New York: Wiley, 1969.

10. MacGregor, R. J. and Lewis, E. R. *Neural Modeling: Electrical Signal Processing in the Nervous System.* New York: Plenum Press, 1977.

11. Moore, W. J. *Physical Chemistry*, 4th edition. Englewood Cliffs, NJ: Prentice-Hall, 1972.

12. Nicholls, J. G., Martin, A. R., and Wallace, B. G. *From Neuron to Brain*, 3rd edition. Sunderland, MA: Sinauer, 1992.

13. Peusner, L. *Concepts in Bioenergetics.* Englewood Cliffs, NJ: Prentice-Hall, 1974.

14. Reif, F. *Fundamentals of Statistical and Thermal Physics.* New York: McGraw-Hill, 1965.

15. Tuckwell, H. C. *Introduction to Theoretical Neurobiology.* Vol. 1, *Linear Cable Theory and Dendritic Structure.* Cambridge: Cambridge University Press, 1988a.

16. Tuckwell, H. C. *Introduction to Theoretical Neurobiology.* Vol. 2, *Nonlinear and Stochastic Theories.* Cambridge: Cambridge University Press, 1988b.

17. Woodbury, J. W. Action potential: Properties of excitable membranes. In *Physiology and Biophysics,* edited by T. C. Ruch and H. D. Patton. Philadelphia: Saunders, 1965, pp. 26-57.

18. Woodbury, J. W. Eyring rate theory model of the current-voltage relationships of ion channels in excitable membranes. In *Chemical Dynamics: Papers in Honor of Henry Eyring,* edited by J. O. Hirschfelder. New York: Wiley, 1971, pp. 601-617.

19. Wylie, C. R. *Advanced Engineering Mathematics.* New York: McGraw-Hill, 1975.

Original Articles

1. Hille, B. Ionic selectivity, saturation, and block in sodium channels. A four-barrier model. *J. Gen. Physiol.* 66: 535-560, 1975a.

2. Hille, B. The receptor for tetrodotoxin and saxitoxin: A structural hypothesis. *Biophys. J.* 15: 615-619, 1975b.

3. Hodgkin, A. L. and Huxley, A. F. Currents carried by sodium and potassium ions through the membrane of the giant axon of *Loligo. J. Physiol. (Lond.)* 116: 449-472, 1952a.

4. Hodgkin, A. L. and Huxley, A. F. The components of membrane conductance in the giant axon of *Loligo. J. Physiol. (Lond.)* 116: 473-496, 1952b.

5. Hodgkin, A. L. and Huxley, A. F. The dual effect of membrane potential on sodium conductance in the giant axon of *Loligo. J. Physiol. (Lond.)* 116: 497-506, 1952c.

6. Hodgkin, A. L. and Huxley, A. F. A quantitative description of membrane current and its application to conduction and excitation in nerve. *J. Physiol. (Lond.)* 117: 500-544, 1952d.

7. Hodgkin, A. L., Huxley, A. F., and Katz, B. Ionic currents underlying activity in the giant axon of the squid. *Arch. Sci. Physiol.* 3: 129-150, 1949.

8. Hodgkin, A. L., Huxley, A. F., and Katz, B. Measurements of current-voltage relations in the membrane of the giant axon of *loligo. J. Physiol. (Lond.)* 116: 424-448, 1952.

9. Hodgkin, A. L. and Katz, B. The effect of sodium ions on the electrical activity of the giant axon of the squid. *J. Physiol. (Lond.)* 108: 37-77, 1949.

Chapter 6

Books and Reviews

1. Aidley, D. J. *The Physiology of Excitable Cells*, 3rd edition. Cambridge: Cambridge University Press, 1989.

2. Armstrong, C. M. Sodium channels and gating currents. *Physiol. Rev.* 61: 644–683, 1981.

3. Cole, K. S. *Membranes, Ions and Impulses: A Chapter of Classical Biophysics.* Berkeley: University of California Press, 1968.

4. Eggers, D. F., Jr., Gregory, N. W., Halsey, G. D., Jr., and Rabinovitch, B. S. *Physical Chemistry.* New York: Wiley, 1964.

5. Ferreira, H. G. and Marshall, M. W. *The Biophysical Basis of Excitability.* Cambridge: Cambridge University Press, 1985.

6. Hille, B. *Ion Channels of Excitable Membranes*, 2nd edition. Sunderland, MA: Sinauer, 1992.

7. Jack, J. J. B., Noble, D., and Tsien, R. W. *Electric Current Flow in Excitable Cells.* Oxford: Clarendon Press, 1975.

8. Jewett, D. L. and Rayner, M. D. *Basic Concepts of Neuronal Function.* Boston, MA: Little, Brown, 1984.

9. Junge, D. *Nerve and Muscle Excitation*, 3rd edition. Sunderland, MA: Sinauer, 1992.

10. Kittel, C. *Thermal Physics.* New York: Wiley, 1969.

11. MacGregor, R. J. and Lewis, E. R. *Neural Modeling: Electrical Signal Processing in the Nervous System.* New York: Plenum Press, 1977.

12. Moore, W. J. *Physical Chemistry*, 4th edition. Englewood Cliffs, NJ: Prentice-Hall, 1972.

13. Nicholls, J. G., Martin, A. R., and Wallace, B. G. *From Neuron to Brain*, 3rd edition. Sunderland, MA: Sinauer, 1992.

14. Noble, D. Applications of Hodgkin-Huxley equations to excitable tissues. *Physiol. Rev.* 46: 1–50, 1966.

15. Peusner, L. *Concepts in Bioenergetics.* Englewood Cliffs, NJ: Prentice-Hall, 1974.

16. Reif, F. *Fundamentals of Statistical and Thermal Physics.* New York: McGraw-Hill, 1965.

17. Tuckwell, H. C. *Introduction to Theoretical Neurobiology.* Vol. 2, *Nonlinear and Stochastic Theories.* Cambridge: Cambridge University Press, 1988b.

Original Articles

1. Armstrong, C. M. and Bezanilla, F. Currents related to movement of the gating particles of the sodium channels. *Nature (Lond.)* 242: 459-461, 1973.

2. Armstrong, C. M. and Bezanilla, F. Charge movement associated with the opening and closing of the activation gates of the Na channels. *J. Gen. Physiol.* 63: 533-552, 1974.

3. Armstrong, C. M. and Bezanilla, F. Inactivation of the sodium channel. II: Gating current experiments. *J. Gen. Physiol.* 70: 567-590, 1977.

4. Bezanilla, F. Gating of sodium and potassium channels. *J. Membrane Biol.* 88: 97-111, 1985.

5. Bezanilla, F. and Armstrong, C. M. Inactivation of the sodium channel. I: Sodium current experiments. *J. Gen. Physiol.* 70: 549-566, 1977.

6. Bezanilla, F., Perozo, E., Papazian, D. M., and Stefani, E. Molecular basis of gating charge immobilization in *Shaker* potassium channels. *Science* 254: 679-683, 1991.

7. Cole, K. S. and Moore, J. W. Ionic current measurements in the squid giant axon membrane. *J. Gen. Physiol.* 44: 123-167, 1960.

8. Frankenhaeuser, B. Quantitative description of sodium currents in myelinated nerve fibres of *Xenopus laevis. J. Physiol. (Lond.)* 151: 491-501, 1960a.

9. Frankenhaeuser, B. Sodium permeability in toad nerve and in squid nerve. *J. Physiol. (Lond.)* 152: 159-166, 1960b.

10. Frankenhaeuser, B. A quantitative description of potassium currents in myelinated nerve fibres of *Xenopus laevis. J. Physiol. (Lond.)* 169: 424-430, 1963.

11. Frankenhaeuser, B. and Hodgkin, A. L. The action of calcium on the electrical properties of squid axons. *J. Physiol. (Lond.)* 137: 218-244, 1957.

12. Hodgkin, A. L. and Huxley, A. F. Currents carried by sodium and potassium ions through the membrane of the giant axon of *Loligo. J. Physiol. (Lond.)* 116: 449-472, 1952a.

13. Hodgkin, A. L. and Huxley, A. F. The components of membrane conductance in the giant axon of *Loligo. J. Physiol. (Lond.)* 116: 473-496, 1952b.

14. Hodgkin, A. L. and Huxley, A. F. The dual effect of membrane potential on sodium conductance in the giant axon of *Loligo. J. Physiol. (Lond.)* 116: 497-506, 1952c.

15. Hodgkin, A. L. and Huxley, A. F. A quantitative description of membrane current and its application to conduction and excitation in nerve. *J. Physiol. (Lond.)* 117: 500-544, 1952d.

16. Hodgkin, A. L., Huxley, A. F., and Katz, B. Ionic currents underlying activity in the giant axon of the squid. *Arch. Sci. Physiol.* 3: 129-150, 1949.

17. Hodgkin, A. L., Huxley, A. F., and Katz, B. Measurements of current-voltage relations in the membrane of the giant axon of *Loligo. J. Physiol. (Lond.)* 116: 424-448, 1952.

18. Hodgkin, A. L. and Katz, B. The effect of sodium ions on the electrical activity of the giant axon of the squid. *J. Physiol. (Lond.)* 108: 37-77, 1949.

19. Hodgkin, A. L. and Keynes, R. D. Active transport of cations in giant axons from *Sepia* and *Loligo. J. Physiol. (Lond.)* 128: 28-60, 1955a.

20. Hodgkin, A. L. and Keynes, R. D. The potassium permeability of a giant nerve fibre. *J. Physiol. (Lond.)* 128: 61-88, 1955b.

21. Keynes, R. D. and Rojas, E. Kinetics and steady state properties of charge systems controlling sodium conductance in the squid giant axon. *J. Physiol. (Lond.)* 239: 393-434, 1974.

22. White, M. M. and Bezanilla, F. Activation of squid axon K channels. *J. Gen. Physiol.* 85: 539-554, 1985.

Chapter 7

Books and Reviews

1. Brown, D. A., Gähwiler, B. H., Griffith, W. H., and Halliwell, J. V. Membrane currents in hippocampal neurons. *Prog. Brain Res.* 83: 141-160, 1990.

2. Hille, B. *Ionic Channels of Excitable Membranes.* Sunderland, MA: Sinauer, 1992.

3. Jack, J. J. B., Noble, D., and Tsien, R. W. *Electric Current Flow in Excitable Cells.* Oxford: Clarendon Press, 1975.

4. Llinás, R. R. The intrinsic electrophysiological properties of mammalian neurons: Insights into central nervous system function. *Science* 242: 1654-1664, 1988.

5. Storm, J. F. Potassium currents in hippocampal pyramidal cells. *Prog. Brain Res.* 83: 161-187, 1990.

Original Articles

1. Beck, H., Ficker, E., and Heinemann, U. Properties of two voltage-activated potassium currents in acutely isolated juvenile rat dentate gyrus granule cells. *J. Neurophysiol.* 68: 2086-2099, 1992.

2. Brown, D. A. and Adams, P. R. Muscarinic suppression of a novel voltage-sensitive K^+ current in a vertebrate neurone. *Nature (Lond.)* 283: 673-676, 1980.

3. Brown, D. A. and Griffith, W. H. Calcium-activated outward current in voltage-clamped hippocampal neurones of the guinea pig. *J. Physiol. (Lond.)* 337: 287-301, 1983.

4. Carbone, E. and Lux, H. D. A low voltage-activated calcium conductance in embryonic chick sensory neurons. *Biophys. J.* 46: 413-418, 1984.

5. Connor, J. A. and Stevens, C. F. Inward and delayed outward membrane currents in isolated neural somata under voltage clamp. *J. Physiol. (Lond.)* 213: 1-19, 1971a.

6. Connor, J. A. and Stevens, C. F. Voltage clamp studies of a transient outward membrane current in gastropod neural somata. *J. Physiol. (Lond.)* 213: 21-30, 1971b.

7. Fisher, R. E., Gray, R., and Johnston, D. Properties and distribution of single voltage-gated calcium channels in adult hippocampal neurons. *J. Neurophysiol.* 64: 91-104, 1990.

8. Fox, A. P., Nowycky, M. C., and Tsien, R. W. Kinetic and pharmacological properties distinguishing three types of calcium currents in chick sensory neurones. *J. Physiol. (Lond.)* 394: 149-172, 1987a.

9. Fox, A. P., Nowycky, M. C., and Tsien, R. W. Single-channel recordings of three types of calcium channels in chick sensory neurones. *J. Physiol. (Lond.)* 394: 173-200, 1987b.

10. Gähwiler, B. H. Facilitation by acetylcholine of tetrodotoxin-resistant spikes in rat hippocampal pyramidal cells. *Neuroscience* 11: 381-388, 1984.

11. Guharay, F. and Sachs, F. Stretch-activated single ion channel currents in tissue-cultured embryonic chick skeletal muscle. *J. Physiol. (Lond.)* 352: 685-701, 1984.

12. Hagiwara, S., Miyazaki, S, and Rosenthal, N. P. Potassium current and the effect of cesium on this current during anomalous rectification of the egg cell membrane of a starfish. *J. Gen. Physiol.* 67: 621-638, 1976.

13. Halliwell, J. V. and Adams, P. R. Voltage-clamp analysis of muscarinic excitation in hippocampal neurones. *Brain Res.* 250: 71-92, 1982.

14. Hansen, A. J., Hounsgaard, J., and Jahnsen, H. Anoxia increases potassium conductance in hippocampal nerve cells. *Acta Physiol. Scand.* 115: 301-310, 1982.

15. Lancaster, B. and Adams, P. R. Calcium-dependent current generating the afterhyperpolarization of hippocampal neurons. *J. Neurophysiol.* 55: 1268-1282, 1986.

16. Llinás, R. and Yarom, Y. Electrophysiology of mammalian inferior olivary neurones *in vitro.* Different types of voltage-dependent ionic conductances *J. Physiol. (Lond.)* 315: 549-567, 1981.

17. Madison, D. V., Malenka, R. C., and Nicoll, R. A. Phorbol esters block a voltage-sensitive chloride current in hippocampal pyramidal cells. *Nature (Lond.)* 321: 695-697, 1986.

18. Mintz, I. M., Adams, M. E., and Bean, B. P. P-type calcium channels in rat central and peripheral neurons. *Neuron* 9: 85-95, 1992.

19. Owen, D. G., Segal, M., and Barker, J. L. A Ca^{2+}-dependent Cl^- conductance in cultured mouse spinal neurons. *Nature (Lond.)* 311: 567-570, 1984.

20. Regan, L. J., Sah, D. W. Y., and Bean, B. P. Ca^{2+} channels in rat central and peripheral neurons: High-threshold current resistant to dihydropyridine blockers and ω-conotoxin. *Neuron* 6: 269–280, 1991.

21. Rushton, W. A. H. A theory of the effects of fibre size in medullated nerve. *J. Physiol. (Lond.)* 115: 101–122, 1951.

22. Schwindt, P. C., Spain, W. J., and Crill, W. E. Long-lasting reduction of excitability by a sodium-dependent potassium current in cat neocortical neurons. *J. Neurophysiol.* 61: 233–244, 1989.

23. Segal, M. and Barker, J. L. Rat hippocampal neurons in culture: Potassium conductances. *J. Neurophysiol.* 51: 1409–1433, 1984.

24. Stafstrom, C. E., Schwindt, P. C., Chubb, M. C., and Crill, W. E. Properties of persistent sodium conductance and calcium conductance of layer V neurons from cat sensorimotor cortex in vitro. *J. Neurophysiol.* 53: 153–170, 1985.

25. Storm, J. F. Temporal integration by a slowly inactivating K^{+} current in hippocampal neurons. *Nature (Lond.)* 336: 379–381, 1988.

26. Williams, S. and Johnston, D. Muscarinic depression of synaptic transmission at the hippocampal mossy fiber synapse. *J. Neurophysiol.* 64:1089–1097, 1990.

Chapter 8

Books and Reviews

1. Aidley, D. J. *The Physiology of Excitable Cells*, 3rd edition. Cambridge: Cambridge University Press, 1989.

2. Catterall, W. A. Cellular and molecular biology of voltage-gated sodium channels. *Physiol. Rev.* 72: 515–548, 1992.

3. Hall, Z. W. *Introduction to Molecular Neurobiology*. Sunderland, MA: Sinauer, 1992.

4. Hille, B. *Ion Channels of Excitable Membranes*, 2nd edition. Sunderland, MA: Sinauer, 1992.

5. Junge, D. *Nerve and Muscle Excitation*, 3rd edition. Sunderland, MA: Sinauer, 1992.

6. Kandel, E. R., Schwartz, J. H., and Jessell, T. M. *Principles of Neural Science*, 3rd edition. Norwalk, CT: Appleton and Lange, 1991.

7. Nicholls, J. G., Martin, A. R., and Wallace, B. G. *From Neuron to Brain*, 3rd edition. Sunderland, MA: Sinauer, 1992.

8. Sakmann, B. and Neher, E. *Single Channel Recording*. New York: Plenum Press, 1983.

Original Articles

1. Catterall, W. A. Voltage-dependent gating of sodium channels: correlating structure and function. *Trends Neurosci.* 9: 7-10, 1986.

2. Fisher, R. E., Gray, R., and Johnston, D. Properties and distribution of single voltage-gated calcium channels in adult hippocampal neurons. *J. Neurophysiol.* 64: 91-104, 1990.

3. Gray, R. and Johnston, D. Rectification of single GABA-gated chloride channels in adult hippocampal neurons. *J. Neurophysiol.* 54: 134-142, 1985.

4. Hamill, O. P., Marty, A., Neher, E., Sakmann, B., and Sigworth, F. J. Improved patch-clamp techniques for high-resolution current recording from cells and cell-free membrane patches. *Pflügers Arch.* 391: 85-100, 1981.

5. Hille, B. Ionic selectivity, saturation, and block in sodium channels. A four-barrier model. *J. Gen. Physiol.* 66: 535-560, 1975a.

6. Hille, B. The receptor for tetrodotoxin and saxitoxin: A structural hypothesis. *Biophys. J.* 15: 615-619, 1975b.

7. Neher, E. and Sakmann, B. Single-channel currents recorded from membrane of denervated frog muscle fibres. *Nature (Lond.)* 260: 779-802, 1976.

8. Neher, E., Sandblom, J., and Eisenman, G. Ion selectivity, saturation, and block in gramicidin A channels. II. Saturation behavior of single channel conductances and evidence for the existence of multiple binding sites in the channel. *J. Membrane Biol.* 40: 97-116, 1978.

9. Sigworth, F. J. The variance of sodium current fluctuations at the node of Ranvier. *J. Physiol. (Lond.)* 307: 97-129, 1980a.

10. Sigworth, F. J. The conductance of sodium channels under conditions of reduced current at the node of Ranvier. *J. Physiol. (Lond.)* 307: 131-142, 1980b.

Chapter 9

Books and Reviews

1. Bendat, J. S. and Piersol, A. G. *Random Data: Analysis and Measurement Procedures*, 2nd edition. New York: Wiley, 1986.

2. DeFelice, L. J. *Introduction to Membrane Noise.* New York: Plenum Press, 1981.

3. Devore, J. L. *Probability and Statistics for Engineering and Sciences.* Pacific Grove, CA: Brooks/Cole, 1991.

4. Godfrey, M. G., Roebuck, E. M., and Sherlock, A. J. *Concise Statistics.* London: Edward Arnold, 1988.

5. Hogg, R. V. and Tanis, E. A. *Probability and Statistical Inference*, 4th edition. New York: Macmillan, 1993.

6. Kasai, M., Yoshioka, T., and Suzuki, H. *Biosignal Transduction Mechanisms.* Tokyo: Japan Scientific Society Press, 1989.

7. Leon-Garcia, A. *Probability and Random Processes for Electrical Engineering.* Reading, MA: Addison-Wesley, 1989.

8. Marmarelis, P. Z. and Marmarelis, V. Z. *Analysis of Physiological Systems.* New York: Plenum Press, 1978.

9. Mosteller, F., Rourke, R. E., and Thomas, G. B. *Probability With Statistical Applications*, 2nd edition. Reading, MA: Addison-Wesley, 1970.

10. Neher, E. and Stevens, C. F. Conductance fluctuations and ionic pores in membranes. *Annu. Rev. Biophys. Bioeng.* 6: 345–381, 1977.

11. Ochi, M. K. *Applied Probability and Stochastic Processes in Engineering and Physical Sciences.* New York: Wiley, 1990.

12. Peebles, P. Z., Jr. *Communication System Principles.* Reading, MA: Addison-Wesley, 1976.

13. Press, W. H., Tenkolsky, S. A., Vettering, W. T., and Flannery, B. P. *Numerical Recipes in C*, 2nd edition. Cambridge: Cambridge University Press, 1992.

14. Sakmann, B. and Neher, E. *Single Channel Recording.* New York: Plenum Press, 1983.

15. Tuckwell, H. C. *Stochastic Processes in the Neurosciences.* Philadelphia: Society for Industrial and Applied Mathematics, 1989.

16. Wylie, C. R. *Advanced Engineering Mathematics.* New York: McGraw-Hill, 1975.

17. Ziemer, R. E. and Tranter, W. H. *Principles of Communications: Systems, Modulation and Noise.* Boston: Houghton Mifflin, 1976.

Original Articles

1. Colquhoun, D. and Sakmann, B. Fluctuations in the microsecond time range of the current through single acetylcholine receptor ion channels. *Nature (Lond.)* 294: 464–466, 1981.

2. Colquhoun, D. and Sakmann, B. Fast events in single-channel currents activated by acetylcholine and its analogues at the frog muscle end-plate. *J. Physiol. (Lond.)* 369: 501–557, 1985.

3. Neher, E. and Sakmann, B. Single-channel currents recorded from membrane of denervated frog muscle fibres. *Nature (Lond.)* 260: 779–802, 1976.

4. Sigworth, F. J. The variance of sodium current fluctuations at the node of Ranvier. *J. Physiol. (Lond.)* 307: 97–129, 1980a.

5. Sigworth, F. J. The conductance of sodium channels under conditions of reduced current at the node of Ranvier. *J. Physiol. (Lond.)* 307: 131–142, 1980b.

6. Sigworth, F. J. Covariance of nonstationary sodium current fluctuations at the node of Ranvier. *Biophys. J.* 34: 111–133, 1981.

Chapter 10

Books and Reviews

1. Bendat, J. S. and Piersol, A. G. *Random Data: Analysis and Measurement Procedures*, 2nd edition. New York: Wiley, 1986.

2. Catterall, W. A. Cellular and molecular biology of voltage-gated sodium channels. *Physiol. Rev.* 72: 515–548, 1992.

3. DeFelice, L. J. *Introduction to Membrane Noise*. New York: Plenum Press, 1981.

4. Devore, J. L. *Probability and Statistics for Engineering and Sciences*. Pacific Grove, CA: Brooks/Cole, 1991.

5. Godfrey, M. G., Roebuck, E. M., and Sherlock, A. J. *Concise Statistics*. London: Edward Arnold, 1988.

6. Hille, B. *Ion Channels of Excitable Membranes*, 2nd edition. Sunderland, MA: Sinauer, 1992.

7. Hoffman, K. and Kunze, R. *Linear Algebra*. Englewood Cliffs, NJ: Prentice-Hall, 1961.

8. Hogg, R. V. and Tanis, E. A. *Probability and Statistical Inference*, 4th edition. New York: Macmillan, 1993.

9. Junge, D. *Nerve and Muscle Excitation*, 3rd edition. Sunderland, MA: Sinauer, 1992.

10. Kasai, M., Yoshioka, T., and Suzuki, H. *Biosignal Transduction Mechanisms*. Tokyo: Japan Scientific Society Press, 1989.

11. Kuhn, T. S. *The Structure of Scientific Revolutions*, 2nd edition. Chicago: University of Chicago Press, 1970.

12. Leon-Garcia, A. *Probability and Random Processes for Electrical Engineering*. Reading, MA: Addison-Wesley, 1989.

13. Nicholls, J. G., Martin, A. R., and Wallace, B. G. *From Neuron to Brain*, 3rd edition. Sunderland, MA: Sinauer, 1992.

14. Noble, B. *Applied Linear Algebra*. Englewood Cliffs, NJ: Prentice-Hall, 1969.

15. Ochi, M. K. *Applied Probability and Stochastic Processes in Engineering and Physical Sciences*. New York: Wiley, 1990.

16. Patlak, J. B. Molecular kinetics of voltage-dependent Na^+ channels. *Physiol. Rev.* 71: 1047–1080, 1991.

17. Peebles, P. Z., Jr. *Communication System Principles*. Reading, MA: Addison-Wesley, 1976.

18. Sakmann, B. and Neher, E. *Single Channel Recording*. New York: Plenum Press, 1983.

19. Segel, I. H. *Enzyme Kinetics*. New York: Wiley, 1975.

20. Tuckwell, H. C. *Introduction to Theoretical Neurobiology.* Vol. 2, *Nonlinear and Stochastic Theories.* Cambridge: Cambridge University Press, 1988b.

21. Tuckwell, H. C. *Stochastic Processes in the Neurosciences.* Philadelphia: Society for Industrial and Applied Mathematics, 1989.

22. Wylie, C. R. *Advanced Engineering Mathematics.* New York: McGraw-Hill, 1975.

23. Ziemer, R. E. and Tranter, W. H. *Principles of Communications: Systems, Modulation and Noise.* Boston: Houghton Mifflin, 1976.

Original Articles

1. Anderson, C. R. and Stevens, C. F. Voltage clamp analysis of acetylcholine produced end-plate current fluctuations at frog neuromuscular junction. *J. Physiol. (Lond.)* 235: 655–691, 1973.

2. Colquhoun, D. How fast do drugs work? *Trends Pharmacol. Sci.* 2: 212–217, 1981.

3. Colquhoun, D., Dionne, V. E., Steinbach, J. H., and Stevens, C. F. Conductance of channels opened by acetylcholine-like drugs in muscle endplate. *Nature (Lond.)* 253: 204–206, 1975.

4. Colquhoun, D. and Hawkes, A. G. Relaxation and fluctuations of membrane currents that flow through drug-operated channels. *Proc. R. Soc. Lond. [Biol.]* 199: 231–262, 1977.

5. Colquhoun, D. and Hawkes, A. G. On the stochastic properties of single ion channels. *Proc. R. Soc. Lond. [Biol.]* 211: 205–235, 1981.

6. Colquhoun, D. and Hawkes, A. G. On the stochastic properties of bursts of single ion channel openings and of clusters of bursts. *Phil. Trans. R. Soc. Lond. [Biol.]* 300: 1–59, 1982.

7. Colquhoun, D. and Hawkes, A. G. The principles of stochastic interpretation of ion-channel mechanisms. In *Single Channel Recording,* edited by B. Sakmann and E. Neher. New York: Plenum Press, 1983, pp. 135–175.

8. Colquhoun, D. and Sakmann, B. Fluctuations in the microsecond time range of the current through single acetylcholine receptor ion channels. *Nature (Lond.)* 294: 464–466, 1981.

9. Colquhoun, D. and Sakmann, B. Fast events in single-channel currents activated by acetylcholine and its analogues at the frog muscle end-plate. *J. Physiol. (Lond.)* 369: 501–557, 1985.

10. Colquhoun, D. and Sheridan, R. E. The modes of action of gallamine. *Proc. R. Soc. Lond. [Biol.]* 211: 181–203, 1981.

11. Hille, B. Ionic selectivity, saturation, and block in sodium channels. A four-barrier model. *J. Gen. Physiol.* 66: 535–560, 1975a.

12. Hille, B. The receptor for tetrodotoxin and saxitoxin: A structural hypothesis. *Biophys. J.* 15: 615–619, 1975b.

13. Hodgkin, A.L. Beginning: some reminiscences of my early life (1914–1917). *Annu. Rev. Physiol.* 45: 1–16, 1983.

14. Hodgkin, A. L. and Huxley, A. F. A quantitative description of membrane current and its application to conduction and excitation in nerve. *J. Physiol. (Lond.)* 117: 500–544, 1952d.

15. Horn, R. and Vandenberg, C. A. Statistical properties of single sodium channels. *J. Gen. Physiol.* 84: 505–534, 1984.

16. Katz, B. and Miledi, R. Membrane noise produced by acetylcholine. *Nature (Lond.)* 226: 962–963, 1970.

17. Katz, B. and Miledi, R. The statistical nature of the acetylcholine potential and its molecular components. *J. Physiol. (Lond.)* 224: 665–699, 1972.

18. Neher, E. and Sakmann, B. Single-channel currents recorded from membrane of denervated frog muscle fibres. *Nature (Lond.)* 260: 779–802, 1976.

19. Patlak, J. and Horn, R. Effect of N-bromoacetamide on single sodium channel currents in excised membrane patches. *J. Gen. Physiol.* 79: 333–351, 1982.

20. Puia, G., Santi, M. R., Vicine, S., Pritchett, D. B., Purdy, R. H., Paul, S. M., Seeburg, P. H., and Costa, E. Neurosteroids act on recombinant human GABA$_A$ receptors. *Neuron* 4: 759–765, 1990.

Chapter 11

Books and Reviews

1. Aidley, D. J. *The Physiology of Excitable Cells.* Cambridge: Cambridge University Press, 1989.

2. Bracewell, R. N. *The Fourier Transform and Its Applications.* New York: McGraw-Hill, 1978.

3. Busch, C. and Sakmann, B. Synaptic transmission in hippocampal neurons: Numerical reconstruction of quantal IPSCs. *Cold Spring Harbor Symp. Quant. Biol.* 55: 69–80, 1990.

4. Cooper, J. R., Bloom, F. E., and Roth, R. H. *The Biochemical Basis of Neuropharmacology*, 5th edition. New York: Oxford University Press, 1986.

5. Ferreira, H. G. and Marshall, M. W. *The Biophysical Basis of Excitability.* Cambridge: Cambridge University Press, 1985.

6. Hall, Z. W. *An Introduction to Molecular Neurobiology.* Sunderland, MA: Sinauer, 1992.

7. Jack, J. J. B., Kullmann, D. M., Larkman, A. U., Major, G., and Stratford, K. J. Quantal analysis of excitatory synaptic mechanisms in the mammalian central nervous system. *Cold Spring Harbor Symp. Quant. Biol.* 55: 57–67, 1990.

8. Katz, B. *Nerve, Muscle and Synapse.* New York: McGraw-Hill, 1966.

9. Korn, H. and Faber, D. S. Regulation and significance of probabilistic release mechanisms at central synapses. In *Synaptic Function,* edited by G. M. Edelman, W. E. Gall, and W. M. Cowan. New York: Wiley, 1987, pp. 57-108.

10. Korn, H. and Faber, D. S. Quantal analysis and synaptic efficacy in the CNS. *Trends Neurosci.* 14: 439-445, 1991.

11. Larkman, A., Stratford, K. and Jack, J. Quantal synaptic transmission? *Nature (Lond.)* 353: 396, 1991.

12. Magleby, K. L. Short-term changes in synaptic efficacy. In *Synaptic Function,* edited by G. M. Edelman, W. E. Gall, and W. M. Cowan. New York: Wiley, 1987, pp. 21-56.

13. Martin, A. R. Junctional transmission II. Presynaptic mechanisms. In *Handbook of Physiology.* Sec. 1, *The Nervous System,* vol. 1. Bethesda, MD: Am. Physiol. Soc., 1977, pp. 329-355.

14. McLachlan, E. M. The statistics of transmitter release at chemical synapses. In *International Review of Physiology, Neurophysiology III,* edited by R. Porter. Baltimore, MD: University Park Press, 1978, pp. 49-117.

15. Purves, D. and Lichtman, J. W. *Principles of Neural Development.* Sunderland, MA: Sinauer, 1985.

16. Redman, S. Quantal analysis of synaptic potentials in neurons of the central nervous system. *Physiol. Rev.* 70: 165-198, 1990.

17. Sakmann, B. and Neher, E. *Single Channel Recording.* New York: Plenum Press, 1983.

18. Selkoe, D. J. Regulation and significance of probabilistic release mechanisms at central synapses. *Trends Neurosci.* 10: 57-108, 1987.

19. Silinsky, E. M. The biophysical pharmacology of calcium-dependent acetylcholine secretion. *Pharmacol. Rev.* 37: 81-132, 1985.

20. Stein, R. B. *Nerve and Muscle.* New York: Plenum Press, 1980.

21. Steinbach, J. H. and Stevens, C. F. Neuromuscular transmission. In *Frog Neurobiology,* edited by R. Llinás and W. Prech. New York: Springer, 1976, pp. 33-92.

22. Stevens, C. F. Quantal release of neurotransmitter and long-term potentiation. *Neuron* 10 (Suppl.): 55-63, 1993.

Original Articles

1. Baxter, D., Bittner, G. D., and Brown, T. H. Quantal mechanism of long-term synaptic potentiation. *Proc. Natl. Acad. Sci. USA* 82: 5978-5982, 1985.

2. Bekkers, J. M. and Stevens, C. F. Presynaptic mechanism for long-term potentiation in the hippocampus. *Nature (Lond.)* 346: 724-729, 1990.

3. Boyd, I. A. and Martin, A. R. The end-plate potential in mammalian muscle. *J. Physiol. (Lond.)* 132: 74-91, 1956.

4. Brown, T. H., Perkel, D. H., and Feldman, M. W. Evoked neurotransmitter release: Statistical effects of nonuniformity and nonstationarity. *Proc. Natl. Acad. Sci. USA* 73: 2913-2917, 1976.

5. DEL Castillo, J. and Katz, B. Quantal components of the end-plate potential. *J. Physiol. (Lond.)* 124: 560-573, 1954a.

6. DEL Castillo, J. and Katz, B. Statistical factors involved in neuromuscular facilitation and depression. *J. Physiol. (Lond.)* 124: 574-585, 1954b.

7. Edwards, F. LTP is a long term problem. *Nature (Lond.)* 350: 271-272, 1991.

8. Edwards, F., Konnerth, A., and Sakmann, B. Quantal analysis of inhibitory synaptic transmission in the dentate gyrus of rat hippocampal slices: A patch-clamp study. *J. Physiol. (Lond.)* 430: 213-249, 1990.

9. Fatt, P. and Katz, B. Spontaneous subthreshold activity at motor nerve endings. *J. Physiol. (Lond.)* 117: 109-128, 1952.

10. Jack, J. J. B., Redman, S. J., and Wong, K. The components of synaptic potentials evoked in cat spinal motoneurones by impulses in single group Ia afferents. *J. Physiol. (Lond.)* 321: 65-96, 1981.

11. Katz, B. and Miledi, R. The role of calcium in neuromuscular facilitation. *J. Physiol. (Lond.)* 195: 481-492, 1968.

12. Kullmann, D. M. and Nicoll, R. A. Long-term potentiation is associated with increases in quantal content and quantal amplitude. *Nature (Lond.)* 357: 240-244, 1992.

13. Larkman, A., Stratford, K., and Jack, J. Quantal analysis of excitatory synaptic action and depression in hippocampal slices. *Nature (Lond.)* 350: 344-347, 1991.

14. Liao, D., Jones, A., and Malinow, R. Direct measurement of quantal changes underlying long-term potentiation in CA1 hippocampus. *Neuron* 9: 1089-1097, 1992.

15. Malinow, R. Transmission between pairs of hippocampal slice neurons: Quantal levels, oscillations, and LTP. *Science* 252: 722-724, 1991.

16. Malinow, R. and Tsien, R. W. Presynaptic enhancement shown by whole-cell recordings of long-term potentiation in hippocampal slices. *Nature (Lond.)* 346: 177-180, 1990.

17. Manabe, T., Renner, P., and Nicoll, R. A. Postsynaptic contribution to long-term potentiation revealed by the analysis of miniature synaptic currents. *Nature (Lond.)* 355: 50-55, 1992.

18. Martin, A. R. A further study of the statistical composition of the endplate potential. *J. Physiol. (Lond.)* 130: 114-122, 1955.

19. Stevens, C. F. A comment on Martin's relation. *Biophys. J.* 16: 891-895, 1976.

20. Zengel, J. E. and Magleby, K. L. Differential effects of Ba^{2+}, Sr^{2+}, and Ca^{2+} on stimulation-induced changes in transmitter release at the frog neuromuscular junction. *J. Gen. Physiol.* 76: 175-211, 1980.

21. Zengel, J. E. and Magleby, K. L. Changes in miniature end-plate potential frequency during repetitive nerve stimulation in the presence of Ca^{2+}, Ba^{2+}, and Sr^{2+} at the frog neuromuscular junction. *J. Gen. Physiol.* 77: 503–529, 1981.

Chapter 12

Books and Reviews

1. Augustine, G. J., Adler, E. M., and Charlton, M. P. The calcium signal for transmitter secretion from presynaptic nerve terminals. *Ann. N. Y. Acad. Sci.* 635: 365–380, 1991.

2. Augustine, G. J., Charlton, M. P., and Smith, S. J. Calcium action in synaptic transmitter release. *Annu. Rev. Neurosci.* 10: 633–693, 1987.

3. Magleby, K. L. Short-term changes in synaptic efficacy. In *Synaptic Function*, edited by G. M. Edelman, W. E. Gall, and W. M. Cowan. New York: Wiley, 1987, pp. 21–56.

4. Monck, J. R. and Fernandez, J. M. The exocytotic fusion pore and neurotransmitter release. *Neuron* 12: 707–716, 1994.

5. Pappas, G. D. and Purpura, D. P. *Structure and Function of Synapses.* New York: Raven Press, 1972.

6. Peters, A., Palay, S. L., and Webster, H. *The Fine Structure of the Nervous System*, 3rd edition. New York: Oxford University Press, 1991.

7. Smith, S. J. and Augustine, G. J. Calcium ions, active zones and synaptic transmitter release. *Trends Neurosci.* 11: 458–464, 1988.

8. Südhof, T. C. and Jahn, R. Proteins of synaptic vesicles involved in exocytosis and membrane recycling. *Neuron* 6: 665–677, 1991.

9. Zucker, R. S. The role of calcium in regulating neurotransmitter release in the squid giant synapse. In *Presynaptic Regulation of Neurotransmitter Release: A Handbook*, edited by J. Feigenbaum and M. Hanani. London: Freund, 1991, pp. 153–195.

10. Zucker, R. S. Calcium regulation of ion channels in neurons. In *Intracellular Regulation of Ion Channels*, edited by M. Morad and Z. Agus. Berlin: Springer, 1992, pp. 191–201.

11. Zucker, R. S., Delaney, K. R., Mulkey, R., and Tank, D. W. Presynaptic calcium in transmitter release and posttetanic potentiation. *Ann. N. Y. Acad. Sci.* 635: 191–207, 1991.

Original Articles

1. Adrian, R. H., Chandler, W. K., and Hodgkin, A. L. Voltage clamp experiments in striated muscle fibres. *J. Physiol. (Lond.)* 208: 607–644, 1970.

2. Augustine, G. J. and Charlton, M. P. Calcium dependence of presynaptic calcium current and post-synaptic response at the squid giant synapse. *J. Physiol. (Lond.)* 381: 619-640, 1986.

3. Augustine, G. J., Charlton, M. P., and Smith, S. J. Calcium entry into voltage-clamped presynaptic terminals of squid. *J. Physiol. (Lond.)* 367: 143-162, 1985a.

4. Augustine, G. J., Charlton, M. P., and Smith, S. J. Calcium entry and transmitter release at voltage-clamped nerve terminals of squid. *J. Physiol. (Lond.)* 367: 163-181, 1985b.

5. Augustine, G. J. and Neher, E. Calcium requirements for secretion in bovine chromaffin cells. *J. Physiol. (Lond.)* 450: 247-271, 1992.

6. Blundon, J. A., Wright, S. N., Brodwick, M. S., and Bittner, G. D. Residual free calcium is not responsible for facilitation of neurotransmitter release. *Proc. Natl. Acad. Sci. USA* 90: 9388-9392, 1993.

7. Breckenridge, L. J. and Almers, W. Currents through the fusion pore that forms during exocytosis of a secretory vesicle. *Nature (Lond.)* 328: 814-817, 1987.

8. Charlton, M. P., Smith, S. J., and Zucker, R. S. Role of presynaptic calcium ions and channels in synaptic facilitation and depression at the squid giant synapse. *J. Physiol. (Lond.)* 323: 173-193, 1982.

9. Delaney, K., Tank, D. W., and Zucker, R. S. Presynaptic calcium and serotonin-mediated enhancement of transmitter release at crayfish neuromuscular junction. *J. Neurosci.* 11: 2631-2643, 1991.

10. DEL Castillo, J. and Katz, B. The effect of magnesium on the activity of motor nerve endings. *J. Physiol. (Lond.)* 124: 553-559, 1954.

11. Dodge, F. A., and Rahamimoff, R. Co-operative action of calcium ions in transmitter release at the neuromuscular junction. *J. Physiol. (Lond.)* 193: 419-432, 1967.

12. Jenkinson, D. H. The nature of the antagonism between calcium and magnesium ions at the neuromuscular junction. *J. Physiol. (Lond.)* 138: 434-444, 1957.

13. Katz, B. and Miledi, R. The role of calcium in neuromuscular facilitation. *J. Physiol. (Lond.)* 195: 481-492, 1968.

14. Llinás, R. and Nicholson, C. Calcium role in depolarization-secretion coupling: an aequorin study in squid giant synapse. *Proc. Natl. Acad. Sci. USA* 72: 187-190, 1975.

15. Llinás, R., Steinberg, I. Z., and Walton, K. Presynaptic calcium currents and their relation to synaptic transmission: Voltage clamp study in squid giant synapse and theoretical model for the calcium gate. *Proc. Natl. Acad. Sci. USA* 73: 2918-2922, 1976.

16. Llinás, R., Steinberg, I. Z. and Walton, K. Presynaptic calcium currents in squid giant synapse. *Biophys. J.* 33: 289-322, 1981a.

17. Llinás, R., Steinberg, I. Z. and Walton, K. Relationship between presynaptic calcium current and postsynaptic potential in squid giant synapse. *Biophys. J.* 33: 323–352, 1981b.

18. Miledi, R. Transmitter release induced by injection of calcium ions into nerve terminals. *Proc. R. Soc. Lond. [Biol.]* 183: 421–425, 1973.

19. Neher, E. and Augustine, G. J. Calcium gradients and buffers in bovine chromaffin cells. *J. Physiol. (Lond.)* 450: 273–301, 1992.

20. Simon, S. M. and Llinás, R. Compartmentalization of the submembrane calcium activity during calcium influx and its significance in transmitter release. *Biophys. J.* 48: 485–498, 1985.

21. Swandulla, D., Hans, M., Zipser, K., and Augustine, G. J. Role of residual calcium in synaptic depression and posttetanic potentiation: Fast and slow calcium signaling in nerve terminals. *Neuron* 7: 915–926, 1991.

22. Winslow, J. L., Duffy, S. N., and Charlton, M. P. Short-term homosynaptic facilitation of transmitter release is not caused by residual ionized calcium. *J. Neurophysiol.* (in press).

23. Yamada, W. M. and Zucker, R. S. Time course of transmitter release calculated from simulations of a calcium diffusion model. *Biophys. J.* 61: 671–682, 1992.

Chapter 13

Books and Reviews

1. Aidley, D. J. *The Physiology of Excitable Cells,* 3rd edition. Cambridge: Cambridge University Press, 1989.

2. Brown, T. H., Chang, V. C., Ganong, A. H., Keenan, C. L., and Kelso, S. R. Biophysical properties of dendrites and spines that may control the induction and expression of long-term synaptic potentiation. In *Long-Term Potentiation: From Biophysics to Behavior,* edited by P. W. Landfield and S. Deadwyler. New York: Liss, 1988, pp. 201–264.

3. Claiborne, B. J., Zador, A. M., Mainen, Z. F., Brown, T. H. Computational models of hippocampal neurons. In *Single Neuron Computation,* edited by T. McKenna, J. Davis, and S. F. Zornetzer. San Diego: Academic Press, 1992, pp. 61–80.

4. Cooper, J. R., Bloom, F. E., and Roth, R. H. *The Biochemical Basis of Neuropharmacology,* 5th edition. New York: Oxford University Press, 1986.

5. Hall, Z. W. *An Introduction to Molecular Neurobiology.* Sunderland, MA: Sinauer, 1992.

6. Jonas, P. and Spruston, N. Mechanisms shaping glutamate-mediated excitatory postsynaptic currents in the CNS. *Curr. Opin. Neurobiol.* 4: (in press), 1994.

7. Kandel, E. R., Schwartz, J. H., and Jessell, T. M. *Principles of Neural Science,* 3rd edition. Norwalk, CT: Appleton and Lange, 1991.

8. Koch, C. and Segev, I. *Methods in Neuronal Modeling.* Cambridge, MA: MIT Press, 1989, chapters 2, 3, and 13.

9. Lisman, J. E. and Harris, K. M. Quantal analysis and synaptic anatomy: Integrating two views of hippocampal plasticity. *Trends Neurosci.* 16: 141-147, 1993.

10. Nicholls, J. G., Martin, A. R., and Wallace, B. G. *From Neuron to Brain.* Sunderland, MA: Sinauer, 1992.

11. Pappas, G. D. and Purpura, D. P. *Structure and Function of Synapses.* New York: Raven Press, 1972.

12. Peters, A., Palay, S. L., and Webster, H. *The Fine Structure of the Nervous System,* 3rd edition. New York: Oxford University Press, 1991.

13. Rall, W. and Segev, I. Space-clamp problems when voltage clamping branched neurons with intracellular microelectrodes. In *Voltage and Patch Clamping with Microelectrodes,* edited by T.G. Smith, H. Lecar, S. J. Redman, and P. W. Gage. Bethesda, MD: Am. Physiol. Soc., 1985, pp. 191-215.

14. Sakmann, B. Elementary steps in synaptic transmission revealed by currents through single ion channels. *Neuron* 8: 613-629, 1992.

15. Siegel, G. J., Agranoff, B. W., Albers, R. W., and Molinoff, P. B. *Basic Neurochemistry.* New York: Raven Press, 1989.

16. Spruston, N., Jaffe, D. B., and Johnston, D. Dendritic attenuation of synaptic potentials and currents: the role of passive membrane properties. *Trends Neurosci.* 17: 161-166, 1994.

17. Steinbach, J. H. and Stevens, C. F. Neuromuscular transmission. In *Frog Neurobiology,* edited by R. Llinás and W. Prech. New York: Springer, 1976, pp. 33-92.

18. Takeuchi, A. Junctional transmission I. Postsynaptic mechanisms. In *Handbook of Physiology.* Sec. 1, *The Nervous System,* vol. 1. Bethesda, MD: Am. Physiol. Soc., 1977, pp. 295-328.

19. Weight, F. F. Physiological mechanisms of synaptic modulation. In *The Neurosciences: Third Study Program,* edited by F. O. Schmitt and F. G. Worden. Cambridge, MA: MIT Press, 1974, pp. 929-941.

20. Zador, A. M. Biophysics of computation in single hippocampal neurons. Ph.D. dissertation, Yale University, 1992.

Original Articles

1. Carnevale, N. T. and Johnston, D. Electrophysiological characterization of remote chemical synapses. *J. Neurophysiol.* 47: 606-621, 1982.

2. Clements, J. D., Lester, R. A. J., Tong, G., Jahr, C. E., and Westbrook, G. L. The time course of glutamate in the synaptic cleft. *Science* 258: 1498-1501, 1992.

3. Colquhoun, D., Jonas, P., and Sakmann, B. Action of brief pulses of glutamate on the AMPA/kainate receptors in patches from different neurones of rat hippocampal slices. *J. Physiol. (Lond.)* 458: 261-287, 1992.

4. Colquhoun, D., Large, W. A., and Rang, H. P. An analysis of the action of a false transmitter at the neuromuscular junction. *J. Physiol. (Lond.)* 266: 361–395, 1977.

5. Gage, P. W. and McBurney, R. N. Effects of membrane potential, temperature and neostigmine on the conductance change caused by a quantum of acetylcholine at the toad neuromuscular junction. *J. Physiol. (Lond.)* 244: 385–407, 1975.

6. Hartzell, H. C., Kuffler, S. W., and Yoshikami, D. Post-synaptic potentiation: Interaction between quanta of acetylcholine at the skeletal neuromuscular synapse. *J. Physiol. (Lond.)* 251: 427–463, 1975.

7. Johnston, D. and Brown, T. H. Interpretation of voltage-clamp measurements in hippocampal neurons. *J. Neurophysiol.* 50:464–486, 1983.

8. Katz, B. and Miledi, R. The binding of acetylcholine to receptors and its removal from the synaptic cleft. *J. Physiol. (Lond.)* 231: 549–574, 1973.

9. Kordas, M. On the role of junctional cholinesterase in determining the time course of the end-plate current. *J. Physiol. (Lond.)* 270: 133–150, 1977.

10. Magleby, K. L. and Stevens, C. F. The effect of voltage on the time course of end-plate currents. *J. Physiol. (Lond.)* 223: 151–171, 1972a.

11. Magleby, K. L. and Stevens, C. F. A quantitative description of end-plate currents. *J. Physiol. (Lond.)* 223: 173–197, 1972b.

12. Mayer, M. L. and Westbrook, G. L. Mixed-agonist action of excitatory amino acids on mouse spinal cord neurones under voltage clamp. *J. Physiol. (Lond.)* 354: 29–53, 1984.

13. Spruston, N., Jaffe, D. B., Williams, S. H., and Johnston, D. Voltage- and space-clamp errors associated with the measurement of electronically remote synaptic events. *J. Neurophysiol.* 70: 781–802, 1993.

14. Vyklicky, L., Patneau, D. K., and Mayer, M. L. Modulation of excitatory synaptic transmission by drugs that reduce desensitization at AMPA/kainate receptors. *Neuron* 7: 971–984, 1991.

Chapter 14

Books and Reviews

1. Hubbard, J. I., Llinás, R., and Quastel, D. M. J. *Electrophysiological Analysis of Synaptic Transmission.* Baltimore, MD: Williams and Wilkins, 1969.

2. Langmoen, I. A. and Andersen, P. The hippocampal slice *in vitro.* A description of the technique and some examples of the opportunities it offers. In *Electrophysiology of Isolated Mammalian CNS Preparations,* edited by G. A. Kerkut and H. V. Wheal, London: Academic Press, 1981, pp. 51–105.

3. Mitzdorf, U. Current source-density method and application in cat cerebral cortex: Investigation of evoked potentials and EEG phenomena. *Physiol. Rev.* 65: 37–100, 1985.

4. Plonsey, R. *Bioelectric Phenomena.* New York: McGraw-Hill, 1969.

5. Ruch, T. C. and Patton, H. D. *Physiology and Biophysics,* 19th edition. Philadelphia: Saunders, 1965.

6. Stevens, C. F. *Neurophysiology: A Primer.* New York: Wiley, 1966.

Original Articles

1. Freeman, J. A. and Nicholson, C. Experimental optimization of current source-density technique for anuran cerebellum. *J. Neurophysiol.* 38: 369–382, 1975.

2. Haberly, L. B. and Shepherd, G. M. Current-density analysis of summed evoked potentials in opossum prepyriform cortex. *J. Neurophysiol.* 36: 789–803, 1973.

3. Holsheimer, J. Electrical conductivity of the hippocampal CA1 layers and application to current-source-density analysis. *Exp. Brain Res.* 67: 402–410, 1987.

4. Lorente de Nó, R. A study of nerve physiology. *Stud. Rockefeller Inst. Med. Res.* 132: 384–482, 1947.

5. Mauro, A. Properties of thin generators pertaining to electrophysiological potentials in volume conductors. *J. Neurophysiol.* 23: 132–143, 1960.

6. Mitzdorf, U. and Singer, W. Prominent excitatory pathways in the cat visual cortex (A17 and A18): A current source density analysis of electrically evoked potentials. *Exp. Brain Res.* 33: 371–394, 1978.

7. Miyakawa, H. and Kato, H. Active properties of dendritic membrane examined by current source density analysis in hippocampal CA1 pyramidal neurons. *Brain Res.* 399: 303–309, 1986.

8. Nicholson, C. and Freeman, J. A. Theory of current source-density analysis and determination of conductivity tensor for anuran cerebellum. *J. Neurophysiol.* 38: 356–368, 1975.

9. Rall, W. and Shepherd, G. M. Theoretical reconstruction of field potentials and dendrodendritic synaptic interactions in olfactory bulb. *J. Neurophysiol.* 31: 884–915, 1968.

10. Richardson, T. L., Turner, R. W., and Miller, J. J. Action-potential discharge in hippocampal CA1 pyramidal neurons: Current source-density analysis. *J. Neurophysiol.* 58: 981–996, 1987.

11. Rodriguez, R. and Haberly, L. B. Analysis of synaptic events in the opossum piriform cortex with improved current source-density techniques. *J. Neurophysiol.* 61: 702–718, 1989.

Chapter 15

Books and Reviews

1. Baudry, M. and Davis, J. L. *Long-Term Potentiation: A Debate of Current Issues.* Cambridge, MA: MIT Press, 1991.

2. Bliss, T. V. P. and Lynch, M. A. Long-term potentiation of synaptic transmission in the hippocampus: properties and mechanisms. In *Long-Term Potentiation: From Biophysics to Behavior,* edited by P. W. Landfield and S. Deadwyler. New York: Liss, 1988, pp. 3–72.

3. Brown, T. H., Chang, V. C., Ganong, A. H., Keenan, C. L., and Kelso, S. R. Biophysical properties of dendrites and spines that may control the induction and expression of long-term synaptic potentiation. In *Long-Term Potentiation: From Biophysics to Behavior,* edited by P. W. Landfield and S. Deadwyler. New York: Liss, 1988, pp. 201-264.

4. Brown, T. H., Ganong, A. H., Kairiss, E. W., Keenan, C. L., and Kelso, S. R. Long-term potentiation in two synaptic systems of the hippocampal brain slice. In *Neural Models of Plasticity,* edited by J. H. Byrne and W. O. Berry, San Diego: Academic Press, 1989, pp. 266–306.

5. Brown, T. H., Kairiss, E. W., and Keenan, C. L. Hebbian synapses: Biophysical mechanisms and algorithms. *Annu. Rev. Neurosci.* 13: 475–511, 1990.

6. Buzsáki, G. and Vanderwolf, C. H. *Electrical Activity of the Archicortex.* Budapest: Akadémiai Kiadó, 1985.

7. Byrne, J. H. and Berry, W. O., eds. *Neural Models of Plasticity.* San Diego: Academic Press, 1989.

8. Churchland, P. S. and Sejnowski, T. J. *The Computational Brain.* Cambridge, MA: MIT Press, 1992.

9. Cohen N. J. and Eichenbaum, H. *Memory, Amnesia, and the Hippocampal System.* Cambridge, MA: MIT Press, 1993.

10. Eichenbaum, H., Otto, T., and Cohen, N. J. The hippocampus: What does it do? *Behav. and Neural Biol.* 57: 2-36, 1992.

11. Gustafsson, B. and Wigström, H. Basic features of long-term potentiation in the hippocampus. *Semin. Neurosci.* 2: 321–333, 1990.

12. Hebb, D. O. *Organization of Behavior.* New York: Wiley, 1949.

13. Johnston, D., Williams, S. W., Jaffe, D., and Gray, R. NMDA-receptor independent long-term potentiation. *Annu. Rev. Physiol.* 54: 489–505, 1992.

14. Ito, M. Long-term depression. *Annu. Rev. Neurosci.* 12: 85–102, 1989.

15. Kandel, E. R., Schwartz, J. H., and Jessell, T. M. *Principles of Neural Science.* Norwalk, CT: Appleton and Lange, 1991.

16. Kandel, E. R. and Spencer, W. A. Cellular neurophysiological approaches to the study of learning. *Physiol. Rev.* 48: 65–134, 1968.

17. Koch, C. and Segev, I. *Methods in Neuronal Modeling.* Cambridge, MA: MIT Press, 1989, chapters 2, 3, and 13.

18. Kohonen, T. *Associative Memory.* Berlin: Springer-Verlag, 1978.

19. Landfield, P. W. and Deadwyler, S., eds. *Long-Term Potentiation: From Biophysics to Behavior.* New York: Liss, 1988.

20. Lisman, J. E. and Harris, K. M. Quantal analysis and synaptic anatomy: Integrating two views of hippocampal plasticity. *Trends Neurosci.* 16: 141-147, 1993.

21. Madison, D. V., Malenka, R. C., and Nicoll, R. A. Mechanisms underlying long-term potentiation of synaptic transmission. *Annu. Rev. Neurosci.* 14: 379-397, 1991.

22. Malenka, R. C. The role of postsynaptic calcium in the induction of long-term potentiation. *Molec. Neurobiol.* 5: 289-295, 1991.

23. McNaughton, B. L. Neuronal mechanisms for spatial computation and information storage. In *Neural Connections, Mental Computations,* edited by L. Nadel, L. A. Cooper, P. Culicover, and R. M. Harnish, Cambridge, MA: MIT Press, 1989, pp. 285-350.

24. McNaughton, B. L. and Morris, R. G. M. Hippocampal synaptic enhancement and information storage within a distributed memory system. *Trends Neurosci.* 10: 408-415, 1987.

25. Morris, R. G. M. Toward a representational hypothesis of the role of hippocampal synaptic plasticity in spatial and other forms of learning. *Cold Spring Harbor Symp. Quant. Biol.* 55: 161-173, 1990.

26. Nicoll, R. A., Kauer, J. A., Malenka, R. C. The current excitement in long-term potentiation. *Neuron* 1: 97-103, 1988.

27. Rolls, E. T. Functions of neuronal networks in the hippocampus and neocortex in memory. In *Neural Models of Plasticity,* edited by J. H. Byrne and W. O. Berry, San Diego: Academic Press, 1989a, pp. 240-265.

28. Rolls, E. T. The representation and storage of information in neuronal networks in the primate cerebral cortex and hippocampus. In *The Computing Neuron,* edited by R. Durbin, C. Miall, and G. Mitchison. Wokingham, U.K.: Addison-Wesley, 1989b, pp. 125-159.

29. Sejnowski, T. J. Skeleton filters in the brain. In *Parallel Models of Associative Memory,* edited by G. E. Hinton and J. A. Anderson. Hillsdale, NJ: Erlbaum, 1981, pp. 189-212.

30. Sejnowski, T. J. Induction of synaptic plasticity by Hebbian covariance in the hippocampus. In *The Computing Neuron,* edited by R. Durbin, C. Miall, and G. Mitchison. Wokingham, U.K.: Addison-Wesley, 1989.

31. Sejnowski, T. J. and Tesauro, G. The Hebb rule for synaptic plasticity: algorithms and implementations. In *Neural Models of Plasticity,* edited by J. H. Byrne and W. O. Berry, San Diego: Academic Press, 1989, pp. 94-103.

32. *Seminars in The Neurosciences.* 2: 317-420, 1990.

33. Squire, L. R. and Zola-Morgan, S. The medial temporal lobe memory system. *Science* 253: 1380–1386, 1991.

34. Swanson, L. W., Teyler, T. J., and Thompson, R. F. Hippocampal long-term potentiation: Mechanisms and implications for memory. *Neurosci. Res. Program Bull.* 20: 613–769, 1982.

35. Teyler, T. J. and DiScenna, P. Long-term potentiation as a candidate mnemonic device. *Brain Res. Rev.* 7: 15–28, 1984.

36. Teyler, T. J. and DiScenna, P. The role of hippocampus in memory: A hypothesis. *Neurosci. Biobehav. Rev.* 9: 377–389, 1985.

37. Thompson, R. F. *Introduction to Physiological Psychology.* New York: Harper and Row, 1975.

38. *Trends in Neuroscience* 15, 1992.

Original Articles

1. Alger, B. E. and Teyler, T. J. Long-term and short-term plasticity in the CA1, CA3, and dentate regions of the rat hippocampal slice. *Brain Res.* 110: 463–480, 1976.

2. Aniksztejn, L. and Ben-Ari, Y. Novel form of long-term potentiation produced by a K^+ channel blocker in the hippocampus. *Nature (Lond.)* 349: 67–69, 1991.

3. Barrionuevo, G. and Brown, T. H. Associative long-term potentiation in hippocampal slices. *Proc. Natl. Acad. Sci. USA* 80: 7347–7351, 1983.

4. Bienenstock, E. L., Cooper, L. N., and Munro, P. W. Theory for the development of neuron selectivity: Orientation specificity and binocular interaction in visual cortex. *J. Neurosci.* 2: 32–48, 1982.

5. Buzsáki, G. Two-stage model of memory trace formation: A role for "noisy" brain states. *Neuroscience* 31: 551–570, 1989.

6. Carnevale, N. T. and Johnston, D. Electrophysiological characterization of remote chemical synapses. *J. Neurophysiol.* 47: 606–621, 1982.

7. Grover, L. M. and Teyler, T. J. Two components of long-term potentiation induced by different patterns of afferent activation. *Nature (Lond.)* 347: 477–479, 1990.

8. Grover, L. M. and Teyler, T. J. N-methyl-D-aspartate receptor-independent long-term potentiation in area CA1 of rat hippocampus: Input-specific induction and preclusion in a non-tetanized pathway. *Neuroscience* 49: 7–11, 1992.

9. Holmes, W. R. Is the function of dendritic spines to concentrate calcium? *Brain Res.* 519: 338–342, 1990.

10. Holmes, W. R. and Levy, W. B. Insights into associative long-term potentiation from computational models of NMDA receptor-mediated calcium influx and intracellular calcium concentration changes. *J. Neurophysiol.* 63: 1148–1168, 1990.

11. Kelso, S. R. and Brown, T. H. Differential conditioning of associative synaptic enhancement in hippocampal brain slices. *Science* 232: 85-87, 1986.

12. Kelso, S. R., Ganong, A. H., and Brown, T. H. Hebbian synapses in hippocampus. *Proc. Natl. Acad. Sci. USA* 83: 5326-5330, 1986.

13. Kirkwood, A., Dudek, S. M., Gold, J. T., Aizenman, C. D., and Bear, M. F. Common forms of synaptic plasticity in the hippocampus and neocortex in vitro. *Science* 260: 1518-1521, 1993.

14. Koch, C. and Zador, A. The function of dendritic spines: Devices subserving biochemical rather than electrical computation. *J. Neurosci.* 13: 413-422, 1993.

15. Koch, C., Zador, A., and Brown, T. H. Dendritic spines: Convergence of theory and experiment. *Science* 256: 973-974, 1992.

16. Kullmann, D. M., Perkel, D. J., Manabe, T., and Nicoll, R. A. Ca^{2+} entry via postsynaptic voltage-sensitive Ca^{2+} channels can transiently potentiate excitatory synaptic transmission in the hippocampus. *Neuron* 9: 1175-1183, 1992.

17. Malinow, R. and Miller, J. P. Postsynaptic hyperpolarization during conditioning reversibly blocks induction of long-term potentiation. *Nature (Lond.)* 320: 529-530, 1986.

18. Marr, D. Simple memory: A theory for archicortex. *Philos. Tran. R. Soc. Lond.* 262: 23-81, 1971.

19. McNaughton, B. L., Douglas, R. M., and Goddard, G. V. Synaptic enhancement in fascia dentata: Cooperativity among coactive afferents. *Brain Res.* 157: 277-293, 1978.

20. Schulz, P.E., Cook, E., and Johnston, D. Changes in paired-pulse facilitation suggest presynaptic involvement in long-term potentiation. *J. Neurosci.* (in press), 1994.

21. Spruston, N., Jaffe, D. B., Williams, S. H., and Johnston, D. Voltage- and space-clamp errors associated with the measurement of electronically remote synaptic events. *J. Neurophysiol.* 70: 781-802, 1993.

22. Treves, A. and Rolls, E. T. A computational analysis of the role of the hippocampus in memory. *Hippocampus* (in press), 1994.

23. Wigström, H., Gustafsson, B., Huang, Y.-Y., and Abraham, W. C. Hippocampal long-term potentiation is induced by pairing single afferent volleys with intracellularly injected depolarizing current pulses. *Acta Physiol. Scand.* 126: 317-319, 1986.

24. Zador, A., Koch, C., and Brown, T. H. Biophysical model of a Hebbian synapse. *Proc. Natl. Acad. Sci. USA* 87: 6718-6722, 1990.

Appendix A

Books and Reviews

1. Brown, P. B., Maxfield, B. W., and Moraff, H. *Electronics for Neurobiologists.* Cambridge, MA: MIT Press, 1973.

2. Diefenderfer, A. J. *Principles of Electronic Instrumentation.* Philadelphia: Saunders, 1972.

3. Halliday, D. and Resnick, R. *Fundamentals of Physics.* New York: Wiley, 1993.

4. Hoenig, S. A. *How to Build and Use Electronic Devices Without Frustration, Panic, Mountains of Money, or an Engineering Degree.* Boston: Little, Brown, 1980.

5. Kettenmann, H. and Grantyn, R. *Practical Electrophysiological Methods.* New York: Wiley, 1992.

6. Malmstadt, H. V., Enke, C. G., and Crouch, S. R. *Electronic Measurements for Scientists.* Menlo Park, CA: W. A. Benjamin, 1974.

7. Purves, R. D. *Microelectrode Methods for Intracellular Recording and Iontophoresis.* London: Academic Press, 1981.

8. Sakmann, B. and Neher, E. *Single-Channel Recording.* New York: Plenum Press, 1983.

9. Smith, T. G., Lecar, H., Redman, S. J., and Gage, P. W. *Voltage and Patch Clamping with Microelectrodes.* Bethesda, MD: Am. Physiol. Soc., 1985.

10. Standen, N. B., Gray, T. A., and Whitaker, M. J. *Microelectrode Techniques.* Cambridge: The Company of Biologists, 1987.

Appendix B

Books and Reviews

1. Bradbury, S. *An Introduction to the Optical Microscope.* Oxford: Oxford University Press, 1989.

2. Cebulla, D.-M. W. *Handbook of Incident Light Microscopy.* Germany: Carl Zeiss.

3. Cherry, R. J. *New Techniques of Optical Microscopy and Microspectroscopy.* Boca Raton, FL: CRC Press, 1991.

4. Cohen, L., Höpp, H.-P., Wu, J.-Y., and Xiao, C. Optical measurement of action potential activity in invertebrate ganglia. *Annu. Rev. Physiol.* 51: 527–541, 1989.

5. Delly, J. D. *Photography Through the Microscope.* Rochester, NY: Eastman Kodak Company, 1980.

6. De Weer, P. and Salzberg, B. M. *Optical Methods in Cell Physiology.* New York: Wiley, 1986.

7. Grinvald, A. Real-time optical mapping of neuronal activity: From single growth cones to the intact mammalian brain. *Annu. Rev. Neurosci.* 8: 263–305, 1985.

8. Halliday, D. and Resnick, R. *Fundamentals of Physics.* New York: Wiley, 1993.

9. Haugland, R. P. *Handbook of Fluorescent Probes and Research Chemicals.* Eugene, OR: Molecular Probes Inc., 1992.

10. Hecht, E. *Optics.* Reading, MA: Addison-Wesley, 1987.

11. Herman, B. and Lemasters, J. J. *Optical Microscopy: Emerging Methods and Applications.* San Diego: Academic Press, 1993.

12. Holz, H. M. *Worthwhile Facts About Fluorescence Microscopy.* Germany: Carl Zeiss.

13. Inoué, S. *Video Microscopy.* New York: Plenum Press, 1986.

14. Lieke, E. E., Frostig, R. D., Arieli, A., Ts'o, D. Y., Hildesheim, R., and Grinvald, A. Optical imaging of cortical activity. *Annu. Rev. Physiol.* 51: 543–581, 1989.

15. Ross, W. N. Changes in intracellular calcium during neuron activity. *Annu. Rev. Physiol.* 51: 491–506, 1989.

16. Rost, F. W. D. *Fluorescence Microscopy,* vol. 1. Cambridge: Cambridge University Press, 1992.

17. Salzberg, B. M. Optical recording of voltage changes in nerve terminals and in fine neuronal processes. *Annu. Rev. Physiol.* 51: 507–526, 1989.

18. Sears, F. W. *Optics.* Cambridge, MA: Addison-Wesley, 1949.

19. Tsien, R. Y. Fluorescent probes of cell signalling. *Annu. Rev. Neurosci.* 12: 227–254, 1989.

Original Articles

1. Grynkiewicz, G., Poenie, M., and Tsien, R. Y. A new generation of Ca^{2+} indicators with greatly improved fluorescence properties. *J. Biol. Chem.* 260: 3440–3450, 1985.

2. Hoffman, R. The modulation contrast microscope: Principles and performance. *J. Microscopy* 110: 205–222, 1977.

3. Lev-Ram, V., Miyakawa, H., Lasser-Ross, N., and Ross, W. N. Calcium transients in cerebellar Purkinje neurons evoked by intracellular stimulation. *J. Neurophysiol.* 68: 1167–1177, 1992.

4. Ross, W.N., Salzberg, B.M., Cohen, L.B., and Davila, H.V. A large change in dye absorption during the action potential. *Biophys. J.* 14: 983–986, 1974.

5. Wallén, P., Carlsson, K., Liljeborg, A., and Grillner, S. Three-dimensional reconstruction of neurons in the lamprey spinal cord in whole-mount, using a confocal laser scanning microscope. *J. Neurosci. Methods* 24: 91–100, 1988.

Index